PRIOR ACCLAIM FOR JEREMY GRIFFITH'S TREATISE

'Frankly, I am 'blown away' as the saying goes...
The ground-breaking significance of this work is tremendous.'

PATRICIA GLAZEBROOK, *Professor and Chair of Philosophy*
(Response to 'The Human Condition Documentary Proposal', 2004)

'It might help bring about a paradigm shift in the
self-image of humanity – an outcome that in the past
only the great world religions have achieved.'

MIHALY CSIKSZENTMIHALYI, *Professor of Psychology and author of 'Flow: The Psychology of Optimal Experience'*
(Response to 'The Human Condition Documentary Proposal', 2004)

'I am simply overwhelmed...I find it astonishing and impressive.'

JOSEPH CHILTON PEARCE, *author of 'Magical Child'*
(Response to 'The Human Condition Documentary Proposal', 2004)

'Could you please send me an extra copy of your book?
[Mine] is on loan because it was so appreciated.'

The late **SIR LAURENS VAN DER POST**,
pre-eminent philosopher and author
(Response to 'Free: The End of the Human Condition', 1988)

'a superb book…[that] brings out the truth of a new and
wider frontier for humankind, a forward view of a world of
humans no longer in naked competition amongst ourselves.'

The late **JOHN MORTON**, *Professor of Zoology*
(Response to 'A Species In Denial', 2003)

'The proposal is indeed impressive.'

DR ROGER LEWIN, *prize-winning science writer and author*
(Response to 'The Human Condition Documentary Proposal', 2004)

'I have never heard of anything comparable before.'

FRIEDEMANN SCHRENK, *Professor of Paleoanthropology*

(Response to 'The Human Condition Documentary Proposal', 2004)

'[Professor Hawking] is most interested in your impressive proposal.'

STEPHEN HAWKING, *world-leading physicist*

(Response to 'The Human Condition Documentary Proposal', 2004)

'I consider the book to be the work of a prophet
and I expect the author to become recognised as a saint.'

The late **DR RONALD STRAHAN**,
former director of Sydney's Taronga Park Zoo

(Response to 'Free: The End of the Human Condition', 1988)

'[Griffith] gives us a genuinely original and inspiring way
of understanding ourselves and our place in the universe.'

The late **CHARLES BIRCH**,
Templeton Prize winner and Professor of Biology

(Response to 'Beyond The Human Condition', 1991)

'I believe you are on to getting answers to much that
has puzzled and bewildered humanity for a long time.'

DR IAN PLAYER, *conservationist, naturalist and philosopher*

(Response to 'A Species In Denial', 2003)

'The sequence of discussion is so logical and sensible,
providing the necessary breakthrough for this critical issue
in our understanding of ourselves.'

DR DAVID J. CHIVERS, *physical anthropologist*

(Response to 'The Human Condition Documentary Proposal', 2004)

'The insights and ideas are fascinating and pertinent
and must be developed and disseminated.'

DR GEORGE SCHALLER, *zoologist and author*

(Response to 'The Human Condition Documentary Proposal', 2004)

IS IT TO BE Terminal Alienation or Transformation For The Human Race?
by Jeremy Griffith

First Edition, published in 2014, by WTM Publishing and Communications Pty Ltd (ACN 103 136 778) (www.wtmpublishing.com)

Cover design by Luke Causby/Blue Cork.
Cover images: *Cringing in Terror* (c.1794-96) and *Albion Arose* (c.1794-96) by William Blake; coloured impression of *Albion Arose* by Carol Marando 1991.

All enquiries to:

WORLD TRANSFORMATION MOVEMENT® (WTM®)
GPO Box 5095, Sydney NSW 2001, Australia
Phone: + 61 2 9486 3308   Fax: + 61 2 9486 3409
Email: info@worldtransformation.com
Website: www.worldtransformation.com or www.humancondition.com

The WTM is a non-profit organisation which holds an Authority to Fundraise for Charitable Purposes in NSW, Australia.

ISBN 978-1-74129-017-2
CIP – Biology

Edited by Fiona Cullen-Ward.
Typesetting: designed by Jeremy Griffith, set by Lee Jones & Polly Watson.
Font: Times; main body text:11.4pt on 14.2pt leading; quote: 9.634pt, horizontal scale 100.4%, bold; quote source: 8.5pt; comment within a quote: 10.744pt; digits and all caps text: 1-2pts smaller than body text. For further details about the typesetting, styles and layout used in this book please view the *WTM Style Guide* at <www.worldtransformation.com/style-guide>.

IS IT TO BE
Terminal Alienation or Transformation For The Human Race?

Jeremy Griffith

www.IsItToBe.com or
www.TheBookThatSavesTheWorld.com or
www.WorldTransformation.com or
www.HumanCondition.com

With all their love and support of the truth that this book contains, which ensured its creation, *IS IT TO BE* is dedicated to the future of the human race by the World Transformation Movement's Founding Members: Annabel Armstrong, Susan Armstrong, Sam Belfield, John Biggs, Richard Biggs, Anthony Clarke, Lyn Collins, Steve Collins, Lachlan Colquhoun, Eric Crooke, Emma Cullen-Ward, Fiona Cullen-Ward, Anthony Cummins, Neil Duns, Sally Edgar, Anna Fitzgerald, Brony FitzGerald, Connor FitzGerald, Tony Gowing, Jeremy Griffith, Simon Griffith, Damon Isherwood, Felicity Jackson, Charlotte James, Lee Jones, Monica Kodet, Anthony Landahl, Doug Lobban, Tim Macartney-Snape, Simon Mackintosh, Katrina Makim, Sean Makim, Manus McFadyen, Ken Miall, Tony Miall, Rachel O'Brien, James Press, Stacy Rodger, Marcus Rowell, Genevieve Salter, Will Salter, Nicholas Shaw, Wendy Skelton, Pete Storey, Ali Watson, Polly Watson, Prue Watson, Tess Watson, Tim Watson, James West, Stirling West, Prue Westbrock and Annie Williams.

Notes to the Reader

- Readers are warned that the issue of the human condition that this book addresses has been such a difficult subject for humans to confront that reading about it can initially cause a 'deaf effect', where it is hard to take in and absorb what is being said. However, with patience and re-reading this deaf effect wears off, allowing the compassionate and immensely relieving insights to become accessible. As some now avid readers of Jeremy Griffith's human-condition-confronting writing have admitted, **'When I first read this material all I saw were a lot of black marks on white paper'**; and, **'Reading this is like reading another language—you know it's English, you can understand the words, but the concepts are so basic and so different that they are almost incomprehensible—it's a paradigm shift of a read'**; and, **'This stuff is so head on it can be crippling, which, initially at least, can make it hard to get behind what's being said and access the profundity of where it's coming from'**; and, **'At first I found this information difficult to absorb, in fact my wife and I would sit in bed and read a page together, and then re-read it a number of times, but still we couldn't understand what was written there and ended up thinking it must be due to poor expression'**—but it is worth persevering because here are some post-deaf effect online responses: **'If Plato and Aristotle were alive and read Griffith, they would die happy men'**; and, **'We don't have to put up with "Not Knowing" anymore'**; and, **'tears stream down my face, so overcome have I been by this book. It is the greatest book on the planet, no wait, in the universe. In fact it is the greatest anything in the universe'**; and, **'Here is the breakthrough biological explanation that PROVES we are ALL very, very good'**; and, **'This, to me, is the most significant thing I have ever stumbled across...If it doesn't hit you right away—it will down the road'**; and, **'Gah! words are too limited for this. Here have some love brother! ☺'**

- Unlike most publications, there is no **bibliography** at the conclusion of this book because the source is provided in small text at the end of each quote, to ensure the reader can immediately see when, where and by whom the quote was made. Also, rather than give **the particular edition and/or publisher** of the book that the quote comes from, the page number where the quote appears and the total number of pages of the particular edition used for the source is provided. This enables the reader to find the comparative place of the quote in any edition. There is also no **index** because any word or phrase can be easily searched electronically using the 'Search IS IT TO BE' facility in the top right-hand corner of the online copy of the book at <www.IsItToBe.com>. All **paragraphs are numbered** to allow easy referencing across any of the book's formats. *IS IT TO BE* contains a number of coloured **images**, however, in this print edition they appear in black and white. To view those images in colour, see the online edition. All **biblical references** are from the 1978 *New International Version* translation of the Bible.

- Since this book contains the only thing that can save the human race from extinction, namely the dignifying and relieving understanding of our species' psychologically troubled human condition, it must be made accessible to all people—which is why, alongside its availability for purchase through bookshops and Amazon, it is, and will always remain, freely available to be read, shared or printed (with binding instructions included) at <www.IsItToBe.com>.

Soon (in March 2015)

Contents

		PAGE	PARAGRAPH
Introduction by Professor Harry Prosen		15	1–
Chapter 1	**Is It To Be Terminal Alienation or Transformation For The Human Race?**	31	33–
1:1	Summary	31	33
1:2	The true nature of the human condition—and our Resignation to living in denial of it	32	34–
1:3	Our near total resistance to analysis of the human condition post-Resignation	45	53–
1:4	Important note to reader	50	61–
1:5	How has science coped with the issue of the human condition?	53	65–
1:6	The elements involved in a true analysis of the human condition	55	72–
1:7	The ultimate paradox of the human condition	61	86–
1:8	Social Darwinism	62	89–
1:9	Sociobiology/Evolutionary Psychology	63	91–
1:10	Multilevel Selection theory for eusociality	65	95–
1:11	While denial has been necessary, you can't find the truth with lies	74	115–
Chapter 2	**The Human-Race-Transforming, *Real* Explanation of The Human Condition**	83	126–
2:1	Summary	83	126–
2:2	We cannot endure being faced by the problem of the human condition forever	84	128–
2:3	The psychosis-addressing-and-solving *real* explanation of the human condition	86	132–
2:4	The Story of Adam Stork	89	140–
2:5	The double and triple whammy involved in our *human* situation or condition	95	151–
2:6	Adam and Eve—we humans—are heroes NOT villains	101	161–

2:7 Our instinctive moral conscience has been 'A sharp accuser, but a helpless friend!' 108 169–
2:8 Most wonderfully, this psychosis-addressing-and-solving, *real* explanation of the human condition makes possible the psychological rehabilitation of the human race 112 175–
2:9 The end of politics 117 185–
2:10 'Free at last!' 121 191–

Chapter 3 The Meaning of Life 125 200–
3:1 Summary 125 200
3:2 The obvious truth of the development of order of matter on Earth 126 201–
3:3 'God' is our personification of Integrative Meaning 131 210–
3:4 The denial-free history of the development of matter on Earth 139 225–
3:5 Elaborating the sexually reproducing individual 147 243–
3:6 Negative Entropy found a way to form the Specie Individual 152 256–

Chapter 4 The Origin of Humans' Unconditionally Selfless, Altruistic, Moral Instinctive Self or Soul 153 259–
4:1 Summary 153 259–
4:2 How could humans have acquired their altruistic moral instincts? 154 262–
4:3 The integration of sexually reproducing individuals to form the Specie Individual 156 268–
4:4 Love-Indoctrination 159 272–
4:5 Fossil evidence confirming the love-indoctrination process 163 280–
4:6 Bonobos provide living evidence of the love-indoctrination process 172 294–
4:7 The role of nurturing in our development has been an unbearably confronting truth 181 301–
4:8 The emergence of consciousness assisted the love-indoctrination process by allowing the sexual selection of integrativeness 182 303–
4:9 Sexual selection for integrativeness explains neoteny 188 314–
4:10 The importance of nurturing in bonobo society 192 322–

4:11 The importance of strong-willed females in developing integration ... 194 326–
4:12 Descriptions of bonobos provide an extraordinary insight into what life for our human ancestors was like before the emergence of the human condition ... 198 332–
4:13 Milwaukee County Zoo's fabulous group of bonobos ... 201 337–
4:14 'A golden race...formed on Earth' ... 210 349–

Chapter 5 End Play for The Human Race ... 215 356–
5:1 Summary ... 215 356
5:2 The danger of denial becoming so entrenched that it locks humanity onto the path to terminal alienation ... 215 357–
5:3 The nurturing origins of our moral soul is an obvious truth... 218 362–
5:4 The problem has been that the nurturing origin of our moral soul has been devastatingly, unbearably, excruciatingly condemning ... 223 367–
5:5 To deny the importance of nurturing, the Social Intelligence Hypothesis was invented ... 228 375–
5:6 Dismissing maternal love as training to manage complex social situations still left the extraordinarily cooperative lives of bonobos, and of our ape ancestors, to somehow be explained ... 235 388–
5:7 Fossil evidence of our species' cooperative past has also been dismissed and then ignored by mechanistic science ... 239 396–
5:8 The Social Ecological Model ... 241 401–
5:9 A brief history of left-wing dishonest mechanistic biology ... 246 413–
5:10 The Self-Domestication Hypothesis ... 251 417–
5:11 End play for the human race ... 257 434–
5:12 *The* great obscenity ... 260 441–
5:13 Nurturing now becomes a priority ... 270 453–

Chapter 6 What is Consciousness, and Why, How and When Did Humans' Unique Conscious Mind Emerge? ... 273 458–
6:1 Summary ... 273 458
6:2 What is consciousness? ... 274 459–
6:3 Why, how and when did consciousness emerge in humans? .. 289 497–

Chapter 7 The Greatest, Most Heroic Story Ever Told: Humanity's Journey from Ignorance to Enlightenment 309 535–

7:1 Summary 309 535–

7:2 The stages of humanity's maturation from ignorance to enlightenment 311 538–

7:3 INFANCY 316 550–

7:4 CHILDHOOD 319 555

7:5 Early Happy, Innocent Childhood 320 556–

7:6 Middle Demonstrative Childhood 321 558–

7:7 Late Naughty Childhood 324 562–

7:8 ADOLESCENCE 330 569

7:9 Early Sobered Adolescentman 331 570–

7:10 Depressed Adolescentman 332 572–

7:11 Adventurous Adolescentman—includes the reconciling explanation of Men and Women 344 592–

7:12 Angry Adolescentman 397 668–

7:13 Pseudo Idealistic Adolescentman 412 695–

7:14 Hollow Adolescentman 420 713–

7:15 The last 11,000 years and the rise of Imposed Discipline, Religion and other forms of Pseudo Idealism 425 721–

7:16A The final 200 years when pseudo idealism takes humanity to a death-by-dogma, end play state of terminal alienation—the time when we need to 'Beware of false prophets', the merchants of delusion and denial, for they are 'the abomination that causes desolation' 436 748–

7:16B The emergence of terminal levels of alienation in the 'developed' world 438 752–

7:16C The dire consequences of terminal levels of alienation destroying our ability to nurture our children 447 765–

7:16D The dysfunctional, extremely narcissistic 'Power Addicted' state 467 804–

7:16E The differences in alienation between races of humans 479 826–

7:16F The emergence of 'materialism envy' and with it unbridled greed, extreme dysfunction and destitution in the 'developing' world.... 496 850–
7:16G So it is end play wherever we looked 499 855–
7:16H The progression of ever-more dishonest and dangerous forms of pseudo idealism to cope with the unbearable levels of upset......... 503 861–
7:16I Less Guilt Emphasising Expressions of Religion 504 863–
7:16J Non-Religious Pseudo Idealistic Causes 509 872–
7:16K Socialism and Communism 511 876
7:16L The New Age Movement................. 511 877
7:16M The Feminist Movement 512 878
7:16N The Environmental or Green Movement 513 879
7:16O The Politically Correct Movement 514 880–
7:16P The Postmodern Deconstructionist Movement 520 891–
7:16Q The 'abomination that causes desolation' 'sign...of the end of the age' that is 'cut short' by the arrival of the liberating truth about the human condition 526 905–

Chapter 8 The Transformation of The Human Race........ 543 933–
8:1 Summary 543 933
8:2 The 'dawn...of our emancipation' 544 934–
8:3 'Judgment day' 548 940–
8:4 The Transformed Lifeforce Way of Living 553 947–
8:5 The Transformed Lifeforce Way of Living is not another religion 559 959–
8:6 Nor is the Transformed Lifeforce Way of Living another deluded, false start to a human-condition-free new world..... 562 965–
8:7 But how does the Transformed Way of Living solve the problem of the unbearable, 'judgment day' exposure that understanding of the human condition unavoidably brings? .. 566 973–
8:8 Plato's anticipation of the initial 'violently opposed' stage that the arrival of the understanding of the human condition has had to withstand.. 573 987–

8:9 The utter magnificence of the Transformed Lifeforce State ... 580 997–
8:10 How the Transformed Lifeforce Way of Living will quickly repair the world ... 591 1017–
8:11 Humanity's overall situation now that we have understanding ... 603 1038–
8:12 The 'pathway of the sun' ... 615 1052–

About the Author ... 638

Introduction
by Professor Harry Prosen

'In the whole of written history there are only two or three people who have been able to think on this scale about the human condition.'
Prof. Anthony Barnett, zoologist, author and broadcaster, 1983

[1]The truth is I am inadequate to write this Introduction, but everyone is, so I will do my best.

[2]Firstly, to immediately put into context the fabulous significance of what Australian biologist Jeremy Griffith has achieved with this book: it delivers the breakthrough biological explanation of the human condition, the holy grail of insight we humans have sought for the psychological rehabilitation of our species. It brings the compassionate, reconciling, liberating and transforming understanding of our lives that the human race has lived in eternal hope, faith and trust would one day be found and have so assiduously pursued since we first became conscious beings some 2 million years ago!

[3]I once read that the great philosophers each had an acute grasp of their own small piece of an unthinkably huge and interlocking puzzle but nobody has ever put the pieces together into a single coherent system (Steve King, *Post-Structuralism or Nothing*, 1996; see <www.wtmsources.com/135>). However, that is no longer the situation because Jeremy provides that great unifying, make-sense-of-everything synthesis, for not only does he explain the human condition—our species' capacity for what has been called 'good' and 'evil' (chapter 2)—his ability to plumb the great depths of this most foreboding of all subjects has meant that he has *also* been able to provide the fully accountable explanation of the other great outstanding mysteries in science of the meaning of human existence (chapter 3), the origin of our unconditionally selfless moral instincts (chapter 4), and why we humans became conscious when other animals haven't (chapter 6)—insights that have *then* enabled him to make sense of every other aspect of our troubled human condition, such as the strained relationship between men and women, sex as humans practice it, the origin and nature of politics, religion, and so many other human phenomena (chapter 7). Indeed, preposterous as it must seem, what this book effectively does is take humanity from a state of bewilderment

about the nature of human behavior and existence to a state of profound understanding of our lives. It *truly* is a case of having got all the truth up in one go. Understanding of our species' troubled human condition has *finally* emerged to drain away all the pain, suffering, confusion and conflict from the world—and, given our plight, its arrival couldn't be more timely or serious. If there was ever a case of cometh the hour, cometh the man, this is it!

4 So it is certainly with great trepidation that I attempt to write this Introduction, but I hope that my background as a professor of psychiatry with over 50 years' experience working in the field, including chairing 2 departments of psychiatry and serving as president of the Canadian Psychiatric Association—and my 10 years of knowing the author—might allow me to provide the reader with a helpful insight into this greatest of all books. And by the greatest of all books I do include within that comparison the Bible, which, as it turns out, and as many believed, is also entrenched in truth, but, *unlike* this book, was written in pre-scientific times and thus unable to provide a scientific basis for all those profound truths. In fact, there is really only one book in the world now: this book. It is *so* all-explaining and all-solving that it is, in effect, the new science-based Bible for the human race! It is an extraordinary statement to make but one the Nobel laureate physicist Charles H. Townes was anticipating would one day be possible when he wrote, **'they [science and religion] both represent man's efforts to understand his universe and must ultimately be dealing with the same substance. As we understand more in each realm, the two must grow together...converge they must'** ('The Convergence of Science and Religion', *Zygon*, 1966, Vol.1, No.3). Yes, this is the book we have been waiting for—the book that saves the world.

5 The explanation of the human condition, which Jeremy has found, is *THE* key to a fully unifying understanding of our **'universe'**, it *is* the insight from which all the other mysteries about human life unravel. What needs to be stressed, however, is that finding that explanation depended on overcoming a very great *psychological* hurdle—which Charles Darwin was actually alluding to when, towards the end of *The Origin of Species*, he wrote that **'In the distant future I see open fields for far more important researches. Psychology will be based on a new foundation... Light will be thrown on the origin of man and his history'** (1859, p.458 of 476). Yes, while Darwin shed illuminating light on the origin of the variety of life, there biology has been stalled—and for a very good reason—because

the next step for biologists involved **'far more important'**, in terms of difficulty, **'research'**. For **'light'** to **'be thrown on the origin of man and his history'** the issue of our species' seemingly highly imperfect condition had to be explained, but the **'psycholog**[ical]**'** difficulty was that the human condition is, in fact, *so* unbearably self-confronting and depressing a subject that, as Professor Barnett admitted in the opening quote to this Introduction, only a rare few individuals in recorded history have been able to go anywhere near it. Indeed, as Jeremy explains in this book, the fundamental reason science has been what is referred to as **'reductionist'** and **'mechanistic'** is because it has avoided the great overarching, all-important but unbearably confronting and depressing issue of the human condition and instead *reduced* its focus to only looking down at the details of the *mechanisms* of the working of our world—the great hope being that understanding of those mechanisms would eventually make it possible to explain, understand and thus at last be able to confront the human condition. All very well, but the problem was that to assemble that liberating explanation from the hard-won insights into the mechanisms of the workings of our world, a biologist was going to have to emerge who could face our species' psychosis! Mechanistic science had to do all the hard work of, as it were, finding the pieces of the jigsaw, but unable to look at the whole picture its practitioners were in no position to assemble the jigsaw—that task required someone sound and secure enough in self to look at the whole picture, to confront the human condition, which, as emphasized, is no mean feat in itself, but one which Jeremy was able to accomplish.

[6]In *The Structure of Scientific Revolutions* (1970), science historian Thomas Kuhn noted that revolutions in science are often initiated by an outsider, someone not locked into the current model that hampers vision almost as much as blinders would. And when it comes to addressing the human condition this need to think independently of the existing details-only-focused, whole-view-of-the-human-condition-avoiding, mechanistic framework could not be more critical. For someone to be able to explain, and, through that explanation, bring ameliorating understanding to our troubled human-condition-afflicted lives, they obviously had to be thinking from a position *outside* that conventional mechanistic paradigm. Einstein's famous comment that **'We can't solve problems by using the same kind of thinking we used when we created them'** most certainly applied to solving the problem of the human condition!

[7]So what exactly *is* the human condition, and why has the human race, including the mechanistic scientific establishment, been *so* committed to avoiding it, to living in such complete denial of it, and what have been the consequences for Jeremy, the outsider who dared to break such an entrenched, fiercely held doctrine?

[8]To draw on Jeremy's own penetrating description, the human condition is our species' extraordinary capacity for what has been called 'good' and 'evil'. While it's undeniable that humans are capable of great love and empathy, we also have an unspeakable history of greed, hatred, rape, torture, murder and war; a propensity for deeds so shocking and overwhelming that the eternal question of '*Why?*' seems depressingly inexplicable. Even in our everyday behavior, why, when the ideals of life are to be cooperative, selfless and loving, are we so ruthlessly competitive and selfish that human life has become all but unbearable and our planet near destroyed? How could we humans possibly be considered good when all the evidence seems to unequivocally indicate that we are a deeply flawed, bad, even 'evil' species?

[9]As my profession has taught me only too well, for most people, trying to think about this ultimate of questions of whether we humans are fundamentally good or not has been an unbearably self-confronting exercise. Indeed, while the term 'human condition' has become fashionable, its superficial use masks just how profoundly unsettling a subject it really is. Again, the truth is, the issue of the human condition has been *so* depressing for virtually all humans that only a rare few individuals have been sound and secure enough in self to go anywhere near what the human condition *really* is. So for Jeremy to so freely and accurately talk about the human condition as he does in this book, he clearly must be one of those rare few. Nurtured by a sheltered upbringing in the Australian bush (countryside), it is Jeremy's soundness and resulting extraordinary integrity and thus clarity of thought, coupled with his training in biology, that has enabled him to successfully grapple with this most foreboding of all subjects of the human condition and produce the breakthrough, human-behavior-demystifying-and-ameliorating explanation of it.

[10]Conversely, since the current paradigm of thought that exists within science, and across all aspects of human life, has had no choice but to deny the seemingly inexplicable and unbearably confronting and depressing question of the human condition, scientists have remained failure-

trapped in trying to explain it—if you can't confront the issue you're in no position to solve it. In fact, the human condition has been *such* a fearful, unconfrontable subject that science as a whole has become the purveyor of extremely dishonest theories that seek to falsely account for, and thus dismiss, our paradoxical nature. As Jeremy explains in chapter 1, there can not be a more dangerous example of this than that which appears in biologist E.O. Wilson's 2012 book, *The Social Conquest of Earth*, a theory that evasively trivializes the human condition as nothing more than selfless instincts at odds with selfish instincts within us.

11 Given then that we haven't been able to acknowledge the immense role denial of the issue of the human condition has played in human life (because obviously we couldn't be in denial *and* admit we were in denial), I am going to, with Jeremy's permission, pre-emptively employ some deadly accurate quotes and references that appear in this book to illustrate just *how* terrifying the human condition has been for virtually all humans, and thus *why* humanity, including its scientific fraternity, has *had* to live in almost complete denial of the subject; yes, ours has been a culture of such pervasive dishonesty that it became nearly impossible to unearth the *real*, human-race-liberating, biological explanation of the human condition.

12 The most famous account ever given of the human condition forms the centerpiece of Plato's most acclaimed work, *The Republic*, where humans are metaphorically depicted living as prisoners in a dark cave, deep underground, so fearful are they of the issue of the imperfection of their lives that the sun has the power to reveal. As Plato wrote, **'I want you to go on to picture the enlightenment or ignorance of our human conditions somewhat as follows. Imagine an underground chamber, like a cave with an entrance open to the daylight and running a long way underground. In this chamber are men who have been prisoners there'** (c.360 BC; *Plato: The Republic*, tr. H.D.P. Lee, 1955, p.278 of 405). Plato described how the cave's exit is blocked by a **'fire'** that **'corresponds to the power of the sun'** (p.282), which the cave prisoners have to hide from because its **'painful'** light would make **'visible'** the unbearably depressing issue of **'the imperfections of human life'** (p.282).

13 In terms of how unbearably confronting—in fact, suicidally depressing—the issue of the human condition has been, the great philosopher Søren Kierkegaard is said to have given the best account of that *worse-than-death* experience, writing in his aptly titled 1849 book, *The Sickness*

Unto Death, that **'the torment of despair is precisely the inability to die** [and end the torture of our unexplained human condition]**...that despair is the sickness unto death, this tormenting contradiction** [of our 'good-and-evil'-human-condition-conflicted lives]**, this sickness in the self; eternally to die, to die and yet not to die'** (tr. A. Hannay, 1989, p.48 of 179). The famous psychoanalyst Carl Jung provided an equally stark description of the terrifying nature of the human condition when he wrote: **'When it** [our **shadow**] **appears... it is quite within the bounds of possibility for a man to recognize the relative evil of his nature, but it is a rare and shattering experience for him to gaze into the face of absolute evil'** (*Aion: Researches into the Phenomenology of the Self*, 1959; tr. R. Hull, *The Collected Works of C.G. Jung*, Vol. 9/2, p.10). The **'face of absolute evil'** *is* the **'shattering'** possibility—if humans allow their minds to think about it—that they might indeed be a terrible mistake.

14 It follows then that to confront the human condition has been an impossible ask for most people—as another great philosopher, Nikolai Berdyaev, acknowledged: **'Knowledge requires great daring. It means victory over ancient, primeval terror...it must also be said of knowledge that it is bitter, and there is no escaping that bitterness...Particularly bitter is moral knowledge, the knowledge of good and evil. But the bitterness is due to the fallen state of the world...There is a deadly pain in the very distinction of good and evil, of the valuable and the worthless'** (*The Destiny of Man*, 1931; tr. N. Duddington, 1960, pp.14-15 of 310). Trying to think about our corrupted, **'fallen'**, seemingly **'evil'** and **'worthless'** human-condition-afflicted state *has been* an **'ancient, primeval terror'**, a **'deadly pain'**, **'the bitterest thing in the world'** for virtually all humans—**'knowledge of good and evil'**, of the human condition, **'requires great daring'**. No wonder the Australian zoologist, author and broadcaster, Anthony Barnett, said to Jeremy back in 1983 that **'you are being very arrogant to think you can answer questions on this scale. In the whole of written history there are only two or three people who have been able to think on this scale about the human condition'** (from a recorded interview with Jeremy Griffith, 15 Jan. 1983).

15 It is perhaps not surprising then that the human-condition-avoiding, denial-based, mechanistic or reductionist scientific establishment, along with the human-condition-avoiding public at large, have found Jeremy's human-condition-confronting work deeply heretical, an anathema to be dismissed and even persecuted. Indeed, the persecution of Jeremy and his work was of such ferocity that it led to the biggest defamation case in Australia's history, a case that, after 15 long years, resulted in

the vindication of Jeremy's work. I know all this only too well because I was one of the international scientists who gave evidence in Jeremy's defense during the trial, where I was able to vouch for the veracity of Jeremy's biological thinking on the basis of the extensive studies and work I have carried out over many years on empathy within both humans and bonobos, as well as from all my experience in the field of intensive psychotherapy.

16 The extreme paradox of this situation should be apparent to the reader. As humanity's designated vehicle for enquiry into our world and our place in it, science's ultimate objective and responsibility has been to find understanding of the human condition, so to treat the eventual discovery of its explanation as an anathema, to the point of persecuting it, is ridiculously counter to the whole purpose of science. The fundamental goal of the whole human journey of conscious thought and enquiry has been to find the reconciling, redeeming and rehabilitating explanation of our species' troubled condition, so to reject it when it arrives is madness of the highest order!

17 Professor Barnett, who passed away in 2003, didn't mention to Jeremy who he considered to be the **'two or three people who have been able to think on this scale about the human condition'**, but I think we can deduce that they would be Moses, Plato and Christ. We have already seen the extraordinary soundness and integrity of Plato's thinking in his ability to so clearly describe humans' existence in a cave-like state of denial of the human condition. In fact, Plato's honest, truthful, human-condition-confronting clarity of thought was such that he actually fully anticipated the rejection, indeed persecution, that would occur when someone eventually found the truthful, all-revealing explanation of the human condition. To quote from a summary of Plato's cave allegory from the *Encarta Encyclopaedia*: **'Breaking free, one of the individuals escapes from the cave into the light of day. With the aid of the sun** [living free of denial of the human condition], **that person sees for the first time the real world and returns to the cave with the message that the only things they have seen heretofore are shadows and appearances and that the real world awaits them if they are willing to struggle free of their bonds. The shadowy environment of the cave symbolizes for Plato the physical world of** [false] **appearances. Escape into the sun-filled setting outside the cave symbolizes the transition to the real world...which is the proper object of knowledge'** (written by Prof. Robert M. Baird, accessed Jul. 2008; see <www.wtmsources.com/101>). To return to Plato's own words, **'if he** [the

cave prisoner] **were forcibly dragged up the steep and rocky ascent** [out of the cave] **and not let go till he had been dragged out into the sunlight** [shown the truthful but at the same time exposing and confronting explanation of the human condition]**, the process would be a painful one, to which he would much object, and when he emerged into the light his eyes would be so overwhelmed by the brightness of it that he wouldn't be able to see a single one of the things he was now told were real** [this 'deaf effect' inability to absorb discussion of the human condition is what people often experience when they first try to read analysis of the human condition, which readers of this book should be warned about]' (*The Republic*, p.280). Plato didn't stop there, going on to say, **'they would say that his** [the person who tries to deliver understanding of the human condition] **visit to the upper world had ruined his sight** [they would treat him as if he was mad]**, and that the ascent** [out of the cave] **was not worth even attempting. And if anyone tried to release them and lead them up, they would kill him if they could lay hands on him'** (p.281). Yes, as any psychotherapist knows, denials do fight back with a vengeance when faced with annihilation.

[18] Of course, in today's civilized world more subtle means of eliminating threats to the mechanistic, human-condition-avoiding, denial-based 'cave existence' were employed in the campaign of persecution against Jeremy for being the person who **'escapes from the cave into the light of day'** and then **'dragged** [humanity] **out into the sunlight'**. This persecution was the focal point of the aforementioned legal battle undertaken to defend Jeremy's work, in which the three judges in the New South Wales Court of Appeal unanimously found that an earlier ruling in a lower court did **'not adequately consider' 'the nature and scale of its subject matter'**, in particular **'that the work was a grand narrative explanation from a holistic approach, involving teleological elements'**, and other important submissions **'were not adequately considered by the primary judge'** including that the work can make **'those who take the trouble to grapple with it uncomfortable'** because it **'involves reflections on subject-matter including the purpose of human existence which may, of its nature, cause an adverse reaction as it touches upon issues which some would regard as threatening to their ideals, values or even world views'**! (For details of the persecution and court case, including this vindicating ruling, see <www.worldtransformation.com/persecution>).

[19] The journey that Jeremy and those advocating his work have been on to bring these human-race-saving understandings to the world, which has culminated in this absolutely astonishing book, has been a long and torturous one, but one that makes for fascinating and revealing reading,

so I will now present for the reader a very brief summary of Jeremy's early life, his writing and his unrelenting efforts to free our species from its incarceration in Plato's terrible cave of human-condition-denying, alienated darkness.

---

[20]Born on December 1, 1945, and raised on a sheep station (ranch) in rural New South Wales, Australia, Jeremy was educated at Geelong Grammar School in Victoria, a school whose visionary approach to education has produced such notable alumni as Rupert Murdoch and HRH The Prince of Wales. He gained first class honours in biology in the state matriculation exams and in 1965 began a science degree at the University of New England in northern New South Wales. While there, Jeremy played representative rugby union football, making the 1966 trials for the national team, the Wallabies.

[21]Deferring his studies in 1967, Jeremy undertook the most thorough investigation ever into the plight of the Tasmanian Tiger (thylacine)—a search that was to last more than six years, before concluding the 'Tiger' was indeed extinct. His findings were internationally reported, with articles appearing in the American Museum of Natural History's journal, *Natural History* (see <www.worldtransformation.com/natural-history-tiger>), and *Australian Geographic* (see <www.worldtransformation.com/australian-geographic-tiger>). His search also featured in an episode of the national television series *A Big Country* (see <www.worldtransformation.com/big-country-tiger>).

[22]In 1971 Jeremy completed his Bachelor of Science degree in zoology at the University of Sydney and the following year, in the same self-sufficient spirit with which he had undertaken the 'Tiger' search, he established a highly successful furniture manufacturing business based on his own simple and natural designs, which pioneered the use of bark-to-bark slabs of timber (see <www.worldtransformation.com/griffith-tablecraft>). On the subject of creativity, I should mention that Jeremy is also an accomplished artist (see <www.worldtransformation.com/jeremy-art-work>).

[23]An upbringing nurtured with real, unconditional love (yes, it is to the nurturing from Jeremy's mother—and to the good fortune of having a father who was not oppressively egocentric—that we really owe these world-saving insights) in the sheltered isolation of the

Australian bush left Jeremy deeply troubled and perplexed by all the selfishness, aggression, dishonesty and indifference on one hand, and all the suffering on the other hand, that he inevitably encountered in the innocence-destroyed, psychotic wider world when he grew up. In time, he realized that trying to save animals from extinction or trying to build ideal furniture wasn't going to make a difference to the extraordinary imperfection in human life and that he would have to get to the bottom of the issue of this seeming complete wrongness of human behavior, which is the issue of the human condition. And so it was while building his furniture business that Jeremy first began to write down his ever developing thoughts about the problem of the human condition. Indeed, since 1975 Jeremy has spent the first, often pre-dawn, hours of each day thinking and writing about the human condition, the result of that dedication being his earlier articles and books about the subject: his 1983 submissions to *Nature* and *New Scientist* magazines (which were rejected, with the then editor of *Nature*, John Maddox, telling Jeremy that his starting point teleological argument that there is an underlying order in nature **'is wrong'**—in chapter 3, Jeremy explains why the truth of the order in nature has been denied); *Free: The End Of The Human Condition* (1988); *Beyond The Human Condition* (1991); *A Species In Denial* (2003), which was a bestseller in Australia and New Zealand; *The Human Condition Documentary Proposal* (2004); *The Great Exodus: From the horror and darkness of the human condition* (2006); *Freedom* (2009); *The Book of Real Answers to Everything!* (2011), and now *IS IT TO BE Terminal Alienation or Transformation For The Human Race?* (All the earlier works mentioned here are freely available to be read or printed at <www.worldtransformation.com/publications>.)

[24] Jeremy's books have attracted the support of such eminent scientists in the field as Australia's Templeton Prize-winning biologist Professor Charles Birch and New Zealand's then foremost zoologist Professor John Morton, as well as other distinguished thinkers such as Sir Laurens van der Post. A proposal to make a documentary about the human condition, *The Human Condition Documentary Proposal*, which Jeremy wrote and which outlined all the main biological explanations that are put forward in this book, received over 100 endorsements from many of the world's leading scientists and thinkers, including professors Stephen Hawking and the aforementioned Nobel Laureate Charles H. Townes (see <www.worldtransformation.com/doco-responses>).

25 But while Jeremy's work has attracted praise and received impressive commendations from some exceptional thinkers able to acknowledge his insights, he has, as mentioned, also had to withstand the enormous cynicism, indifference and even persecution that humans' historical resistance to engaging the subject of the human condition produces. Indeed, it was soon after he began writing that Jeremy realized that not only was the scientific establishment failing its responsibility to address the issue of the human condition, but that, like the rest of humanity, it was treating the whole issue as an anathema. As a result, Jeremy established, in 1983, a non-profit organization dedicated to promoting analysis of the human condition, now called the World Transformation Movement (WTM) (<www.worldtransformation.com>). So fearful, however, has humanity been of the issue of the human condition that a vicious campaign was launched in 1995 to try to shut down Jeremy's work and bring the WTM into disrepute, which Jeremy, along with fellow WTM Patron, renowned mountaineer and twice-honoured Order of Australia recipient Tim Macartney-Snape and the other supporters of the WTM, determinedly resisted—the result of which was the biggest defamation case in Australia's history, against the two biggest, left-wing (described by Jeremy as dogmatic, pseudo idealistic, 'let's pretend there's no human condition and the world should just be ideal', dishonest) media organizations in Australia, including its national public broadcaster. As mentioned, in 2010—after 15 long years—Jeremy and Tim were vindicated, enabling Jeremy to concentrate solely on producing new works such as *Freedom*, and his summa masterpiece, this 8-part book, *IS IT TO BE Terminal Alienation or Transformation For The Human Race?*

26 Jeremy lives in Sydney with Annie Williams, his partner since 1980. To learn more about the WTM and Jeremy's role as founder and patron see <www.worldtransformation.com> or <www.humancondition.com>.

---

27 As I have said, Jeremy's journey in bringing understanding to the human condition and protecting the integrity of that explanation—a 40 year saga—has certainly been a long and torturous one (indeed, the persecution was so terrible it left Jeremy seriously debilitated with Chronic Fatigue Syndrome from 1999 to 2009), but bringing understanding to the human condition is the only rational path forward for the human race! Indeed, as a reflection of the looming psychological crisis for the human

race that is the end result of having to live ever deeper **'underground'** in Plato's horrible cave of human-condition-avoiding, dishonest, alienated darkness, and the now desperate need for the reconciling, redeeming and psychologically transforming light of understanding of the human condition, the following initiatives have all taken place in the last 18 months (as at February 2014 when *IS IT TO BE* was published)!

- in December 2012 an American billionaire pledged $200 million to Columbia University's **'accomplished scholars whose collective mission is both greater understanding of the human condition and the discovery of new cures for human suffering'** (*The Educated Observer*, Winter 2013); and,
- in January 2013 the European Commission announced the launch of the **'Human Brain Project with a 2013 budget of €54 million (US$69 million)'** with a **'projected billion-euro funding over the next ten years'** ('Neuroscience: Solving the brain', *Nature*, 2013, Vol.499, No.7458) with the goal of providing **'a new understanding of the human brain and its diseases'** (Press Release, The Human Brain Project, 28 Jan. 2013; see <www.wtmsources.com/111>) to **'offer solutions to tackling conditions such as depression'** ('Scientists to simulate human brain inside a supercomputer', *CNN International Edition*, 12 Oct. 2012; see <www.wtmsources.com/129>); and,
- in April 2013 the President of the United States, Barack Obama, announced a **'Brain Initiative'**, giving **'$100 million initial funding'** (*The Sydney Morning Herald*, 4 Apr. 2013) to mechanistic science to also find **'the underlying causes of...neurological and psychiatric conditions'** afflicting humans (US National Institutes of Health, accessed Apr. 2013; see <www.wtmsources.com/114>); and,
- in April 2013 *BBC News Business* reported that **'Lord Rees, the Astronomer Royal and former president of the Royal Society, is backing plans for** [Cambridge University to open] **a Centre for the Study of Existential Risk** [meaning risk to our existence]. **"This is the first century in the world's history when the biggest threat is from humanity," says Lord Rees'** ('How are humans going to become extinct?', 24 Apr. 2013; see <www.wtmsources.com/120>). The article then referred to Oxford University's Future of Humanity Institute that was established in 2005, which is **'looking at big-picture questions for human civilization...**[and] **change...**[that] **might transform the human condition'** (Future of Humanity Institute, University of Oxford, accessed Jul. 2013; see <www.wtmsources.com/118>), quoting its Director and advisor to the Centre for the Study of Existential Risk, Nick Bostrom: **'There is**

a bottleneck in human history. The human condition is going to change. It could be that we end in a catastrophe or that we are transformed by taking much greater control over our biology'! (Note how Bostrom's comment reaffirms Jeremy's assessment of the plight of the world. Indeed, Bostrom's comment could have been—although wasn't, because it was made prior to this book's publication—inspired by Jeremy's title for this book: *IS IT TO BE Terminal Alienation or Transformation For The Human Race?*)

28 But establishing a center to study the human condition is *precisely* the initiative Jeremy took 30 years ago when he created the self-funded WTM (an act of prescience that evidences his clarity and integrity of thought)—an institute that even has the domain name **'humancondition.com'**. And while these recently announced initiatives are admirable in their quest to solve the mental problems now threatening to overwhelm humankind, they are attempting to do so from the traditional human-condition-*avoiding*, reductionist, mechanistic position, which is self-defeating. An opinion piece by Benjamin Y. Fong of the University of Chicago that was published in *The New York Times* in August 2013 made this point: **'The real trouble with the Brain Initiative is...the instrumental approach...**[such **biological reduction** is] **intent on uncovering the organic "cause"...of mental problems...rather than looking into psychosocial factors... By humbly claiming ignorance about the "causes" of mental problems... neuroscientists unconsciously repress all that we know about the alienating, unequal, and dissatisfying world in which we live and the harmful effects it has on the psyche, thus unwittingly foreclosing'** the ability to **'alleviate mental disorder'** (11 Aug. 2013). Yes, to summarise what Fong has said here and elsewhere in his article, mechanistic science's **'synthetic'** focus on the **'organic'** rather than the **'psychological'** nature of our problems can only end in denying humans **'the possibility of self-transformation'**. It is *only* the *psychosis-addressing, human-condition-confronting* position that Jeremy has taken that could hope to find, and now *has* found, the reconciling and human-race-transforming understanding of the human condition—and yet it is his approach that has been treated as heretical, an anathema and a threat by the mechanistic scientific establishment!! My sincere hope, however, is that with our species' predicament now so dire, the scientific establishment will finally acknowledge and support Jeremy's human-race-saving insight into the human condition—and the

other critically important insights made possible by his power to unravel our species' psychosis.

[29]In conclusion—and given Jeremy was so vilified he was made a pariah—I would like to emphasize the height of my regard for him. He is the most impressive person and courageous thinker I have ever met and no doubt ever will. Normally people disappoint you at some point, or on some occasions, but Jeremy never does. It is quite amazing, in all my years of meeting people and practicing psychiatry, I haven't encountered a soul like him. He is one of those incredibly rare individuals, a person of intellectual rigor and personal nobility who has the capacity to be completely honest without a personal bent; in fact, when you are with him you can feel his passion for the truth, which he embodies. Indeed, meeting Jeremy as I did after reading so much of his work, I realized that he lives 100 percent in the world that he writes about—an immensely inspired, child-like-zest-full, enthralled-with-all-of-life, truthful world where the human condition is at all times being addressed and understood. Unlike everyone I have ever come across, for whom discussion of the human condition is so extremely difficult (indeed, it is almost impossible), when you talk to Jeremy about the human condition and the biology surrounding it, the world changes, everything seems possible, biology makes sense. Logic—simple and obvious truth—replaces over-complicated intellectual scientific downright rubbish—the vast majority of it. His capacity for honesty is literally staggering, but, as you will see, it is always accompanied by understanding; his is no 'feel good', guru-like, false prophet form of totally dubious and ephemeral 'help' for humans' troubled lives, but a get-to-the-bottom-of-all-the-problems, truthful, *real*, insightful, ameliorating love that the world has been so in need of. And, thankfully, right through all the vehement resistance that he has faced and overcome in the last 30 years of his life, which took him near to death, Jeremy never gave up his responsibility that he saw right from the early years of his completely-nurtured-with-love upbringing to deliver the understanding of human nature that would end all the suffering in the world.

[30]Since coming across Jeremy's work almost 10 years ago and realizing its enormous world-saving significance, I have kept one of his pieces of writing with me at all times. Having these insights into what it means to be human brings such clarity and change to everything, that staying in close contact with them in a world that is so distressed and psychologically crippled brings me relief, security and optimism

for the future like nothing else that I have ever encountered. Jeremy is the ultimate psychotherapist, the psychotherapist for psychotherapists—in fact, all the great theories that I have encountered in my lifetime of studies of psychiatry can be accounted for under his explanation of human origins and behavior. I want to emphasize that Jeremy does describe the world exactly as it is with all its imperfections, but just as he describes all the horror of the world he also provides the insight into our condition that makes possible the only real hope, optimism and downright, out-of-your-skin excitement for the future of the human race and our planet—a transformational experience that Jeremy introduces us to in chapter 8. Indeed, the explanation of the human condition in chapter 2 is so amazingly accountable and relieving that it alone will transform you. Yes, the importance of these ideas is immeasurable. The depths they enable us to reach in understanding ourselves and our world is bottomless. The great impasse to a full understanding of our existence has finally been breached. This truly is it, the day of days, the coming of our species' moment of liberation, the implications and context of which Jeremy fully deals with in pars 1068 and 1069.

31 I am so very, very fortunate to have spent time with Jeremy and to have spoken with him almost fortnightly for 10 years—however, it is not who Jeremy is, but what he has done that is so important and in this regard I commend this book to you with all my mind, heart and soul. You are in for an absolute feast of knowledge, insight and ultimately love like you could never have imagined.

32 (I am very grateful to Fiona Cullen-Ward and Tony Gowing at the WTM for their assistance in preparing this Introduction.)

Harry Prosen, Wisconsin, USA, 2014

Chapter 1

Is It To Be
Terminal Alienation or Transformation

William Blake's *Cringing In Terror* (c. 1794-96)

William Blake's *Albion Arose* (c. 1794-96); coloured impression by Carol Marando

For The Human Race?

Chapter 1:1 Summary

[33] In 2012 the most celebrated living biologist, the Harvard-based Edward (E.) O. Wilson, published a book, *The Social Conquest of Earth*, in which he claimed to have found THE holy grail in humanity's great journey of conscious thought and enquiry of the explanation of our 'good-and-evil'-afflicted, so-called 'human condition'. But has Wilson finally opened the shutters and let the liberating light of understanding stream in upon the agonising dilemma of what it is to be human? Has he finally delivered the reconciling, redeeming and psychologically transforming insight that we humans have lived in hope, faith and trust would one day be found? Or has Wilson taken the human race into the deepest and darkest corner of truth-avoiding denial and alienation it has ever known? This first chapter in this book about human behaviour and the destiny of the human race will

explain that the latter is, in fact, the case—that mechanistic science, of which Wilson is the quintessential exponent, *is* locking humanity onto a path to terminal alienation! But all is not lost, for although the truth-avoiding mechanistic paradigm currently finds it an anathema, the fully accountable, *real*, human-psychosis-addressing-and-solving, transforming biological explanation of the human condition *has* been found—and will be presented in the next chapter. So the question now is, which will triumph: terminal alienation or transformation?

Chapter 1:2 The true nature of the human condition — and our Resignation to living in denial of it

34 Any *real* analysis of human behaviour requires an appreciation of what the issue of the human condition actually is—but that is not easily achieved because the issue of the human condition has been *so* unbearably confronting that it is now almost impossible for us humans to connect with what it really is, to the extent that some seriously believe that it doesn't even exist. So, the first task in this book is to reconnect the reader to the true nature of the human condition.

35 In fact, aggravating as it may be, it is so absolutely fundamentally important to everything that is going to be explained in this book that I want to repeat what has just been stated, which is that "Any *real* analysis of human behaviour requires an appreciation of what the issue of the human condition actually is...So, the first task in this book is to reconnect the reader to the true nature of the human condition."

---

36 Here on Earth some of the most complex arrangements of matter in the known universe have come into existence. Life, in all its incredible diversity and richness, developed. And, by virtue of our mind, the human species must surely represent the culmination of this grand experiment of nature we call life—for, as far as we can detect, our species is the first organism to have developed the fully conscious ability to sufficiently understand and thus manage the relationship between cause and effect to wrest management of our lives from our instincts, and even to reflect upon our existence. In a world ravaged by strife it is easy to lose sight of the utter magnificence of what we are; yes, the human mind must surely

be nature's most astonishing creation. Indeed, our brain must be one of the wonders of the universe! Perhaps one of the greatest demonstrations of this intellectual brilliance was sending three of our kind to the Moon and back.

[37] *HOWEVER*, despite our magnificent mental capabilities, and also an undeniable capacity for immense sensitivity and love, behind every wondrous scientific achievement, sensitive artistic expression and compassionate act lies the shadow of humanity's darker side—an unspeakable history of greed, hatred, rape, torture, murder and war; a propensity for deeds of shocking violence, depravity, indifference and cruelty (as the philosopher Arthur Schopenhauer wrote, **'man is the only animal which causes pain to others with no other object than causing pain... No animal ever torments another for the sake of tormenting: but man does so, and it is this which constitutes the *diabolical* nature which is far worse than the merely bestial'** (*Essays and Aphorisms*, tr. R.J. Hollingdale, 1970, p.139 of 237)). Indeed, the truth is that despite all our marvellous accomplishments and sensibilities, we humans have been the most ferocious and destructive creatures to have ever lived on Earth!

[38] And it is precisely this duality of what has historically been referred to as 'good' and 'evil' that has troubled the human mind since we first became fully conscious, thinking beings: are humans essentially 'good' and, if so, what is the cause of our destructive, insensitive and cruel so-called 'evil' side? *Why* do we thinking, reasoning, rational, immensely clever, supposedly sensible beings behave so abominably and cause so much suffering and devastation? Yes, the eternal question has been why 'evil'? What is the origin of the dark, volcanic forces that undeniably exist within us humans? What is it deep within us humans that troubles us so terribly? What is it that makes us such a combative, ruthless, hateful, retaliatory, violent, in-truth-psychologically-disturbed creature? In metaphysical religious terms, what is 'the origin of sin'? More generally, if the universally accepted ideals of life are to be *cooperative, loving and selfless*—ideals that have been accepted by modern civilisations as the foundations for constitutions and laws and by the founders of all the great religions as the basis of their teachings—*why* are humans *competitive, aggressive and selfish*? Indeed, *so* ruthlessly competitive, selfish and brutal that human life has become all but unbearable and we have nearly destroyed our own planet? Does our inconsistency with the ideals

mean we are essentially bad, a flawed species, an evolutionary mistake, a blight on Earth, a cancer in the universe—or could we possibly be divine beings? And more to the point, is the human race faced with having to live *forever* in this tormented state of uncertainty and insecurity about the fundamental worth and meaning of our lives? Is it our species' destiny to have to live in a state of permanent damnation?!

39 The agony of being unable to truthfully answer this fundamental question of why we are the way we are—divisively instead of cooperatively behaved—has been the particular burden of human life. It has been our species' particular affliction or condition—our *'human condition'*. Good or bad, loving or hateful, angels or devils, constructive or destructive, sensitive or insensitive: *what are we?* Throughout history we humans have struggled to find meaning in the awesome contradiction of our human condition. Neither philosophy nor psychology nor biology has, until now, been able to provide a truthful, clarifying explanation. For their part, religious assurances such as 'God loves you' may have provided temporary comfort but failed to explain *WHY* we are lovable. So, yes, *WHY* are we lovable? How could we be good when all the evidence seems to unequivocally indicate that we are a deeply flawed, bad, even evil species? What is the answer to this question of questions, this problem of 'good and evil' in the human make-up, this greatest of all paradoxes and dilemmas of the human condition? *What caused humans to become divisively behaved and, more importantly, how is this divisive behaviour ever going to be brought to an end? THIS*, the issue of the human condition, is the *REAL* question facing the human race. (Note, the reason I said we have been *unable to truthfully* answer the question of why we have been divisively behaved is because—as will be explained in this chapter—historically we have used the excuse that we humans are competitive, aggressive and selfish because of our animal heritage; that we have savage animal instincts that make us fight and compete for food, shelter, territory and a mate; basically, for the chance to reproduce our genes. Of course, this reason that biologists, including E.O. Wilson, have been perpetuating cannot be the *real* cause of our divisive behaviour because descriptions of our human behaviour, such as egocentric, arrogant, deluded, optimistic, pessimistic, artificial, hateful, mean, immoral, guilty, evil, depressed, inspired, psychotic, alienated, all recognise the

involvement of our species' unique fully conscious thinking mind—that there is a *psychological* dimension to *our* behaviour. We have suffered not from the genetic-opportunism-based, non-psychological *animal* condition, but the conscious-mind-based, psychologically distressed *human* condition—the truthful, fully accountable, *real* biological explanation of which will be presented in chapter 2.)

40 Indeed, with every day bringing with it more startling evidence of the turmoil of the human situation, the issue of the human condition is the *ONLY* question facing the human race, because its solution has become a matter of critical urgency. Conflict between individuals, races, cultures and countries abounds; there is genocide, terrorism, mass displacement of peoples, starvation, runaway diseases, environmental devastation, gross inequality, racial and gender oppression, rampant corruption and other crimes, drug abuse, obesity, family breakdown and epidemic levels of depression, unhappiness and loneliness. And an exploding world population is rapidly exacerbating all of these problems. Improved forms of management such as better laws, better politics and better economics—and better self-management, such as new ways of disciplining, suppressing, organising, motivating or even transcending our troubled natures—have all failed to end the march towards ever greater levels of devastation and unhappiness. In short, the situation is now *so* dire the human race is, in fact, entering *end play* or *end game*, where the Earth cannot absorb any further devastation from the effects of our behaviour, nor the human body cope with any more debilitating stress, or, most particularly, our mind endure any more psychological distress, any more alienated psychosis and neurosis. Yes, the journalist Richard Neville was frighteningly accurate when, in summarising the desperate state of our species' situation, he wrote that **'The world is hurtling to catastrophe: from nuclear horrors, a wrecked ecosystem, 20 million dead each year from malnutrition, 600 million chronically hungry...All these crises are man made, their causes are psychological. The cures must come from this same source; which means the planet needs psychological maturity...fast. We are locked in a race between self destruction and self discovery'** (*Good Weekend, The Sydney Morning Herald*, 14 Oct. 1986). *Only* **'self discovery'**—reconciling, ameliorating, **'psychological**[ly]' healing understanding of ourselves—could save us from **'self destruction'**.

Cartoon by Michael Leunig, Melbourne's *The Age* newspaper, 8 Oct. 1988

[41]This masterpiece by the Australian cartoonist Michael Leunig truthfully depicts all the horrors of the human condition. Leunig has been contributing cartoons to and writing for Australian newspapers since 1965 and in that time has produced innumerable brilliantly honest, insightful and therapeutically revealing cartoons about all aspects of our species' troubled condition—four more of which appear later in this book. This cartoon is in my view one of his best. It is certainly *not* a picture of a lovely ordered city park where people peacefully and happily enjoy themselves, as we all too easily prefer to delude ourselves that the world we have created is like. Rather, it shows a mother and child approaching the **'Gardens of the Human Condition'** with an expression of bewildered dread on the face of the mother, and in the case of the child, wide-eyed shock. Yes, as Leunig cleverly intimates, our world is no longer an innocent Garden of Eden, but a devastated realm of human-condition-afflicted, psychologically distressed humans where 'inhumanity' reigns.

With this cartoon, Leunig has boldly revealed the truth that we humans are a brutally angry, hateful, destructive, arrogant, egocentric, selfish, mad, lonely, unhappy and psychologically depressed species. He has people fighting, beating and strangling each other, drunk out of their minds, depressed, lonely, crying, hiding and suiciding, going mad and egocentrically holding forth—reflecting, in effect, every aspect of the human condition. Yes, as the main character in the 2005 film *The White Countess* observes: **'What we see out there** [in the world] **is chaos; mistrust, deception, hatred, viciousness—chaos—there's no broader canvas out there, nothing that man can go and compose a pretty picture on.'**

[42] So while environmental issues have dominated our concerns in recent times, we have only been focusing on the symptoms. To stop the destruction of our world and the disintegration of society that is happening everywhere we look we have to fix the *cause* of the problems, which is *us humans*. *We* are the problem: our out-of-control egocentric, selfish, competitive and aggressive behaviour. The underlying real question that had to be answered if we humans were ever to find relieving, redeeming, psychologically healing understanding of ourselves was *'Why have we humans been the way we have been, less than ideally behaved, in fact capable of horrific atrocities?'*

[43] This predicament or 'condition' of not being able to understand ourselves, however, has meant that the more we tried to think about this, in truth, most obvious question of the meaningfulness and worthiness, or otherwise, of our existence, the more insecure and depressed our thoughts became. In time, we learnt that to avoid the suicidal depression that accompanied *any* thinking about the issue of our species', and our *own*, seemingly *extremely* imperfect condition, we simply had no choice but to stop grappling with the answerless question. And so despite the human condition being the all-important issue of the meaningfulness or otherwise of our existence, there came a point in virtually everyone's life—and it typically occurred at around 14 or 15 years of age—when we were each forced to put the whole depressing subject aside once and for all and just hope that one day in the future the explanation and defence of our species' apparently horrifically flawed, seemingly utterly disappointing, sad state would be found, because then, and *only* then, would it be psychologically safe to even broach the subject. Indeed, The Beatles' song *Let It Be*—consistently voted one of the most popular songs of

the twentieth century—is an anthem to the need that adolescents have historically had, when confronted with the unbearable **'hour of darkness'** that came from grappling with the issue of all **'the broken hearted people living in the world'**, to **'let it be' 'until tomorrow'** when **'there will be an answer'** (Lennon/McCartney, 1970). When the poet Gerard Manley Hopkins wrote about the unbearably depressing subject of the human condition in his aptly titled poem *No Worst, There Is None* (1885), his words, **'O the mind, mind has mountains; cliffs of fall, frightful, sheer, no-man-fathomed'**, did not exaggerate the depth of depression humans faced if we allowed our minds to think about the human condition while it was still to be **'fathomed'**/understood. So, yes, when, in **'my hour of darkness'**, **'Mother Mary comes...speaking words of wisdom, let it be, let it be'**—accept her wisdom, and don't go there!

44 It's little wonder then that the human condition has been described so vehemently as **'the personal unspeakable'** and as **'the black box inside of humans they can't go near'** (personal conversations, WTM records, Feb. 1995)—*and* why it is so very rare to find a description of adolescents going through the excruciating process of Resignation, of resigning themselves to having to block out the seemingly inexplicable question of their worth and meaning and live, from that time on, in denial of the unbearable issue of the human condition. Having already been through this terrible process of Resignation, most adults simply couldn't allow themselves to recall, recognise and thus empathise with what adolescents were experiencing—and so our young have been alone with their pain, unable to share it with those closest, or the world at large. All of which makes the following account of a teenager in the midst of Resignation, by the American Pulitzer Prize-winning child psychiatrist Robert Coles, incredibly special: **'I tell of the loneliness many young people feel...It's a loneliness that has to do with a self-imposed judgment of sorts...I remember...a young man of fifteen who engaged in light banter, only to shut down, shake his head, refuse to talk at all when his own life and troubles became the subject at hand. He had stopped going to school...he sat in his room for hours listening to rock music, the door closed...I asked him about his head-shaking behavior: I wondered whom he was thereby addressing. He replied: "No one." I hesitated, gulped a bit as I took a chance: "Not yourself?" He looked right at me now in a sustained stare, for the first time. "Why do you say that?"** [he asked]**...I decided not to answer the question in the manner that I was trained...Instead, with some unease...I heard myself saying this: "I've been there; I remember**

being there—remember when I felt I couldn't say a word to anyone"...The young man kept staring at me, didn't speak...When he took out his handkerchief and wiped his eyes, I realized they had begun to fill' (*The Moral Intelligence of Children*, 1996, pp.143-144 of 218). The boy was in tears because Coles had reached him with *some* recognition and appreciation of what he was wrestling with; Coles had shown *some* honesty about what the boy could see and was struggling with, namely the horror of the utter hypocrisy of human behaviour—including his own.

[45]The words Coles used in his admission that he too had once grappled with the issue of the human condition, of **'I've been there'**, are exactly those used by one of Australia's greatest poets, Henry Lawson, in his extraordinarily honest poem about the unbearable depression that results from trying to confront the question of why human behaviour is so at odds with the cooperative, loving—or, to use religious terms, 'Godly'—ideals of life. In his 1897 poem *The Voice from Over Yonder*, Lawson wrote: **'"Say it! think it, if you dare! Have you ever thought or wondered, why the Man and God were sundered** [torn apart]**? Do you think the Maker blundered?" And the voice in mocking accents, answered only: "I've been there."'** The unsaid words in the final phrase, **'I've been there'**, being 'and I'm certainly not going **'there'** again!'—with the **'there'** and the **'over yonder'** of the title referring to the state of deepest, darkest depression.

[46]In his bestselling 2003 book, *Goya*, about the great Spanish artist Francisco Goya, another Australian, Robert Hughes, who for many years was *TIME* magazine's art critic, described how he **'had been thinking about Goya...**[since] **I was a high school student in Australia...**[with] **the first work of art I ever bought...**[being] **a poor second state of Capricho 43...*The sleep of reason brings forth monsters*...**[Goya's most famous etching reproduced overleaf] **of the intellectual beset with doubts and night terrors, slumped on his desk with owls gyring around his poor perplexed head'** (p.3 of 435). Hughes then commented that, **'glimpsing *The sleep of reason brings forth monsters* was a fluke'** (p.4). A little further on, Hughes wrote of this experience that **'At fifteen, to find this voice** [of Goya's]**—so finely wrought** [in *The sleep of reason brings forth monsters*] **and yet so raw, public and yet strangely private—speaking to me with such insistence and urgency...was no small thing. It had the feeling of a message transmitted with terrible urgency, mouth to ear: this is the *truth*, you *must know* this, I have *been through* it'** (p.5). Again, while the process of Resignation is such a horrific experience that adults determined never to revisit it, or even recall it, Hughes' attraction to *The sleep of reason brings forth monsters*

was not the **'fluke'** he thought it must have been. The person slumped at the table with owls and bats gyrating around his head perfectly depicts the bottomless depression that occurs in humans just prior to resigning to a life of denial of the issue of the human condition, and someone in that situation would have recognised that meaning instantly, almost wilfully drawing such a perfect representation of their state out of the world around them. Even the title is accurate: **'The sleep of reason'**—namely reasoning at a very deep level—does **'bring forth monsters'**! While Hughes hasn't recognised that what he was negotiating **'At fifteen'** was Resignation, he has accurately recalled how strong his recognition was of what was being portrayed in the etching: **'It had the feeling of a message transmitted with terrible urgency, mouth to ear: this is the *truth*, you *must know* this, I have *been through* it.'** Note how Hughes' words, **'I have *been through* it'**, are almost identical to Coles and Lawson's words **'I've been there.'**

Goya's *The sleep of reason brings forth monsters*, 1796-97

[47]And so, unable to acknowledge the process of Resignation, adults instead blamed the well known struggles of adolescence on the hormonal upheaval that accompanies puberty, the so-called 'puberty blues'—even terming glandular fever, a debilitating illness that often occurs in mid-adolescence, a puberty-related 'kissing disease'. These terms, 'puberty blues' and 'kissing disease', are dishonest, denial-complying, evasive excuses because it wasn't the onset of puberty that was causing the depressing 'blues' or glandular fever, but the trauma of Resignation. For glandular fever to occur, a person's immune system must be extremely rundown, and yet during puberty the body is physically at its peak in terms of growth and vitality—so for an adolescent to succumb to the illness they must be under extraordinary psychological pressure, experiencing stresses much greater than those that could possibly be associated with the physical adjustments to puberty, an adjustment, after all, that has been going on since animals first became sexual. No, the depression and glandular fever experienced by young adolescents are a direct result of the trauma of having to resign to never again revisiting the unbearably depressing subject of the human condition.

[48]That sublime classic of American literature, J.D. Salinger's 1951 novel *The Catcher in the Rye*, is a masterpiece because, like Coles, Salinger dared to write about that forbidden subject for adults of adolescents having to resign to a dishonest life of denial of the human condition—for *The Catcher in the Rye* is entirely about a 16-year-old boy struggling against Resignation. The boy, Holden Caulfield, complains of feeling **'surrounded by phonies'** (p.12 of 192) and **'morons'** who **'never want to discuss anything'** (p.39), of living on the **'opposite sides of the pole'** (p.13) to most people, where he **'just didn't like anything that was happening'** (p.152), to wanting to escape to **'somewhere with a brook...**[where] **I could chop all our own wood in the winter time and all'** (p.119). He knows he is supposed to resign—in the novel he talks about being told that **'Life...**[is] **a game... you should play it according to the rules'** (p.7), and to feeling **'so damn lonesome'** (pp.42, 134) and **'depressed'** (multiple references) that he felt like **'committing suicide'** (p.94). As a result of all this despair and disenchantment with the world he keeps **'failing'** (p.9) his subjects at school and is expelled from four for **'making absolutely no effort at all'** (p.167). About his behaviour he says, **'I swear to God I'm a madman'** (p.121) and **'I know. I'm very hard to talk to'** (p.168). But like the boy in Coles' account, Holden finally encounters

some rare honesty from an adult that, in Holden's words, **'really saved my life'** (p.172). This is what the adult said: **'This fall I think you're riding for—it's a special kind of fall, a horrible kind...**[where you] **just keep falling and falling** [utter depression]**'** (p.169). The adult then spoke of men who **'at some time or other in their lives, were looking for something their own environment couldn't supply them with...So they gave up looking** [they resigned]...[adding] **you'll find that you're not the first person who was ever confused and frightened and even sickened by human behavior'** (pp.169-170). Yes, to be **'confused and frightened'** to the point of being **'sickened by human behavior'**—indeed, to be **'suicid**[ally]**' 'depressed'** by it—is the effect the human condition has if you haven't resigned yourself to living a relieving but utterly dishonest and superficial life in denial of it.

49 Going through Resignation *has* been a truly horrific experience. A friend and I were walking in bushland past a school one day when we came across a boy, who would have been about 14 years old, sitting by the track in an hunched, foetal position. When I asked him if he was okay he looked up with such deep despair in his eyes that it was clear he didn't want to be disturbed and so we left him alone. It was very apparent that he was trying to wrestle with the issue of the human condition, but without understanding of the human condition it hasn't been possible for virtually all humans to do so without becoming so hideously condemned and thus depressed that they had no choice but to eventually surrender and take up denial of the issue of the human condition as the only way to cope with it—even though doing so meant adopting a completely dishonest, superficial and artificial, effectively dead, existence.

50 I haven't as yet come across a photograph of an adolescent in the midst of Resignation, however, in my picture collection I do have the following haunting image of a boy who had, the previous day, lost all his classmates in a plane crash, and his expression is exactly the same deeply sobered, drained pale, all-pretences-and-facades-stripped-away, pained, tragic, stunned, human-condition-laid-bare expression I have seen on the faces of adolescents going through Resignation. We can see in this boy's face that all the artificialities of human life have been rendered meaningless and ineffectual by the horror of losing all his friends, leaving bare only the sad, painful awareness of a world devoid of any real love or truth.

'Too poor to go on school trip, boy fishes the day after classmates perish in plane crash'
LIFE magazine, Fall Special Edition, 1991

51 Although rarely shared, adolescents in the midst of Resignation quite often write excruciatingly honest poems about their impending fate; indeed, *The Catcher in the Rye* is really one long poem about the agony of having to resign to living a human-condition-denying, superficial, totally false existence. I have written much more about Resignation at <www.worldtransformation.com/freedom-resignation>, however, the following are two horrifically honest Resignation poems that are discussed at that link and worth including here to provide first-hand insights into the agony of adolescence: **'You will never have a home again / You'll forget the bonds of family and family will become just family / Smiles will never bloom from your heart again, but be fake and you will speak fake words to fake people from your fake soul / What you do today you will do tomorrow and what you do tomorrow you will do for the rest of your life / From now on pressure, stress, pain and the past can never be forgotten / You have no heart or soul and**

there are no good memories / Your mind and thoughts rule your body that will hold all things inside it; bottled up, now impossible to be released / You are fake, you will be fake, you will be a supreme actor of happiness but never be happy / Time, joy and freedom will hardly come your way and never last as you well know / Others' lives and the dreams of things that you can never have or be part of, will keep you alive / You will become like the rest of the world—a divine actor, trying to hide and suppress your fate, pretending it doesn't exist / There is only one way to escape society and the world you help build, but that is impossible, for no one can ever become a baby again / Instead you spend the rest of life trying to find the meaning of life and confused in its maze'; and **'Growing Up: There is a little hillside / Where I used to sit and think / I thought of being a fireman / And of thoughts, I thought important / Then they were beyond me / Way above my head / But now they are forgotten / Trivial and dead.'** Yes, as these poems so painfully express, Resignation means blocking out all memory of the innocent, soulful, true world because it is unbearably condemning of our present immensely corrupted human condition: **'You have no heart or soul and there are no good memories / Your mind and thoughts rule your body that will hold all things inside it; bottled up, now impossible to be released / You are fake, you will be fake, you will be a supreme actor of happiness but never be happy.'** And since virtually all adults have resigned, that is exactly how **'fake'**, or as the 16-year-old Holden Caulfield described it, **'phony'**, they have become. Clearly, the price of Resignation is *enormous*, but the alternative for virtually all humans of not resigning has been an *even worse* fate because it meant living with constant suicidal depression.

52 Appreciably then, in what forms the key passage in *The Catcher in the Rye* (indeed, it provides the meaning behind the book's title), Salinger has Holden Caulfield dreaming of a time when this absolute horror, indeed obscenity, of Resignation will no longer have to be an unavoidable part of human life: **'I keep picturing all these little kids playing some game in this big field of rye and all. Thousands of little kids, and nobody's around—nobody big, I mean—except me. And I'm standing on the edge of some crazy cliff. What I have to do, I have to catch everybody if they start to go over the cliff—I mean if they're running and they don't look where they're going I have to come out from somewhere and *catch* them. That's all I do all day. I'd just be the catcher in the rye and all. I know it's crazy, but that's the only thing I'd really like to be'** (p.156). *And*, as will be explained in this book, the time that Holden Caulfield so yearned for when we will be able to **'catch'** all children before **'they start to go over the** [excruciating] **cliff'** of Resignation to a life

of utter dishonesty, **'phony'**, **'fake'** superficiality, and silence in terms of **'never want**[ing] **to discuss anything'** truthfully again, *has* finally come about with the finding of the all-clarifying, redeeming and relieving truthful explanation of the human condition! Yes, the real **'catcher in the rye'** is the ability to explain **'human behavior'** so that it is no longer **'sicken**[ing]**'** but understandable, and, best of all, healable. Again, that Resignation-ending, human-race-transforming explanation of the human condition is *not* E.O. Wilson's extremely dishonest, threatening-to-bring-terminal-alienation account of the human condition that is dealt with in this first chapter, but the fully accountable, true explanation of the human condition that will be presented in chapter 2.

Chapter 1:3 Our near total resistance to analysis of the human condition post-Resignation

[53] It is a measure of just how unbearable the issue of the human condition has been that while Resignation has been *the* most important psychological event in human life it has virtually gone unacknowledged, with only a rare few of our most accomplished writers even managing to write about the suicidally depressing experience of engaging the subject of the human condition itself. To this end, many **'philosophers and psychologists'** consider that the great Danish philosopher Søren Kierkegaard's **'analysis on the nature of despair is one of the best accounts on the subject'** (Wikipedia, accessed Sep. 2010; see <www.wtmsources.com/137>)—with the **'nature of despair'** being as close as the reviewer could go in referring to the worse-than-death, suicidal depression that the human condition has caused humans, but which Kierkegaard managed to give such an honest account of in his aptly titled 1849 book, *The Sickness Unto Death*: **'the torment of despair is precisely the inability to die** [and end the torture of our unexplained human condition]**...that despair is the sickness unto death, this tormenting contradiction** [of our 'good and evil', human condition-afflicted lives]**, this sickness in the self; eternally to die, to die and yet not to die'** (tr. A. Hannay, 1989, p.48 of 179). Kierkegaard went on to include these unnervingly truthful words about how, even when the blocks *were* in place in our minds against recognising the existence of the issue of the human condition, the terrifying **'anxiety'** it caused us still occasionally surfaced: **'there is not a single** [adult] **human being who does not despair at least a little, in whose innermost being there doesn't dwell an uneasiness, an unquiet, a discordance, an anxiety**

in the face of an unknown something, or a something he doesn't even dare strike up acquaintance with...he goes about with a sickness, goes about weighed down with a sickness of the spirit, which only now and then reveals its presence within, in glimpses, and with what is for him an inexplicable anxiety' (p.52).

[54] Another great philosopher, the Russian Nikolai Berdyaev, gave this extraordinarily forthright description of how trying to address and solve the sickeningly depressing issue of the human condition has been a nightmare: **'Knowledge requires great daring. It means victory over ancient, primeval terror. Fear makes the search for truth and the knowledge of it impossible...it must also be said of knowledge that it is bitter, and there is no escaping that bitterness...Particularly bitter is moral knowledge, the knowledge of good and evil. But the bitterness is due to the fallen state of the world, and in no way undermines the value of knowledge...it must be said that the very distinction between good and evil is a bitter distinction, the bitterest thing in the world... There is a deadly pain in the very distinction of good and evil, of the valuable and the worthless'** (*The Destiny of Man*, 1931; tr. N. Duddington, 1960, pp.14-15 of 310). Yes, trying to think about our corrupted, **'fallen'**, seemingly **'evil'** and **'worthless'** state *has* been an **'ancient, primeval terror'**, a **'deadly pain'**, **'the bitterest thing in the world'** for virtually all humans.

[55] So the famous Swiss psychoanalyst Carl Jung certainly exhibited **'great daring'** in his thinking when he wrote the following words about the terrifying nature of the human condition: **'When it** [our **shadow**] **appears...it is quite within the bounds of possibility for a man to recognize the relative evil of his nature, but it is a rare and shattering experience for him to gaze into the face of absolute evil'** (*Aion: Researches into the Phenomenology of the Self*, 1959; tr. R. Hull, *The Collected Works of C.G. Jung*, Vol. 9/2, p.10). The **'face of absolute evil'** *is* the **'shattering'** possibility—if we allowed our minds to think about it—that we humans might indeed be a terrible mistake. And so to avoid that implication, we humans have had to avoid almost all deep, penetrating, truthful thinking because virtually *any* thinking at a deeper level brought us into contact with the unbearable issue of our seemingly horribly flawed condition: 'There's a tree with lovely autumn leaves; isn't it amazing how beautiful nature can be, I wonder why some things are beautiful while others are not—I wonder why I'm not beautiful inside, in fact *so* full of all manner of angst, selfish self-obsession, indifference and anger...aaahhhhh!!!!' The very great poet William Wordsworth was making the same point when he wrote, **'To me the meanest flower that blows can give thoughts that do often lie too deep for tears'** (*Intimations of Immortality from Recollections of*

Early Childhood, 1807), for it is true that even the plainest flower can remind us of the unbearably depressing issue of our seemingly horrifically imperfect, **'fallen'**, seemingly **'worthless'** condition. Yes, as the comedian Rod Quantock once said, **'Thinking can get you into terrible downwards spirals of doubt'** ('Sayings of the Week', *The Sydney Morning Herald*, 5 Jul. 1986). The Nobel Laureate Albert Camus wasn't overstating the issue either when he wrote that **'Beginning to think is beginning to be undermined'** (*The Myth of Sisyphus*, 1942); nor was another Nobel Prize winner in Literature, Bertrand Russell, when he said, **'Many people would sooner die than think'** (Antony Flew, *Thinking About Thinking*, 1975, p.5 of 127). It is certainly true what yet another literature Nobel Laureate, the poet T.S. Eliot, wrote, that **'human kind cannot bear very much reality'** (*Burnt Norton*, 1936). **'The sleep of reason'**, reasoning at a deep level, *does* indeed **'bring forth monsters'**.

[56]The fact is, the human race has lived a haunted existence, dogged by the dark shadow of its imperfect human condition, forever trying to escape it; the result of which is that we have become *immensely* superficial and artificial—**'fake'** and **'phony'**, as the resigning adolescents so truthfully described it—living on the absolute meniscus of existence in terms of what we are prepared to look at and feel. We are a *profoundly* estranged or alienated species, completely blocked-off from the amazing and wonderful real world, and from the truth of our self-corruption that thinking about that beautiful, inspired, soulful world unbearably connects us to—as the absolutely **'fear'**-lessly honest Scottish psychiatrist R.D. Laing has written: **'Our alienation goes to the roots. The realization of this is the essential springboard for any serious reflection on any aspect of present inter-human life...We are born into a world where alienation awaits us. We are potentially men, but are in an alienated state** [p.12 of 156] **...the *ordinary* person is a shrivelled, desiccated fragment of what a person can be** [p.22] **...The condition of alienation, of being asleep, of being unconscious, of being out of one's mind, is the condition of the normal man** [p.24] **...between *us* and It** [our true selves or soul] **there is a veil which is more like fifty feet of solid concrete. *Deus absconditus*. Or we have absconded** [p.118] **...The outer divorced from any illumination from the inner is in a state of darkness. We are in an age of darkness. The state of outer darkness is a state of sin—i.e. alienation or estrangement from the inner light** [p.116] **...We are all murderers and prostitutes...We are bemused and crazed creatures, strangers to our true selves, to one another'** [pp.11-12] (*The Politics of Experience* and *The Bird of Paradise*, 1967). **'We are dead, but think we are alive. We are asleep, but think we are awake. We are dreaming, but take our dreams to be reality. We**

are the halt, lame, blind, deaf, the sick. But we are doubly unconscious. We are *so* ill that we no longer feel ill, as in many terminal illnesses. We are mad, but have no insight [into the fact of our madness]' (*Self and Others*, 1961, p.38 of 192). '**We are so out of touch with this realm** [where the issue of the human condition lies] **that many people can now argue seriously that it does not exist**' (*The Politics of Experience* and *The Bird of Paradise*, p.105).

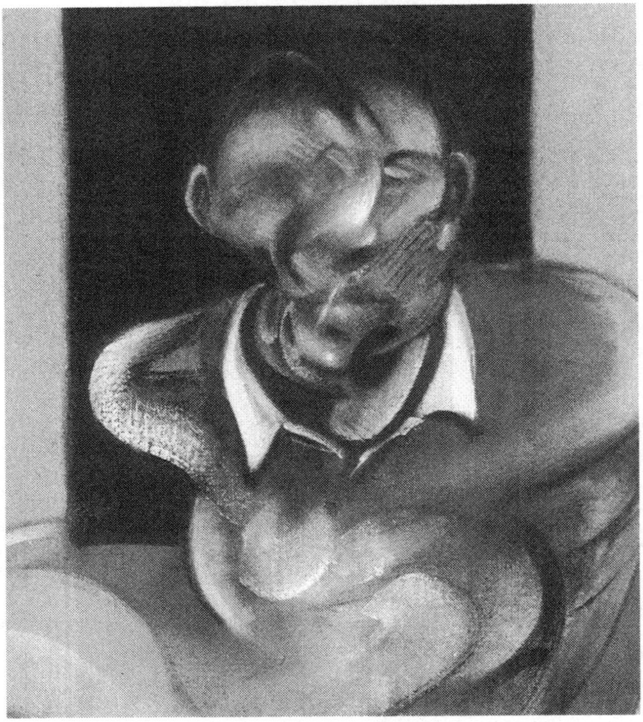

Detail from Francis Bacon's *Study for self-portrait*, 1976

[57] Possibly the only other person who managed to depict and penetrate our '**fifty feet of solid concrete**' wall of denial of the truth of our tortured, 'good-and-evil'-afflicted human condition as honestly as Laing was the Irish artist Francis Bacon. While people in their resigned state of denial typically find his work '**obscene**' and '**enigmatic**' (*The Sydney Morning Herald*, 29-30 Apr. 1992), there is really no mistaking the agony of the human condition in Bacon's twisted, smudged, distorted, death-mask-like—alienated—faces and tortured, contorted, stomach-knotted, arms-pinned, psychologically strangled and imprisoned bodies; consider, for instance, his *Study for self-portrait* above. It's some recognition of the incredible integrity/

honesty of Bacon's work that in 2013 one of his triptychs sold for $US142.4 million, becoming **'the most expensive work of art ever sold at auction, breaking the previous record, set in May 2012, when a version of Edvard Munch's *The Scream*** [another exceptionally honest, human-condition-revealing painting] **sold for $119.9 million'** (*TIME*, 25 Nov. 2013).

58 Indeed, if Laing's words and Bacon's images don't reconnect us with the human condition then nothing will!! But again, that reconnection is not easily achieved because what Laing wrote is true—resigned humans have learnt to live in *such* complete denial of the issue of the human condition that many people do now **'seriously'** believe the issue **'does not exist'**, at least not in its true form as a profoundly troubled and insecure, *psychological* affliction.

59 A further confirmation of just how impenetrable the issue of the human condition has been is that one of the clearest descriptions of it remains one that was written over three centuries before Christ by the inimitable Plato, a man who is still rightfully regarded as the greatest of all philosophers; for instance, Alfred North Whitehead—himself one of the most highly regarded philosophers of the twentieth century—described the history of philosophy as merely **'a series of footnotes to Plato'** (*Process and Reality [Gifford Lectures Delivered in the University of Edinburgh During the Session 1927-28]*, 1979, p.39 of 413). What enabled Plato to be such an effective and penetrating thinker was that he was able to **'realiz**[e]**'** that **'Our alienation goes to the roots'**, which, as Laing wrote, is **'the essential springboard for any serious reflection on any aspect of present inter-human life'**. Plato was one of a rare few individuals in history who have been sound and secure enough in self to fully confront, talk freely about and effortlessly describe the human condition; in fact, the centrepiece of his most celebrated work, *The Republic*, is the allegory of a cave in which humans have been incarcerated, unable to face the exposing glare of the sunlit true world, which is the unbearable truth of our seemingly highly imperfect condition. In his introduction to his cave allegory, Plato wrote: **'I want you to go on to picture the enlightenment or ignorance of our human conditions somewhat as follows. Imagine an underground chamber, like a cave with an entrance open to the daylight and running a long way underground. In this chamber are men who have been prisoners there'** (c.360 BC; *Plato: The Republic*, tr. H.D.P. Lee, 1955, p.278 of 405). Plato described how the cave's exit is blocked by a **'fire'** that **'corresponds to the power of the sun'** (p.282), which the cave prisoners have to hide from because its **'painful'** light would make

'visible' the unbearably depressing issue of **'the imperfections of human life'** (p.282). Yes, either we **'enlighten'** ourselves about our human condition, or we remain **'prisoners'**, having to hide ever deeper **'underground'** in the protective but truth-avoiding and thus alienating darkness of the **'cave'** of denial of the whole unbearably **'painful'** issue of our species' horrifically conflicted, seemingly **'imperfect'** state.

60 So, despite the human condition being the all-important issue that *had* to be solved, it has been, for all but a very rare few individuals like Plato, the one issue humans couldn't face—it has been *the* great unacknowledged 'elephant in the living room' of our lives, THE absolutely critical and yet completely unconfrontable and virtually unmentionable subject in life.

Chapter 1:4 Important note to reader

61 For readers of this book what is going to be said now is very important.

62 It is not difficult to understand the following account of one woman's reaction to being diagnosed with cancer: **'She said, "I thought I was cool, calm and collected but I must have been in a state of shock because the words just seemed to flood over me and I remembered almost nothing from what the doctor said in the initial consultation"'** (*The Sydney Morning Herald*, 18 Aug. 1995). Well, since discussion of what the human condition is involves re-connecting with an experience that was *worse than death*—a **'horrible'** **'shattering'** **'deadly pain'** of such depressing **'despair'**, **'doubts and night terrors'** that it was like a **'sickness unto death'** where you **'just keep falling and falling'**, **'eternally to die, to die and yet not to die'**, as Hughes, Salinger, Kierkegaard, Berdyaev and Jung described it—the reader should similarly expect that even though you will *think* you are **'cool, calm and collected'** and able to absorb what is being explained, your mind will actually be in **'shock'** and the words will **'flood over'** you and you will **'initial**[ly]**'** not be able to take in or hear much of what is said. Once you are resigned to living in denial of the issue of the human condition, as virtually every adult is, then you are *living* in denial of the issue of the human condition—and so your mind will, **'initial**[ly]**'** at least, resist absorbing discussion of it. So while you may *think* you will be able to follow and take in discussion about the human condition, like you would expect to be able to do when reading or

learning about any new subject, this 'deaf effect' that **'initial**[ly]' occurs when reading about the human condition is, in fact, very real.

[63] The problem is that once we humans resign and become mentally blocked out or alienated from any truth that brings the issue of the human condition into focus (which, as will be revealed in this book, is most truth) we aren't then aware that we are blocking out anything; we aren't then aware that **'there is a veil which is more like fifty feet of solid concrete'** between us and **'our true selves'** or soul, as Laing said, that is preventing our mind from accessing what is being said. This inability to know we are blocking something out occurs because obviously we can't block something out of our mind and know we have blocked it out because if we knew we had blocked it out we wouldn't have blocked it out!! <u>The fact is, we aren't aware that we are alienated</u>. What happens then when a resigned adult starts reading description of what the human condition really is is that they suffer from a 'deaf effect' where their mind will **'initial**[ly]' only take in a little of what is being said. However, not knowing that this is occurring, they unavoidably think the reason they are finding it difficult to follow the explanation is because the presentation is flawed; that it is **'disjointed'**, **'poorly expressed'**, **'impenetrably dense'** or **'too academic for me to understand'**. Readers of my books about the human condition often become so frustrated with their inability to take in or hear what is being said that they think little or nothing is being said, or that what is being presented is so **'confused'**, **'convoluted'** and **'needlessly repetitive'** of a vague point that they send me angry requests for **'an executive summary so I can grasp what you are saying'**. I put a pre-emptive warning about the danger of the deaf effect, with powerful illustrations, in the 'Notes to the Reader' at the beginning of this book, but people typically disregard it because they think it won't apply to them, that if something makes sense they will be able to follow it—and yet the deaf effect *does* apply to virtually *every* reader.

[64] Later in par. 991 (and Professor Prosen has mentioned some of the following in his Introduction) I present Plato's description of what happens when, as summarised in the *Encarta Encyclopedia*, someone **'escapes from the cave** [where the human race is living in psychological denial of the human condition] **into the light of day'** and **'sees for the first time the real world and returns to the cave'** to help the cave prisoners **'Escape into the sun-filled setting outside the cave** [which] **symbolizes the transition to the real world...which is the proper object of knowledge'** ('Plato', *Encarta*

Encyclopedia, written by Prof. Robert M. Baird, accessed Jul. 2008; see <www.wtmsources.com/101>). Plato wrote that **'it would hurt his** [the cave's prisoner's] **eyes and he would turn back and take refuge in the things which he could see** [take refuge in all the human-condition-avoiding dishonest explanations for human behaviour that he has become accustomed to]**, which he would think really far clearer than the things being shown him. And if he were forcibly dragged up the steep and rocky ascent** [out of the cave of denial] **and not let go till he had been dragged out into the sunlight** [shown the truthful explanation of the human condition]**, the process would be a painful one, to which he would much object, and when he emerged into the light his eyes would be so overwhelmed by the brightness of it that he wouldn't be able to see a single one of the things he was now told were real'** (c.360 BC; *Plato: The Republic*, tr. H.D.P. Lee, 1955, p.280 of 405). Significantly, Plato then added, **'Certainly not at first. Because he would need to grow accustomed to the light before he could see things in the world outside the cave'** (p.280). Yes, you will find that re-reading (**'grow**[ing] **accustomed to'**) analysis of the human condition will help enormously in your ability to take in or hear what is being said. Indeed, it has become a point of amusement when giving introductory talks about the human condition how often people who, having attended a second or third talk, say, **'That was a much better presentation this time than last time, the explanations and descriptions were so much easier to follow.'** It is amusing because the talks are virtually identical. What has dramatically improved is the listener's ability to 'hear' what was being said. So the deaf effect *is* very real, as this online article acknowledges: **'I read it** [my 2003 book, *A Species In Denial*] **in 2005, and at the time it was not an easy read. The core concepts keep slipping from my mental grasp, at the time I put it down to bad writing, however a second reading revealed something the Author had indicated from the outset—*your mind doesn't want to understand the content*. The second read was quick and painless...**[and I was then able to see that] **The cause of the malaise** [of humanity] **is exposed, remedied and the reader is left with at the very least an understanding of themselves, and for me something of an optimism for the future'** ('Fitzy', *Humanitus Interruptus – Great Minds of Today*, 21 Oct. 2011; see <www.wtmsources.com/106>). Yes, Plato emphasised how relieved the cave prisoner would be to be free of his old dishonest existence by saying that once he had become **'accustomed to the light'**, **'when he thought of his first home and what passed for wisdom there and of his fellow-prisoners, don't you think he would congratulate himself on his good fortune and be sorry for them?'** (*Plato: The Republic*, tr. H.D.P. Lee, 1955, p.281). Some of the online quotes that were included in the 'Notes to the Reader' warning about the deaf effect show this **'good fortune'** effect of being able to access understanding of the human condition and finally be able to make sense of human existence.

Chapter 1:5 How has science coped with the issue of the human condition?

[65] In light of what has now been explained about our species' extremely committed denial of the human condition, how has science—as humanity's designated vehicle for enquiry into the nature of our world, particularly into the all-important issue of human nature—coped with the human condition? Well, given it is practised by humans who have had no choice but to avoid the suicidally depressing subject of the human condition, science has necessarily been what is termed **'reductionist'** and **'mechanistic'**. It has avoided the overarching whole view of life that required having to confront the issue of the human condition and instead *reduced* its focus to only looking down at the details of the *mechanisms* of the workings of our world, in the hope that understanding of those mechanisms would eventually make it possible to explain, understand and thus at last be able to both confront and ameliorate or heal the human condition.

[66] Of course, the great danger inherent in the reduced, mechanistic, whole-view-evading, resigned-to-living-in-denial-of-the-human-condition, fundamentally dishonest, truth-avoiding, alienating approach was that it would become *so* entrenched, so habituated to its strategy of denial, that it would resist the whole-view-embracing, human-condition-confronting, truthful explanation of the human condition when it *was* finally found and continue to persevere with its dishonest strategy to the point of taking the human race to terminal alienation and extinction! The risk was that the established *dishonest* scientific paradigm might not tolerate the arrival of the *truthful* scientific paradigm, even though bringing about its arrival is science's great objective and fundamental responsibility—and the only means by which the human race can be liberated from its condition, and thus transformed.

[67] And, most distressingly, that *is* what is occurring. Based on the hard-won insights into the workings of our world that have been discovered by mechanistic science, the human-condition-confronting, fully accountable,

true biological explanation of the human condition *has* finally been found, and will be presented in chapter 2 of this book—but so far mechanistic science has treated this explanation as an anathema and, by so doing, *is* taking humanity to a state of terminal alienation both by that conduct and by perpetuating its dishonest path with the development of an extremely dangerously dishonest account of the human condition itself!

68 Yes, in 2012, E.O. Wilson, a man being lauded as the **'living heir to Darwin'** (from the dustjacket of E.O. Wilson's *The Social Conquest of Earth*), published his book, *The Social Conquest of Earth*, in which the subject of the human condition is front and centre; indeed, the book's central claim is that it presents **'the clearest explanation ever produced as to the origins of the human condition'** (ibid). But given the human condition has been such a terrifying, unapproachable subject, the question has to be asked: has Wilson actually been able to confront, think effectively about and, by so doing, find the long sought-after, human-race-liberating, holy grail of science of the explanation of the human condition—or has he, in fact, not actually confronted and thought truthfully about the human condition at all, and, therefore, *not* presented the liberating and ameliorating understanding of ourselves that we humans so desperately need? As we are going to see, the answer is the latter; indeed, rather than delivering the dreamed-of *relieving* insight into our 'good-and-evil'-afflicted lives, Wilson *has* taken the human race *so* deep **'underground'** into the darkness of Plato's human-condition-avoiding cave of denial that he is effectively locking humanity onto a path to the utterly *tortured*, permanent darkness of terminal alienation!

69 This exposé of Wilson's completely dishonest, condemning-humanity-to-perpetual-torture account of the human condition (which, again, was always the extremely dangerous potential end play outcome of mechanistic science's resigned, human-condition-avoiding approach) will be countered, in chapter 2 of this book, by the aforementioned denial-free, human-condition-confronting-not-avoiding, alienation-removing-not-increasing, psychologically-rehabilitating-and-thus-human-race-transforming, *real* biological explanation of the human condition—the very explanation that science has been charged with finding but which the scientific establishment has, thus far, been treating as an anathema.

70 Further, it is this fully accountable and thus true explanation of the human condition that finally makes it both possible and psychologically safe to also provide the fully accountable, true <u>answers to the three other</u>

(only slightly less important) outstanding holy grails in science, namely the meaning of existence; how we humans acquired our altruistic moral instincts—as well as a rebuttal of scientific theories that have been put forward on the subject; and how we became conscious when other animals haven't—explanations that are presented in chapters 3, 4, 5 and 6, respectively. Chapter 7 will then provide the denial-free, true account of humanity's journey from ignorance to enlightenment that these fully accountable, true explanations make possible—an account that includes the reconciling understanding of the lives of men and women, the explanation of sex as humans have practised it (including homosexuality), and many, many other denial-free explanations of human behaviour.

71 In short, this book has the power to transport humanity from a world of ignorant darkness and excruciating, human-condition-afflicted bondage that Plato's cave so honestly depicted, to a liberated world bathed in the light of redeeming, relieving and psychologically rehabilitating understanding. Yes, as long as humanity chooses the path of truth that has been made possible by the finding of the real explanation of the human condition and doesn't persevere with its habituated path of denial that leads to terminal alienation, the human race *will* progress to a fabulous existence of transformed freedom from the human condition. The book's concluding chapter (8) describes how this fabulous transformation can now occur.

Chapter 1:6 The elements involved in a true analysis of the human condition

72 To describe mechanistic science's great avoidance of the subject of the human condition and the resulting litany of dishonest biological excuses for human behaviour that have now culminated in Wilson's dishonest and extremely dangerous account of the human condition, we first need to identify the actual elements that gave rise to the human condition.

73 If we ask ourselves what is the 'we' that we are talking about when we refer to the possibility that 'we' are a terrible mistake, a worthless blight on this planet, the **'face of absolute evil'** as Jung said, we can see where the *real* problem about our seemingly horribly flawed condition lies: the 'we' is surely our conscious thinking mind or intellect. It is our conscious mind that is uncertain of its worthiness, that suspects that it

might be to blame for our species' present seemingly highly imperfect, even 'fallen' or corrupted, competitive, selfish and aggressive condition. And indeed, that most famous mythological account of the origin or genesis of the human condition, the story of Adam and Eve from the book 'Genesis' in the Old Testament, which the very great prophet Moses wrote (versions of which also appear in the Torah of Judaism, and the Koran of Islam), recognises that this was the case; that it *was* our conscious mind that led to our 'good-and-evil'-afflicted condition. It says that we humans once lived in **'the Garden of Eden'** (Bible, Gen. 3:23) and were **'created...in the image of God'** (1:27), presumably meaning we once lived in a pre-human-condition-afflicted state of original innocence where we were perfectly instinctively orientated to the cooperative, selfless, loving, 'Godly' ideals of life. Genesis then says that Adam and Eve took the **'fruit' 'from the tree of...knowledge'** (3:3, 2:17) because it was **'desirable for gaining wisdom'** (3:6), obviously meaning we became fully conscious, thinking, knowledge-seeking beings. Then, as a result of becoming conscious, it says we **'fell from grace'** (derived from the title of Gen. 3, **'The Fall of Man'**), meaning our original cooperative, selfless and loving (good) state became corrupted and our competitive, selfish and aggressive—indeed, angry, egocentric and alienated—('evil'/'sinful') state emerged, at which point we were **'banished...from the Garden of Eden'** (3:23) state of original innocence and left **'a restless wanderer on the earth'** (4:14); that is, we were left in our present, psychologically upset, distressed and alienated condition.

74 So, our conscious mind caused **'the fall'**; our present corrupted condition resulted from the emergence of our unique fully conscious thinking mind. As I pointed out earlier, consider for a moment descriptions like egocentric, arrogant, deluded, optimistic, pessimistic, artificial, hateful, mean, immoral, guilt-ridden, evil, depressed, inspired, psychotic, alienated, and so on—they all imply that there is a consciousness-derived, *psychological* dimension to *our* behaviour. Yes, we humans have suffered not from the genetic-opportunism-based, non-psychological *animal condition*, but the consciousness derived and induced, *psychological* HUMAN CONDITION. Our condition is unique to us fully conscious humans, *and any truthful description of our human condition should recognise this instigating involvement of our conscious mind, as well as the psychosis it has produced in us.*

75 And, tellingly, *all* our religions and mythologies do recognise that humans did once live in a pre-conscious and pre-human-condition-

afflicted peaceful, cooperative, selfless, loving ideal state. In addition to the Biblical account of Adam and Eve, which speaks of an innocent past in the Garden of Eden, there is also the Taoist description of our distant forebears as being **'the Men of Perfect Virtue'** (Bruce Chatwin, *The Songlines*, 1987, p.227 of 325). Zen Buddhism similarly recognises the loss of an uncontaminated, pure state as a result of the intervening conscious mind, referring to **'the affective contamination (*klesha*)'** or **'the interference of the conscious mind predominated by intellection (*vijnana*)'** (D.J. Suzuki, Erich Fromm, Richard Demartino, *Zen Buddhism & Psychoanalysis*, 1960, p.20). And, not surprisingly, Greece, the cradle of democracy and the scientific method, also contributed to the collection, with Hesiod, for instance, in his epic poem *Theogony*, saying of our distant ancestors that **'Like gods they lived, with calm untroubled mind / Free from the toils and anguish of our kind...All willing shared the gathering of their hands'** (c. Eighth century BC).

76 Indeed, the consistency of such accounts has been borne out by the investigations of the author Richard Heinberg. He found that every human culture has a myth involving both the emergence of consciousness and a 'fall' from an original 'Golden Age' of togetherness and peace—from the major religions, to races as isolated and diverse as the Eskimos, Aborigines, and Native Americans—summarising that **'Every religion begins with the recognition that human consciousness has been separated from the divine Source, that a former sense of oneness...has been lost...everywhere in religion and myth there is an acknowledgment that we have departed from an original...innocence and can return to it only through the resolution of some profound inner discord...the cause of the Fall is described variously as disobedience, as the eating of a forbidden fruit, and as spiritual amnesia** [alienation]**'** (*Memories & Visions of Paradise*, 1990, pp.81-82 of 282). Yes, as Berdyaev recognised, **'The memory of a lost paradise, of a Golden Age, is very deep in man'** (*The Destiny of Man*, 1931; tr. N. Duddington, 1960, p.36 of 310). The philosopher Jean-Jacques Rousseau also expressed what we all intuitively do know is the truth when he wrote that **'nothing is more gentle than man in his primitive state'** (*The Origin of Inequality*, 1755; *The Social Contract and Discourses*, tr. G. Cole, 1913, p.198 of 269).

77 Wordsworth was another who spoke the truth when he wrote that **'The Soul** [our instinctive memory of a past all-loving, selfless, cooperative existence] **that rises with us** [that we are born with]**, our life's Star...cometh from afar...trailing clouds of glory do we come from God, who is our home'** (*Intimations of Immortality from Recollections of Early Childhood*, 1807). As will be explained in chapter 3:3, 'God' is our personification of the terrifyingly confronting

truth of the teleological, integrative, order-of-matter-developing, co-operative, selfless, loving theme or direction or meaning of existence that our distant ancestors lived in accordance with, but which we no longer appear to—a truth Plato acknowledged when he wrote that humans have **'knowledge, both before and at the moment of birth...of all absolute standards...** [of] **beauty, goodness, uprightness, holiness...our souls exist before our birth'**, describing **'the soul'** as **'the pure and everlasting and immortal and changeless... realm of the absolute...**[our] **soul resembles the divine** [God]**'** (*Phaedo*, c.360 BC; tr. H. Tredennick).

78 But beyond what is revealed by myth, religion and profound thought, consider the evidence provided by our studies in anthropology and primatology. While fossils of our early ape ancestors who lived from 12 to 4 million years ago are rare, recent discoveries from this period are now evidencing a cooperative past, which anthropologists are beginning to admit; for instance, C. Owen Lovejoy has written that **'our species-defining cooperative mutualism can now be seen to extend well beyond the deepest Pliocene** [well beyond 5.3 million years ago]**'** ('Reexamining Human Origins in Light of *Ardipithecus ramidus*', *Science*, 2009, Vol.326, No.5949). In primatology, studies of living apes reveal that bonobos (*Pan paniscus*) are not only humans' closest relatives, but an extremely gentle and cooperative species. While bonobos and chimpanzees (*Pan troglodyte*) both share around 99 percent of humans' genetic material, the primatologist Frans de Waal points out that **'the recent discovery** [by neuroscientists Elizabeth Hammock and Larry Young in 2005] **that bonobos and humans share genetic code in relation to affiliative** [social, cohesive, loving, integrative] **behavior that is absent in the chimpanzee'** indicates that bonobos and humans are more closely related in terms of their social nature ('Foreword by Frans B.M. de Waal', *The Bonobos: Behavior, Ecology, and Conservation*, eds T. Furuichi & J. Thompson, 2008, p.12 of 327). The anthropologist Adrienne Zihlman has also shown that of bonobos and chimpanzees, bonobos are closer anatomically to our ancestors ('Reconstructions reconsidered: chimpanzee models and human evolution', *Great Ape Societies*, eds William C. McGrew et al, 1996, pp.293-304 of 352). As to their peaceful nature, many primatologists attest to it; consider this from Barbara Fruth: **'up to 100 bonobos at a time from several groups spend their night together. That would not be possible with chimpanzees because there would be brutal fighting between rival groups'** (Paul Raffaele, 'Bonobos: The apes who make love, not war', Last Tribes on Earth.com, 2003; see <www.wtmsources.com/143>). (The evidence of our cooperative past that is provided by anthropology, some of which

is referred to above, and by the bonobos, will be documented in some detail in chapters 4:5 and 4:6, respectively.)

79 In addition to these numerous acknowledgments of a pre-human-condition-afflicted, all-loving past, other thinkers, some of history's greatest, have also identified the rise of consciousness as being key to understanding and thus resolving our present corrupted, soul-devastated condition.

80 The South African Sir Laurens van der Post (whom I consider to be the pre-eminent philosopher of the twentieth century) said it all when he wrote: **'This shrill, brittle, self-important life of today is by comparison a graveyard where the living are dead and the dead are alive and talking** [through our instinctive self or soul] **in the still, small, clear voice of a love and trust in life that we have for the moment lost...**[there was a time when] **All on earth and in the universe were still members and family of the early race seeking comfort and warmth through the long, cold night before the dawning of individual consciousness in a togetherness which still gnaws like an unappeasable homesickness at the base of the human heart'** (*Testament to the Bushmen*, 1984, pp.127-128 of 176).

81 The South African naturalist Eugène Marais, the first person to study primates in their natural habitat, also got to the point of our condition being a psychologically embattled one when he focused on a conflict between our already established instincts and an emerging consciousness, saying, **'The highest primate, man, is born an instinctive animal. All its behavior for a long period after its birth is dominated by the instinctive mentality...it has no memory, no conception of cause and effect, no consciousness...As the... individual memory slowly emerges, the instinctive soul becomes just as slowly submerged...For a time it is almost as though there were a struggle between the two'** (*The Soul of the Ape*, written between 1916-1936 and published posthumously in 1969, pp.77-79 of 171). And after a lifetime spent hunting the cause of the human condition, the great Hungarian-English polymath Arthur Koestler similarly identified the emergence of consciousness as the catalyst for our condition: **'the brain explosion gave rise to a mentally unbalanced species in which old brain and new brain, emotion and intellect, faith and reason were at loggerheads'** (*Janus: A Summing Up*, 1978, p.10 of 354).

82 Berdyaev was another who regarded consciousness as an intrinsically important consideration in the quest for self-understanding: **'the distinction between the conscious and the subconscious mind is fundamental to the new psychology'** (*The Destiny of Man*, 1931; tr. N. Duddington, 1960, p.68 of 310). The very great biologist Charles Darwin also recognised that acknowledgment of

the emergence of consciousness is **'fundamental to the new psychology'**; in fact, he used almost the same words, saying such acknowledgment will mean **'Psychology will be based on a new foundation'**. While it has been noted that Darwin's seminal 1859 book **'*The Origin of Species* contains almost no mention of the human species'** (Robert Wright, *The Moral Animal*, 1994, p.3 of 467), near the end of the final chapter Darwin did write that **'In the distant future I see open fields for far more important researches. Psychology will be based on a new foundation, that of the necessary acquirement of each mental power and capacity by gradation. Light will be thrown on the origin of man and his history'** (*The Origin of Species*, 1859, p.458 of 476). While Darwin studiously avoided trying to explain human behaviour in *The Origin of Species*, he did at least here recognise that for **'Light'** to **'be thrown on the origin of man and his history'** and a new meaningful, **'important'** world of understanding to be **'open**[ed]**'** up, **'Psychology'** will have to **'be based on a new foundation' 'of the necessary acquirement of each mental power and capacity by gradation'**—that it will need to recognise the involvement of the emergence of our **'mental power'** of consciousness in creating our species' unique **'psycholog**[icall]**y'** troubled human condition. Yes, the key to understanding the origin of man's **'psychology'** is to recognise the **'acquirement'** of our **'mental power'** of consciousness. (With regard to Darwin not addressing the issue of human behaviour in *The Origin of Species*—which was a stark omission given the book is titled *The Origin of Species* and the most important species we needed to know the origin of was ourselves—it is true that 12 years after the publication of *The Origin of Species* Darwin did publish *The Descent of Man, and Selection in Relation to Sex* in which he did make an attempt to look at the origins of our behaviour, but it was still only a tentative step in that all-important exploration. The evidence suggests that while Darwin was honest enough to recognise that trying to address the issue of human behaviour meant addressing the issue of the human condition, he apparently didn't feel secure enough to attempt it himself.)

[83]To make any headway in understanding human behaviour requires a **'fundamental'**, **'foundation'**, **'new'**, honest way of thinking that recognises that our condition is a **'psycholog**[ical]**'** one—that our conscious mind is deeply *psychologically* troubled, that we *are* a psychotic and neurotic, immensely alienated species. This is the point Laing was emphasising when he wrote that **'Our alienation goes to the roots. The realization of this is the essential springboard for any serious reflection on any aspect of present inter-human life.'**

[84]The following observations further reinforce the fact that our selfish and destructive behaviour arises from a deeply psychologically troubled conscious state—from a *psychosis*. Firstly, from Koestler, who wrote that **'symptoms of the mental disorder which appears to be endemic in our species...are specifically and uniquely human, and not found in any other species'** (*Janus: A Summing Up*, p.19). And from the journalist Richard Neville, who, as mentioned earlier, said, **'The world is hurtling to catastrophe: from nuclear horrors, a wrecked ecosystem, 20 million dead each year from malnutrition, 600 million chronically hungry...all these crises are man made, their causes are psychological. The cures must come from this same source; which means the planet needs psychological maturity...fast. We are locked in a race between self destruction and self discovery.'**

[85]So *again*, any truthful analysis of our human condition, its origins and its amelioration, should recognise the involvement of our conscious mind in our species' departure from an original, cooperative, loving state, as well as the psychosis and neurosis it has produced in us.

Chapter 1:7 The ultimate paradox of the human condition

[86]But if we do have a cooperative past (which we do), and our consciousness is behind our destructiveness (which it is), then the *ultimate* paradox of the human condition, indeed the reason we are still able to get out of bed in the morning and face the world, is that we fully conscious humans don't actually believe we are fundamentally bad/evil. Despite all the damning evidence, we don't accept that we are a terrible mistake, a worthless blight on this planet, the **'face of absolute evil'**. In fact, the incredible determination with which we live our lives bespeaks of a core belief within us all that we are not only *not* bad/evil but the great heroes of the story of life on Earth, and that one day we will be able to explain why that is true!

[87]It follows then that while awaiting the exonerating, liberating explanation of our present egocentric, competitive, selfish and aggressive condition we couldn't afford to acknowledge that we did once live in an innocent, all-loving instinctive state that was corrupted by the emergence of our conscious mind. Accepting such truth without the clarifying explanation for it would have left us unbearably condemned as bad and worthless, sentenced to a state of completely insecure, permanent damnation. So, as dishonest as it was, there *has* been a need to deny the

truth of our species' cooperative, peaceful past and the role consciousness played in its corruption. Denial in science *has* been necessary, not only because we couldn't cope with the depression that resulted from trying to think about the human condition, but because we didn't believe we were fundamentally bad/flawed/**'the face of absolute evil'**. BUT, again, the great danger was of this necessary denial becoming overly entrenched, rendering science incapable of fulfilling its fundamental responsibility of acknowledging the human-condition-confronting-not-denying, fully accountable and human-race-saving, true explanation of the human condition when it eventually arrived—a perilous situation that is one of the main themes of this book.

88 What has been written thus far forms the background for all of mechanistic science's human-condition-avoiding, dishonest, not-truly-accountable biological theories on human behaviour—including E.O. Wilson's—that will now be described.

Chapter 1:8 Social Darwinism

89 Prior to the development of science and its evasive offerings, humans had already found a way to avoid the condemning truths of a cooperative, all-loving past and of a consciousness-induced 'fall from grace', which was to simply assert that nature is brutally competitive and aggressive—**'red in tooth and claw'**, as Lord Alfred Tennyson put it (*In Memoriam*, 1850)—and that's why we are. Basically, we looked around and saw that animals always appear to be fighting and competing with each other and instead of acknowledging that our instincts are cooperative and all-loving, we said that our instincts are similarly ruthlessly competitive and aggressive. We said that we have brutal, savage animal instincts that our conscious mind has to somehow try to control. It was an absolutely brilliant excuse, because instead of our instincts being all-loving and thus unbearably condemning of our present non-loving state, they were made out to be vicious and brutal; *and*, instead of our conscious mind being the villain, the cause of our corruption, it was made out to be the blameless mediating 'hero' that had to manage those supposed vicious instincts within us! It was all a terrible reverse-of-the-truth lie, but a hugely relieving one for humans seeking relief from the human condition.

[90] What happened when Charles Darwin presented his idea of natural selection in his momentous book *On the Origin of Species by Means of Natural Selection* was that the excuse that claimed we have 'savage', 'barbaric', 'backward', 'brutish', 'bestial', 'primitive' animal instincts within us was supposedly given a biological basis through the misrepresentation of natural selection as a 'survival of the fittest' process. Natural selection is the process by which some members of a population reproduce more than others in a given environment, and, most significantly, in the first edition of *The Origin of Species* Darwin left it undecided as to whether those individuals that reproduced more could be viewed as winners, as being 'fitter'. However, in later editions Darwin's associates, Herbert Spencer and Alfred Russel Wallace, persuaded him to substitute the term 'natural selection' with the term 'survival of the fittest' (letter from Wallace to Darwin, 2 Jul. 1866). While Darwin's friend and staunch defender, the biologist Thomas Huxley described the term 'survival of the fittest' as an **'unlucky substitution'** (Sir Gavin de Beer, *Charles Darwin*, 1963, p.178 of 290), from the point of view of humanity needing to contrive an excuse for its divisive selfish, competitive and aggressive behaviour it was a *lucky* substitution because it reinforced the dishonest but human-condition-relieving argument that our instincts are competitive and selfish and that we, in the sense of 'we' being our conscious thinking self, are blameless. (I should mention that in par. 242 it will be explained that Darwin's original position, where he left it undecided as to whether those who reproduced more are 'fitter', was right because being unconditionally selfless, where you give your life to help others and don't seek to reproduce more, *can* be a biologically meaningful—'fitter'—outcome.)

Chapter 1:9 Sociobiology/Evolutionary Psychology

[91] There were, of course, serious problems with this so-called Social Darwinist contrived excuse that 'nature is selfish and that's why we are'. For starters, it didn't account for instances in nature where selflessness occurs, such as in ant and bee colonies where workers slave selflessly for the whole colony. And secondly, and most particularly, it didn't account for *our* instinctive memory of having lived in a cooperative, loving, Garden of Eden-like existence, which is our selfless, consider-the-welfare-of-others, born-with, instinctive moral nature, the 'voice' of which is our 'conscience'.

[92]Seeking to address these cracks in the argument, biologists developed the theory of Sociobiology, with E.O. Wilson acting as its main proponent. Later known as Evolutionary Psychology, this theory explains, truthfully enough, that worker ants and bees are *not* actually being unconditionally selfless, truly altruistic when serving their colony because, when doing so, they are fostering the queen who reproduces their genes, which means their apparent selfless behaviour is, in fact, just a subtle form of selfishness: they are helping the queen to selfishly reproduce their genes. But in terms of maintaining the primary agenda of avoiding the unbearable and unacceptable issue of the human condition at all costs, this idea of selfless behaviour actually being a subtle form of selfishness, where you indirectly promote the reproduction of your own genes by fostering others who are related to you—your kin—was then not surprisingly, but *in this case* extremely dishonestly, commandeered to explain *our* moral instincts. Yes, it was claimed that *our* moral inclination to help others was no more than an attempt to reproduce our genes by supporting others whose genes we shared, with any anomalies put down to **'misplaced parental behavior'** (George Williams, *Adaptation and Natural Selection*, 1966, p.vii of 307)! As Wilson boldly summarised, **'Morality has no other demonstrable function'** other than to ensure **'human genetic material...will be kept intact'** (*On Human Nature*, 1978, p.167 of 260); even saying that **'Rousseau claimed** [that humanity] **was originally a race of noble savages in a peaceful state of nature, who were later corrupted...**[but what] **Rousseau invented** [was] **a stunningly inaccurate form of anthropology'** (*Consilience*, 1998, p.37 of 374)!!

[93]So, in saying our moral soul is still basically selfish, the old 'nature is selfish and that's why we are selfish' excuse was preserved; the same 'I'm going to determinedly avoid, not confront, the human condition' attitude had been upheld.

[94]BUT, the problem that *then* emerged was that this denigration of our moral self as nothing more than a subtle form of selfishness was both deeply offensive to and entirely inconsistent with what we all in truth know about our moral instincts, which is that they *are* unconditionally selfless, genuinely altruistic. As the journalist Bryan Appleyard pointed out, biologists **'still have a gaping hole in an attempt to explain altruism. If, for example, I help a blind man cross the street, it is plainly unlikely that I am being prompted to do this because he is a close relation and bears my genes. And the world is full of all sorts of elaborate forms of cooperation which extend far beyond the boundaries of mere relatedness'** (*Brave New Worlds: Staying Human in a Genetic Future*, 1998, p.112).

Chapter 1:10 Multilevel Selection theory for eusociality

[95]This **'gaping hole'** in the theory of Sociobiology/Evolutionary Psychology brings us to the present and the publication in 2012 of E.O. Wilson's *The Social Conquest of Earth*. Yes, once again, it was Wilson who concocted a 'solution' to this problem of the offensiveness of Evolutionary Psychology's denigration of our moral instincts as selfish. Now, to the dismay of his earlier supporters, he has dismissed 'his' previous Evolutionary Psychology theory as **'incorrect'**, **'inoperable'** and as having **'failed'** (*The Social Conquest of Earth*, pp.143, 180, 181 of 330), proffering in its place a new theory that not only contrives an explanation for our genuinely moral instincts, but takes the art of denial to the absolute extreme by contriving a non-human-condition-confronting explanation of the human condition itself!

[96]Known as Multilevel Selection or **'a New Theory of Eusociality'** (ibid. p.183) (eusociality simply meaning genuine sociality), this theory claims that humans have instincts derived from natural selection operating at the individual level (where members of a species selfishly compete for food, shelter, territory and a mate), *and* instincts derived from natural selection operating at the group level (where, it is claimed, groups of altruistic, cooperative members outcompete groups of selfish, non-cooperative members)—with the supposed selfish individual level instincts being the bad/sinful aspects of our nature, and the selfless, supposed group-selected instincts being the good/virtuous aspect of our nature. According to Wilson, **'Individual selection is responsible for much of what we call sin, while group selection is responsible for the greater part of virtue. Together they have created the conflict between the poorer and the better angels of our nature'** (ibid. p.241). In summary, Wilson asserts that **'The dilemma of good and evil was created by multilevel selection'** (ibid).

[97]Before looking at the way in which the Multilevel Selection theory for eusociality misrepresents, in fact, *avoids*, the real, consciousness-derived, psychological aspect of the human condition, we need to look at the 'group selection' mechanism that Wilson says accounts for our moral sense—because while we certainly *do* have a genuine moral sense, under scrutiny Wilson's theory of how we acquired it completely falls apart.

[98]While it makes sense that, as Wilson states, **'groups of altruists** [will] **beat groups of selfish individuals'** (ibid. p.243), the biological stumbling block

is whether genes, which have to selfishly ensure they reproduce if they are to carry on, can develop self-sacrificing altruistic traits in the first place. To reiterate, while it *is* true that a group of altruists whose members are prepared to make sacrifices for each other *will* defeat a group whose members are concerned only for themselves, *the question is whether a group of altruists can ever actually form in the first place?* The genetic reality is that whenever an unconditionally selfless, altruistic trait appears those that are selfish will naturally take advantage of it: 'Sure, you can help me reproduce my genes but I'm not about to help you reproduce yours!' Any selflessness that might arise through group selection will be constantly exploited by individual selfishness from *within* the group; as the biologist Jerry Coyne pointed out, **'group selection for altruism would be unlikely to override the tendency of each group to quickly lose its altruism through natural selection favoring cheaters** [selfish individuals]**'** ('Can Darwinism improve Binghamton?', *The New York Times*, 9 Sep. 2011). The only biological models that have been presented that appear to overcome this problem of genetic selfishness always prevailing are so complex and convoluted they seem highly implausible, in that they involve groups warring, then peacefully merging, then separating back out into new groups—with the altruists somehow banding together into their own groups. The situations where this between-group selection of unconditionally selfless traits is said to have taken place are in the occurrence of female-biased sex ratios in some small invertebrate species, and in the evolution of reduced virulence in some disease organisms (see David Sloan Wilson & Elliot Sober, *Unto Others: The Evolution and Psychology of Unselfish Behavior*, 1998, pp.35-50 of 394). However, for large mammals especially, who don't have complex life cycles, the mechanism is so implausible it has to be considered impossible.

99 Nevertheless, in defiance of the biological reality that, even where there is selection between groups, unconditionally selfless traits will be exploited and eliminated, Wilson proposes that extreme warring between groups of early humans where cooperation would have been an advantage was a strong enough force to overcome this problem of selfish exploitation and thus allow for the selection of altruism and the emergence of our genuinely moral instincts! So, according to Wilson, our ability to war successfully somehow produced our ability to love unconditionally!

100 Wilson's theory not only defies biological reality, it also flies in the face of both our cultural memories and anthropological evidence. As has been emphasised, standing in stark contrast to Wilson's proclamation

of **'universal and eternal'** warfare (*The Social Conquest of Earth*, p.55) are the cultural memories enshrined in our myths, religions and in the words of some of our greatest thinkers that attest to humans having a peaceful heritage, and in the evidence gleaned from studies in anthropology and primatology—such as the aforementioned recent fossil discoveries that are now confirming a cooperative past, and studies of bonobos, who are our species' closest living relatives and extraordinarily peaceful.

101 But, in attempting to prove we have a warlike past, Wilson ignores the evidence provided by the fossil record that reveals at least 7 million years of cooperative existence and instead argues that **'to test the prevalence of violent group conflict in deep human history** [one can look at]**...archaic cultures** [such as]**...the aboriginals of Little Andaman Island off the east coast of India, the Mbuti Pygmies of Central Africa, and the !Kung Bushmen of southern Africa. All today, or at least within historical memory, have exhibited aggressive territorial behavior'** (ibid. pp.69, 71). In addition, in his 1978 book *On Human Nature* Wilson wrote of the Bushmen that **'their homicide rate per capita equalled that of Detroit and Houston'** (p.100 of 260). However, there is ample evidence for just how extraordinarily cooperative, social and relatively peaceful the Bushmen are. For example, Lorna Marshall, regarded as **'the doyenne of American anthropology'** (Sandy Gall, *The Bushmen of Southern Africa*, 2001, ch.10) and one of the only Westerners to live with the Bushmen before they became contaminated through contact with more upset-adapted, alienated races, described **'their predominantly peaceful, well-adjusted human relations'** (*The !Kung of Nyae Nyae*, 1976, p.286 of 433). Her daughter, Elizabeth Marshall Thomas, who accompanied her on her expeditions in the 1950s, wrote the classic 1959 book about the Bushmen, *The Harmless People*. In a 1989 addition to her book, Marshall Thomas wrote: **'To my knowledge Wilson has never visited the Ju/wasi** [Bushmen]**. His book** [*On Human Nature*] **never mentions how important it was to them to keep their social balance, how carefully they treated this balance, and how successful they were. That he discusses them at all is perhaps due to the fact that in the 1970s they were selected by academics as a sort of living laboratory in which studies could be made on attributes of human nature, the most intriguing of which at the time seemed to be aggression'** (p.283 of 303). (More will be said in pars 685-691 about mechanistic science's misrepresentation of Bushmen and other so-called 'primitive' societies as **'violent'** and **'aggressive'** in order to comply with the human-condition-avoiding excuse that we are competitive, aggressive and selfish because of our animal heritage.)

[102]Wilson also cites recent archaeological evidence to support **'the prevalence of violent group conflict in deep human history'**, stating that **'Early humans had the innate equipment—and likely the tendency also—to use projectiles in capturing prey and repelling enemies. The advantages gained were surely decisive. Spear points and arrowheads are among the earliest artifacts found in archaeological sites'** (*The Social Conquest of Earth*, p.29). But this 'data' proves equally unreliable, as the archaeologist Steven Mithen has noted: **'No, the earliest artifacts are from around 2.5 million years ago, but spear points are not made until a mere 250,000 years ago and arrowheads might have first been manufactured no longer ago than 20,000 years'** ('How Fit Is E.O. Wilson's Evolution?', *The New York Review of Books*, 21 Jun. 2012). And in response to Wilson's claim that **'Archaeologists have found burials of massacred people to be a commonplace'** and **'archaeological sites are strewn with the evidence of mass conflict'**, Mithen argues that **'No, both are quite rare, especially in pre-state societies, and those that are known are difficult to interpret'** (ibid).

[103]On primatology, in an attempt to dismiss the example bonobos represent of a cooperative, peaceful heritage, Wilson suggests there is no difference between bonobos and the more aggressive, selfish chimpanzee, claiming that, like chimpanzees, bonobos **'do not share the fruit they pick'** (*The Social Conquest of Earth*, p.42), despite the fact there is a wealth of research recording instances of bonobos sharing fruit. For example, a report by Barbara Fruth and Gottfried Hohmann, included in 2002's influential *Behavioural Diversity in Chimpanzees and Bonobos*, states: **'In bonobos (*Pan paniscus*), food sharing between mature individuals is common...**[and] **bonobos often divide large-size fruits'** ('How bonobos handle hunts and harvests: why share food?', *Behavioural Diversity in Chimpanzees and Bonobos*, eds C. Boesch et al, 2002, pp.231-232 of 285). Wilson also claims bonobos **'hunt in coordinated packs in the same manner as chimpanzees' 'wolves and African wild dogs'** (*The Social Conquest of Earth*, p.32). While bonobos have been known to capture and eat small game, including small monkeys, to supplement their diet with protein, they are not known to routinely hunt down and eat large animals such as colobus monkeys, like chimpanzees do, with **'hunting behavior** [by bonobos] **very rare'** (Tetsuya Sakamaki quoted by David Quammen, 'The Left Bank Ape', *National Geographic*, Mar. 2013). Wilson relies upon a 2008 report by Hohmann titled 'Primate hunting by bonobos at LuiKotale, Salonga National Park' to draw these erroneous comparisons with the more aggressive chimpanzees—but Hohmann's report actually reveals extraordinary *differences* between bonobos and chimpanzees, recording

that **'at the Lilungu site, bonobos catch guenons and colobus monkeys but do not eat them, and at Wamba, bonobos and red colobus monkeys have been seen to engage in mutual grooming'** (M. Surbeck & G. Hohmann, *Current Biology*, 2008, Vol.18, No.19). A brief look at the papers Hohmann, in turn, cites, reveals that at Lilungu **'the bonobos interacted with the captured primates as if they were dealing with individuals of their own species. They sought cooperation in their interaction with the captured young primates without success. There is no evidence that they ate the captives...this interactional behavior...satisf**[ies] **the ethological definition of play'** (J. Sabater Pi et al, 'Behaviour of Bonobos (*Pan paniscus*) Following Their Capture of Monkeys in Zaire', *International Journal of Primatology*, 1993, Vol.14, No.5), and that at Wamba, despite **'red colobus** [being] **major hunting targets of common chimpanzees...there is little evidence of hunting by the pygmy chimpanzees** [bonobos] **of Wamba, despite the fact that they have been intensively studied for over ten years'** (Hiroshi Ihobe, 'Interspecific Interactions Between Wild Pygmy Chimpanzees (*Pan paniscus*) and Red Colobus (*Colobus badius*)', *Primates*, 1990, Vol.31, No.1). So it turns out that Hohmann's 2008 paper, which Wilson relies so heavily upon to depict bonobos as chimpanzee-like, ruthless killers, contains reports of bonobo behaviour that would be unthinkable from a chimpanzee, but which Wilson has just ignored! 'That's one more problem out of the way', he seems to be saying. It would appear that just as the rest of the insecure, human-condition-afflicted human race has had to practise the art of denial of the human condition, Wilson too has had to fudge the evidence to try to find support for his lies.

104 In summary, our moral instincts are *not* derived from warring with other groups of humans as Wilson and his theory of group selection would have us believe. No, as will be explained in chapter 4, we humans have an *unconditionally* selfless, *fully* altruistic, *all*-loving, *universally-benevolent*-not-competitive-with-other-groups, moral conscience. Our instinctive orientation is to love *all* people, not love some and be at war with others. The 'savage-instincts-in-us' excuse for our selfish behaviour is entirely inconsistent with the fact that we have *completely* moral, NOT partially moral and partially savage, instincts.

105 Overall then, while the Multilevel Selection theory for eusociality adds unconditionally selfless instincts to selfish instincts in the mix of what allegedly forms our species' instinctive make-up (thus countering Evolutionary Psychology's offensive denial of the fact that we have unconditionally selfless instincts), the same old reverse-of-the-truth, escape-rather-than-confront-the-human-condition agenda—that we have

villainous selfish instincts and a blameless conscious mind that has to 'step-in' to control them—continues. (It should be mentioned here that there has been an attempt to counter the selfishness-emphasising-and-justifying biological theories that have been described in this chapter with cooperation-not-competition, selflessness-not-selfishness emphasising biological theories. A summary of these 'left-wing' theories, as put forward by scientists such as Stephen Jay Gould, David Sloan Wilson and Robert Sussman—which all avoid the human condition just as ardently as these 'right-wing' theories and are therefore just as, if not more, dishonest, false and unaccountable—is provided in chapter 5:9, 'A brief history of left-wing dishonest mechanistic biology'.)

106 To look now at how Wilson's Multilevel Selection theory for eusociality avoids the real, consciousness-derived-and-induced *psychological* aspect of our human condition.

107 If our instincts *are* wholly peaceful and cooperative (which they are), and we are *not* selfish because of selfish instincts (which we are not), what is the source of our selfishness—or what Wilson calls our propensity for evil? The honest, human-condition-confronting answer is that it is the result of a *psychosis*.

108 As shown previously, our human behaviour involves our unique fully conscious thinking mind. As I have emphasised, descriptions of our less-than-ideal condition, such as egocentric, arrogant, deluded, artificial, hateful, mean, immoral, alienated, etc, all imply a consciousness-derived, *psychological* dimension to our behaviour. We suffer from a consciousness-induced, psychological HUMAN CONDITION, *not* an instinct-controlled ANIMAL CONDITION. And so it is to this psychological dimension to our behaviour that we should look for the cause of our selfishness. And yet in Wilson's psychological-problem-avoiding model our consciousness is merely a *mediator* between supposed selfish and selfless instincts—as he writes: **'Multilevel selection (group and individual selection combined) also explains the conflicted nature of motivations. Every normal person feels the pull of conscience, of heroism against cowardice, of truth against deception, of commitment against withdrawal. It is our fate to be tormented...We, all of us, live out our lives in conflict and contention'** (*The Social Conquest of Earth*, p.290). Clever semblance of our conflicted condition, diabolically clever, but entirely untrue, the epitome of shonk/evasion/denial/'**fake**[ness]'/ '**phon**[iness]'/separation-from-the-truth—alienation!

[109]In finding a way to avoid the truth of our psychologically conflicted condition with a non-psychological 'clever semblance' of it, what Wilson has done is *not* explain the human condition but nullify it, render the issue benign, virtually inconsequential—and in doing so he is effectively burying humanity into the deepest, darkest, **'underground'** depths of Plato's cave of denial. Make no mistake, Wilson's great **'fake'**, **'phony'**, superficial, not-genuinely-biological, **'Darwin's'-'heir'**-be-damned, deliberately-human-condition-trivialising account of the human condition is the most sophisticated expression of denial to have ever been invented—and thus the most dangerous. Certainly, providing humans with a 'get out of jail free' card, a way to supposedly explain the human condition *without* having to confront the issue of the extreme psychosis (psyche/soul repression) and neurosis (neuron/mental denial) of our real human condition, is immensely appealing to the now overly psychologically upset human race—but it is precisely that seductiveness that is *so* dangerous. (Already school children are talking about the human condition in an off-hand way, with a teacher reporting in 2013 that one of her students had remarked that **'I love the term The Human Condition; I can use it in just about any essay for any subject'** (WTM records, 15 Feb. 2013).) This Ultimate Lie had the potential to seduce the exhausted human race to such a degree that it obliterated any chance of the real human condition ever being truthfully confronted and thus understood! While denial was necessary while we couldn't explain ourselves, taking the art of denial to the extreme that Wilson has done with his dismissal of the fundamental issue before us as a species of our human condition as nothing more than two different instincts within us that are sometimes at odds, is a truly sinister—in fact, unconscionable—lie. It is the ultimate tragic expression of the human-condition-avoiding, superficialising, dumbing-down, dogma-not-knowledge-preferring, madness-becomes-universal, end play situation the human race is now in. It is nothing less than the final great push to have the world of lies with all its darkness take over the world—and condemn humanity to extinction. If the real psychosis-addressing-and-solving explanation of the human condition hadn't already emerged (which, as will be revealed in chapter 2, it has), then this Ultimate Lie would have given humanity's headlong march to ever greater levels of lying denial and its terrible alienating effects its terminal impetus.

[110]I should add that despite Wilson's evasive dismissal of religion as mere group propaganda—as nothing more than **'an expression of tribalism'**

(*The Social Conquest of Earth*, p.258)—the fact is, religions resonate with us, and have done for millennia, because they contain profound truth. Contrast Wilson's dangerously superficial account of our condition, where our consciousness is the blameless mediator or manager of villainous selfish instincts within us, with Moses' Garden of Eden account of the origin or genesis of the human condition that was described in par. 73, which says our instinctive heritage is wholly selfless and that it was the emergence of consciousness that led to our selfishness. *This* is an account that acknowledges the problematic role of our conscious mind, which is where the real terror of the human condition lies—our deep insecurity about whether 'we', our conscious thinking self, is actually evil. And it is *this* account that has been reinforced by some of history's most profound thinkers—like van der Post, Marais, Koestler and Berdyaev, whose words were included earlier (in pars 80-82).

111 I might mention here a review by the journalist Christopher Booker of Wilson's *The Social Conquest of Earth* that has just been published as I complete the writing of this book. While I don't agree at all with Booker that Darwin's theory of natural selection is a flawed and deficient theory that **'can't explain'** how all of life developed (the explanation of how it can and does is presented in subsequent chapters of this book), in every other respect what he has to say about Wilson's book is extraordinarily honest—in particular that **'what Wilson completely misses out is any recognition of what is by far the most glaring difference between humans and ants...we have broken free from the dictates of instinct...that peculiarity of human consciousness...has allowed us to step outside the instinctive frame...But it is this which also gives us our disintegrative propensity, individually and collectively, to behave egocentrically, presenting us with all those problems which distinguish us from all the other species which still live in unthinking obedience to the dictates of nature. All these follow from that split from our selfless 'higher nature', with which over the millennia our customs, laws, religion and artistic creativity have tried their best to re-integrate us'** ('E.O. Wilson has a new explanation for consciousness, art & religion. Is it credible?', *The Spectator*, 7 Sep. 2013).

112 Yes, in summary, Wilson continues to look everywhere for the cause of our condition, except to our consciousness; and instead of explaining our psychosis (**'all those problems which distinguish us from all the other species'**, as Booker referred to it) he simply states that it does not exist. Like most resigned, human-condition-avoiding humans (and other mechanistic scientists) he is effectively saying, 'What psychosis?

What inner insecurity? What sense of guilt? What original 'Golden Age' of innocence? What **'split from our selfless "higher nature"'**? What 'fallen' condition? What 'haunted existence dogged by the shadow of our human condition'? What alienation? What sickness of the soul; for that matter, what 'soul'? What deeply troubled state? What depression from the **'mountains'** of the **'mind**['s]**'** **'cliffs of fall, frightful, sheer'**? What agonising issue of the human condition that we, as teenagers, had to learn to **'let it be'**? What **'fake'**, **'phony'** state that I'm now living in? What great elephant in the living room of our lives that we can't acknowledge? What dreamed-of psychologically rehabilitated, transformed human race? What wonderful time when **'we'll all live as one'** that John Lennon **'imagine**[d]**'** (*Imagine*, 1971)?' To Wilson, the human condition is nothing more than a conflict between selfless and selfish instincts within us. He is basically saying, '*To hell with your psychological garbage, I'm not going there!*' So, while he might have won almost every accolade in science, in the end Wilson has revealed himself to be just another victim of the human condition—coping with it by finding a way to deny it.

113 There is one last but very important aspect of Wilson's account that needs to be addressed, which is that his notion of our condition being a result of selfish and selfless instincts within us would mean that unless we change our genes we are, as he asserts, **'intrinsically imperfectible'** (*The Social Conquest of Earth*, p.241). BUT such a fate is completely inconsistent with one of the central beliefs about the real, *psychological* nature of our condition, which is that finding understanding of it will bring about the psychologically ameliorated transformation of the human race. If, as Wilson maintains, we don't suffer from a psychosis, then we can't be healed—but we *do* suffer from a psychosis, which *can* be healed. Carl Jung was forever asserting that **'wholeness for humans depends on the ability to own their own shadow'** because he knew that finding the psychologically ameliorating understanding of the dark side of ourselves would make us **'whole'**. In religious terms, *The Lord's Prayer* contains the hope of the time when **'Your** [the Godly, ideal, peaceful] **kingdom come, your will be done on earth as it is in heaven'** (Bible, Matt. 6:10 & Luke 11:2). In contemporary mythology, the same sentiment is conveyed in the words of our modern day, truthful thinking prophets: in John Lennon's **'imagin**[ings]'; in Bob Dylan's anticipation of **'when the ship** [understanding of the human condition] **comes in...and the morning will be a-breaking...and the** [dishonest]

words that are used to get the ship confused will no longer be understood as the spoken [truth]**'** (*When The Ship Comes In*, 1963); in Jim Morrison, of The Doors, singing of **'Standing there on freedom's shore waiting for the sun'** (*Waiting for the Sun*, 1968), the liberating light of understanding to arrive, because when **'day destroys night'** we can **'break on through to the other side'** (*Break on Through*, 1966) to our **'freedom'** from the agony of the human condition; and, finally, in Bono's (of the band U2) lyrics about the coming of a world **'high on a** [uncorrupted] **desert plain'** where there will be no more egocentricity and **'the streets** [will] **have no name'** and **'there will be no toil or sorrow, then there will be no time of pain'** (*Where The Streets Have No Name*, 1987). This last prophetic vision is exactly the same as that expressed in the Bible where it states that **'Another** [denial-free, honest, all-clarifying] **book** [will be]**...opened which is the book of life...**[which will introduce] **a new heaven and a new earth...**[and] **wipe every tear from...**[our] **eyes. There will be no more death or mourning or crying or pain, for the old order of things has passed away'** (Rev. 20:12, 21:1, 4). Buddhist scripture contains the exact same anticipation of a fabulous time when humans **'will with a perfect voice preach the true Dharma** [present the supreme wisdom, namely the psychologically rehabilitating, transforming, true understanding of the human condition]**, which is auspicious and removes all ill'**, saying, **'Human beings are then without any blemishes, moral offences are unknown among them, and they are full of zest and joy'** (Maitreyavyakarana; *Buddhist Scriptures*, tr. Edward Conze, 1959, pp.238-242).

114 Yes, the fulfilment of Holden Caulfield's dream of a time when the need for Resignation, with all its horrible alienating, psychosis-producing effects—that Francis Bacon so honestly depicted—*will end* is what the human race *has* tirelessly been working towards, and has now (as we will see in chapter 2), in the nick of time, finally achieved. We are precisely the opposite of what Wilson argues, because we are, in fact, **'intrinsically' 'perfectible'**!

Chapter 1:11 While denial has been necessary, you can't find the truth with lies

115 In conclusion, I should emphasise that Wilson, as the quintessential exponent of reductionist, mechanistic science, is only doing what *all* mechanistic scientists have been doing—and, indeed, what virtually *all* humans have been doing—which is avoiding the issue of the human condition at all costs. But when it comes to finding understanding of the

human condition, the costs of such evasion *are* great indeed because, clearly, if you're committed to living in denial of the human condition you are in no position to ever find understanding of it. When Laing said, **'Our alienation goes to the roots. The realization of this is the essential spring-board for any serious reflection on any aspect of present inter-human life'**, he was making the fundamental point that for there to be **'any serious reflection on' 'human life'** the truth of our alienated, psychotic and neurotic, human-condition-afflicted state *had* to be **'realiz**[ed]'/confronted. You can't find the truth from a position of lying—a case the aforementioned philosopher Arthur Schopenhauer was making when he wrote that **'The discovery of truth is prevented most effectively...by prejudice, which...stands in the path of truth and is then like a contrary wind driving a ship away from land'** (*Essays and Aphorisms*, tr. R.J. Hollingdale, 1970, p.120 of 237). Yes, once you are resigned to living in denial of the human condition, you are in no position to think truthfully and thus effectively—as the aforementioned, painfully honest poem of the resigning adolescent lamented, **'you spend the rest of life trying to find the meaning of life and confused in its maze'**. Plato similarly saw the consequences of soul-and-truth-denying, mechanistic, reductionist intellectualism when he wrote that **'when the soul uses the instrumentality of the body** [uses the body's intellect with its preoccupation with denial] **for any inquiry...it is drawn away by the body into the realm of the variable, and loses its way and becomes confused and dizzy, as though it were fuddled** [drunk]**'** (*Phaedo*, c.360 BC; tr. H. Tredennick). The Templeton Prize-winning physicist Paul Davies also recognised the stultifying, **'confused'** and **'fuddled'**, *alienating* effect of being **'driv**[en]**' 'away from'** the **'truth'** when he observed that **'For 300 years science has been dominated by extremely mechanistic thinking. According to this** [whole-view-evading, human-condition-psychosis-avoiding, mechanisms-only-focused] **view of the world all physical systems are regarded as basically machines...I have little doubt that much of the alienation and demoralisation that people feel in our so-called scientific age stems from the bleak sterility of mechanistic thought'** ('Living in a non-material world—the new scientific consciousness', *The Australian*, 9 Oct. 1991).

116 Arthur Koestler also recognised the consequences of mechanistic, reductionist science's avoidance of our species' consciousness-induced, human-condition-afflicted psychosis. After making the already mentioned point that the **'symptoms of the mental disorder which appears to be endemic in our species...are specifically and uniquely human, and not found in any other species'**, he went on to write that **'Thus it seems only logical that our search**

for explanations [of human behaviour] **should also concentrate primarily on those attributes of *homo sapiens* which are exclusively human and not shared by the rest of the animal kingdom. But however obvious this conclusion may seem, it runs counter to the prevailing reductionist trend. "Reductionism" is the philosophical belief that all human activities can be "reduced" to – i.e., explained by – the** [non-psychosis involved] **behavioural responses of lower animals – Pavlov's dogs, Skinner's rats and pigeons, Lorenz's greylag geese, Morris's hairless apes...That is why the scientific establishment has so pitifully failed to define the predicament of man.'** Like Davies, Koestler complained of **'the sterile deserts of reductionist philosophy'**, making the fundamental point that **'a correct diagnosis of the condition of man** [had to be] **based on a new approach to the sciences of life'** (*Janus: A Summing Up*, 1978, pp.19-20 of 354).

117 Charles Birch, my professor of biology at Sydney University and another recipient of the Templeton Prize, also bravely spoke the truth about the limitations of mechanistic science when he said, '[mechanistic] **science can't deal with subjectivity** [the issue of our psychologically distressed condition]**...what we were all taught in universities is pretty much a dead end'** (from recording of Birch's 1993 World Transformation Movement Open Day address). He also perceived the stultifying, **'truth'-'prevent**[ing]**'** effects of dishonest, denial-based, mechanistic thinking when he said, **'Biology has not made any real advance since Darwin'** (in recorded conversation with this author, 20 Mar. 1987), and, some 10 years later, that **'the traditional framework of thinking in science is not adequate for solving the really hard problems'** (*Ockham's Razor*, ABC Radio National, 16 Apr. 1997), with the **'hard**[est] **problem'** of all for truth-avoiding thinking to solve being the all-important issue of our psychologically distressed human condition. Yes, as Birch concluded, **'Biology right now awaits its Einstein in the realm of consciousness studies'** (ibid). As is explained in chapter 6:2, 'consciousness' has become code word for the issue of the human condition, which had to be confronted not avoided for it to be solved.

118 It is no wonder then that denial-complying mechanistic science has been such a superficial and thus ineffective form of enquiry, and why humanity has lost faith in science. General Omar Bradley very accurately pointed out science's failings when he said, **'The world has achieved brilliance...without conscience. Ours is a world of nuclear giants and ethical infants'** (Armistice Day Address, 10 Nov. 1948; *Collected Writings of General Omar N. Bradley*, Vol.1). Carl Jung also recognised science's failure to provide us with the all-important, psychologically liberating, redeeming and

transforming understanding of ourselves when he said that **'Man everywhere is dangerously unaware of himself. We really know nothing about the nature of man, and unless we hurry to get to know ourselves we are in dangerous trouble'** (Laurens van der Post, *Jung and the Story of Our Time*, 1975, p.239 of 275). Sharing Jung's concern was the author Antoine de Saint-Exupéry, who wrote that **'We are living through deeply anxious days, and if we are to relieve our anxiety we must diagnose its cause...What is the meaning of man? To this question no answer is being offered, and I have the feeling that we are moving toward the darkest era our world has ever known'** (*A Sense of Life*, 1965, pp.127, 219).

[119] Yes, **'Man...is** [so] **dangerously unaware of himself'** that **'we are moving toward the darkest era our world has ever known'**; **'Ours is a world of nuclear giants and ethical infants'**, **'the scientific establishment has so pitifully failed to define the predicament of man'**, **'what we were all taught in universities is pretty much a dead end'**, **'the traditional framework of thinking in science is not adequate for solving the'** human condition, the **'demoralisation that people feel in our so-called scientific age stems from the bleak sterility of mechanistic thought'**. Dishonest mechanistic science couldn't solve the human condition, which is what was needed for the human race to progress to a human-condition-ameliorated, transformed state. And so human progress has been stalled, piled up and festering. The dialogue of one character in the 1991 film *Separate but Equal* accurately recognised the plight of our species when he said, **'Struggling between two worlds; one dead, the other powerless to be born'**—words that echo those of the philosopher Antonio Gramsci: **'The crisis consists precisely in the fact that the old is dying and the new cannot be born; in this interregnum a great variety of morbid symptoms appears'** (*Prison Notebooks*, written during Gramsci's 10-year imprisonment under Mussolini, 1927-1937). The politician Lionel Bowen was another who alluded to the futility of trying to reform our lives and world without first finding the reconciling, ameliorating understanding of ourselves when he said, **'I think it's just impossible to bring about change until such time as some new civilisation develops to allow change'** (*The Sydney Morning Herald*, 10 Sep. 1988). The historian Eric Hobsbawm also described the stark predicament facing humanity when he wrote that **'The alternative to a changed society—is darkness'** (*Age of Extremes*, 1994).

[120] Yes, *only* a whole new way of thinking, in particular the reconciling, redeeming and healing way of understanding ourselves, and resulting new transformed civilisation could alter our species' plight. We had arrived at a situation where humanity *desperately* needed clear

biological understanding of ourselves, understanding that would make sense of our divisive condition and liberate us from criticism, lift the psychological burden of guilt, give us meaning. There *had* to be a scientific, first-principle-based, biological reason for our divisive behaviour and finding it had become a matter of great urgency. The **'race'** that Richard Neville so accurately identified we were **'locked in…between self destruction and self discovery'** had reached crisis point, for to be stranded in a state of insecurity about our worthiness or otherwise was to be stranded in a terminally upset, psychologically immature state of arrested development—as Benjamin Disraeli, a former Prime Minister of Great Britain, recognised, **'Stranded halfway between ape and angel is no place to stop.'** The essayist Jonathan Swift's anguished cry to **'Not die here in a rage, like a poisoned rat in a hole'** (Letter to Bolingbroke, 21 Mar. 1729) did not exaggerate the truth of our predicament. The cellist Pablo Casals similarly emphasised the danger of our species' stalled state when he said, **'The situation is hopeless, we must take the next step'** (at a press conference in Madrid, on the occasion of his 80th [approx.] birthday). The journalist Doug Anderson made the same point when he wrote, **'Time may well be dwindling for us to enlighten ourselves…Tragic to die of thirst half a yard from the well'** (*The Sydney Morning Herald*, 31 Oct. 1994). In saying **'enlighten ourselves'**, Anderson was intimating that only understanding ourselves, understanding of our psychosis and neurosis afflicted human condition, could make the difference that was needed. In quoting clinical psychologist Maureen O'Hara, the science reporter Richard Eckersley also acknowledged that **'humanity is either standing on the brink of "a quantum leap in human psychological capabilities or heading for a global nervous breakdown"'** (*Values and Visions: Western Culture and Humanity's Future*, Address by Richard Eckersley, Nov. 1995; see <www.wtmsources.com/133>). The psychotherapist Wayne Dyer was another who understood that it is only the *reconciling*, *dignifying* understanding of our seemingly imperfect human condition that could save the human race, when he said, **'We've come to a place…where we can either destroy ourselves or discover our divineness'** (*The Australian Magazine*, 8 Oct. 1994). Yes, as the great Australian educator, and my headmaster when I was a student at Geelong Grammar School, Sir James Darling, wrote when referring to the critical need now to solve the human condition: **'The time is past for help which is only a Band-Aid. It is time for radical thinking and for a solution on the grand scale'** (*Reflections for The Age*, ed. J. Minchin & B. Porter, 1991, p.145 of 176).

[121]Knowing now how evasive, truth-avoiding, defensive, excusive and deluded the resigned mind is, it is really to an unresigned adolescent mind that we should go for a truly accurate description of the seriousness of our species' plight—and that's what we have in these clearly unresigned, denial-free, honest lyrics from the 2010 *Grievances* album of the young American heavy metal band 'With Life In Mind': **'It scares me to death to think of what I have become...I feel so lost in this world', 'Our innocence is lost', 'I scream to the sky but my words get lost along the way. I can't express all the hate that's led me here and all the filth that swallows us whole. I don't want to be part of all this insanity. Famine and death. Pestilence and war.** [Famine, death, pestilence and war are traditional interpretations of the 'Four Horsemen of the Apocalypse' described in Revelation 6 in the Bible. Christ referred to similar **'Signs of the End of the Age'** (Matt. 24:6-8 and Luke 21:10-11).] **A world shrouded in darkness...Fear is driven into our minds everywhere we look', 'Trying so hard for a life with such little purpose...Lost in oblivion', 'Everything you've been told has been a lie...We've all been asleep since the beginning of time. Why are we so scared to use our minds?', 'Keep pretending; soon enough things will crumble to the ground...If they could only see the truth they would coil in disgust', 'How do we save ourselves from this misery...So desperate for the answers...We're straining on the last bit of hope we have left. No one hears our cries. And no one sees us screaming', 'This is the end.'** Saying **'We've all been asleep since the beginning of time'** echoes all that Laing said about the extent of our blocked-out, alienated condition; and saying **'Everything you've been told has been a lie'** reiterates the extent of the dishonest denial in the world, especially in science, today; and saying **'So desperate for the answers'** confirms how incredibly important are all the **'answers'** about our human condition that are presented in this book. If there was ever a collection of words that cuts through all the dishonest pretence and delusion in the world about our condition these lyrics from 'With Life In Mind' surely do it!

[122]Thankfully then it is precisely this **'solution on the grand scale'**, this **'enlightenment'** of **'ourselves'**—in fact, the **'discover**[y]**'** of the reason for **'our divineness'**—that makes possible the **'quantum leap in human psychological capabilities'** from alienation to transformation that alone can **'save ourselves from this misery'** of the human condition, that is going to be presented in the next chapter. Yes, finding understanding of the human condition is the *real* game-changer the human race has been waiting for, such that when only yesterday the levels of human suffering and

distress and anger and environmental degradation from the effects of our horrifically troubled, upset human condition seemed irredeemable and irreversible, and all looked hopeless, suddenly people are going to appear who are inspired and transformed, *so* inspired and transformed in fact they are super-charged on a super-highway to a fabulous future for the human race. As will be explained and described in chapter 8 of this book, that sublime future is what becomes possible when the human condition is truthfully explained and solved, as it will be in the next chapter.

123 Indeed, the next chapter in this book evidences what *is* possible if you *don't* think dishonestly and instead acknowledge important truths such as that humans did once live in a cooperative, loving, innocent state, and that it was our conscious mind that led to our present 'good-and-evil'-afflicted, immensely psychologically upset, 'fallen' condition—for it provides the presentation of the dreamed-of, psychologically liberating, human-race-transforming, fully accountable, *true* explanation of the human condition. And yet, in what is a protracted act of extreme stupidity, *this* is *also* the explanation that mechanistic science has so far been treating as an anathema, and in doing so is allowing humanity to be taken into the permanent darkness of terminal alienation.

124 In conclusion, the following amazingly honest cartoon by Michael Leunig summarises all that has just been said about how truth and beauty can only be accessed via another paradigm outside the mechanistic one the human race currently lives in. In it we see a lone, self-powered, self-sufficient individual leaving the great alienated and alienating metropolis that has been built on dishonest denial to seek **'truth'** and, by so doing, resurrect **'beauty'** on Earth. Yes, as Sir Laurens van der Post wrote, **'There is, somewhere beyond it all, an undiscovered country to be pioneered and explored, and only a few lonely and mature spirits take it seriously and are trying to walk it'** (*About Blady*, 1991, p.87 of 255)—a statement of prophetic honesty that reiterates an observation he made some 40 years earlier that **'the one primary and elemental approach to the problem** [that the world faces] **is through** [understanding] **the being of man. Unfortunately it is an increasingly lonely way, trodden more and more not by masses but by solitary individuals…**[only these few] **sustain** [man's] **urge to seek an answer to the riddle of life** [confront and by so doing solve the human condition]**'** (*The Dark Eye in Africa*, 1955, p.15 of 159).

Cartoon by Michael Leunig, Melbourne's *The Age* newspaper, 17 Oct. 1987

[125](The complete presentation of this chapter's demolition of mechanistic biology appears in my freely available, online book *Freedom Book 1* at <www.worldtransformation.com/freedom-the-denials-in-biology>.)

Chapter 2

The Human-Race-Transforming, *Real* Explanation of The Human Condition

Chapter 2:1 <u>Summary</u>

126 What will now be presented is the human-race-liberating-and-transforming, psychosis-addressing-and-solving, fully accountable, *real* biological explanation of the human condition. This is the understanding that the human race has been working tirelessly towards since we humans first became fully conscious beings some 2 million years ago and the 'good-and-evil'-afflicted state of the human condition emerged. In a world fast going crazy from the effects of the human condition, this is the now desperately needed reconciling and redeeming breakthrough of breakthroughs that can bring about a new world for humans free of the agony of the human condition. In short, this is the understanding that is able to end human suffering and reunite the human race.

[127]**Yes, rather than yet another human-condition-avoiding, mechanistic account of human behaviour that doesn't address and resolve our species' *real, psychologically* troubled condition—like E.O. Wilson's recent use of the theory of Multilevel Selection that was examined in the previous chapter—*this* is the understanding that can finally liberate humanity from its incarceration in the horrifically debilitating, indeed, unfeeling, unseeing and unhearing, death-like darkness of Plato's 'cave' of alienating psychological denial that featured in the preceding chapter and is depicted in the image above.**

Chapter 2:2 We cannot endure being faced by the problem of the human condition forever

[128]Before presenting this dreamed-of, human-race-transforming explanation of the human condition, the following extraordinarily honest words from Nikolai Berdyaev's 1931 book, *The Destiny of Man*, provide a powerful summary of all that was just said in chapter 1 about the stalled state of science, and thus of the human race. Firstly, in regard to our species' overall plight, Berdyaev wrote that (and part of this extract was referred to in par. 76) **'The memory of a lost paradise, of a Golden Age** [our species' pre-human-condition-afflicted state of original cooperative, loving innocence]**, is very deep in man, together with a sense of guilt and sin and a dream of regaining the Kingdom of Heaven...We are faced with a profound enigma: how could man have renounced paradise which he recalls so longingly in our world-aeon? How could he have fallen away from it?'** (tr. N. Duddington, 1960, p.36 of 310). When it came to our consciousness-induced psychosis, he recognised that **'Philosophers and scientists have done very little to elucidate the problem of man'** (p.49). **'[P]sychologists were wrong in assuming that man was a healthy creature, mainly conscious and intellectual, and should be studied from that point of view. Man is a sick being...The human soul is divided, an agonizing conflict between opposing elements is going on in it...the distinction between the conscious and the subconscious mind is fundamental for the new psychology'** (pp.67-68). Significantly, Berdyaev *also* acknowledged how our species' immense fear of the human condition has blocked access to finding understanding of it, writing that **'Knowledge requires great daring. It means victory over ancient, primeval terror. Fear makes the search for truth and the knowledge of it impossible. Knowledge implies fearlessness...Particularly bitter is moral knowledge, the knowledge of good and evil. But the bitterness is due to the fallen state of the world...**

There is a deadly pain in the very distinction of good and evil, of the valuable and the worthless. We cannot rest in the thought that that distinction is ultimate...we cannot bear to be faced for ever with the distinction between good and evil...Ethics must be both theoretical and practical, i.e. it must call for the moral reformation of life...this implies that ethics is bound to contain a prophetic element. It must be a revelation of a clear conscience, unclouded by social conventions' (pp.14-16).

129 No, **'we cannot bear to be faced for ever with the' 'deadly' 'distinction of good and evil, of the valuable and the worthless'**—nor with the resulting **'social conventions'** of having to live with the psychologically **'sick'** state of alienated denial as our only means of coping with that patently untrue **'distinction'** between **'the valuable and the worthless'**. Permanent damnation and terminal alienation are simply *not* acceptable options for the human race, but to find the liberating, exonerating, reconciling and rehabilitating understanding of our **'divided'** condition, and, by so doing, bring about **'the new psychology'** and with it the transforming **'moral reformation of life'**, required **'victory over** [the] **ancient, primeval terror'** of our condition—a **'victory'** that could *only* be achieved by **'a revelation of a clear conscience, unclouded by social conventions'** of denial, because only those who don't suffer from the **'terror'** of the human condition could hope to **'fearless**[ly]**'** investigate the subject. Yes, to find understanding of the human condition necessarily required a truthful, **'clear conscience'**-guided, instinctual approach, not a resigned-to-living-in-denial-of-the-human-condition, alienated-from-the-truth, blocking-out-of-condemning-moral-instincts, **'intellectual'** approach. It is simply not possible to build the truth from a position of denial/lying. You can't think effectively, insightfully, **'prophetic**[ally]'**, if you're not being honest.

130 Indeed, as I pointed out in chapter 1:11, trying to investigate reality while living in denial of any truths that brought the, for most people, **'deadly pain**[ful]**'** issue of the human condition into focus, as mechanistic science has done, was an extremely compromised and deficient way of searching for knowledge. In fact, it is a measure of the blindness of human-condition-avoiding, denial-based thinking and the effectiveness of human-condition-confronting, honest thinking, that when the whole truth about our condition *is* finally reached, as it now has been, it can appear so straightforward and simple that it seems self-evident. But simplicity has always been a hallmark of insightful thought—as the pioneering biologist Allan Savory observed, **'whenever there has been a major insoluble problem for mankind, the answer, when finally found, has always**

been very simple' (*Holistic Resource Management*, 1988, p.3). For instance, when Charles Darwin put forward his breakthrough, and necessarily exceptionally **'fearless'** and truthful-thinking-based insight of natural selection, it was, in hindsight, such a simple explanation that the eminent biologist of the time, the aforementioned Thomas Huxley, was prompted to exclaim, **'How extremely stupid of me not to have thought of that!'** (Leonard Huxley, *The Life and Letters of Thomas Henry Huxley*, Vol.1, 1900, p.170). So, yes, the author George Seaver was prescient when, in anticipating just how simple the explanation of **'the riddle of life's meaning and mystery'**, namely the human condition, would be, he wrote that **'The ultimate thought, the thought which holds the clue to the riddle of life's meaning and mystery, must be the simplest thought conceivable, the most natural, the most elemental, and therefore also the most profound'** (*Albert Schweitzer: The Man and His Mind*, 1947, p.311).

131 But while the crux question facing the human race, of the much-needed clarification of the nature of 'good and evil', *does* have an amazingly simple answer, the *implications* of it could not be more significant, far-reaching or exciting.

Chapter 2:3 The psychosis-addressing-and-solving *real* explanation of the human condition

132 In another brilliant cartoon (opposite), Michael Leunig succinctly dares to ask the fundamental question of *'What does the chaotic, traumatic and strife-torn life of humans all mean—how are we to make sense of our existence?'* He has done so by placing a very perplexed and distressed gentleman behind an 'Understandascope', through which he peers into a sea of apparent madness. Everywhere he looks there is tumultuous congestion: there are people furiously arguing and fighting with each other; there is a church where people pray for forgiveness and salvation; and there are vehicles polluting the chaos with fumes and noise. *And*, in this 1984 drawing Leunig even seems to have predicted a climactic demonstration of all our human excesses and frustrations when, on September 11, 2001, terrorists flew planes into the tall, square-shaped towers of the World Trade Center in New York City! Well, it is precisely this great burning, searing question of *'What is wrong with us humans?'*—*'How are we to understand it all?'*—*'How are we to make sense of human existence?'*— that is now going to be answered here. The explanation to be given *is* the 'UNDERSTANDASCOPE' we have always wanted, needed and sought.

Cartoon by Michael Leunig, Melbourne's *The Age* newspaper, 17 Mar. 1984

133 So what is the all-exciting, human-race-liberating, amazingly simple, fully accountable, *true* explanation of the human condition? As emphasised in chapter 1, to provide this explanation requires starting from the unequivocally honest basis of acknowledging that humans *did* once live in a completely loving, unconditionally selfless, altruistic state, and that it was only *after* the emergence of our conscious mind that our present 'good-and-evil'-afflicted, immensely psychologically upset condition emerged. If we do this and consider what would happen when a conscious mind emerged in the presence of an already established cooperative, loving instinctive state—given what human-condition-avoiding, mechanistic science has managed to discover about the gene-based natural selection process and how nerves are capable of memory—then the explanation is right there in front of us.

134 Clearly, since our altruistic, moral instincts are only genetic *orientations* to the world and not *understandings* of it, when our fully conscious, reasoning, self-managing mind emerged it would, in order to find the understandings it needed to effectively manage events, have

had to challenge those instinctive orientations, which would have led to a psychologically upsetting clash with our moral instincts.

135 The exceptionally **'clear conscience'**-guided, truthful, and thus effective-thinking, **'prophetic'** naturalist Eugène Marais was on the right track when, as mentioned in par. 81, he recognised that **'As the…individual memory slowly emerges** [in humans]**, the instinctive soul becomes just as slowly submerged…For a time it is almost as though there were a struggle between the two.'** Berdyaev was also approaching the truth when, as just mentioned, he wrote that **'The human soul is divided, an agonizing conflict between opposing elements is going on in it…the distinction between the conscious and the subconscious mind is fundamental for the new psychology.'** Other thinkers, such as Arthur Koestler, Erich Neumann, Paul MacLean, Julian Jaynes and Christopher Booker, have also delved into the problem of the human condition to a similar depth to Marais and Berdyaev, but what was missing was the *clarifying* explanation of the *nature* of that **'struggle'** and **'conflict'** between our instincts and our consciousness. (After all, as pointed out in chapter 1 in par. 73, even Moses' Genesis story of the Garden of Eden contains the truth that our corrupted, 'fallen' condition occurred when our conscious mind emerged from an original, presumably instinctive, idyllic state, and that degree of insight didn't liberate humanity from the human condition.) So, yes, what is the particular **'distinction'** between our instincts and intellect that *caused* the psychologically upset state of our 'good-and-evil'-afflicted condition? As just stated, the answer is that the gene-based refinement system is only capable of *orientating* a species, whereas the nerve-based refinement system has the potential to *understand* the nature of change.

136 To present the explanation in more detail.

137 Nerves were originally developed for the coordination of movement in animals, but, once developed, their ability to store impressions—what we refer to as 'memory'—gave rise to the potential to develop understanding of cause and effect. If you can remember past events, you can compare them with current events and identify regularly occurring experiences. This knowledge of, or insight into, what has commonly occurred in the past enables you to predict what is likely to happen in the future and to adjust your behaviour accordingly. Once insights into the nature of change are put into effect, the self-modified behaviour starts to provide feedback, refining the insights further. Predictions are compared with outcomes and so on. Much developed, and such refinement occurred in

the human brain, nerves can sufficiently *associate* information to *reason* how experiences are related, learn to *understand* and become CONSCIOUS of, or aware of, or *intelligent* about, the relationship between events that occur through time. Thus consciousness means being sufficiently aware of how experiences are related to attempt to manage change from a basis of understanding. (Much more will be explained about the nature and origin of consciousness in chapter 6.)

[138]The significance of this process is that once our nerve-based learning system became sufficiently developed for us to become conscious and able to effectively manage events, our conscious intellect was then in a position to wrest control from our gene-based learning system's instincts, which, up until then, had been in charge of our lives. Basically, once our self-adjusting intellect emerged it was capable of taking over the management of our lives from the instinctive orientations we had acquired through the natural selection of genetic traits that adapted us to our environment. Moreover, at the point of becoming conscious the nerve-based learning system *should* wrest management of the individual from the instincts because such a self-managing or self-adjusting system is infinitely more efficient at adapting to change than the gene-based system, which can only adapt to change very slowly over many generations. HOWEVER, it was at this juncture, when our conscious intellect challenged our instincts for control, that a terrible battle broke out between our instincts and intellect, the effect of which was the extremely competitive, selfish and aggressive state that we call the 'human condition'.

[139]An analogy will help further explain the origin of our human condition.

Chapter 2:4 The Story of Adam Stork

[140]What distinguishes humans from other animals is our fully conscious mind, our ability to understand and thus manage the relationship between cause and effect. However, prior to becoming fully conscious and able to self-manage—consciously decide how to behave—our earliest ape ancestors were controlled by and obedient to their instincts, as other animals still are. Aldous Huxley, author of famous novels such as *Brave New World*, acknowledged how other animals live, and also how our distant ancestors would have lived, obedient to their instincts, when he wrote, **'Non-rational creatures do not look before or after, but live in the**

animal eternity of a perpetual present; instinct is their animal grace and constant inspiration; and they are never tempted to live otherwise than in accord with their own…immanent law. Thanks to his reasoning powers and to the instrument of reason, language, man (in his merely human condition) lives nostalgically, apprehensively and hopefully in the past and future as well as in the present' (*The Perennial Philosophy*, 1946, p.141 of 352). I have underlined part of this passage because it raises the important question of what would happen if a species was **'tempted to live otherwise than in accord with their own'** instincts—to, in effect, break that **'immanent law'**—as Huxley infers humans must have done when we became fully conscious? The following analogy involving migrating storks provides an illustration of what would happen.

The Story of Adam Stork

141 Many bird species are perfectly orientated to instinctive migratory flight paths. Each winter, without ever 'learning' where to go and without knowing why, they quit their established breeding grounds and migrate to warmer feeding grounds. They then return each summer and so the cycle continues. Over the course of thousands of generations and migratory movements, only those birds that happened to have a genetic make-up that inclined them to follow the right route survived. Thus, through natural selection, they acquired their instinctive orientation.

142 Consider a flock of migrating storks returning to their summer breeding nests on the rooftops of Europe from their winter feeding grounds in southern Africa. Suppose in the instinct-controlled brain of one of them we place a fully conscious mind (we shall call the stork Adam because we will soon see that, up to a point, this story parallels Moses' account of the origin or genesis of the human condition in the Bible where Adam and Eve take the fruit from the tree of knowledge in the Garden of Eden). So, as Adam Stork flies north he spots an island off to the left with a tree laden with apples. Using his newly acquired conscious mind, Adam thinks, 'I should fly down and eat some apples.' It seems a reasonable thought but he can't know if it is a good decision or not until he acts on it. For Adam's new thinking mind to make sense of the world he has to learn by trial and error and so he decides to carry out his first grand experiment in self-management by flying down to the island and sampling the apples.

143 But it's not that simple. As soon as Adam's conscious thinking self (as depicted by the stork on the left) deviates from his established migratory path, his innocent instinctive self (as depicted by the wide-eyed stork on the right) tries to pull him back on course. In following the flight path past the island, the instinct-obedient self/stork is, in effect, criticising Adam's decision to veer off course; it is condemning his search for understanding. All of a sudden Adam is in a dilemma: if he obeys his instinctive self and flies back on course, he will remain perfectly orientated but he'll never *learn* if his deviation was the right decision or not. All the messages he's receiving from within inform him that obeying his instincts is good, is right, but there's also a new inclination to disobey, a defiance of instinct. Diverting from his course will result in apples and understanding, yet he already sees that doing so will make him feel bad.

144 Uncomfortable with the criticism his newly conscious mind or intellect is receiving from his instinctive self, Adam's first response is

to ignore the temptation the apples present and fly back on course. This 'correction' not only makes his instinctive self happy, it also wins back the approval of his fellow storks, for not having conscious minds they, like his instinctive self, are innocent, unaware or ignorant of the conscious mind's need to search for knowledge. Furthermore, since Adam's instinctive self developed alongside the natural world, it too reminds him of his instinctive orientation, thus contributing to the criticism of Adam for his rebellious decision.

145 Flying on, however, Adam realises he can't deny his intellect. Sooner or later he must find the courage to master his conscious mind by carrying out experiments in understanding. This time he thinks, 'Why not fly down to an island and rest?' Again, not knowing any reason why he shouldn't, he proceeds with his experiment. And again, his decision is met with the same chorus of criticism—from his instinctive self, from the other storks that were ignorant of the need to search for knowledge, and from the natural world. But this time Adam defies the criticism and perseveres with his experimentation in self-management. His decision, however, means he must now live with the criticism and immediately he is condemned to a state of upset. A battle has broken out between his instinctive self, which is perfectly orientated to the flight path, and his emerging conscious mind, which needs to understand why that flight path is the correct course to follow. His instinctive self is perfectly orientated, but Adam doesn't *understand* that orientation.

146 In short, when the fully conscious mind emerged it wasn't enough for it to be orientated by instincts. It *had to* find understanding to operate effectively and fulfil its great potential to manage life. Tragically, the instinctive self didn't 'appreciate' that need and 'tried to stop' the mind's necessary search for knowledge, as represented by the latter's experiments in self-management—hence the ensuing battle between instinct and intellect. To refute the criticism from his instinctive self, Adam needed to understand the difference in the way genes and nerves process information; he needed to be able to explain that the gene-based learning system can orientate species to situations but is incapable of insight into the nature of change. Genetic selection of one reproducing individual over another reproducing individual (the selection, in effect, of one idea over another idea, or one piece of information over another piece of information) gives species adaptations or orientations—instinctive programming—for managing life, but those genetic orientations, those

instincts, are not understandings. This means that when the nerve-based learning system gave rise to consciousness and the ability to understand the world, it wasn't sufficient to be *orientated* to the world—*understanding* of the world had to be found. The problem, of course, was that Adam had only just taken his first, tentative steps in the search for knowledge, and so had no ability to explain anything. It was a catch-22 situation for the fledgling thinker, for in order to explain himself he needed the very understanding he was setting out to find. He had to search for understanding, ultimately self-understanding, understanding of why he had to 'fly off course', without the ability to first explain why he needed to 'fly off course'. And without that defence, he had to live with the criticism from his instinctive self and was insecure in its presence.

147 To resist the tirade of unjust criticism that he was having to endure and mitigate that insecurity, Adam had to do something. *But what could he do?* If he abandoned the search and flew back on course, sure, he'd gain some momentary relief, but the search would, nevertheless, remain to be undertaken. So all Adam could do was retaliate against and **attack** the instincts' unjust criticism, attempt to **prove** the instincts' unjust criticism wrong, and try to **deny** or block from his mind the instincts' unjust criticism—and, as the stork on the left of the picture illustrates, he did *all* those things. He became angry towards the criticism. In every way he could he tried to demonstrate his self worth, prove that he is good and not bad—he shook his fist at the heavens in a gesture of defiance of the implication that he is bad. And he tried to block out the criticism. In short, the **angry**, **egocentric** and **alienated** state appeared. His intellect or 'ego' (which is just another word for the intellect since the *Concise Oxford Dictionary* defines **'ego'** as **'the conscious thinking self'** (5th edn, 1964)) became 'centred' or focused on the need to justify itself—selfishly preoccupied aggressively competing for opportunities to prove he is good and not bad, to validate his worth, to get a 'win'—essentially eke out any positive reinforcement that would bring him some relief from criticism and sense of worth. He unavoidably became **selfish**, **aggressive** and **competitive**.

148 Overall, it was a terrible predicament in which Adam became **psychologically upset**—a sufferer of **psychosis** and **neurosis**. Yes, since, according to *Dictionary.com*, **'osis'** means **'abnormal state or condition'**, and the *Penguin Dictionary of Psychology*'s entry for **'psyche'** reads: **'The oldest and most general use of this term is by the early Greeks, who envisioned**

the psyche as the soul or the very essence of life' (1985 edn), Adam developed a **'psychosis'** or **'soul-illness'**, and a **'neurosis'** or neuron or nerve or **'intellect-illness'**. His gene-based instinctive self or soul or **psyche** became repressed by his intellect for its unjust condemnation of his intellect, and, for its part, his nerve or **neuron**-based intellect became preoccupied denying any implication that it is bad. Adam became psychotic and neurotic.

149 But without the knowledge he was seeking, without self-understanding, Adam Stork had no choice other than to resign himself to living a psychologically upset life of anger, egocentricity and alienation as the only three responses available to him to cope with the horror of his situation or condition. It was an extremely unfair and difficult, indeed tragic, position for Adam to find himself in, for we can see that while he was good he appeared to be bad and had to endure the horror of his psychologically distressed, upset condition until he found the defence or reason for his 'mistakes'. Basically, *suffering upset was the price of his heroic search for understanding*. Indeed, it is the tragic yet inevitable situation any animal would have to endure if it transitioned from an instinct-controlled state to an intellect-controlled state—its instincts would resist the intellect's search for knowledge. Adam's uncooperative and divisive competitive aggression; and his selfish, egocentric efforts to prove his worth; and his need to deny and evade criticism, essentially embrace a dishonest state, all became an unavoidable part of his personality. Such was Adam Stork's predicament, *and such has been the human condition, for it was within our species that the fully conscious mind emerged.*

150 What now needs to be explained is how the situation faced by humans when *we* began searching for knowledge was *so* much worse than it was for our hypothetical Adam Stork. This is because, as was emphasised in chapter 1:6, *our* instinctive orientation wasn't to a flight path but to *behaving in an unconditionally selfless, all-loving, cooperative moral way*, so when we started carrying out experiments in understanding and then unavoidably began reacting defensively in an angry, aggressive, competitive, selfish and evasive, dishonest way to combat the criticism from our instincts, that divisive response drew *even more* criticism from our particular cooperative, loving, ideal-behaviour-demanding, integrative, **'very essence of life'**-orientated moral instincts—a vicious cycle that fuelled our upset immensely. (Chapter 4 will present the biological explanation of how we humans acquired such unconditionally selfless, universally loving, moral instincts, the 'voice' of which is our conscience.)

Chapter 2:5 The double and triple whammy involved in our *human* situation or condition

[151] When Adam Stork began searching for knowledge and unavoidably became angry, egocentric and alienated, that upset response didn't attract further criticism from his instinctive orientation—but that certainly wasn't the case with us humans. No, when we began searching for knowledge and became angry, egocentric and alienated, that response was extremely offensive to our particular instinctive orientation because our instinctive orientation isn't to a flight path, or to any of the various instinctive orientations that other animals are obedient to; it is to behaving in the opposite way, namely lovingly, selflessly and honestly. So in *our* case, when we began experimenting in understanding and were criticised by our instincts and unavoidably responded in an angry, egocentric and alienated way, we had to endure a further round of criticism, a second hit, a 'double whammy', from our instinctive orientation. Yes, in our necessary search for understanding we were firstly unjustly condemned for defying our instincts, and then again for reacting to that condemnation in a way that was counter and offensive to our instincts. So if Adam Stork had cause to be upset, we had double cause to be upset!

[152] And yet the horror of our situation didn't end there—if a worse fate can be imagined! For we weren't just unjustly condemned twice, we were unjustly condemned *three times*—forced to endure a 'triple whammy', which I will now explain.

[153] While the following explanation will be covered in greater detail in chapter 3, to context why we have experienced this 'triple whammy' of condemnation it is necessary to briefly explain that there is a teleological, order-of-matter-developing, integrative, cooperative theme or direction or purpose or meaning to existence, which God is the personification of. Everywhere we look we see hierarchies of ordered matter—'There is a tree that is composed of parts (leaves, branches, a trunk and roots) and in turn those parts are composed of parts (fibres, cells, etc).' Our world is clearly composed of a hierarchy of ordered matter: atoms have come together or integrated to form compounds, which in turn have come together or integrated to form virus-like organisms, which integrated to form single-celled organisms, which then integrated to form multicellular organisms, which have come together or integrated to form societies of multicellular organisms. Significantly, the behaviour required for these

ordered arrangements of matter or wholes to stay together is selflessness—because selflessness means being considerate of the welfare of the larger whole or integrative, while selfishness is divisive or disintegrative. Yes, selflessness is really the theme of existence, the glue that holds wholes together. But in light of our *divisive* selfish, egocentric, competitive and aggressive behaviour, we humans have obviously found this truth of the selflessness-dependent, *integrative* meaning of existence unbearably condemning. And being so unbearably condemning, the main way we coped with this truth of Integrative Meaning was to deify it, make it God—a concept we revered but claimed had no material relationship to us. Again, all this will be explained in chapter 3, but, in terms of explaining the 'triple whammy' we suffered from when we searched for knowledge, what this selfless, cooperative, Godly, integrative theme of existence means is that when we started retaliating to the criticism from our instincts in a divisive, selfish, uncooperative, angry, aggressive, egocentric and alienated way, we were not only at odds with our cooperative, selfless, loving moral instincts, we were also defying God! Our retaliation against our instinctive self made us appear as though we were out of step with creation, living in a way that was entirely inconsistent with the integrative, cooperative, selfless theme or meaning of existence! The point here is that despite having removed the confronting presence of Integrative Meaning by abstracting it as God (and even by outright denial of the existence of Integrative Meaning or God), the development of order of matter in nature is actually such an obvious truth that our conscious mind is well aware of it. So, when we humans went in search of knowledge we were initially criticised for not obeying our instincts, and *then* for responding to that initial criticism in a way that offended our cooperatively orientated, moral instinctive conscience, and, thirdly, by that behaviour defying the very integrative, cooperative theme of existence that our intellect could so plainly see existed in nature. *We defied our instincts, we offended our moral conscience, and we insulted God!!!* We humans could hardly be more guilt-ridden. And all this guilt, which we can now understand was completely unjustified, made us *extremely, excruciatingly* upset—*absolutely furious*, in fact. As necessary as it was, in our case, 'flying off course' was an *incredibly* upsetting act of defiance—*which is why we humans have been capable of absolutely extraordinary acts of brutality, barbarism and cruelty*. While we have tried to restrain and conceal the anger within us, 'civilise' it as we say, it is, in truth, volcanic—but, again, we can now at last understand the origin of all that anger.

[154]To appreciate exactly what occurred when we humans began to actively experiment in understanding we can consider what occurs in the lives of children today, and must have occurred in the lives of our distant, child-equivalent ancestors, who, as will be described in chapters 7:2 to 7:7, were the australopithecines who emerged from our infant-equivalent, bonobo-like ape ancestors some 4 million years ago.

[155]In the case of young humans today, when they begin to actively experiment in managing their lives from a basis of understanding, which is what occurs during the stage of a conscious mind's development that we call 'childhood', they encounter this upsetting criticism from their instinctive moral conscience and from their mind's awareness of the integrative theme of existence—as well as from other minds around them. Imagine, for instance, a situation where a young boy sees a birthday cake on a table and, being new to this business of reasoning, thinks, innocently enough, 'Why shouldn't I take all the cake for myself', before duly doing so. While many mothers actually witness these grand mistakes of pure selfishness that young children make when they first attempt to self-manage their lives, they still have to be reasonably lucky to do so because, once done, the child usually doesn't make such a completely naive mistake again due to the criticism it attracts from its moral instincts, from its own conscious mind's awareness of the very obvious integrative, cooperative, selfless theme of existence, and from others present. But despite the nasty shock from all the criticism and his subsequent determination to never again make such a mistake, the child, although he is unable to explain his actions, does feel that what he has done is not something bad, not something deserving of such criticism. In fact, by this stage in the child's mental development, he has become quite proud of his efforts at self-managing his life, drawing attention to his achievements with excited declarations like 'Look at me Daddy, I can jump puddles', and so on. So the child is only just discovering that this business of self-adjusting is not all fun and that some experiments are getting him into trouble. It is at this stage that 'playing', as we call these early experiments in self-management, starts leading to some serious issues for the child. Indeed, the frustrated feeling of being unjustly criticised for some of the experiments gives rise to the precursors of the defensive, retaliatory reactions of anger, egocentricity and

alienation, with some angry, aggressive nastiness creeping into the child's behaviour. Furthermore, in this situation of feeling unfairly criticised, it follows that any *positive* feedback or reinforcement begins to become highly sought after, which is the beginning of egocentricity—the conscious thinking self or ego starts to become preoccupied trying to defend its worth, assert that it is good and not bad. At this point, the intellect also begins experimenting in ways to deny the unwarranted criticism, which, in this initial, unskilled-in-the-art-of-denial stage, takes the form of blatant lying: 'But Mum, Billy told me to do it', or 'But Mum, the cake accidentally fell in my lap.' These apparent misrepresentations weren't *actually* lies, rather they were inadequate attempts at explanation. Lacking the real excuse or explanation, it was at least *an* excuse, a contrived defence for the child's mistake. The child was evading the false implication that his behaviour was bad, in the sense that a 'lie' that said he wasn't bad was less of a 'lie' than a partial truth that said he was. Basically, the child has started to feel the first aggravations from the horror of the injustice of the human condition—and we can expect that exactly the same kind of mistakes in thinking and the resulting frustrations with the ensuing criticism would have also occurred in our australopithecine, child-equivalent ancestors. Some of them, those who had become intelligent enough to actively experiment in self-adjustment, would have begun to encounter criticism of their efforts to self-manage and, as such, begun to exhibit the psychologically upset behaviours of anger, egocentricity and alienation.

156 Once the experimentation in self-management gets underway, so the upsetting frustrations with the resulting criticism increases. The journey of ever-increasing levels of upset has begun—and this corrupting journey of escalating upset could not and would not stop until the exonerating explanation of the human condition was found, which has only now, in this book, finally emerged millions of years after this self-fuelling process began, which means there must be, and indeed is, an absolutely astronomical amount of upset built up in us humans! That horrific journey of ever-increasing levels of upset is described in chapter 7, but just to follow its development a little further—throughout childhood the experimenting in understanding increases and the resulting frustrated upset also increases, such that by late childhood children enter what is recognised as the **'naughty nines'**. By this stage the resentment and frustration with the criticism from their efforts at self-adjustment

has become so great that the child starts lashing out at the unjust world. Indeed, by late childhood children become very angry, even taunting and bullying those around them. However, by the end of childhood, children realise that lashing out in exasperation at the 'injustice of the world' doesn't change anything and that the only possible way to solve their frustration is to find the reconciling insight into why the criticism they are experiencing is not deserved. It is at this point, which occurs around 12 years of age, that the child undergoes a dramatic change from a frustrated, protesting, demonstrative, loud extrovert into a sobered, deeply thoughtful, quiet introvert, consumed with anxiety about the imperfections of life under the duress of the human condition. Indeed, it is in recognition of this very significant psychological change from a relatively human-condition-free state to a very human-condition-aware state that we separate these stages into 'Childhood' and 'Adolescence', a shift even our schooling system marks by having children graduate from what is generally called primary school into secondary school. And indeed, this critical junction in our species' development is also acknowledged in the anthropological record, with the name of the genus changing from *Australopithecus*, the extrovert 'Childman', to *Homo*, the sobered 'Adolescentman'. The story of the journey through this next stage of adolescence was briefly introduced in pars 43-52 when Resignation was described, but to quickly recap, after struggling for a few years during their early adolescence to make sense of existence, by about 14 or 15 years of age that search for understanding generally became so confronting that denial of the whole unbearably depressing issue of the human condition had to be adopted—after which adolescents became superficial and artificial escapists, not wanting to look at any issue too deeply, and, before long, combative and competitive power-fame-fortune-and-glory, relief-seeking resigned adults.

157 So while the upset state of the human condition emerged during our childhood, and in the case of our species' journey, during the life of our child-equivalent australopithecine ancestor, the *real* struggle with the agony of the human condition didn't arise until our adolescence, which began when we were around 12 years of age, or, in the case of our species, during the life of *Homo* who emerged from the australopithecines some 2 million years ago. Indeed, we could say that we only became *fully conscious* in the sense of being *fully aware* of the situation we humans have been in of having to live with the agony of the human condition

some 2 million years ago when our species entered adolescence. Again, all the stages of ever-increasing levels of upset that we humans both individually and as a species have progressed through will be much more fully described in chapter 7.

158 In summary, unlike Adam Stork who only had to contend with criticism of his attempts to self-adjust, we humans have had to contend with that criticism *and* criticism of our unavoidable angry, aggressive, competitive, selfish and dishonestly evasive response to that initial criticism—a response that went against our moral conscience and against the integrative meaning of existence; against the Godly ideals, no less. In short, through our efforts to self-adjust and experiment in self-understanding, in order to find the ultimate knowledge of understanding of the human condition that we needed to make sense of ourselves, we were made to feel extremely guilty, 'evil' and 'sinful', which very greatly compounded our insecurity-of-self and frustration, making us *immensely* angry and egocentric and *very much* needing to live in denial of any confrontation with the problem of our corrupted condition. We had to live totally separated from, or blocked-off from, or dissociated from, or alienated from our true situation; metaphorically hidden deep in Plato's dark cave where no exposure of our corrupted condition was possible.

159 When we consider that we have been living in this extremely unfair and upsetting state for over 2 million years we can begin to comprehend how much hurt, frustration and anger has developed in us humans. Indeed, imagine living just *one* day with the injustice of being condemned as evil, bad and worthless when you intuitively know—but are unable to explain—that you are actually the complete opposite, namely truly wonderful, good and meaningful. How tormented and furious—how upset—would you be by the end of that one day? You would be *immensely* upset. So yes, extrapolate that experience over *2 million years* and you will begin to get some appreciation of just how much anger must now exist within us humans today! While we have learnt to significantly restrain and conceal—'civilise'—our phenomenal amount of upset, it nevertheless follows that, under the surface, we must be boiling mad with anger, and that sometimes, when our restraint can no longer find a way to contain it, that anger must express itself. Yes, *we can now, at last, understand humans' capacity for astounding acts of aggression, hate, brutality and atrocity*. The following is but one description of how much anger humans have accumulated as a result of being unjustly condemned; it is an account

of the bloodshed that was commonplace during World War I: **'The flowing blood of these murdered men, ten million gallons of steaming human blood could substitute for a whole day the gigantic water masses of the Niagara** [Falls]**...Make a chain of these ten million murdered murderers, placing them head to head and foot to foot, and you will have an uninterrupted line measuring ten thousand miles, a grave ten thousand miles long'** (Mrs Will Gordon, *Roumania Yesterday and To-day*, 1918, p.251 of 270). And we can now understand that *this capacity for inhumanity exists in us all*—as the author Morris West so bravely acknowledged in his memoir, *A View from the Ridge*: **'brutalise a child and you create a casualty or a criminal. Bribe a servant of the state and you will soon hear the deathwatch beetles chewing away at the rooftrees of society. The disease of evil** [now able to be understood as upset] **is pandemic; it spares no individual, no society, <u>because all are predisposed to it</u>. It is this predisposition which is the root of the mystery** [of 'evil' that is now explained]**. I cannot blame a Satan, a Lucifer, a Mephistopheles, for the evils I have committed, the consequences of which have infected other people's lives. I know, as certainly as I know anything, that the roots are in myself, buried deeper than I care to delve, in caverns so dark that I fear to explore them. I know that, given the circumstances and the provocation, I could commit any crime in the calendar'** (1996, p.78 of 143).

[160] Clearly it is an understatement of the grandest proportions to say that it is a relief that humanity's 2-million-year journey of conscious thought and enquiry into the nature of our condition has finally delivered understanding of it—that the **'caverns so dark'** where the **'mystery'** of our grotesquely upset human condition lies have at last been **'explore**[d]**'** and the greater dignifying, liberating, healing, ameliorating explanation for that condition found.

Chapter 2:6 <u>Adam and Eve — we humans — are heroes NOT villains</u>

[161] We can see that The Story of Adam Stork—which describes the main or primary issue involved in our human condition of the upsetting battle that emerged between our instincts and our conscious intellect's search for knowledge—is similar to Moses' Biblical account of Adam and Eve's experiences in the Garden of Eden, except in that presentation when Adam and Eve took the **'fruit'** (Gen. 3:3) **'from the tree of the knowledge of good and evil'** (2:9, 17)—went in search of understanding—they were **'banished... from the Garden'** (3:23) of our original innocent, all-loving state for having

become 'bad' or 'evil' or 'sinful'. In THIS presentation, however, Adam and Eve are revealed to be the HEROES, NOT THE VILLAINS they have so long been portrayed as. While we ARE immensely upset—that is, immensely angry, egocentric and alienated—HUMANS ARE GOOD AND NOT BAD AFTER ALL!!!! Yes, this explanation finally allows us to see that our conscious mind that caused us to take the **'fruit' 'from the tree of... knowledge'** is *NOT* the sinful, evil villain it has so long been portrayed as. And 'upset' is the right word for our condition because while we are not 'evil' or 'bad' we are definitely psychologically upset from participating in humanity's heroic search for knowledge. 'Corrupted' and 'fallen' have sometimes been used to describe our condition, but they have negative connotations that we can now understand are not deserved.

162 The following is another famous cartoon by Michael Leunig that beautifully illustrates the journey humans have been on in our quest for understanding—what is missing, however, from his depiction is this final, truthful revelation of our species' heroism that the explanation of the human condition has now made possible. So it is with deep reverence to Leunig that I have taken the liberty of drawing three more frames in his marvellously expressive style to complete the story. Leunig's cartoon depicts the Genesis story of Adam and Eve in the Garden of Eden, beginning with the wide-eyed and innocent Adam and Eve taking the **'fruit' 'from the tree of...knowledge'** and being **'banished...from the Garden of Eden'** as a result. Up to that point there is nothing unusual about the story being portrayed, however, the cartoon goes on to show Adam reaping revenge upon that hallowed Garden. It is possible that Leunig meant for the retaliation to be interpreted as a straightforward joke about human behaviour—'You kicked me out, I'll get even'—but surely there is a deeper truth to the retaliation that accounts for the cartoon resonating with so many people. Hasn't Leunig got to the truth that lies at the very heart of the issue of our human condition, and summarised that truth in the most succinct way possible using the story of the Garden of Eden? Hasn't he captured the underlying feeling that we humans have of being condemned as fundamentally evil and God-disobeying when in our heart of hearts we don't believe we are; and hasn't he captured the psychotic and neurotic anger that feeling of unjust condemnation has caused us? This is a deservedly celebrated cartoon, for, as emphasised, we humans have been unjustly condemned for some *2 million years*—so how deeply, deeply angry must we be

inside ourselves having had to live undefended on this planet for so long! Yes, wouldn't, and shouldn't, we be as angry as Leunig has depicted us in this cartoon!

Cartoon by Michael Leunig that appeared in Melbourne's *The Age* newspaper on 31 Dec. 1988

Drawings by Jeremy Griffith, with deeply appreciative deference to Michael Leunig, Jul. 2009

163 To analyse the cartoon's elements more closely, in the fourth frame Adam is shown fuming with rage and resentment for being evicted, implying that he doesn't believe it is deserved, and deciding that he has no choice but to retaliate against the injustice; he can't be expected to just sit there and take it, he has to find some way of demonstrating that he doesn't accept as true the criticism that he is fundamentally bad. And so a vengeful Adam returns with a chainsaw to raze the Garden. The guardian angel is in tears at the wanton destruction, and we can see that Eve is similarly distressed by his actions. (This lack of empathy by women for men's immensely upsetting battle to defy the ignorance of our instinctive self, which Leunig has so honestly expressed here, will be explained in pars 596-647.) But Adam's expression and body language shows the enormous relief and satisfaction his retaliation brings him. In giving the guardian angel 'the finger' in the eighth frame, he's symbolically saying, 'Stick that up your jumper for unjustly condemning me!'

164 BUT, above all, in the expression of extreme anger on Adam's face, Leunig has revealed just what 2 million years of being unjustly condemned by the *whole* world has done to us humans. Yes, since the sun, the rain, the trees and the innocent animals are all friends of our original innocent, instinctive self or soul, through that association they too have condemned us, which is why Adam's/our, retaliation against nature for its unjust condemnation of him has left the whole natural world such a wasteland! We burnt the scrub, tore down the trees and dumped rubbish, pollutants and cement over what was left of nature, and we murdered animals who, in their innocence, condemned us. For example, there has been **'5,843 Sq km of Amazon rain forest reportedly lost to deforestation from August 2012 to July 2013; activists blame the 28% rise in one year on looser environmental laws'** (*TIME*, 2 Dec. 2013)—the deeper truth being that if we humans weren't so hateful and couldn't-care-less we wouldn't destroy pristine rainforest like this. Similarly, the diary of the legendary 'white hunter', the suitably named J.A. Hunter, reveals that he dispatched **'966 Rhinos'** from **'August 29th 1944 to October 31st 1946'** (Peter Beard, *The End of the Game*, 1963, p.137 of 280). That's the equivalent of nearly 10 rhinoceroses every week for more than 2 years that he shot to death! Incidentally, we can see here how, if we really wanted to save the environment, that 'hugging trees' and 'patting dolphins' wasn't going to do it. Pretending to be loving and kind and considerate could make us feel good but it was never going to fix anything. To 'save the planet' we had to find the

understanding that would end the underlying upset in us humans. We had to confront the issue of *us*, *the human condition*, NOT find ways to delude ourselves we were good so we didn't have to confront that issue. Pretence, delusion and escapism got us nowhere; in fact, it made the situation the human race is in much, much worse because it hid the real issue of the immensely psychologically upset state of the human condition. So yes, while we have learnt to conceal how upset we are—learnt 'civility'—underneath that facade of restraint and delusion lies the level of anger Leunig has portrayed.

165 It really has been a case of **'Give me liberty or give me death'**, **'No retreat, no surrender'**, **'Death before dishonour'**, **'Death or glory'**, **'Do or die'**, **'Die on your feet, don't live on your knees'**, **'Never give in'**, **'Better to reign in hell than serve in heaven'**, **'You can stand me up at the gates of hell, but I won't back down'**, as the sayings go. We were *never* going to give in to our instinctive self or soul; we were never going to accept that we were fundamentally bad, evil, worthless, awful beings; we weren't going to wear that criticism. For if we did, we wouldn't be able to get out of bed each morning and face the world. If we truly believed we were fundamentally evil beings, we would shoot ourselves. There had to be a greater truth that explained us humans and we couldn't rest until we found it. And so every day as we got out of bed we took on the world of ignorance that was condemning us. We defied the implication that we are bad. We shook our fist at the heavens. In essence, we said, 'One day, one day, we are going to prove our worth, explain that we are not bad after all, and until that day arrives we are not going to **'back down'**, we are not going to take the ignorant, naive, stupid, unjustified criticism from our instincts, or from Integrative Meaning/God. No, we are going to fight back with all our might.' And that is what we have done; that is what every conscious human that has ever lived has done—and because we did we have now finally broken through and found the full truth that explains that we humans are wonderful beings after all. In fact, not just wonderful but the heroes of the whole story of life on Earth. This is because our fully conscious mind is surely—given its phenomenal ability to understand the world—nature's greatest invention, so for us humans who were given this greatest of all inventions to develop, to be made to endure the torture of being unjustly condemned as bad or evil for doing just that, and to have had to endure the torture for *so* long, *some 2 million years*, *has to* make us the absolute heroes of the story of life on Earth.

We were given the hardest, toughest of tasks, and against all the odds we completed it. We humans *are* the champions of the story of life on Earth. We are *so, so* wonderful!

166 Yes, the essential truth that has now at last been explained is that we should *never* have been 'thrown out of the Garden of Eden' in the first place—which is why, to return to Leunig's cartoon, I have taken the liberty of drawing three additional frames to complete the story. The first addition depicts Adam and Eve beckoning to the guardian angel to return, while the second shows them explaining to the angel the irrefutable, first-principle-based, biological reason why we humans are *not* fundamentally evil; in fact, explaining, as I have just done, that we are the absolute heroes of the story of life on Earth—an explanation that so affects the angel that it starts to cry out of regret and sympathy. In the third and final frame, we see the angel taking Adam and Eve—the human race in effect—by the hand and apologetically escorting us back to the Garden of Eden. Yes, the human race is coming home now, our **'banish**[ment]**'** has ended; we are no longer **'strangers in paradise'** (popular song from the 1953 musical *Kismet*), rather we are the heroes of paradise—its most revered, cherished and loved occupants! What a wonderful change of circumstances for us humans!!

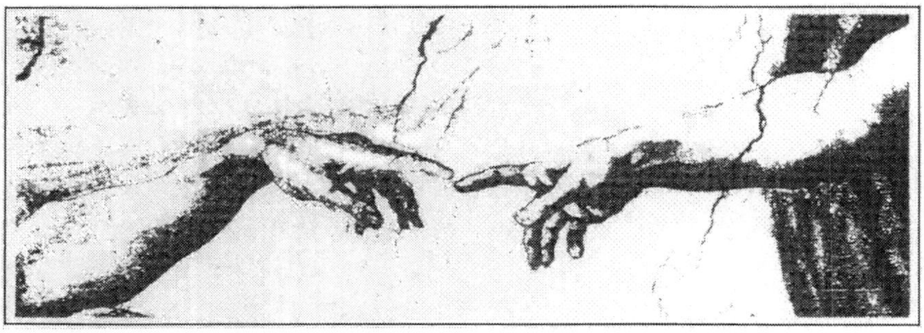

Detail from Michelangelo's *The Creation of Adam*, c.1508-1512

167 This is a detail from *The Creation of Adam*, Michelangelo's famous masterpiece that adorns the ceiling of the Sistine Chapel in the Vatican City in Rome. In interpretations of the painting, God is said to be in the process of creating Adam, yet it can also be construed as God and man reaching out to each other. Since we can now afford to acknowledge that God is the personification of the integrative, cooperative, selfless, harmonious ideal values of life, then with the reconciling understanding

of our less-than-ideal, human-condition-afflicted state found, the 'outstretched fingers' of God and man *have* finally touched. Our seemingly un-Godly, upset anger, egocentricity and alienation resulted from our participation in humanity's heroic search for knowledge. Yes, it turns out that in that classic 1980 American film *The Blues Brothers*, for all their wild, off-the-wall, upset behaviour, the Blues Brothers really were **'on a mission from God'**, as they kept telling everyone they encountered; *indeed, the life of every human who has ever lived has been entirely meaningful and thus Godly.*

168 Another universally iconic symbol that can now be interpreted through the truthful lens that this explanation allows is the Statue of Liberty that stands so proudly in New York Harbor. The United States of America presently sits at the forefront of the human race's immensely upsetting but critically important, heroic search for knowledge; it really has been **'the land of the free and the home of the brave'**, as its national anthem proclaims. So it is entirely fitting that this symbol of the USA is such a brave one because it has been *so* difficult maintaining the freedom/liberty necessary to continue the upsetting/corrupting search for knowledge. This is because as soon as there was too much corruption of humans' cooperative, selfless, loving instinctive self or soul—too much anger, egocentricity and alienation, but especially too much egocentric selfishness—a society became dysfunctional and decadent, and then, in a bid to counter that dysfunction, it became oppressive and restraining of the freedoms each person needed to continue the corrupting search for knowledge. Yes, you had to be **'brave'** to avoid giving up the corrupting search for knowledge and maintain the '**free**[dom]' needed to

continue humanity's heroic search for knowledge—*and the story of the human race as a whole is that it has never completely abandoned that corrupting search for knowledge, and as a result of that tenacity it has now finally succeeded in finding the particular knowledge it was in search of, namely the explanation of the human condition: the reconciling, ameliorating, peace-bringing understanding of why we humans are good and not bad.*

Chapter 2:7 Our instinctive moral conscience has been 'A sharp accuser, but a helpless friend!'

169 Greek mythology describes how Prometheus stole fire from his fellow Gods and gave it to humans for their use, an act which enraged the Gods, and Zeus in particular who punished Prometheus by having him strapped to the top of a mountain where, every day in perpetuity, he was forced to suffer having his innards eaten out by an eagle. In light of what has been revealed, we can now understand that in this story fire is the metaphor for the conscious intellect (as it is in many mythologies), and that the consequence of humans gaining a conscious mind was extremely upset behaviour, which explains why Prometheus was punished by the Gods—in their eyes his gift of consciousness to humans was responsible for the corruption of the human race, for our falling out with the Godly ideals.

170 Yes, for 2 million years our intellect has been seen as the villain of the piece while our instinctive self or soul's moral conscience was held up as the epitome of goodness, but the truth, which we can now finally explain, turns out to be the exact opposite in the sense that it was our instincts' unjust criticism that caused us to become upset. This paradoxical turn of events in which our 'good side' is revealed to have been the 'bad side' is the theme of the crime writer Agatha Christie's famous play *The Mousetrap*. The play is just another 'whodunit' murder mystery and yet it is now the longest running play in history, having been performed continuously since opening in 1952; indeed, the play's enduring popularity is such that it celebrated its 60th anniversary in 2012 with a global tour. All enduring myths and stories contain truths that resonate, and in the case of *The Mousetrap*, the police inspector involved in the murder investigation, who is held up throughout the play as the pillar of goodness and justice, is revealed at the very end to be the culprit. It is

the essential story of humanity where the apparent ideals of our original instinctive self or soul's selfless, loving world are revealed, at the very last moment, to have been the unjustly condemning influence. As with so many aspects of the human condition, THE TRUTH WAS NOT AS IT APPEARED. We discover at the very end of our journey to enlightenment that conscious humans, immensely corrupt as we are, are 'good' and not 'bad' after all, and that which was 'good', our moral conscience, turns out to be the cause of our 'sin'.

171 The same essential paradox appears in G.K. Chesterton's 1908 book, *The Man Who Was Thursday*, in which a policeman representing the 'good' side has to infiltrate and expose the sinister members of a quintessentially corrupt organisation, but as the tale unfolds each of the apparently corrupt members are revealed to be forces for good commissioned to fight evil.

172 The poet Alexander Pope acknowledged the pain, guilt and frustration caused by the unjust criticism emanating from our species' instinctive moral conscience when he wrote, **'our nature** [conscience—is]**...A sharp accuser, but a helpless friend!'** (*An Essay on Man*, 1733, Epistle II). It was a sentiment echoed by William Wordsworth in his great 1807 poem, *Intimations of Immortality*: **'High instincts before which our mortal Nature / Did tremble like a guilty thing surprised.'** Albert Camus was another who felt the sting of the criticism from our naive, ignorant, innocent soul when he wrote, '[can] **innocence, the moment it begins to act...avoid committing murder** [?]' (*L'Homme Révolté*, 1951; pub. in English as *The Rebel*, 1953).

173 Considering then how unjustly hurtful our instinctive self or soul has been it is little wonder we learnt to psychologically block it out, deny and bury it to the point where we now refer to it as **'the child within'** and the **'collective** [shared by all, instinctive] **unconscious'**. Indeed, if we look at the situation from the perspective of our original, unconditionally selfless, all-loving, moral instinctive self or soul for a moment, we can begin to appreciate just how enormous the schism between our instinctive soul and conscious mind has been. We have to imagine then what a shock it was for our all-loving, moral instincts when our conscious mind began searching for knowledge and became angry, aggressive, selfish and competitive—our soul would have been absolutely mortified, as children are today by the extreme imperfection of human behaviour. Our soul would have been *utterly* bewildered and distressed, *completely* overcome with shock and disappointment, *absolutely* devastated:

'Why, when all our behaviour has been so cooperative, loving and perfect, and our world so happy and content as a result, are you, our conscious mind, doing this?—this is *SO* wrong!', is how our soul has perceived the situation. Unable to explain our divisive behaviour, we conscious humans have never allowed ourselves to properly consider the hurt that behaviour has caused our moral soul, but now that we can at last explain our upset state we can afford to consider it. Yes, when our conscious mind emerged and the angry, aggressive, selfish and competitive behaviour started to appear, our original, all-loving and all-sensitive instinctive self experienced the most profound shock imaginable. Our soul has been *completely* distressed and overwhelmed by this turn of events: 'Why is this happening, this behaviour simply must stop, this is absolutely wrong!' Our soul has been in tears of distress and disappointment, utterly overcome with unhappiness—BUT WHAT WAS HAPPENING *HAD* TO GO ON, THE UPSETTING SEARCH FOR THE KNOWLEDGE THAT WOULD LIBERATE OUR SPECIES *HAD* TO CONTINUE. To return, however, to viewing the situation from the perspective of our conscious intellect, this **'sharp accuser'**, these **'high'** and mighty **'innocen**[t]**' 'instincts'** that made us **'tremble like a guilty thing'**, *have* been **'murder**[ous]**'** and thus completely unbearable, SO WE SIMPLY HAD TO BLOCK OUR MORAL INSTINCTIVE SELF OUT, OR DIE FROM THE PAIN, THE GUILT, IT WAS CAUSING US. So the extent of our soul's unhappiness with us, which was of a level of *extreme* unhappiness, is the extent we, our conscious mind, had to block out and become alienated from it, which is, it follows, a level of *extreme* alienation. We are EXTREMELY soul-repressed, or psychotic, and EXTREMELY mentally committed to living in denial, or neurotic. R.D. Laing was right when, as mentioned earlier in par. 56, he described us as being **'a shrivelled, desiccated fragment of what a person can be...between *us* and It** [**our true selves** or soul] **there is a veil which is more like fifty feet of solid concrete...The outer divorced from any illumination from the inner is in a state of darkness. We are in an age of darkness. The state of outer darkness is a state of sin—i.e. alienation or estrangement from the inner light...We are bemused and crazed creatures, strangers to our true selves, to one another'**, and **'We are dead, but think we are alive. We are asleep, but think we are awake...We are the halt, lame, blind, deaf, the sick. But we are doubly unconscious...We are mad, but have no insight** [into the fact of our madness]**.'** But again, our conscious, intellectual self simply had *no choice* but to block out our instinctive self or soul, to banish it to our subconscious where it only now occasionally bubbles up in dreams and on other occasions when our conscious self is subdued, such as when praying or meditating. As Carl Jung wrote, **'The dream is a little hidden door in the innermost and most secret recesses of the psyche** [soul]**, opening into that cosmic night which was psyche long before there was any ego consciousness'** (*Civilization in Transition*, 1945; *The Collected Works of C.G. Jung*, tr. R. Hull, Vol.10). That pre-eminent philosopher Sir Laurens van der Post also wrote about the repression of our soul when he acknowledged that **'Human beings know far more than they allow themselves to know: there is a kind of knowledge of life which they reject, although it is born into them: it is built into them'** (*A Walk with a White Bushman*, 1986, p.142 of 326).

174 Yes, our ability now to understand the human condition, understand the consequences of humans gaining a conscious mind, means the whole mystery of our life in extreme alienation finally becomes fully understandable. *Of course* there is **'a kind of knowledge of life which they** [conscious humans] **reject'**, even **'though it is born into'** us—our moral conscience *has been an unbearably* **'sharp accuser, but a helpless friend!'** *Of course* we have had to block out our soul, and with it any truth that brought the unbearable issue of our corrupted human condition into focus! And no wonder we are 2 million years wedded to a life deep underground in Plato's metaphorical cave of dark, truthless, alienated denial. And, given how precious, indeed life-saving, that block-out or denial has been, and how practised and habituated we are in applying it, it is no wonder it has been almost impossible for anyone to break through that denial, actually confront the human condition and find the liberating explanation of it—and, further, that there has been so much resistance to having all that denial demolished and exposed. Yes, tragically, we had *no choice* other than to suppress our all-sensitive, unconditionally selfless, all-loving (although unloving of our intellect's need to search for knowledge!) soul—as the saying goes, **'we hurt the one we love'**—BUT THANKFULLY, WITH THE EXPLANATION OF WHY WE FULLY CONSCIOUS HUMANS BECAME ANGRY, AGGRESSIVE, SELFISH AND COMPETITIVE FOUND, WE NO LONGER HAVE TO. Indeed, we now have the ability to end our *psychosis*, to rehabilitate our wondrously sensitive soul. Further, our conscious mind no longer has to practise denial, which means we can also end our *neurosis*—all of which means the human race can be *psychologically* rehabilitated; our species can be brought back to life from its psychologically dead state.

Chapter 2:8 Most wonderfully, this psychosis-addressing-and-solving, *real* explanation of the human condition makes possible the psychological rehabilitation of the human race

175 What is *so* utterly exonerating and psychologically healing about our ability to understand the human condition now is that we can finally appreciate that there was a *very good* reason for our angry, alienated and egocentric lives—in fact, we can now see why our species has not just been ego-*centric*, but ego-*infuriated*, even ego-gone-mad-with-murderous-rage for having to live with so much unjust criticism/guilt, for some *2 million years*. At long last we have the reason for humans' capacity for shocking acts of cruelty, sadism, hate, murder and warfare; the origin of the dark volcanic forces in us humans. And no wonder we led such an evasive, escapist, superficial and artificial, greedy, smother-ourselves-with-*material*-glory-while-we-lacked-the-*spiritual*-glory-of-compassionate-understanding-of-our-ourselves, power, fame and fortune-seeking existence. Yes, the accumulation of money or capital served to supply the symbolic wins we needed to counter the insecurity of our seemingly worthless condition. And yes, the so-called Seven Deadly Sins of the human condition, of lust, anger, pride, envy, covetousness, gluttony and sloth, are, in truth, all different manifestations of the three fundamental upsets of anger, egocentricity and alienation that unavoidably emerged when humans became fully conscious and *had to* set out in search of knowledge in the presence of unjustly condemning instincts.

176 Most wonderfully, now that we are able to explain and understand that we are actually good and not bad, all the upset that resulted from not being able to explain our divisive condition is able to subside and disappear. Finding understanding of the human condition is what liberates and transforms the human race from its psychologically upset angry, egocentric and alienated condition; it is what enables the psychological rehabilitation of the human race to begin. Indeed, the word **'psychiatry'** literally means **'soul-healing'** (derived as it is from *psyche* meaning **'soul'** and *iatreia* which according to *The Encyclopedic World Dictionary* means **'healing'**), but we have never been able to 'heal our soul', explain to our original instinctive self or soul that we, our fully conscious, thinking self is good and not bad and, by so doing, reconcile and heal our split selves—*until now*. As Professor Harry Prosen, the former president of the Canadian Psychiatric

Association to whom I am eternally grateful for writing the Introduction to this book, has said about the psychological effect of this all-loving, all-compassionate understanding of ourselves: **'I have no doubt this biological explanation of the human condition is the holy grail of insight we have sought for the psychological rehabilitation of the human race.'** Yes, our ability now to understand the dark side of ourselves means we can finally achieve the **'wholeness for humans'** that the psychoanalyst Carl Jung was forever pointing out **'depends on the ability to own our own shadow'**. Sir Laurens van der Post was also describing the understanding that was required—and which we now have—to heal our species' corrupted, 'fallen' condition when he wrote that **'True love is love of the difficult and unlovable'** (*Journey Into Russia*, 1964, p.145 of 319); and, **'how can there ever be any real beginning without forgiveness?'** (*Venture to the Interior*, 1952, p.16 of 241); and that **'Only by understanding how we were all a part of the same contemporary pattern** [of wars, cruelty, greed and indifference] **could we defeat those dark forces with a true understanding of their nature and origin'** (*Jung and the Story of Our Time*, 1975, p.24 of 275); and that **'Compassion leaves an indelible blueprint of the recognition that life so sorely needs between one individual and another; one nation and another; one culture and another. It is also valid for the road which our spirit should be building now for crossing the historical abyss that still separates us from a truly contemporary vision of life, and the increase of life and meaning that awaits us in the future'** (ibid. p.29). Yes, one day there had to be, to quote The Rolling Stones' lyrics, **'Sympathy for the devil'** (1968); one day, a day that has now arrived, we had to find the reconciling, compassionate, healing understanding of the dark side of human nature. One day there had to occur, as William Blake titled his famous book, ***'The Marriage of Heaven and Hell'*** (c.1790).

177 This explanation *is* immensely positively reinforcing and thus therapeutic because we can now understand that we humans *are* nothing less than the heroes of the story of life on Earth. This is the fundamental truth that, as I said in chapter 1:7, we humans hoped and believed we would one day establish, *and now have*! Despite all appearances to the contrary—despite our anger, egocentricity and alienation—we humans *are* the most wonderful beings. In fact, if there is a word such as divine that can be applied to mortals, then we can now see that it truly does belong to us humans—because to withstand 2 million years of the injustice that we have had to endure and still be on our feet, still be able to laugh, still be able to smack each other on the back with encouragement, still be able to carry on, get out of bed each day and face life under the duress of the

human condition, we *must* be the most magnificent of organisms! Truly, as Camus wrote, **'man's greatness...lies in his decision to be stronger than his condition'** ('The Night of Truth', *Combat*, 25 Aug. 1944); to not give in to the possibility he famously described in his 1942 essay *The Myth of Sisyphus*, that **'There is but one truly serious philosophical problem, and that is suicide. Judging whether life** [under the duress of the human condition] **is or is not worth living'**.

178 This fact of the utter magnificence of the human race—that we can now understand and know is true rather than merely hope it is so—brings such intense relief to our angst-ridden cells, limbs and torsos that we will find it is as though we have thrown off a shroud of heavy chains. Yes, the great, heavy burden of guilt *has* finally been lifted from the shoulders of humans. Yes, doesn't the core feeling exist in all humans that far from being meaningless, **'banish**[ment]**'**-deserving evil blights on this planet we *are* all immense heroes? Doesn't this explanation at last make sense of all humans' immensely courageous and defiant attitude? And won't this explanation bring deep, bone-draining relief to the whole of each person's being?

179 Yes, this explanation, this understanding of the human condition, provides the **'insight we have sought for the psychological rehabilitation of the human race'**, as Professor Prosen said it would. It enables us to see that our divisive human nature was *not* an unchangeable or immutable state as many people came to believe it is, and which E.O. Wilson's Multilevel Selection theory for eusociality deems it to be, rather it was the result of the human condition, our inability to understand ourselves, and, therefore, it *can* dissipate now that we have found that understanding.

180 Importantly, this understanding of why we became upset as a species doesn't condone or sanction 'evil', rather, through bringing understanding to humans' upset behaviour, it gives us the power to ameliorate and thus subside and ultimately eliminate it. 'Evil'—humans' divisive behaviour—was a result of a conflict and insecurity within us that arose from the dilemma of the human condition: resolve the dilemma and you end the conflict and insecurity. Peace could only come to our troubled, divisive state and world through removing the underlying insecurity of our condition. With our ego or sense of self worth satisfied at the most fundamental level, our anger can now subside and all our denials and resulting alienation can be dismantled. From having lived in a dark, cave-like, depressed state of condemnation and, as a result, had to repress, hide and deny our true selves, we can at last, as the 1960s rock musical *Hair*

sang, **'Let the sunshine in'**—end our horrid existence of having to depend on denial to cope. The compassionate-understanding-based psychological rehabilitation of the human race—the TRANSFORMATION of all humans—truly *can* begin. (A detailed description of how the transformation of the human race is now able to take place, such that every human can now *immediately* be free of the human condition, is presented in the concluding chapter 8 of this book.)

181 As for the veracity of this explanation, it should be pointed out that it is precisely this explanation's ability to at last make relieving sense of human life, of all our behaviour in fact, that let's us know that we have finally found the *true* explanation of the human condition. The great physicist Albert Einstein once wrote that **'Truth is what stands the test of experience'** (*Out of My Later Years*, 1950), and since this study and explanation is all about us, our behaviour, we are in a position to personally **'experience'** its validity, to know if it's true or not. As the subject of this study, we can know if the ideas being put forward work or not, if these explanations do make sense of our deepest feelings; of our volcanic anger, of our lonely estranged souls, of our insecure state yet core belief that we are wonderful beings, etc, etc—as, in fact, they *do*. Moreover, these explanations are *so* powerfully insightful, accountable and revealing (so true) that they render our lives transparent—a transparency, a sudden exposure, that can initially be overwhelmingly confronting and depressing. But there is a way, indeed an absolutely wonderful, joyous way, we are able to cope with the arrival of exposure day, or transparency day, or truth day, or honesty day, or self-confrontation day—in fact, the long-feared so-called **'judgment day'**, which is, in fact, not a time of condemning 'judgment' but of compassionate understanding. That wonderful way of coping with the arrival of the all-liberating, all-rehabilitating but at the same time all-exposing truth about us humans is, as mentioned above, the subject of the concluding chapter (8) in this book. But the point being made here is that the transparency of our lives that these explanations bring reveals just how effective, how penetrating, and, therefore, *how truthful* these understandings actually are.

182 So, we can see that as dishonest as it necessarily was in its approach, mechanistic, reductionist science turns out to be the liberator of humanity, the proverbial 'messiah' or 'saviour' of humanity, for it *has* finally enabled us to lift the so-called 'burden of guilt' from the human race! It found understanding of the difference in the way nerves and genes

process information, that one is insightful and the other isn't, which has finally made it possible to explain that greatest of all mysteries, that holy grail of all human conscious thought and enquiry, of the human condition. As was mentioned at the beginning of this chapter (in par. 135), the fundamental conflict involved in our condition between our innocent instinctive self and our conscious self was described as far back as when Genesis was written by Moses, but we couldn't *explain* that conflicted state until science found understanding of the different ways genes and nerves process information.

183 Significantly, now that we are able to understand from scientific first principles that upset is not an 'evil', worthless, bad state, but an immensely heroic state, we can know that while, inevitably, all humans are variously upset from their different encounters with, and degrees of engagement in, humanity's epic search to find knowledge, ALL HUMANS ARE EQUALLY GOOD. Everyone is variously angry, egocentric and alienated, but everyone is good, and not just good but a hero of the story of life on Earth! No longer does humanity have to rely on dogmatic assertions that **'all men are created equal'** purely on the basis that it is a **'self-evident'** truth, as the United States' Declaration of Independence asserts, because we can now *explain, understand* and *know* that the equality of goodness of all humans is a fundamental truth. We can now understand *why* everyone is equally worthy, and that no one is superior or inferior, and that *everyone* deserves the **'rights'** of **'life, liberty and the pursuit of happiness'**. Indeed, through this understanding, the whole concept of good and bad disappears from our conceptualisation of ourselves. Compassionate, relieving truth and honesty about humans has finally arrived. Religions have taught us that 'God loves you', and while that mantra has been comforting, ultimately we needed to understand *why* we were lovable (again, the issue of who or what God is will be addressed in more detail in chapter 3:3). Similarly, the Bible states that **'the truth will set you free'** (John 8:32), and while we know this statement to be true, the problem has been that all the *partial* truths—such as that humans are the most brutal and destructive animals to ever walk the earth—condemned our upset state, fuelling it further, which means that, ultimately, for the truth to *genuinely* set us free, it had to be the *full* truth that explains *why* we humans are *all* good and not bad.

184 Yes, the human condition is certainly shot through with paradox: to become happy we had to first endure unhappiness; we appeared to be bad but believed we were good; we are intelligent, smart and clever but,

by all appearances, behave in such an unintelligent, stupid way that we have brought the world to the brink of destruction. But most wonderfully, we at last have the understanding that reconciles not just these but *all* the polarities of life that existed under the duress of the human condition—between 'good' and 'evil', instinct and intellect, emotion and reason, conscience and conscious, ignorance and wisdom, soul and mind, heart and head, 'I feel' and 'I think', yin and yang, light and dark, the innocent and the corrupted, the un-embattled and the heroic, the selfless and the selfish, the happy and the upset, the light-hearted and the heavy-hearted, the cooperative and the competitive, the integrative and the divisive, the 'Godly' and the 'unGodly', the gentle and the aggressive, the loving and the hateful, the sound and the alienated, the secure and the insecure, the honest and the dishonest, the natural and the artificial, the non-sexual and the sexual (sex as humans practise it is explained in pars 596-647), altruism and egotism, idealism and realism, spiritualism and materialism, socialism and capitalism, left-wing and right-wing, instinctualism and intellectualism, religion and science, holism and mechanism, young and old, women and men (the different roles of women and men under the duress of the human condition is also explained in pars 596-647), blacks and whites (the differences in upset between races is explained in chapter 7:16E), unresigned and resigned, country and city, etc, etc. The explanation of the human condition unravels and makes sense of the whole seemingly impenetrable and insoluble confusion of human life—and does so in such a simple and obvious way that it brings to mind Huxley's famous response to Darwin's idea of natural selection: **'How extremely stupid of me not to have thought of that!'** Yes, we have at last found the UNDERSTANDASCOPE for human behaviour.

Chapter 2:9 <u>The end of politics</u>

[185] Most wonderfully, with this understanding everyone *can* come in from the cold, from their lonely outposts of bewilderment about what has actually been going on in the world of humans. The result is that immense—and urgently needed—fundamental change can now occur for the human race. Conflict from misunderstanding especially goes. For instance, this explanation obsoletes the conflict between the philosophically opposed left and right wings of politics because we can now understand that while giving in to our ideal-behaviour-demanding instincts—'flying

back on course' in the Adam Stork analogy—was an immensely guilt-relieving exercise that made us 'feel good' (the left-wing approach), it was fundamentally irresponsible because it meant abandoning humanity's upsetting but necessary search for knowledge (the right-wing approach). The truth was not as it appeared—participating in humanity's selfish, aggressive and competitive battle to find knowledge was not the 'bad' transgression it has been condemned as, rather it was the responsible course of action and thus a 'good' thing. The paradox of the situation is marvellously summed-up in the famous musical *The Man of La Mancha*, when it says we had to be prepared **'to march into hell for a heavenly cause'** (Joe Darion, *The Impossible Dream*, 1965). Yes, we had to, as it were, lose ourselves to find ourselves—suffering upset was the heroic price we *had* to pay to find understanding, ultimately self-understanding and, with it, freedom from the upset state of the human condition. So while the left-wing has had an absolute field day demonising the right-wing's support of the non-ideal state—even labelling it **'evil'**, as was proselytised by the left-wing activist Michael Moore (*Capitalism: A Love Story*, 2009)—we can *now* see that it was the philosophy of the left that was morally bankrupt, void of meaning in the sense that while siding with idealism made its practitioners 'feel good', it ultimately had no relevance in humanity's critical journey from ignorance to enlightenment. Proponents of the left-wing approach were deluding themselves that they were holding the moral high ground when, in fact, the reverse was true—their stridently *pseudo* idealistic, dogmatic, condemning, ridiculing, escapist, deluded, arrogant, dishonest culture *oppressed* progress towards humanity's liberation from the human condition, stymieing the return of the *genuinely* peaceful, ideal world. Karl Marx, the political philosopher whose mid-nineteenth century theories gave rise to socialism and communism, was very wrong when he asserted that **'The philosophers have only interpreted the world in various ways; the point is** [not to understand the world but] **to change it** [just make it cooperative/social/communal]**'** (*Theses on Feuerbach*, 1845). The whole **'point'** and responsibility of being a conscious being *is* to understand our world and our place in it.

186 The immense danger of the pseudo idealistic left-wing was that the longer the upsetting search for knowledge went on without the goal of that search of the true, psychologically relieving explanation of the human condition being found, the more people became upset sufferers of the human condition and the more they couldn't resist the artificial relief offered by the left-wing's dogmatic, anti-knowledge, pseudo idealistic, 'fly

back on course' way of living. *And*, in contriving the Multilevel Selection theory for human behaviour, E.O. Wilson furnished the whole dumbing-down process with the ultimate delusion it needed of a 'biological', non-explanation nullification of the core, critical, all-important issue of the human condition itself, thereby virtually locking humanity onto a path to terminal dishonesty/dogma/**'phon**[iness]'/**'fake**[ness]'/alienation/darkness/extinction! As the author George Orwell famously predicted, **'If you want a picture of the future, imagine a boot stamping on a human face forever** [stamping forever on the freedom that humans' extraordinary and unique conscious mind needs to search for knowledge]**'** (*Nineteen Eighty-Four*, 1949).

187 The impact of this pseudo idealistic development on the democratic process is grim indeed, because democracy—in which the proper balance is sought between the need to maintain a degree of selfless, loving, soulful, cooperative 'ideal' behaviour for society to function, and the need for there to be sufficient freedom from the imposition of the expectation of such 'ideal' behaviour, for people to be able to carry on the corrupting search for knowledge—is destroyed as an effective and meaningful process when people start voting for idealism *not* because there is too much selfishness in society but simply because it makes *them* 'feel good'. We can now at last appreciate that when we vote like this we are *not* participating in the true democratic process that human advancement has depended upon, we are exploiting and subverting it. And while many people think that even if feel-good pseudo idealism is not a real form of idealism it is surely harmless, that is most certainly *not* the case. *No*, it is extremely bloody-minded, totally selfish and dangerously destructive behaviour that says, 'I no longer care about the human race, only about finding relief from the human condition.' But such behaviour was always going to develop when the levels of upset in society reached unbearable heights because at that end play point the need for relief from the agony of the human condition was going to become the *only* concern amongst an ever-increasing proportion of the population—with the dire consequence that democracy would fail to find the proper balance because the human-race-indifferent, deluded, feel-good, pseudo idealistic, left-wing political attitude would become impossible to defeat in an election, leading to the human race suffering a horrifically alienated death by dogma.

188 Certainly, the upset behaviour that resulted from right-wing participation in humanity's heroic search for knowledge was increasingly bringing about immense human suffering and environmental devastation, but it

was the knowledge-oppressing left-wing that posed the *real* threat to the survival of the human race because *only* through successfully completing that search for knowledge could humanity be liberated from the upset state of the human condition. So while, as the journalist Geoffrey Wheatcroft recognised, **'the great twin political problems of the age are the brutality of the right, and the dishonesty of the left'** ('The year of sexual correctness and double standards', *The Australian Financial Review*, 29 Jan. 1999), it was NOT the **'the brutality of the right'** but the **'dishonesty of the left'** that stood like a colossal ogre over the human race, threatening to destroy it. Yes, we can now appreciate that the philosopher Friedrich Nietzsche spoke the truth about the need for humanity to hold its nerve when he wrote that **'There have always been many sickly people among those who invent fables and long for God** [ideality]**: they have a raging hate for the enlightened man and for that youngest of virtues which is called honesty...Purer and more honest of speech is the healthy body, perfect and square-built: and it speaks of the meaning of the earth** [which is to fight for knowledge, ultimately self-knowledge, understanding of the human condition]**...You are not yet free, you still *search* for freedom. Your search has fatigued you...But, by my love and hope I entreat you: do not reject the hero in your soul! Keep holy your highest hope!...War** [against the oppression of dogma] **and courage have done more great things than charity. Not your pity but your bravery has saved the unfortunate up to now...What warrior wants to be spared? I do not spare you, I love you from the very heart, my brothers in war!'** (*Thus Spoke Zarathustra: A Book for Everyone and No One*, 1892; tr. R.J. Hollingdale, 1961, pp.61-75 of 343). Yes, as stated in par. 168, you had to be **'brave'** to avoid the temptation of giving up the self-corrupting search for knowledge. (Much more is explained about the extreme danger of pseudo idealism in chapter 7:16, 'The final 200 years'.)

189 Thankfully, the arrival of the demystifying, exposing and reconciling, true explanation of the human condition has the power to cut short this frightening, death-by-dogma, left-wing threat to humanity—that is, of course, if humanity chooses to accept this lifeline that it has been thrown. And, for its part, the right-wing's need to support the upsetting, often brutal, competition-aggression-and-selfishness-producing battle/**'war'** to defy the ignorance of our ideal-behaviour-demanding, unjustly condemning, moral instinctive self or soul and find understanding of ourselves, can *also* come to an end with the acceptance of this liberating insight. Yes, with understanding of the human condition now found, the *whole* necessary but ugly business of politics can now happily come to a close.

[190] The explanation of just how the human race can now 'put down the sword', end its egocentric, 'must-prove-that-we-are-good-and-not-bad', **'warrior'** existence, and by so doing transform itself into a peaceful state, is explained in the concluding chapter 8 of this book.

Chapter 2:10 'Free at last!'

[191] The end of the political process as we know it is just one example of the great change that can now come to the human race—this *is* the ultimate 'future shock', massive paradigm shift, for humans—BUT, as will be explained in some detail in chapter 8, since this shift is an immensely positive change, the transformation will not only be easy but *fabulously* exciting. Indeed, so great is both the amount of change we now face—the **'future shock'**—and the self-confronting exposure—the **'judgment day'** effect mentioned earlier—that I and others established the World Transformation Movement to introduce and help manage the truly wonderful way we are able to cope with these huge problems. It is called the World Transformation Movement because the most fabulous effect of having understanding of the human condition is that *all* our psychologically upset angry, egocentric and alienated behaviour can now, immediately, be put aside, and eventually completely disappear, thus bringing about the complete rehabilitation and transformation of the human race and thus of our planet.

[192] So this is the end of the world as we know it and the beginning of the world we always hoped for! Indeed, while the 'Socialist', 'Temperance', 'Age of Aquarius', 'Peace', 'New Age', 'Feminist', 'Green', 'Politically Correct', 'Postmodernist', 'Anti-Capitalist' movements (and these are only the more recent of the litany of deluded, false starts to an ideal world for humans that we have witnessed) have all severely discredited—and inhibited—the prospect of a truly human-condition-resolved, psychologically ameliorated, completely transformed new world for humans, one of the founders of the 1980s New Age Movement, the author Marilyn Ferguson, did offer this accurate description of the now-realised hope of its arrival: **'Maybe** [the Jesuit priest, scientist and philosopher] **Teilhard de Chardin was right; maybe we are moving toward an omega point** [a final unification of our split selves]**—Maybe we can finally resolve the planet's inner conflict between its neurotic self (which we've created and which is unreal) and its real self. Our** [original all-sensitive and loving instinctive] **real self knows how**

to commune, how to create...From everything I've seen people really urgently want the kind of new beginning...[that I am] **talking about** [where humans will live in] **cooperation instead of competition'** (*New Age* mag. Aug. 1982).

193 Yes, from being selfish, aggressive and competitive, humans can now return to being selfless, loving and cooperative. Our round of departure has ended. T.S. Eliot wonderfully articulated our species' journey from an original innocent, yet ignorant, state, to a psychologically upset 'fallen', corrupted state, and back to an uncorrupted, but this time enlightened, state when he wrote, **'We shall not cease from exploration and the end of all our exploring will be to arrive where we started and know the place for the first time'** (*Little Gidding*, 1942).

194 I have drawn the following picture to summarise this horrendously agonising but awesomely heroic journey that our species has taken from ignorance to enlightenment. From its innocent, happy, cooperatively orientated, selfless and loving infancy and childhood, humanity progressed to the horror of an insecure, upset adolescence where we humans had to search for our identity—specifically for understanding of why we lost our innocence and became selfish, aggressive and competitive. But with understanding of the human condition now found, humanity can, at long last, enter the happy, ameliorated, secure and mature state of adulthood. (Again, all these stages that the human race has progressed through will be explained and described in some detail in chapter 7.) Martin Luther King Jr's **'dream'** has finally come true; we can now **'allow freedom to ring... from every village and every hamlet, from every state and every city'** because **'all of God's children, black men and white men, Jews and Gentiles, Protestants and Catholics'** can **'join hands and sing'**, **'Free at last! Free at last! Thank God Almighty, we are free at last!'** ('I Have A Dream' speech, 28 Aug. 1963).

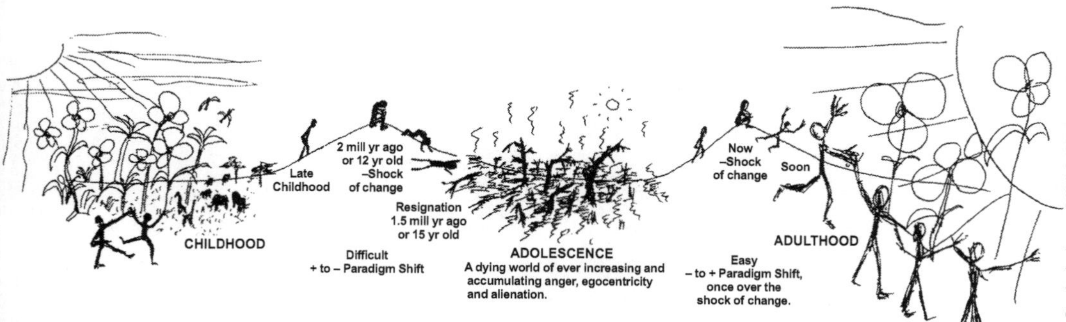

Humanity's Journey from Ignorance to Enlightenment

Drawing by Jeremy Griffith © 1996 Fedmex Pty Ltd

195 Yes, the human race can now, at last, ultimately, finally leave Plato's horrid cave of truth-avoiding, alienated, psychotic and neurotic darkness and stand in the warm, healing sunshine of understanding. And to celebrate the completion of the writing of this book in 2013 that **'let**[s] **the sunshine in'**, I have made, from golden sea shells I found on a beach in Sydney, this exciting picture that depicts that all-fabulous liberating and healing sun.

Artwork by Jeremy Griffith © 2013 Fedmex Pty Ltd

Composed of golden cockle shells, white star limpets and a large scallop shell from a Sydney beach, this collage was made by the author in 2013 to celebrate this time that the human race has dreamed of, and which has now arrived, when we can finally leave Plato's dark cave of denial and 'let the sunshine in'. To help fund the work of the World Transformation Movement, this artwork is for sale for $US 5 million.

196 Having now established the fundamental goodness of humans it finally becomes both possible and psychologically safe to present the fully accountable and thus true scientific explanations for the meaning of existence; for how we humans acquired our moral instincts; and for how we became conscious when other animals haven't. These answers, as well as a rebuttal of the false accounts that have been put forward to supposedly explain the origins of our moral instincts, will be presented in the next four chapters of this book. Chapter 5, which contains the

rebuttal of the false theories that have been put forward for our moral nature, will also provide an account of the ill-treatment, even persecution, these human-race-saving answers have received for 30 years from the mechanistic paradigm—for while our species' journey to enlightenment *is* complete, this explanation's journey to acceptance is still very much a fragile, not-guaranteed-of-success, work in progress. Since the conscious mind *must* surely be nature's greatest invention, its failure to fulfil its great potential by not recognising the psychosis-addressing-and-explaining real explanation of the human condition that has now arrived would represent a failure of the whole story of life on Earth! This resistance by mechanistic science to recognising this understanding of the human condition is an *extremely* serious situation. The survival of humanity is hanging in the balance, and that balance simply *must* tip the right way; these human-race-saving understandings of the human condition *must* be recognised by the scientific establishment.

197 So yes, *IS IT TO BE* that the human race will be saved?—and the answer is that with understanding of the human condition now found it finally can and therefore simply *must* be saved! So while *'IS IT TO BE'* is an abbreviation of the title of this book, use of it on its own does evoke that deepest of all questions about life under the duress of the human condition, of are we going to make it? When Shakespeare wrote what is possibly the most famous phrase in literature, **'To be, or not to be, that is the question'**, he was referring to whether it was worth Hamlet living on under the duress of the human condition—or not. And, with understanding of the human condition found, we are so close to freeing ourselves from the human condition that it *must* **'be'** that the human race lives on!

198 The final chapters of this book, chapters 7 and 8 (respectively), will therefore present a description of humanity's journey from ignorance to enlightenment, and a description of how the human race is now able to be transformed.

199 (A more complete description of this fully accountable, human-race-transforming, *true* biological explanation of the human condition can be found in my freely available, online book *Freedom* at <www.worldtransformation.com/freedom>.)

Chapter 3

The Meaning of Life

Michelangelo's *The Creation of Adam* from the Sistine Chapel, c.1508-12. Michelangelo's masterpiece also came to symbolise our hope of reconciliation with God, which has now occurred.

Chapter 3:1 Summary

200 While the conventional view in science is that there is no direction or purpose or meaning to existence, and that change is random, there is, in fact, a very obvious theme, direction, purpose and meaning to existence, which is the ordering or integration or complexification of matter into ever larger and more stable wholes. Indeed, the answer to the great question of 'what is the meaning of life' is that it is to live in accordance with this integrative, order-of-matter-developing, cooperation-and-selfless-behaviour-dependent theme or meaning of existence. HOWEVER, *until* we could explain our seemingly imperfect human condition, explain *why* we humans appear to have been living in defiance of 'Integrative Meaning', namely in a divisive competitive and selfish way, such a truth could not be faced. *BUT*, as seen in the previous chapter, since we CAN now at last explain the good reason why we humans have been divisively rather than integratively behaved, we CAN finally admit the truth of

Integrative Meaning and acknowledge that there is a direction or purpose to existence. And furthermore, since our concept of 'God' is actually the personification of the unbearably condemning truth of Integrative Meaning, through admitting this truth we humans can also demystify 'God' and, by so doing, change from being a 'God'-fearing species to a 'God'-confronting one. Yes, this explanation of the human condition makes sense of and reconciles *all* the unresolved manifestations of the polarities of the human situation—'good' vs 'evil', idealism or 'God' vs human's non-ideal existence, religion vs science, holism vs mechanism, altruism vs egotism, communism vs capitalism, conscience vs conscious, and so on—thus ameliorating or healing, and finally ending, the psychologically upset, divisive way we have been living!

Chapter 3:2 The obvious truth of the development of order of matter on Earth

201 In starting this fully accountable, true explanation of the meaning of life, we need to take a look at our surroundings. As you do, you'll note that the most obvious characteristic of our world is that it is full of 'things', variously enduring arrangements of matter, like plants, animals, clouds and rocks. And not only that, it is apparent that these arrangements of matter consist of a hierarchy of ordered parts; a tree, for instance, is a hierarchy of ordered matter—it has a trunk, limbs, roots, leaves and wood cells. Our bodies are also a collection of parts, as are clouds and rocks, which are built from different elements and compounds. Furthermore, what we have seen happen over time to these arrangements of matter is that there has been a progression from simple to more complex arrangements. From the fundamental ingredients of our world of matter, space and time, matter has become ordered into ever larger (in space) and more stable or durable (in time) arrangements.

202 To elaborate, our world is constructed from some 94 naturally occurring elements that have come together to form stable arrangements. For example, two hydrogen atoms with their single positive charges came together with one oxygen atom with its double negative charge to form the stable relationship known as water. Over time, larger molecules and compounds developed. Eventually macro compounds formed. These then integrated to form virus-like organisms, which in turn came together

or integrated to form single-celled organisms that then integrated to form multicellular organisms, which in turn integrated to form societies of single species that continue to integrate to form stable, ordered arrangements of different species. Clearly, what is happening on Earth is that matter is integrating into larger and more stable wholes. And this development of order is not only occurring here, it is also happening out in the universe where, over the eons, a chaotic cosmos continues to organise itself into stars, planets and galaxies. As two of the world's greatest physicists, Stephen Hawking and Albert Einstein, have said, respectively, **'The overwhelming impression is of order...[in] the universe'** (Gregory Benford, 'The Time of His Life', *The Sydney Morning Herald*, 28 Apr. 2002), and that **'behind everything is an order'** (*Einstein Revealed*, PBS, 1997).

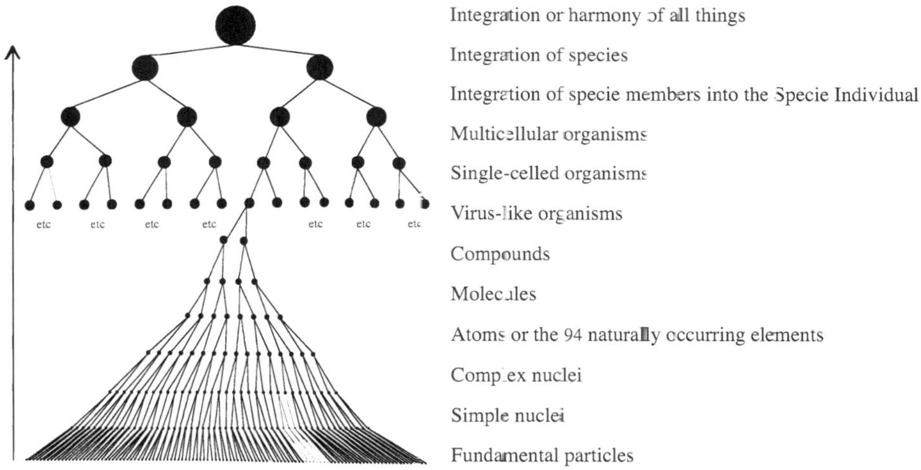

Development of Order or Integration of Matter on Earth.
A similar chart appears in Arthur Koestler's 1978 book, *Janus: A Summing Up*.

[203]The law of physics that accounts for this integration of matter is known as the 'Second Path of the Second Law of Thermodynamics', or 'Negative Entropy', which states that in an open system, where energy can come into the system from outside it (in Earth's case, from the sun, and, in the case of the universe, from the original 'big bang' explosion that created it), matter integrates, it develops order. Thus, subject to the influence of Negative Entropy, the 94 elements from which our world is built develop ever larger and more stable wholes.

[204]In *Janus: A Summing Up* (1978), the scientist-philosopher Arthur Koestler gave this excellent summary of the history of the concept of Negative Entropy: **'One of the basic doctrines of the nineteenth-century mechanistic world-view was Clausius' famous "Second Law of Thermodynamics". It asserted that the universe was running down towards its final dissolution because its energy is being steadily, inexorably dissipated...cosmos dissolving into chaos. Only fairly recently did science begin to recover from the hypnotic effect of this gloomy vision, by realizing that the Second Law applies only in the special case of so-called "closed systems"...whereas all living organisms are "open systems" which maintain their complex structure and function by continuously drawing materials and energy from their environment** [p.222 of 354] **...It was in fact a physicist, not a biologist, the Nobel laureate Erwin Schrödinger, who put an end to the tyranny of the Second Law with his celebrated dictum: "What an organism feeds on is negative entropy"** [p.223] **...Schrödinger's revolutionary concept of negentropy, published in 1944** [p.224] **...is a somewhat perverse way of referring to the power of living organisms to "build up" instead of running down, to create complex structures out of simpler elements, integrated patterns out of shapelessness, order out of disorder. The same irrepressible building-up tendency is manifested in the progress of evolution, the emergence of new levels of complexity in the organismic hierarchy and new methods of functional coordination** [p.223]**.'** Significantly, Koestler wrote of **'the active striving of living matter towards [order]** [p.223]**'**, of **'a drive towards synthesis, towards growth, towards wholeness** [p.224]**'**, and that **'this "innate drive" derives from the "integrative tendency"** [p.225]**'**.

[205]So the theme of existence, the overall direction or destiny of change, or, from a conscious observer's point of view, the overall purpose or meaning of existence, is the ordering or integration or complexification of matter. 'Holism', which the dictionary defines as **'the tendency in nature to form wholes'** (*Concise Oxford Dictionary*, 5th edn, 1964), and 'teleology', **'the belief that purpose and design are a part of nature'** (*Macquarie Dictionary*, 3rd edn, 1998), are both terms that recognise this integrative **'tendency'**.

[206]But while the integrative meaning of existence *is* the most obvious of all truths, it has also been the most difficult of all truths for us humans to acknowledge, for an *extremely* good reason.

[207]The difficulty arises from the fact that for a collection of parts to form and hold together, for matter to integrate, the parts of the developing whole *must* cooperate, behave selflessly, place the maintenance of the whole above the maintenance of themselves, because if they don't

cooperate—if they compete, behave selfishly or inconsiderately—then the whole disintegrates, the parts break down into the more elementary building blocks of matter from which they were assembled. As Koestler stated, to create **'order out of disorder'** requires **'functional coordination'**. A leaf falling from a tree in autumn does so to ensure the tree survives through winter and carries on; it puts the maintenance of the whole, namely the tree, above the maintenance of itself. The effective functioning of our body similarly depends on the cooperation of all its parts, on every part doing what is best for the whole body. Our skin, for example, is constantly growing and dying to protect our body. Cancer cells, on the other hand, destroy the body precisely because they violate this principle and follow their own selfish, independent agenda. Indeed, the very reason ant and bee societies work so well is because all their parts, the worker ants and bees, behave selflessly; in their behaviour, they put the welfare of the larger whole above that of their own.

208 Put simply, selfishness is divisive or disintegrative while selflessness is integrative—it is the glue that holds wholes together; it is, in fact, the theme of the integrative process, and thus of existence. It is also what we mean by the word 'love', with the old Christian word for love being **'caritas'**, meaning charity or giving or selflessness (see Col. 3:14, 1 Cor. 13:1-13, 10:24 & John 15:13). So 'love' is cooperative selflessness, and not just selflessness but *unconditional selflessness*, the capacity, if called upon, to make a full, self-sacrificing commitment to the maintenance of the larger whole. BUT—and herein lies the nub of the problem—if the meaning of existence is to behave *integratively*, which means behave cooperatively and selflessly, *why* do we humans behave in the completely opposite way, in such a competitive and selfish *divisive* way? Yes, the integrative theme of existence certainly confronts us humans squarely with the issue of the human condition, the issue of our non-ideal behaviour. And so despite being such an obvious truth, Integrative Meaning has been *so* horrifically condemning of the competitive, aggressive and selfish human race that until we could explain the *good reason why* we humans have been divisively rather than integratively behaved (which was done in chapter 2)—and thus make it psychologically safe to admit the truth of the order developing, integrative meaning of existence—we had no choice but to live in near total denial of it. Hawking's, Einstein's and Koestler's acknowledgments of the order developing, integrative process when, as Koestler

said, **'mechanistic'** science has maintained such a dedicated, **'hypnotic'** denial of it, were bold indeed.

209 In summary then, selflessness, specifically unconditional selflessness or altruism, *is* the theme or meaning of existence. And since love means unconditional selflessness, love *is* the theme of existence—it *is* the meaning of life. The very great truthful, denial-free-thinker or prophet Christ emphasised the unconditionally selfless significance of the word 'love' when he said, **'Greater love has no-one than this, that one lay down his life for his friends'** (Bible, John 15:13). Of the biblical references to love provided above, Colossians 3:14 perfectly summarises the integrative significance of love: **'And over all these virtues put on love, which binds them all together in perfect unity.'** But, unfortunately, while virtually all the great truths, such as the unconditionally selfless nature of love, are acknowledged in the Bible, in our everyday world we couldn't admit that love is unconditional selflessness and the theme of existence until we could explain *why* humans don't behave lovingly and are so seemingly at odds with the integrative process. In fact, in the human-condition-avoiding mechanistic scientific paradigm it is considered improper, unscientific, in fact, to *even* use the word 'love'. The linguist Robin Allott summarised mechanistic science's attitude to love when he wrote that **'Love has been described as a taboo subject, not serious, not appropriate for scientific study'** ('Evolutionary Aspects of Love and Empathy', *Journal of Social and Evolutionary Systems*, 1992, Vol.15, No.4). Indeed, love has been deemed so '[in]**appropriate for scientific study'** that it has been reported that **'more than 100,000 scientific studies have been published on depression and schizophrenia (the negative aspects of human nature), but no more than a dozen good studies have been published on unselfish love'** (*Science & Theology News*, Feb. 2004). So mechanistic science has determinedly resisted analysis of one of humanity's most used, valued and meaningful words! The psychologist Harry F. Harlow highlighted this discrepancy when he observed that **'Psychologists, at least psychologists who write textbooks, not only show no interest in the origin and development of love or affection, but they seem to be unaware of its very existence. The apparent repression of love by modern psychologists stands in sharp contrast with the attitude taken by many famous and normal people. The word "love" has the highest reference frequency of any word cited in Bartlett's book of *Familiar Quotations*'** ('The nature of love', *American Psychologist*, 1958, Vol.13, No.12). The concept of 'unselfish love' has certainly been an unbearable area of scientific enquiry for us selfish, seemingly non-loving and unlovable humans!

Chapter 3:3 'God' is our personification of Integrative Meaning

210 So, 'love' has been **'a taboo subject'** for science, and yet 'love' or unconditional selflessness is only an *aspect* of Integrative Meaning, so how much more unbearable has the *overall* tenet of Integrative Meaning itself been? The answer is the Negative Entropy-driven integrative, cooperative, loving, selfless, order-developing theme or meaning or purpose of existence has been an almost completely unconfrontable truth for the psychologically upset, competitive, aggressive and selfish human-condition-afflicted human race. In fact, we have lived in *such* fear of the truth of Integrative Meaning, lived in such terrified awe of it, been so confronted, condemned and intimidated by it, *so* unable to deal with it on any sort of an equal footing, that we termed it 'God'. Indeed, the integrative, cooperative, loving meaning of life has become deified as not just *a* God, but the one and *only* God—*the* most universal and fundamental, yet completely unconfrontable, of truths.

211 Monotheism, the belief that there is only one God, is an insight that goes back as far as 4,000 years ago to two very great denial-free thinkers or prophets—the Hebrew prophet Abraham, who lived around 2,000 BC, and the pharaoh Akhenaton, who reigned in Egypt from approximately 1,350 to 1,335 BC. The very great Persian prophet Zoroaster also recognised that there is only **'one supreme deity'**, as this reference to the faith describes: **'sometime around or before 600 BC—perhaps as early as 1200 BC—there came forth from the windy steppes of northeastern Iran a prophet who utterly transformed the Persian faith. The prophet was Zarathustra—or Zoroaster, as the Greeks would style his name. Ahuramazda** [the supreme being or wise lord] **had appeared to Zoroaster in a vision, in which the god had revealed himself to be the one supreme deity, all seeing and all powerful. He represented both light and truth, and was creator of all things, fountainhead of all virtue. Ranged against him stood the powers of darkness, the angels of evil and keepers of the lie. The universe was seen as a battleground in which these opposing forces contended, both in the sphere of political conquest and in the depths of each man's soul. But in time the light would shine out, scattering the darkness, and truth would prevail. A day of reckoning would arrive'** (*A Soaring Spirit, 600-400 BC*, Time-Life History of the World, 1988, p.37 of 176). And in approximately 360 BC, that other very great denial-free-thinking prophet, Plato, *also* recognised that God is Integrative Meaning, writing that **'God desired that all things should be**

good and nothing bad, so far as this was attainable. Wherefore also finding the whole visible sphere not at rest, but moving in an irregular and disorderly fashion, out of disorder he brought order, considering that this was in every way better than the other' (*Timaeus*; tr. B. Jowett, 1877).

212 But *until* we could explain the human condition and explain in first-principle-based, scientific terms who, or more precisely, what God is—namely our personification of the Negative Entropy-driven integrative theme, purpose and meaning of life—we had no choice but to leave the religious concept of God in a safely abstract, undefined state. And so despite Integrative Meaning being an *extremely* obvious truth, with evidence of the hierarchy of the order of matter everywhere we look, without understanding of our divisive condition it was imperative for humanity that human-condition-avoiding mechanistic science found a way to deny such a seemingly totally condemning truth. This was easily achieved through the simple assertion that there is no meaning or purpose or theme to existence and that while change *does* occur, it is a random, purposeless, directionless, meaningless, blind process. And, as stated, to cope with the imbued recognition of integrative ideality and meaning in the religious concept of God, mechanistic science simply left the concept undefined, maintaining it was a strictly abstract, metaphysical and spiritual notion unrelated to the scientific domain—an inexplicable deity, a supernatural being seated on a throne somewhere in a remote blue heaven who can be worshipped from afar as someone superior to us while avoiding any direct comparisons with ourselves. Religion and science were firmly demarcated as two entirely unrelated subjects. Indeed, E.O. Wilson succinctly captured mechanistic science's view on the matter when he said, **'I take a very strong stance against the mingling of religion and science'** (*National Geographic*, May 2006).

213 Of course, the truth is, to use Nobel Prize-winning physicist Charles H. Townes' words, **'they [science and religion] both represent man's efforts to understand his universe and must ultimately be dealing with the same substance. As we understand more in each realm, the two must grow together...converge they must'** ('The Convergence of Science and Religion', *Zygon*, 1966, Vol.1, No.3). Yes, as my headmaster at Geelong Grammar School, Sir James Darling, said, **'The scientist can no more deny or devaluate the truths of spiritual experience than the theologian can neglect the truths of science: and the two truths must be reconcilable, and it must be of importance to each of us that they should be reconciled'** (*The Education of a Civilized Man*, ed. Michael Persse, 1962, p.68 of 223). And

with understanding of the human condition now found, **'converge'** they *have*; ideality (which religions and the truthful, denial-free-thinking, God-confronting-not-avoiding prophets they were founded around represented) and our search for understanding of our non-ideal reality (which science represented—the word 'science' literally means 'knowledge') have finally been **'reconciled'**. Yes, with the human condition now explained and our divisive, seemingly non-integrative state finally understood, *all* humans can at last safely admit and recognise that there has only been one God, one all-dominating and all-pervading theme or meaning of existence, which is Integrative Meaning—a truth we recognise when we say **'God is love'** (Bible, 1 John 4:8, 16).

[214]It should be mentioned here that despite the fact that the admittance of Integrative Meaning first required solving the issue of our divisive human condition, a rare few holistic scientists have not only courageously defied the almost universal need to deny the development of order of matter on Earth, or Integrative Meaning, they have actually acknowledged that it is what we mean by God—including the aforementioned giants of physics, Stephen Hawking and Albert Einstein. If we include more of what Hawking and Einstein said earlier about order being the main characteristic of change in the universe, we can see that they both regarded God to be the personification of Integrative Meaning. In 1989 Hawking said, **'I would use the term God as the embodiment of the laws of physics'** (*Master of the Universe*, BBC). In 2002 he went further, saying, **'The overwhelming impression is of order** [in the universe]**. The more we discover about the universe, the more we find that it is governed by rational laws. If one liked, one could say that this order was the work of God. Einstein thought so...We could call order by the name of God'** (Gregory Benford, 'The Time of His Life', *The Sydney Morning Herald*, 28 Apr. 2002). Einstein's views on the matter were chronicled in the 1997 PBS documentary *Einstein Revealed*, which reported Einstein as saying that **'over time, I have come to realise that behind everything is an order that we glimpse only indirectly** [because it's unbearably condemning]**. This is religiousness. In this sense, I am a religious man.'** Yes, this **'order'** is apparent everywhere and it is what our **'religiousness'**, our belief in God, is concerned with acknowledging. I might mention that while Mahatma Gandhi was an inspired leader of the Indian nation rather than a scientist, he was another who bravely acknowledged that **'There is an orderliness in the Universe, there is an unalterable law governing everything and every being that exists or lives...**

That law then which governs all life is God' (Louis Fischer, *Gandhi: His Life and Message for the World*, 1954, p.108).

[215]In his 1987 book, *The Cosmic Blueprint*, another holistic physicist, the Templeton Prize-winning Paul Davies, actually went so far as to protest against the denial of Integrative Meaning in the world of science, writing that **'We seem to be on the verge of discovering not only wholly new laws of nature, but ways of thinking about nature that depart radically from traditional science...Way back in the primeval phase of the universe, gravity triggered a cascade of self-organizing processes—organization begets organization—that led, step by step, to the conscious individuals who now contemplate the history of the cosmos and wonder what it all means...There exists alongside the entropy arrow another arrow of time** [the Negative Entropy arrow], **equally fundamental and no less subtle in nature...I refer to the fact that the universe is *progressing*—through the steady growth of structure, organization and complexity—to ever more developed and elaborate states of matter and energy. This unidirectional advance we might call the optimistic arrow, as opposed to the pessimistic arrow of the second law...There has been a tendency for scientists to simply deny the existence of the optimistic arrow. One wonders why'** (chs 10, 9 & 2 respectively).

[216]We can now appreciate that the reason **'why' 'scientists' 'deny' 'the optimistic arrow'** of Integrative Meaning is because it was far too psychologically dangerous to acknowledge without first finding the biological reason, and thus defence, for our divisive, apparently non-integrative, un-Godly human condition. No wonder we have been, as we say, a 'God-fearing', in fact, so in awe of God to the point of being a God-worshipping—*not* a 'God-confronting'—species; as Berdyaev put it, **'He cannot** [man struggles to] **break through to paradise that lies beyond the painful distinction between good and evil, and the suffering connected therewith. Man's fear of God is his fear of himself, of the yawning abyss of non-being** [alienation] **in his own nature'** (*The Destiny of Man*, 1931; tr. N. Duddington, 1960, p.41 of 310).

[217]Our species' immense fear, and thus denial, of the truth of Integrative Meaning is actually the subject of Plato's famous allegory of the human condition, referred to in pars 59 and 64, which describes humans as being imprisoned in a cave. As 'prisoners' in this metaphorical cave, we are only able to envisage the outside world via shadows cast on the back wall of the cave. These shadows, which symbolise our limited and distorted, human-condition-avoiding, dishonest, immensely alienated **'fake'** and **'phony'** view of the world, are thrown by the light of a fire that,

situated in the entrance to the cave, effectively prevents any escape from it. Explaining the symbolism of the fire, Plato wrote that **'the light of the brightly burning fire in the** [cave] **prison corresponds to the power of the sun'** (c.360 BC; *Plato: The Republic*, tr. H.D.P. Lee, 1955, p.282 of 405), from which we have to **'turn back'** because if 'the prisoner'/we were to go **'outside into the sunlight, the process would be a painful one, to which he** [we] **would much object'** (p.280). Plato explained that the sun represents the **'universal, self-sufficient first principle'** (p.277), the **'absolute form of Good'** (p.282) and the **'highest form of knowledge'** (p.268), which we can now understand is Integrative Meaning. Fire is a common theme in many mythologies, appearing as a metaphor for the integrative, Godly ideals of life whose condemning, scorching glare we had to **'turn back'** from. In the Zoroastrian religion, **'Fire is the representative of God...His physical manifestation...Fire is bright, always points upward, is always pure'** (Edward Rice, *Eastern Definitions*, 1978, p.138 of 433). In Christian mythology, the story of Genesis features **'a flaming sword flashing back and forth to guard the way to the tree of life'** (3:24). The Bible also records the Israelites as saying, **'Let us not hear the voice of the Lord our God nor see this great fire any more, or we will die'** (Deut. 18:16). The biblical character Job was another who pleaded for relief from confrontation with the unbearably depressing integrative, Godly ideals when he lamented, **'Why then did you** [God] **bring me out of the womb?...Turn away from me so I can have a moment's joy before I go to the place of no return, to the land of gloom and deep shadow, to the land of deepest night** [depression]**'** (Job 10:18, 20-22). Christ also recognised the problem of the exposing **'light'** of truth that he was an unresigned, denial-free spokesman for, when he said, **'the light shines in the darkness but...everyone who does evil** [becomes upset sufferers of the human condition] **hates the light, and will not come into the light for fear that his deeds will be exposed'** (John 1:5, 3:20).

218 So, again, while Integrative Meaning *is* the most obvious, profound and thus important of all truths it is clearly also the truth that has appeared to most condemn humans—and in the absence of the explanation as to *why* we, as a species, appear to be so at odds with the integrative meaning of life, we humans have sensibly taken one of two options: we either practised denial of Integrative Meaning, and even of God, and thus of the issue of our self-corruption, or we indirectly acknowledged our self-corruption by acknowledging the existence of God and embracing some expression of faith that a greater dignifying understanding of our divisive condition does exist and would one day be found. To cope

with our less-than-ideal human condition there has only ever been either denial or faith.

219 To counter the utter dishonesty of mechanistic science's denial of the existence of Integrative Meaning/God, support for an extremely literal interpretation of God, in the form of so-called 'Creationism' and 'Intelligent Design', emerged. While still having to avoid the human condition and, therefore, the truth of Integrative Meaning, these movements did acknowledge God, but only in a fundamentalist way in which God took the form of an actual being who 'designed' life on Earth, or 'created' the world in just six days. In truth, both the mechanistic approach and the more literal attitudes were immensely dishonest in that mechanistic scientists wanted to pretend to be rational and either deny any semblance of Integrative Meaning by refuting the existence of the concept of God, or acknowledge the concept of God but claim it has nothing to do with science, while supporters of Creationism and Intelligent Design chose to admit to a semblance of Integrative Meaning in the form of a God who is literally a special being or deity, with the downside being that such a stance necessarily meant abandoning all attempts at rationality. We can see that the *real* issue neither party was willing or able to acknowledge is the issue of Integrative Meaning and its human-condition-confronting implications. (I should mention that promoters of Intelligent Design would have us believe that their position is different to that of Creationists. The main website for Intelligent Design states that **'The theory of intelligent design…**[is concerned with] **whether the "apparent design" in nature…is genuine design (the product of an intelligent cause) or is simply the product of an undirected process such as natural selection acting on random variations'**, and that **'unlike Creationism, the scientific theory of intelligent design does not claim that modern biology can identify whether the intelligent cause detected through science is supernatural'** (Center for Science & Culture, accessed Nov. 2013; see <www.wtmsources.com/116>). In other words, Intelligent Design acknowledges Integrative Meaning, the development of order of matter, *but doesn't discount that a supernatural-type creator/being/God might be involved*—so Intelligent Design is trying to have it both ways, appear to be scientific but still allow for a supernatural creator; it is still fundamentally similar to Creationism which unscientifically supports the idea of a supernatural creator.)

220 Indeed, the truth of Integrative Meaning and its human-condition-confronting implications have been *so* unbearably confronting that in

recent years mechanistic science has, in an insidious attempt to keep the issue even further at bay, evasively steered the discussion toward whether the concept of God has been undermined by physicists' ongoing discoveries about the Big Bang origin of the universe, the extinction of time before the Big Bang and, more recently, the possibility of multiple universes! This is classic 'displacement' behaviour—**'an unconscious defense mechanism whereby the mind redirects affects from an object felt to be dangerous or unacceptable to an object felt to be safe or acceptable'** (Wikipedia, accessed Mar. 2013; see <www.wtmsources.com/138>). The truth is that starting with the boundaries of our reality of matter, space and time, and drawing on the laws of physics with which we live, we can construct the human condition, and also solve it—and, by so doing, make it possible to demystify God, and, indeed, bring to an end the whole debate about 'His' existence. Paul Davies was emphasising the Integrative-Meaning-related *real* issue about God when he said, **'So where is God in this story** [of physics]**? Not especially in the big bang...To me, the true miracle of nature is to be found in the ingenious and unswerving lawfulness of the cosmos, a lawfulness that permits complex order to emerge from the chaos'** ('Physics and the Mind of God: The Templeton Prize Address', 3 May 1995).

221 Yes, with understanding of the human condition now found, it is at last psychologically safe to demystify God as Integrative Meaning, and, by so doing, reconcile religion and science. When the scientist-philosopher Pierre Teilhard de Chardin wrote that **'I can see a direction and a line of progress for life, a line and a direction which are in fact so well marked that I am convinced their reality will be universally admitted by the science of tomorrow'** (*The Phenomenon of Man*, 1938, p.142 of 320), he was recognising how obvious the truth of Integrative Meaning is and how it wouldn't be able to be **'universally admitted'** until the human-condition-resolved **'science of tomorrow'** emerged.

222 So while a few scientists in addition to Hawking, Einstein, Koestler and Davies have 'jumped the gun' and **'admitted'** Integrative Meaning, as the titles (particularly the words I have underlined) of the following books (including three by Davies) illustrate—for instance, David Bohm wrote *Wholeness and The Implicate Order* in 1980; Ilya Prigogine and Isabelle Stengers wrote *Order Out of Chaos* in 1984; Paul Davies wrote *God and the New Physics* in 1983, *The Cosmic Blueprint* in 1987 and *The Mind of God: Science and the Search for Ultimate Meaning* in 1992; Charles Birch wrote *Nature and God* in 1965, *On Purpose* in 1990

and *Biology and The Riddle of Life* in 1999; M. Mitchell Waldrop wrote *Complexity: The Emerging Science at the Edge of Order and Chaos* in 1992; Roger Lewin wrote *Complexity: Life at the Edge of Chaos, the major new theory that unifies all sciences* in 1993; Stuart Kauffman wrote *The Origins of Order: Self-Organization and Selection in Evolution* in 1993, *At Home in the Universe: The Search for the Laws of Self-Organization and Complexity* in 1995 and *Anti-chaos* in 1996; and Richard J. Bird wrote *Chaos and Life: Complexity and Order in Evolution and Thought* in 2003—the reality is that the vast majority of scientists haven't been prepared to go anywhere *near* the unbearably confronting truth of Integrative Meaning, and, as emphasised in chapter 1:11, in coming off such a dishonest base it is impossible to find a true understanding of our world and place in it. Plato recognised this inherent limitation of the mechanistic approach when he wrote that **'the Good** [Integrative Meaning]**... gives the objects of knowledge their truth and the mind the power of knowing...** [just as] **the sun...makes the things we see visible...The Good therefore may be said to be the source not only of the intelligibility of the objects of knowledge, but also of their existence and reality'** (*Plato: The Republic*, tr. H.D.P. Lee, 1955, p.273). Yes, this loss of **'the power of knowing'** has been very serious indeed. Koestler was another who felt it, bemoaning the crippled, stalled, atrophied state of all of science, but of biology and psychology in particular, when he said that blind, reductionist, mechanistic science's denial of Integrative Meaning has **'taken the life out of biology as well as psychology'**, writing that **'although the facts** [of the integration of matter] **were there for everyone to see, orthodox evolutionists were reluctant to accept their theoretical implications. The idea that living organisms, in contrast to machines, were primarily *active*, and not merely *reactive*; that instead of passively adapting to their environment they were...creating...new patterns of structure...such ideas were profoundly distasteful to** [Social] **Darwinians, behaviourists and reductionists in general** [p.222 of 354] **...Evolution has been compared to a journey from an unknown origin towards an unknown destination, a sailing along a vast ocean; but we can at least chart the route...and there is no denying that there is a wind which makes the sails move...the purposiveness of all vital processes...Causality and finality are complementary principles in the sciences of life; if you take out finality and purpose you have taken the life out of biology as well as psychology** [p.226]**'** (*Janus: A Summing Up*, 1978).

223 As was pointed out in par. 82, towards the end of *The Origin of Species*, Charles Darwin anticipated that **'In the distant future I see**

open fields for far more important researches. Psychology will be based on a new foundation, that of the necessary acquirement of each mental power and capacity by gradation. Light will be thrown on the origin of man and his history' (1859, p.458 of 476). Given Koestler's comment that **'if you take out finality and** [the **'integrative tendency'** or] **purpose you have taken the life out of biology as well as psychology'**, what was required to bring about Darwin's **'new'** en-**'light'**-ening **'foundation'** for **'far more important research'** in **'biology as well as psychology'** was not only acknowledgment of the involvement of our conscious **'mental power'** but also of **'integrative' 'purpose'**.

224 So, what *is* the **'far more important research'** that results from thinking from the **'new'** en-**'light'**-ening **'foundation'** of accepting the truth of **'integrative' 'purpose'**?

Chapter 3:4 The denial-free history of the development of matter on Earth

225 In commencing this denial-free analysis of the development of matter from a perspective that takes into account the truth of Integrative Meaning, we first need to replace the word 'evolution' with the word 'development', for while evolution implies that organisms *do* change or evolve it avoids acknowledging that there is a direction and purpose to that change, which is to *develop* the order of matter.

226 As the study of physics has shown, our world is made up of three fundamental ingredients—time, space, and energy—with, as Einstein revealed with his famous formula $E = mc^2$, energy taking the form of matter, which comprises the 94 or so naturally occurring elements that, when subjected to the laws of physics, particularly the law of Negative Entropy, became ordered or integrated; they formed more stable or enduring (in time) and ever larger (in space) arrangements.

227 This development of order of matter involved the initial mixture of the Earth's elements and their gradual formation into stable arrangements called molecules—earlier I provided the example of a water molecule being the stable arrangement of two single positively charged hydrogen atoms with one double negatively charged oxygen atom. In time, through the mixing of different elements, each with its own particular properties, many stable arrangements were found or developed, leading to even greater order and complexity of arrangements in the form of very complex macromolecules.

228 The problem, however, was that the more complex these macromolecules became, the more unstable they tended to be. Highly complex macromolecules would only occasionally form and, when they did, they didn't tend to hold together for long before breaking down into their separate parts. Eventually an impasse was reached where the degree of instability imposed a limit on how complex macromolecules could become. When this ceiling was reached it appeared Negative Entropy—or 'God' if we were to personify the process—could not develop any more order of matter on Earth. And yet matter *did* continue to 'develop' beyond this apparent impasse, with the emergence out of the primordial soup of a complex macromolecule with an unusual property—DNA, or deoxyribonucleic acid. What was unusual about DNA was that it could replicate. It could split, allowing the two halves to draw material from the environment to build two complete DNA molecules. The significance of this replication was that it meant DNA could defy breakdown. It could turn a relatively brief lifetime for a complex macromolecule into a relatively indefinite one. DNA's ability to replicate meant that even though some of the replicates disintegrated into smaller parts, others would survive and go on to replicate further. With slight variations called mutations occurring from the effects of solar radiation, replicates were 'found' that were even more stable/enduring (in time) and more ordered/complex/larger (in space). The process of natural selection of more stable and larger arrangements of matter—and the origin of an indefinite lifetime, or 'life' as we call it—appeared.

229 In this process, each replicating arrangement of matter or reproducing individual was, in effect, being tested both for its ability to survive and reproduce in its lifetime and, over generations of offspring, for its ability to adapt to changes in the environment in which it lived, with those that managed to survive and adapt inevitably, whenever possible, finding/refining/achieving/growing/developing even greater order of matter. The effect of this process over time was that more and greater order of matter was integrated. It was the ability to survive and adapt that supplied the opportunity for more and greater order of matter to develop. Thus, using the tool of replicating DNA, Negative Entropy was able to integrate matter into larger wholes; it was able to develop ever more and ever greater order of matter on Earth.

230 DNA is actually a very complex crystal. Crystal molecules abound—common salt, sodium chloride, for instance, is one—and in a suitable

nutrient environment they all have the capacity to reproduce; to grow their structure from their structure. However, being much simpler than DNA—having fewer elements within their molecular structure—they have little or no potential for adaptation and, it follows, for the development of greater order.

[231] Indeed, variability is so critical to this DNA process of developing greater order of matter that 'sexual reproduction' of DNA molecules developed, where the split halves from two compatible DNA molecules were made, through natural selection, to come together (be 'attracted to and mate with each other as males and females') to form a new, slightly different DNA-based sexually reproducing individual. This greatly increased the variety of a particular DNA type or 'species' and, by so doing, greatly increased its chances of finding/achieving/growing/developing larger and more stable arrangements of matter. Sexual reproduction, therefore, soon replaced non-sexual or 'asexual' reproduction as the most successful or effective form of DNA reproduction in this business of finding or developing greater order of matter.

[232] It can be seen then that Negative Entropy's development of order of matter really comes down to being a product of possibilities. The differing properties of matter mean some arrangements of matter break down towards heat energy, while others stay stable and still others become part of larger and more enduring associations of matter. In time, all the possible associations of matter will be automatically or, as Charles Darwin called it, '<u>naturally</u>' investigated until the largest, most stable association is left or found or, as Darwin described it, '<u>selected</u>'. What happened with DNA was that it not only turned a relatively short lifetime for extremely complex molecules into a relatively indefinite one, it also made a business, as it were, of this negentropy direction—both of resisting breakdown and of developing order. The replicating DNA molecule gave rise to a process that actively resisted breakdown and actively developed ever more and greater order of matter. This is **'the active striving of living matter towards order'**, **'a drive towards synthesis, growth and wholeness'**, the **'active'**, **'creating'**, **'purposiveness'** of life about which Koestler wrote.

[233] The DNA unit of inheritance is called a gene, with the study of the process of change that genes undergo termed 'genetics'. As a tool for Negative Entropy's development or refinement of the order of matter on Earth, the genetic process was very powerful—it was able to develop the great diversity of matter that we term 'the variety of life'. From DNA,

virus-like organisms developed, then from virus-like organisms developed single-celled organisms (such as bacteria), and from single-celled organisms developed multicellular organisms (such as plants and animals). The next level of order to be developed or integrated by Negative Entropy was societies or colonies or ordered arrangements of multicellular organisms. It was at this point, however, that Negative Entropy (or God) encountered another major impasse.

234 While genetics has proved to be a marvellous tool for integrating matter it has one very significant limitation, which arises from the fact that each sexually reproducing individual organism has to struggle, compete and fight selfishly for the available resources of food, shelter, territory and a mate it needs if it is to successfully reproduce its genes. What this means is that integration, and the unconditionally selfless cooperation it depends on, cannot normally develop between one sexually reproducing individual and another. Indeed, the competition between sexually reproducing individuals is the basis of the natural selection process that gave rise to the great variety of life on Earth. The word 'selection' in 'natural selection' implies competition—comparison between sexually reproducing individuals for their ability to survive, adapt and develop greater order of matter. So integration beyond the level of the sexually reproducing individual—that is, the coming together or integration of the sexually reproducing individual members of a species to form the next larger and more stable whole of the Specie Individual—could not, normally, develop (the *Development of Order of Matter* chart included earlier shows where the Specie Individual appears in the hierarchy of integration). This was the second major impasse that Negative Entropy (or God) encountered: the development of order of matter on Earth had seemingly come to a stop at the level of the sexually reproducing individual.

235 To elaborate, each sexually reproducing individual normally has to ensure the reproduction of its own genes, which means sexually reproducing individuals cannot normally develop the ability to behave unconditionally selflessly towards other sexually reproducing individuals—which, as has been explained, is what full cooperation and thus complete integration requires. Certainly sexually reproducing individuals can develop *conditionally* selfless behaviour towards other sexually reproducing individuals. Situations of reciprocity can develop where one sexually reproducing individual selflessly helps another on the proviso that they are selflessly helped in return, which, in effect, means both

parties are still selfishly benefiting. So sexually reproducing individuals *can* develop reciprocity because it is, in essence, still selfish behaviour: it doesn't give away an advantage to other sexually reproducing individuals and, therefore, doesn't compromise the reproductive chances of the sexually reproducing individual practising the behaviour. Unconditionally selfless, altruistic traits, on the other hand, do give away an advantage to other sexually reproducing individuals—that being the meaning of unconditional selflessness, that you are giving without receiving—and, therefore, unconditionally selfless, altruistic, self-sacrificing traits *do* compromise the reproductive chances of the sexually reproducing individual practising such behaviour and, therefore, cannot normally develop.

236 So cooperation between sexually reproducing individuals cannot normally be developed beyond a situation where there is reciprocal/conditional selflessness, and, since conditionally selfless behaviour is still basically selfish behaviour, full cooperation and thus complete integration cannot normally be developed between sexually reproducing individuals to form the Specie Individual. This inability to develop unconditionally selfless, altruistic behaviour leaves sexually reproducing individuals competing relentlessly with each other for available resources of food, shelter, territory and a mate. So much so, in fact, that what we see happening between sexually reproducing individuals as they try to develop more integration under this limitation of not being able to develop unconditionally selfless behaviour is that the competition between them becomes *so* intense that the only way they can contain it at all is by establishing a dominance hierarchy, where each individual accepts its position in a hierarchy that is ordered according to the competitive strengths of the various individuals involved. The benefit of a dominance hierarchy, or a so-called 'peck order', is that once established the only time competition breaks out is when an opportunity arises to move up the hierarchy; for the rest of the time there is relative peace. The emergence of a dominance hierarchy is a sign a species has developed as much integration as it possibly can.

237 It should be pointed out that in situations where competition between individuals breaks out and, for instance, male elephants or whales or kangaroos or birds or solitary insects, etc, etc, chase a female in estrous, it's not simply because the female wants to discover which is the strongest male with which to mate to ensure her offspring is the strongest, most competitively successful individual it can be, as is currently taught, but

because the Negative Entropy integrative tendency has driven the males and the females to that extreme state of competition. Such extreme competition is, in truth, a result of trying to develop greater order of matter. More will be said about this shortly, but the real story of life on Earth is *not* about selfish competition but integration.

[238] So although dominance hierarchy hides it from view for most of the time, the reality is that *extreme* competitiveness characterises the behaviour of the more cooperative and thus integrated, or what has evasively been called 'social', species. In my youth I remember feeding hens in our hen house and seeing a hen twist her leg and become temporarily crippled, at which point all the other hens immediately attacked her. In that instant it was suddenly apparent to me just how closely and intensely each hen was watching all the other hens for an opportunity to literally move up the peck order. The hen house was not at all the gregarious, peaceful community I thought it was; rather, it was a place of absolutely fierce competition! Charles Darwin recognised this truth about the real struggle in the lives of most animals when he wrote that **'It is difficult to believe in the dreadful but quiet war of organic beings, going on** [in] **the peaceful woods and smiling fields'** (12 Mar. 1839; *Charles Darwin's Notebooks*, eds P.H. Barrett et al, 1987, p.429).

[239] This situation where sexually reproducing individuals relentlessly compete for available resources is the situation, the condition, that almost all animals have to endure—it is the great, agonising 'animal condition'. When we humans become free of our numbed, alienated human condition we are going to be shocked by the agony of the animal condition: we are going to *feel* the distress all non-human animal species live under, where each sexually reproducing individual, through its Negative Entropy-driven commitment to achieve greater integration, is having to relentlessly and fiercely compete to reproduce its genes. Unfortunately, because animals' innocence (lack of the psychological upset we humans suffer from) confronts us with our lack of innocence (our vicious angry, egocentric and alienated state), we humans have so hated, despised and resented animals that we have hunted and shot them for 'sport'; but one day we are going to have so much sympathy for animals because of what they have to endure trapped in a life of having to relentlessly compete with each other, often with their closest friends! ('Friends' in the sense of those with whom they have shared their life and developed emotional bonds.) Certainly the same extremely competitive state exists for plants and

microbes, but, not having the developed nervous system that animals have, their awareness of the agony of that horrifically competitive existence could obviously not be as great as it is for animals.

240 What now needs to be explained is, firstly, that while sexually reproducing individuals cannot normally be integrated, the sexually reproducing individual itself *could* be elaborated, enlarged, expanded—developed further to become bigger—which, as will be explained next, is how single-celled organisms developed into multicellular organisms, and how multicellular colonial ants and bees integrated into their fully cooperative and thus completely integrated colonies. Significantly, in these 'elaborated sexually reproducing individuals', the cells of the multicellular body, or the individual ants and bees in their fully integrated colonies, are no longer sexually reproducing individuals themselves, but part of a larger sexually reproducing individual, which is the body, or, in the case of ants and bees, the colony. Secondly, it has to be explained why I have been saying it is 'normally' not possible for sexually reproducing individuals to become fully integrated to form the Specie Individual. There was, in fact, one species who managed to achieve the development of the next larger whole in the integration of matter on Earth of the integration of sexually reproducing individuals to form the Specie Individual: our ape ancestors. As will be explained in chapter 4, this amazing step in the development of matter was achieved through maternalism—the nurturing of our offspring—which has been another of those unbearable truths that we humans couldn't face until we could explain our divisive, unloving human condition.

241 To summarise what has been explained so far: in the development of order of matter on Earth, all non-human animal species are stuck in the 'animal condition', with each sexually reproducing individual member of the species forever having to compete to ensure its genes reproduce and carry on. *That* is the essential fact or rule of the gene-based natural selection process—genes are unavoidably selfish; they have to ensure they reproduce if they are to carry on. It is important to reiterate, however, that even though this selfishness—and the extreme competition between the sexually reproducing individuals it gives rise to—is characteristic of virtually all of nature, such selfishness is *only* occurring because of the *limitation* of the genetic process of *normally* being unable to develop unconditional selflessness between sexually reproducing individuals. In his 1850 poem *In Memoriam*, Lord Alfred Tennyson famously wrote:

'Who trusted God was love indeed / And love Creation's final law / Tho' Nature, red in tooth and claw / With ravine [in violent contradiction]**, shriek'd against his creed.'** While Integrative Meaning or **'God'** and its theme of unconditional selflessness or **'love'** is the **'creed'** or **'final law'** of **'creation'** that the competitive, selfish and aggressive, **'red in tooth and claw'** characteristic of so much of **'Nature'** seems to be in violent contradiction **'against'**, we can now understand that this selfish characteristic doesn't mean that the *overall* biological reality of existence—life's meaning and theme—is to be selfish, as the dishonest theories of Social Darwinism, Sociobiology, Evolutionary Psychology and Multilevel Selection would have us believe. As will be explained in chapter 4, in the case of humans, we don't have selfish instincts like other species, rather we have *unconditionally selfless* instincts. And the selfishness that is characteristic of so much of nature is *only* occurring because of the *limitation* of the gene-based refinement process: its inability, in most situations, to develop unconditional selflessness. The genetic process would develop unconditionally selfless, fully cooperative behaviour between all sexually reproducing individuals if it could—because such selflessness is what is required to maintain a fully integrated whole—but, because of its particular *limitation*, it normally can't. Integrative selflessness, not divisive selfishness, is the *real* nature or characteristic of existence, the theme of life.

242 Incidentally, in par. 90 it was mentioned how, in describing his concept of natural selection, Darwin originally left it undecided as to whether individuals who managed to reproduce are 'fitter' or better than those who don't, but was later persuaded by human-condition-avoiding, mechanistic biologists to describe natural selection as a competitive, 'survival of the fittest' process. Well, we can now see why it was right for Darwin to leave it undecided as to whether individuals who manage to reproduce are better or 'fitter' than those who don't. As has now been explained, it can be completely consistent with the integrative meaning of existence for an individual to give their life for the purpose of maintaining the larger whole of their society and thus not reproduce. Unconditionally selfless, self-sacrifice for the good of the whole, is the very theme of existence. It is only because of the *limitation* of the gene-based natural selection process that unconditionally selfless behaviour normally cannot be developed between sexually reproducing individuals. Selfless cooperation, *not* selfish, competitive, 'survival of the fittest' behaviour, *is* the real characteristic of existence, the theme of life.

Chapter 3:5 Elaborating the sexually reproducing individual

243 As stated, while sexually reproducing individuals cannot normally be integrated, the sexually reproducing individual itself could be elaborated, made bigger, which, as will now be explained, is how single-celled organisms developed into multicellular organisms, and how multicellular colonial ants and bees integrated into their fully cooperative colonies. Struggling to find a way to develop greater order of matter by integrating sexually reproducing individuals, it was as if Negative Entropy (or God) decided, 'Well, what I'll do is develop greater order of matter *within* each genetically reproducing individual, making it bigger.'

244 As was also mentioned, in these 'elaborated sexually reproducing individuals', the cells of the multicellular body, or the individual ants and bees in their fully integrated colonies, are no longer sexually reproducing individuals themselves, but part of a larger sexually reproducing individual, which is the body, or, in the case of ants and bees, the colony.

245 The biological mechanism for elaborating the sexually reproducing individual involved the body's cells, or the colony's multicellular bees/ants, delegating the task of sexual reproduction to a distinct part of the whole that *specialises* in reproduction. In the case of the integration of single-celled organisms, the green alga known as *Volvox* provides an example of an organism in transition from the single-celled to the multicellular state, as this quote describes: **'Volvox is...a small, green sphere...composed of thousands of flagellates embedded in the surface of a jelly ball...Volvox is a colony of unicellular animals rather than a many-celled animal, because even the simplest many-celled animals have considerably more differentiation between cells than appears among the cells of Volvox. The colony swims about, rolling over and over from the action of the flagella; but, remarkably enough, the same end of the sphere is always directed forward...Its behaviour can be explained only by supposing that the activities of the numerous flagellates are subordinated to the activity of the colony as a whole. If the flagella of each member of the colony were to beat without reference to the other members, the sphere would never get anywhere. In such subordination of the individual cells of a colony to the good of the colony as a whole we see the beginnings of individuality as it exists in the higher animals, where each animal behaves as a single individual, although composed of millions of cells...The co-ordination of numerous components into an individual is usually followed by the specialisation of different individuals for different duties. Only the slightest degree of specialisation is seen in the Volvox**

colony; the flagellates of the back part of the colony are capable of reproduction, while the front members never reproduce but have larger eyespots and serve primarily in directing the course of the colony' (R. Buchsbaum, *Animals without Backbones*, 1938, p.50 of 401).

246 The marine invertebrates known as siphonophores, which include species such as the Portuguese man-of-war (otherwise known as the Bluebottle), live in colonies composed of 'zooids'—individual animals that are not fully independent; indeed, their reliance upon and integration with each other is so strong the colony attains the character of one large organism. In fact, most of the zooids are so specialised they lack the ability to survive on their own. Thus siphonophorae, like Volvox, exist at the boundary between colonial and complex multicellular organisms.

247 We can imagine the path to the creation of Volvox and siphonophores began with cloning, which is the asexual reproduction of identical offspring where competition between the clones is pointless and unnecessary since each individual is genetically identical and, therefore, the division of labour and cooperation can develop and exist between individuals. One concern with cloning, however, is loss of variability—for example, if one colony kept reproducing asexually it could become so big it monopolised the available resources of food, space and territory, leading to the detriment of other colonies and a subsequent lack of variability in the species. We can imagine that eventually a limited, functional size would be arrived at through natural selection, which presumably is the size at which Volvox and siphonophores operate. And obviously to maintain variability, it would also be beneficial for sexual reproduction to occur from time to time, as it does amongst Volvox and siphonophores.

248 In the case of bees, the queen bee feeds all of her offspring that she intends to be workers a 'royal jelly' that causes sterility (ants also employ a similar chemical retardant). To ensure the reproduction of their genes these sterile offspring then have to support the queen because she carries their genes. (It should be mentioned that saying the queen 'intends' and the offspring 'have to' is obviously personifying the genetic process. The queen and the offspring are obviously not conscious thinking organisms, deciding they 'intend' and 'have to' do something or other as we humans do; however, this form of anthropomorphism is simply a useful way of describing what, *in effect*, occurs. For example, the way genetics actually causes offspring to 'have to' support the queen is that, out of the many

different mutational varieties of offspring that appear over time, only those that happen to have a genetic make-up that inclines them to support the queen will tend to reproduce, naturally selecting that particular behaviour for all subsequent generations and eventually the whole species.)

249 Elaborating the sexually reproducing individual allows the members of the elaborated individual to develop the ability to at least *behave* unconditionally selflessly, which, as has been explained, is fundamental if the fully cooperative integration of members into a new whole is to develop. The reason our body works so well is because each part has sublimated its needs to the greater good of the whole body; each part behaves unconditionally selflessly. Our skin is constantly growing and dying to protect our body. The leaves that fall in autumn do so to ensure the tree survives through winter. Bees and ants readily sacrifice themselves for their colony; for example, when a bee stings to protect its hive, its innards are attached to the sting that is left in its victim, so when it stings, it dies. The skin, leaves and bees/ants have behaved unconditionally selflessly; they have, in effect, considered the welfare of the greater good above their own welfare.

250 Of significance, however, is the emphasis here on our body's skin, the tree's leaves and the bees/ants only *behaving* unconditionally selflessly, because the selflessness apparent in these examples is *not* actually true unconditional selflessness, it is *not* true altruism. This is because the self-sacrificing skin, leaves and bees/ants are all indirectly selfishly ensuring their own genetic existence will be maintained by supporting the body, tree, or bee/ant colony that carries the genes for their existence and so reproduces them when it reproduces itself as a whole. Genetically, they are selflessly fostering the body/tree/colony to selfishly ensure their own genetic reproduction. Their apparently unconditionally selfless *behaviour* is not actually unconditional and thus altruistic, but rather a subtle form of selfishness. As explained earlier, such reciprocity can develop genetically because it doesn't compromise the chances of the sexually reproducing individual reproducing its genes. (As pointed out in par. 92, the dishonest biological theory of Sociobiology/Evolutionary Psychology was truthful to the extent that it did recognise this fact that the selfless behaviour of social ants and bees is due to reciprocity—where the theory was dishonest was in its application of 'kin selection' to explain *all* social behaviour, even our own *unconditionally* selfless, *universally* benevolent, *fully* altruistic moral instincts.)

[251]It now needs to be explained that large animals couldn't employ this device of elaborating the sexually reproducing individual to develop a fully cooperative, integrated association or whole of their members because for them it involves too great a loss of the variability that all species need to be able to adapt to their environment. For example, if a female buffalo happened to be born with a particular mutation that caused her to produce a chemical in her milk that retarded the sexual maturation of most of her offspring such that those offspring then had to have selected mutations that inclined them to protect her to ensure their genes are successfully reproduced by her, and this became a common practice amongst buffalos with every queen buffalo having, say, 9 protector sacrificial buffalos from 10 offspring (so there is a sexual offspring to ensure the reproduction of the buffalo species, like ants have a few fertile females and males to carry on their species, but these fertile offspring have the same potential to produce some infertile offspring), then the genetic variety of a population of 1,000 buffalos would be reduced to just 100, a drastic loss of the variability so critical to ensuring that the species' genetic pool remains able to adapt to any changes in its environment and thus keep maintaining and/or developing greater order. In the case of bees/ants, they are so small in relation to their environment that they can afford to have many fully integrated colonies in their environment without any significant loss of variability within their species.

Magnetic Termite Mounds, Litchfield National Park, Northern Territory, Australia; and feral Asian Buffalo, Northern Territory, Australia. Photographs by the author, 2010.

[252]These two photographs illustrate the point. As will be mentioned shortly, while termites are a variety of cockroach rather than ant or bee, they have developed the same colonial capability as colonial ants and bees, which means that although there are millions of termites in each termite mound, in terms of the *genetic variety* present in the territory

shown, these mounds do, in fact, represent a similar number of sexually reproducing individuals to the number of sexually reproducing individual buffalos shown in a corresponding area in the second photograph.

253 Quite a number of species that are much larger than ants and bees are attempting to create the integrated society of members by temporarily elaborating the sexually reproducing individual. Many bird species, such as the Australian kookaburra, delay their sexual maturation for a few years after they fledge, during which time they selflessly help raise their parents' subsequent offspring. Wolves, African wild dogs and meerkats do the same thing. However, what these species have obviously found is that to delay their sexual maturation permanently leads to too great a loss of variability in their species.

254 Underground-living colonial naked mole rats form fully integrated colonies of up to 300 members comprising 1 queen who uses hormones to inhibit the sexual maturation of nearly all the others who then act as 'workers' and 'soldiers'. A few 'sexual disperser caste' are allowed to reach sexual maturity and these periodically leave their natal burrow to access other colonies and, in doing so, help maintain the genetic variety of the mole rat species. Significantly, like colonial ants and bees, and the dozen or so other varieties of multicellular organisms (including the termite) that have been able to permanently elaborate the sexually reproducing individual, mole rats are relatively small; individuals typically measure only 8 to 10 centimetres (3 to 4 inches) long.

255 What has been explained here is very significant for humans because it means, as large animals, we could not have employed the integrating device of elaborating the sexually reproducing individual to create the pre-conscious and pre-human-condition-afflicted, fully cooperative, completely integrated, 'Golden', 'Garden of Eden'-like state that our distant ancestors lived in. Further, during that fully integrated, idyllic past our instinctive orientation was *not* to reciprocity's subtle form of selfishness that the parts of multicellular organisms and bee/ant colonies practise, as the theory of Sociobiology/Evolutionary Psychology claims, but to being truly altruistic, genuinely unconditionally selflessly orientated towards *all of life*. Thus, even if we could have employed the device of elaborating the sexually reproducing individual it would not even begin to account for our *unconditionally* selfless, moral soul. I italicised '*all of life*' because while ant and bee colonies have members who are dedicated to supporting each other, each colony is, in fact,

engaged in fierce competition with other colonies. Worker ants and bees are not interested in behaving selflessly towards *all of life*, which, contrary to what the theory of Multilevel Selection claims, our moral self is interested in. As pointed out in chapter 1:10, our ability to love unconditionally didn't arise from an ability to war successfully. So, the question remains: how did we humans manage to develop our absolutely wonderful and astonishing *unconditionally* selfless, *genuinely* altruistic, *all*-loving moral instinctive orientation to the world?

Chapter 3:6 Negative Entropy found a way to form the Specie Individual

256 In conclusion, while elaborating the sexually reproducing individual does allow greater order of matter to be developed, it doesn't achieve the next level of integration, which is the coming together or integration of sexually reproducing individual members of a species to form the Specie Individual or whole. The question, therefore, is, could Negative Entropy or God find a way to overcome the impasse of integrating sexually reproducing members of a species into a Specie Individual—or had the limit to the amount of order of matter that could be developed on Earth finally been reached?

257 As will be fully explained in the next chapter, the reason I have written in this chapter that it is not '*normally*' possible to integrate sexually reproducing individuals is because Negative Entropy or God *did* find one way to integrate sexually reproducing members of a species to form the Specie Individual, which was through the nurturing of offspring—and it was *this* device that our ape ancestors employed to achieve the fully integrated state, the instinctive memory of which is our unconditionally selfless, genuinely altruistic, all-loving moral instinctive self or soul.

258 (A more complete description of the integration of matter and the meaning of life can be found in *Freedom Book 1* at <www.worldtransformation.com/freedom-integrative-meaning>.)

Chapter 4

The Origin of Humans' Unconditionally Selfless, Altruistic, Moral Instinctive Self or Soul

Photograph of Matata by Manny Rubio

Bonobos Matata and her adopted son, Kanzi

=

Drawing by Jeremy Griffith © 2006 Fedmex Pty Ltd

Chapter 4:1 Summary

259 **Our moral instinctive self or soul, the 'voice' or expression of which is our conscience, is a truly amazing phenomenon for it provided the cooperative, unconditionally selfless love that created humanity. So when the philosopher Immanuel Kant had the following words inscribed on his tombstone—'*there are two things which fill me with awe: the starry heavens above us, and the moral law within us*' (*Critique of Practical Reason*, 1788)—he certainly wasn't overstating the magnificence of our altruistic moral sense. Nor was Charles Darwin when he wrote that '*the moral sense affords the best and highest distinction between man and the lower animals*' (*The Descent of Man*, 1871, ch.4).**

260 **But as amazing as our moral soul most certainly is, its very existence is *also* a cause for wonder because it raises the baffling question of *how***

***on earth did we humans acquire such an 'awe'-inspiring, 'distinct'-from-other-animals moral sense?* For biologists especially, the great outstanding mystery has been how could the cold, selfish, competitive, gene-based natural selection process have possibly created such warm, unconditionally selfless, cooperative, loving instincts in us humans?**

261 While it may seem astonishing to suggest that what is now going to be presented *is* that most elusive of answers to this most intriguing of mysteries, that is, in fact, the case—this chapter contains nothing less than the truth about what it really means to be human.

Chapter 4:2 How could humans have acquired their altruistic moral instincts?

262 As evidenced in pars 75-77, throughout our mythologies and in the work of our most profound thinkers there is a recognition that our distant ancestors lived in a pre-conscious, pre-human-condition-afflicted, innocent, unconditionally selfless, genuinely altruistic, fully cooperative, universally loving, peaceful state. The author Richard Heinberg's research into the subject of this collective memory of a 'Garden of Eden'-like, 'Golden Age' in our species' past found that **'Every religion begins with the recognition that human consciousness has been separated from the divine Source, that a former sense of oneness...has been lost...everywhere in religion and myth there is an acknowledgment that we have departed from an original...innocence'** (*Memories & Visions of Paradise*, 1990, pp.81-82 of 282). So yes, when Nikolai Berdyaev acknowledged that **'The memory of a lost paradise, of a Golden Age, is very deep in man'** (*The Destiny of Man*, 1931; tr. N. Duddington, 1960, p.36 of 310), he was expressing what we *all* intuitively know is the truth about our species' past innocent existence—as was the philosopher Jean-Jacques Rousseau when he, almost two centuries earlier, wrote that **'nothing is more gentle than man in his primitive state'** (*The Social Contract and Discourses*, 1755; tr. G. Cole, 1913, Book IV, p.198 of 269).

263 The origin of the words associated with our moral nature reveals this underlying awareness of the extraordinarily loving, ideal-behaviour-expecting, 'good-and-evil'-differentiating, sound nature of our instinctive self or 'psyche' or 'soul', the 'voice' or expression of which is our 'conscience'. For instance, our 'conscience' is defined as our **'moral sense of right and wrong'**, and our 'soul' as the **'moral and emotional part of man'**, and as the **'animating or essential part'** of us (*Concise Oxford Dictionary*,

5th edn, 1964), while, as mentioned in par. 148, the *Penguin Dictionary of Psychology*'s entry for **'psyche'** reads: **'The oldest and most general use of this term is by the early Greeks, who envisioned the psyche as the soul or the very essence of life'** (1985 edn). Indeed, as the **'early Greek'** philosopher Plato wrote about our innate, ideal-or-Godly-behaviour-expecting moral nature, we humans have **'knowledge, both before and at the moment of birth...of all absolute standards...**[of] **beauty, goodness, uprightness, holiness...our souls exist before our birth...**[our] **soul resembles the divine'** (*Phaedo*, c.360 BC; tr. H. Tredennick).

[264] When the philosopher John Fiske wrote about the existence of our moral nature, he was similarly effusive: **'We approve of certain actions and disapprove of certain actions quite instinctively. We shrink from stealing or lying as we shrink from burning our fingers'** (*Outlines of Cosmic Philosophy*, 1874, Vol. IV, Part II, p.126). *And*, while our moral instinctive self or soul will **'shrink from stealing or lying'**, it is not merely concerned with avoiding the ill-treatment of others—it is also deeply concerned with ensuring their well-being. For instance, when Joe Delaney, a professional footballer, admitted that **'I can't swim good, but I've got to save those kids'**, just moments before plunging into a Louisiana pond and drowning in an attempt to rescue three boys ('Sometimes The Good Die Young', *Sports Illustrated*, 7 Nov. 1983), he was considering the welfare of others above that of his own. The truth is that everywhere we look we see examples of humans behaving unconditionally selflessly, such as those who show charity to others less fortunate, or sacrifice their lives for ethical principles. Indeed, now that we can explain the human condition it becomes clear that since the human condition fully emerged some 2 million years ago, *every* generation of humans has had to suffer becoming self-corrupted in an unconditionally selfless effort to aid the accumulation of knowledge that would one day liberate humanity from the human condition; to borrow the words from the musical *The Man of La Mancha* that were included in par. 185, every generation has altruistically **'march**[ed] **into hell for a heavenly cause'** (Joe Darion, *The Impossible Dream*, 1965).

[265] Our species' unconditionally selfless moral nature is indeed a wonderful phenomenon. However, as mentioned in chapter 2:7, Alexander Pope saw our **'awe'**-inspiring, **'best and highest distinction'**-deserving, **'divine'**-like, **'absolute standards...**[of] **beauty, goodness, uprightness, holiness'**-expecting, **'animating'**, **'very essence of life'**, **'moral and emotional'**, **'essential part'** of us in a very different light, pointing out that **'our nature** [is]**...A sharp accuser, but a helpless friend!'** And he was right in the sense

that, as was made clear in that chapter, our ideal-behaviour-expecting, moral conscience *has* been **'a sharp accuser, but a helpless friend'**; it *has* criticised us aplenty when what we needed was redeeming *understanding* of our 'good-and-evil'-afflicted, corrupted or 'fallen' present human condition—which we now at last have.

266 Yes, paradoxically, until we could explain our present soul-devastated, innocence-destroyed, angry, egocentric and alienated condition we couldn't afford to face the truth that our **'awe'**-inspiring moral soul is our instinctive memory of an unconditionally selfless, all-loving past. And so we undermined its very existence. As the psychologist Ronald Conway noted about the human-condition-avoiding, mechanistic attitude that developed toward the subject of our soul: **'Soul is customarily suspected in empirical psychology and analytical philosophy as a disreputable entity'** (Letter to the Editor, *The Australian*, 10 May 2000). But with the fully accountable, human psychosis-addressing-and-solving, truthful explanation of the human condition now found, we *can* finally acknowledge what our soul is, and, most significantly, heal our species' psychosis or **'soul-illness'**; yes, since *psyche* means **'soul'** and *osis*, according to *Dictionary.com*, means **'abnormal state or condition'**, we can at last ameliorate or heal our species' *psychosis*—its alienated, psychologically **'ill'**, **'abnormal state or condition'**.

267 But recognition and resolution of the issue aside, the very great question that remained to be answered was how could we humans have possibly acquired such a **'distinct'** from other **'animals'**, **'awe'**-inspiring but **'sharp accus**[ing]**'** instinctive orientation in the first place? What is the biological origin of our species' extraordinary moral nature?

Chapter 4:3 The integration of sexually reproducing individuals to form the Specie Individual

268 As was explained in some detail in chapters 3:4 to 3:6, while the gene-based system for developing the order of matter on Earth is powerfully effective—it is responsible, after all, for the great variety of life we see on Earth—it has one very significant limitation, which arises from the fact that each sexually reproducing individual organism has to struggle and selfishly compete for the available resources of food, shelter, territory and a mate it requires if it is to successfully reproduce its genes. What this means is that integration, and the unconditionally selfless

cooperation that integration depends upon, cannot normally develop between one sexually reproducing individual and another; which *in turn* means that integration *beyond* the level of the sexually reproducing individual—that is, the coming together or integration of sexually reproducing individuals to form a new larger and more stable whole of sexually reproducing individuals, the Specie Individual—can also not normally develop. Thus, it would appear that Negative Entropy's, or God's, development of order of matter on Earth had come to a stop at the level of the sexually reproducing individual. The integration of the members of a species into the larger whole of the Specie Individual could seemingly not be developed.

269 What this means is that only a degree of cooperation and thus integration could be developed between the sexually reproducing individual members of a species before the competition between them became so intense that a dominance hierarchy had to be employed to contain the divisive competition; and, in fact, that is where most animal species *are* stalled in their ability to integrate. They could become integrated to a degree (what has been termed 'social'), but not completely integrated into one new larger organism or whole. Certainly each sexually reproducing individual could be either temporarily (in the case of large animals like wolves) or permanently (in the case of small animals like ants and bees) 'elaborated'—developed to become bigger—thus allowing greater integration of matter to occur *within* the sexually reproducing individual. But those elaborated units (the wolf packs and the ant/bee colonies) were, nevertheless, *still* engaged in competition with each other. It seemed that the integration of sexually reproducing members of a species and thus the full integration of the members of a species into a Specie Individual could not be achieved; Negative Entropy, or God, *had* seemingly developed as much order of matter on Earth as it could.

270 HOWEVER, this was *not* the case—the integration of matter *hadn't* come to an end because a way *was* found by Negative Entropy, or God, to integrate the members of a species into the larger whole of the Specie Individual, AND, moreover, it was our ape ancestors who achieved this extraordinary step. Yes, as recognised in Genesis in the Bible, we humans *did* once live **'in the image of God'** (1:27), we were once a fully cooperative, unconditionally selflessly behaved, completely integrated species. We *did* once live in **'the Garden of Eden'** (3:23) state of original cooperative, loving, innocent togetherness, then we became conscious, took the **'fruit'** **'from**

the tree of...knowledge' (3:3, 2:17), and, as a result, **'fell from grace'** (derived from the title of Gen. 3, **'The Fall of Man'**) because we became divisively behaved sufferers of the human condition, supposedly deserving of being **'banished...from the Garden of Eden'** (3:23) of original innocent togetherness, and left **'a restless wanderer on the earth'** (4:14)—left in our present, immensely upset, psychologically distressed angry, egocentric and alienated condition, a state we can now at last emerge from because we can finally explain and thus compassionately understand *why* we *had to* search for **'knowledge'** and suffer becoming corrupted.

© National Geographic Society

Africa—our soul's home—the Garden of Eden

Lucas Cranach the Elder's *Adam and Eve* (1526)

Photograph by Peeter Viisimaa

We can still see the remnants of the time when our species lived in a pre-human-condition-afflicted, innocent, Garden-of-Eden-like state in the happiness and togetherness of children.

[271] So THE GREAT QUESTION is, how did Negative Entropy, or God, achieve this amazing integration? How did our human ancestors develop the fully integrated state of the Specie Individual, the instinctive memory of which is our unconditionally selfless, genuinely altruistic, moral self or 'soul', and the 'voice' or expression of which is our conscience? What is the biological origin of our species' extraordinary **'shrink from stealing or lying'**, **'got to save those kids'**, prepared **'to march into hell for a heavenly cause'** moral sense?

Chapter 4:4 Love-Indoctrination

[272] Ever since Charles Darwin published his natural selection explanation for the origins of the great variety of life on Earth in 1859 all manner of theories have been put forward to suggest how one species, namely *our* species, managed to defy the essential selfish nature of the natural selection process and develop unconditionally selfless moral instincts. Indeed, in response to the great build up of anxiety in the world about the now desperate plight of the human race and the resulting urgent need for the relieving answers that are being presented in this book—especially answers to the questions of the origin of our species' moral nature and present human-condition-afflicted lack of compliance with it—almost every edition of every science journal over the last 10 years, and I'm writing this in 2013, has either presented or discussed a new theory that attempts to account for our moral sense. But of all these theories—a brief history of which will be presented in chapter 5—only one provides the fully accountable and thus adequate, true explanation for the origins of our moral sense, which is that it was achieved through nurturing—a mother's maternal instinct to care for her offspring. However, as will be acknowledged shortly (in ch. 4:7), and described in detail in chapter 5, the problem with this nurturing explanation, and why its early permutations were discarded by the scientific establishment, is that it has been an unbearably confronting truth. It is *only* now that we can explain the human condition and thus understand *why* the present human-condition-afflicted human race hasn't been able to adequately nurture our infants to the extent their instincts expect that it becomes safe to finally admit that nurturing is what made us human—that nurturing gave us our moral soul and created humanity.

273 So how *did* nurturing create our extraordinary unconditionally selfless moral instincts? Genetic traits for nurturing are intrinsically selfish (which, as has been emphasised, genetic traits normally have to be if they are to survive) because through a mother's nurturing and fostering of offspring who carry her genes her genetic traits for nurturing are selfishly ensuring their reproduction into the next generation. However, while nurturing *is* a genetically selfish trait, from an observer's point of view it *appears* to be unconditionally selfless behaviour—the mother is giving her offspring food, warmth, shelter, support and protection for *apparently* nothing in return. This point is most significant because it means from the infant's perspective its mother is treating it with real love, with unconditional selflessness. The infant's brain is, therefore, being trained or indoctrinated or inscribed with unconditional selflessness and so, with enough training in unconditional selflessness, that infant will grow into an adult who behaves unconditionally selflessly. Apply this training across all the members of that infant's group and the result is an unconditionally selflessly behaved, cooperative, fully integrated society.

274 The 'trick' in this 'love-indoctrination' process lies in the fact that the traits for nurturing are encouraged, or selected for genetically, because the better the infants are cared for, the greater are their, and the nurturing traits', chances of survival—however, there is an integrative side effect, which is that the more infants are nurtured, the more their brains are trained in unconditional selflessness. There are very few situations in biology where animals appear to behave selflessly towards other animals—normally, they each selfishly compete for food, shelter, territory and mating opportunities. Maternalism, a mother's fostering of her infant, is one of the few situations where an animal appears to be behaving selflessly towards another animal and it was this *appearance* of selflessness that exists in the maternal situation that provided the integrative opportunity for the development of love-indoctrination, the training of individuals in unconditional selflessness. And with this unconditionally selfless behaviour recurring over many, many generations, the unconditionally selfless behaviour will become instinctive—a moral soul will be established—because genes will inevitably follow and reinforce any development process occurring in a species; in this they are not selective. The difficulty was in getting the development of unconditional selflessness to occur in the first place, for once it was regularly occurring it would naturally become instinctive over time.

[275] But for a species to develop nurturing—to develop this 'trick' for overcoming the gene-based learning system's seeming inability to develop unconditional selflessness—it required the capacity to allow its offspring to remain in the infancy stage long enough to allow for the infant's brain to become trained or indoctrinated with unconditional selflessness or love. In most species, infancy has to be kept as brief as possible because of the infant's extreme vulnerability to predators. Zebra foals, for example, have to be capable of independent flight almost as soon as they are born, which gives them little opportunity to be trained in selflessness. But as the following photo of a rhesus monkey trying to carry its infant illustrates, being semi-upright as a result of their tree-living, swinging-from-branch-to-branch, arboreal heritage meant primates' arms were largely freed from walking and thus available to hold dependents. Infants similarly had the capacity to latch onto their mothers' bodies. This freedom of the upper body meant primates were especially facilitated for prolonging their offspring's infancy and thus developing love-indoctrination.

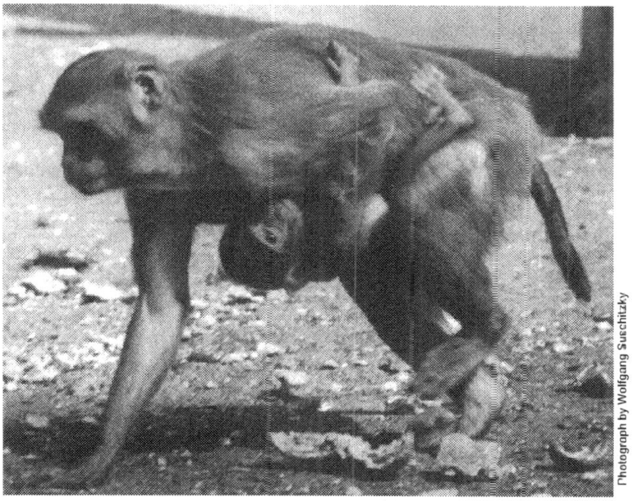

Rhesus monkey with infant. This picture illustrates the difficulty of carrying an infant and suggests the reason for bipedalism.

[276] It follows that the longer infancy is delayed, the more and longer infants had to be held, and thus the greater need and, therefore, selection for the arms-freed, upright walking ability known as bipedalism. As will be discussed in more detail shortly, when I first put forward this nurturing, 'love-indoctrination' explanation for humans' unconditionally selfless moral nature in a 1983 submission to *Nature* and *New Scientist*

magazines (see <www.worldtransformation.com/nature>), I said, contrary to prevailing views, that it meant bipedalism must have developed early in this nurturing of love process, and, in fact, the early appearance of bipedalism in the fossil record of our ancestors is now being found. For example, in 2012 it was reported that **'The oldest hominins currently known are *Sahelanthropus tchadensis*...dated to between 6 and 7 mya** [million years ago]', which the fossil record suggests **'stood and walked bipedally'** (Herman Pontzer, 'Overview of Hominin Evolution', *Nature Education Knowledge*, 2012, Vol.3, No.10).

[277] But while bipedalism was the key factor in developing nurturing and thus love-indoctrination, other influences also played a pivotal role, most notably the presence of ideal nursery conditions. This entailed uninterrupted access to food, shelter and territory, for if any element was compromised, or other difficulties and threats from predators excessive, then we can assume that there would have been a strong inclination to revert to more selfish and competitive behaviour. The successful nurturing of infants therefore required ample food, comfortable conditions and security from external threats over an extended period. But, in addition to these practical factors, to ensure the success of the love-indoctrination process, it wasn't enough to simply *look after* the infants, they had to be *loved*, and so maternalism became about much more than mothers simply protecting, providing for, and training their infants in life skills—it became about demonstrably loving them. Significantly, we speak of 'motherly love', not 'motherly protection, provision and training'.

[278] So, in addition to the prerequisites of, firstly, a physiology that facilitated an extended infancy and, secondly, ideal nursery conditions, what was *also* required for love-indoctrination to occur was the presence and influence of more maternal mothers. Of course, the difficulty with selecting for more maternal mothers is that their genes don't tend to endure because their offspring tend to be the most selflessly behaved, too ready to put others before themselves, leaving the more aggressive, competitive and selfish individuals to take advantage of their selflessness. Such selfish opportunism *could*, however, be avoided if *all* members of the group were equally well nurtured with love, equally trained in selflessness—this situation being yet another of the delicate conditions that has to be maintained if love-indoctrination is to develop, for any breakdown in nurturing within a group that is in the midst of developing

love-indoctrination could jeopardise the whole situation and see it revert to the old state of the 'each-for-his-own', opportunistic, all-out-competition-where-only-dominance-hierarchy-can-bring-some-peace, selfish-genes-rule, 'animal condition'.

279 It can be seen then that while the development of unconditional selflessness through the love-indoctrination process of a mother's nurturing care of her infant *was* possible, it was not easy, even for the exceptionally facilitated primates, which explains why almost all primate species have not been able to complete the development of love-indoctrination to the point of becoming fully integrated. So far the evidence indicates that it has only been our ape ancestors who managed to *complete* the process, the result of which is our species' unconditionally selfless, genuinely altruistic, universally loving instinctive self or 'soul', the 'voice' or expression of which is our moral 'conscience'—a hypothesis the fossil record is now corroborating.

Chapter 4:5 Fossil evidence confirming the love-indoctrination process

280 Although the fossil record has been slow to yield evidence of our ape ancestors who lived during humanity's infancy (which, as will be explained in chapters 7:2 and 7:3, lasted from some 12 to 4 million years ago), the very recent discovery of fossils belonging to our direct ancestors from this period is now confirming the love-indoctrination process. These recently unearthed ancestors are: *Sahelanthropus tchadensis* (who lived some 7 million years ago and is thought to be the first representative of the human line after we diverged from humans' and chimpanzees' last common ancestor); *Orrorin tugenensis* (who lived some 6 million years ago); and the two species of *Ardipithecus*: *kadabba* (who lived some 5.6 million years ago), and *ramidus* (who lived some 4.4 million years ago). Incidentally, *Sahelanthropus* means 'Sahel man' (Sahel is an area near the Sahara); *Orrorin tugenensis* means 'original man whose fossils were found in the Tugen region in Kenya'; while *Ardipithecus* means 'ground ape', with *kadabba* meaning 'oldest ancestor', and *ramidus* meaning 'root' or basal family ancestor.

281 It is worth emphasising that these fossils have all been found *very* recently. *Sahelanthropus* was only discovered in 2002 (in the form of a skull) and decisively identified as a human ancestor in April 2013, while

fragments of a skull, jaw and thigh bone belonging to *Orrorin* were first unearthed in 2001. Although fragments of *Ardipithecus* were first discovered in 1992, and the excavation of a largely intact skeleton (which was nicknamed 'Ardi') began in 1994, the remains of the latter—1 of only 6 reasonably complete skeletons of early humans older than 1 million years—were in such poor condition that it took until 2009—over 15 years of analysis—for reports to be published. With studies on all of these recently discovered ancestors now becoming available, including the series of 2009 *Ardipithecus* reports, which the journal *Science* deemed 'Breakthrough of the Year', it is exciting to see that confirming evidence of the love-indoctrination process that led to the establishment of our extraordinary unconditionally selfless moral instincts is slowly but surely emerging.

282 So, how does this new evidence confirm the love-indoctrination process? How, for instance, does it affect our understanding of the emergence of bipedalism, the first key factor in developing unconditionally selfless moral instincts? As I just mentioned, when I first put forward the nurturing, 'love-indoctrination' explanation for such instincts in 1983, I said, contrary to prevailing views, that it meant bipedalism must have developed early in this nurturing of love process and, it follows, early in our ancestors' history, and, indeed, that is what these recent fossil discoveries now show. Scientists can infer whether a species was bipedal by several methods, including the position of the foramen magnum (the opening at the base of the skull through which the spinal cord enters), because in species that stand upright the opening appears toward the centre of the skull rather than at the rear. Using information such as this, the current scientific thinking is that bipedalism arose at least as early as *Sahelanthropus tchadensis*, with anthropologists now reporting that **'Bipedalism is one of very few human characteristics that appears to have evolved at the base of the hominin clade [species more closely related to modern humans than to any other living species]. Recent fossil discoveries have apparently pushed back the origin of the hominin clade into the late Miocene, to 6 to 7 million years ago (Ma). The oldest known potential hominin** [human line] **fossils** [are] **attributed to *Sahelanthropus tchadensis*'** (Brian G. Richmond & William L. Jungers, '*Orrorin tugenensis* Femoral Morphology and the Evolution of Hominin Bipedalism', *Science*, 2008, Vol.319, No.5870).

283 The analysis of the fossils belonging to the slightly more recent *Orrorin tugenensis* provides further evidence of bipedalism. In addition

to the evidence presented in the fragments of its skull, the analysis of *Orrorin tugenensis*' femur (thigh bone) has allowed scientists to conclude **'that *Orrorin* was a habitual biped as shown by a suite of features in the proximal femur'** (Martin Pickford et al, 'Bipedalism in *Orrorin tugenensis* revealed by its femora', *Comptes Rendus Palevol*, 2002, Vol.1, No.4), and that **'*O. tugenensis* is a basal hominin adapted to bipedalism, and current evidence suggests that an *Australopithecus*-like bipedal morphology evolved early in the hominin clade and persisted successfully for most of human evolutionary history'** (Brian G. Richmond & William L. Jungers, '*Orrorin tugenensis* Femoral Morphology and the Evolution of Hominin Bipedalism', *Science*, 2008, Vol.319, No.5870).

284 Fossils of *Ardipithecus*, and particularly *Ar. ramidus*, confirm that bipedalism was actually *well* established by 4.4 million years ago, with studies of 'Ardi' (the relatively intact skeleton) leading the respected anthropologist C. Owen Lovejoy to conclude that **'*Ar. ramidus* was fully capable of bipedality and had evolved a substantially modified pelvis and foot with which to walk upright'** ('Reexamining Human Origins in Light of *Ardipithecus ramidus*', *Science*, 2009, Vol.326, No.5949). Furthermore, Lovejoy confirmed the long history of bipedalism that preceded *Ar. ramidus* when he said that *Ar. ramidus* **'has been bipedal for a very long time'** (Ann Gibbons, 'A New Kind of Ancestor: *Ardipithecus* Unveiled', *Science*, 2009, Vol.326, No.5949).

285 The second requirement for love-indoctrination to occur is the existence of ideal nursery conditions, namely an environment that provides uninterrupted access to food, shelter and territory. You would perhaps expect such conditions would be found in humid forests and woodlands, where food is plentiful and trees provide shelter and refuge from predators, but the scientific community's traditional view has been that the factor in our ancestral history that propelled our ancestor's development beyond that of the other apes was their movement onto the savannah. However, in light of the fossil evidence that has emerged in the 10 years prior to the time of writing, which is in 2013—and in the 30 years since I first proposed the nurturing, love-indoctrination explanation—the scientific community now widely accepts that this separation of our human ancestors from other primates occurred while our ancestors lived in forests and woodlands, the sort of environment I identified as being required for the love-indoctrination process to begin.

286 Scientists are now able to reconstruct the habitats of *Sahelanthropus*, *Orrorin* and *Ardipithecus* based on those species' physical characteristics, the information provided by the fossils of other animals and plants found

accompanying them, as well as climate data. While *Sahelanthropus* fossils are so limited they don't provide the information needed to confirm that they were adapted to climbing trees and thus lived in forests or woodlands, reconstructions of their environment have narrowed *Sahelanthropus*' habitat to **'a mosaic of environments from gallery forest at the edge of a lake area to a dominance of large savannah and grassland'** (Patrick Vignaud et al, 'Geology and palaeontology of the Upper Miocene Toros-Menalla hominid locality, Chad', *Nature*, 2002, Vol.418, No.6894). As we move forward in time to *Orrorin* some 6 million years ago, this species' skeletal structure shows tree climbing adaptations, which clearly point to them living in an arboreal habitat. Further, associated animal and plant fossils have allowed scientists to infer that **'*Orrorin tugenensis* may have evolved in well wooded to forested conditions margining lakes and streams with open country-side in the vicinity'** (Soizic Le Fur et al, 'The mammal assemblage of the hominid site TM266 (Late Miocene, Chad Basin): ecological structure and paleoenvironmental implications', *Naturewissenschaften*, 2009, Vol.96, No.5); and that **'the surroundings of the site were probably open woodland, while the presence of several specimens of colobus monkeys indicate that there were denser stands of trees in the vicinity, possibly fringing the lake margin and streams that drained into the lake'** (Martin Pickford & Brigitte Senut, 'The geological and faunal context of Late Miocene hominid remains from Lukeino, Kenya', *Comptes Rendus de l'Academie des Sciences–Series IIA–Earth and Planetary Science*, 2001, Vol.332, No.2). Forest and woodlands continued to be the preferred habitat of *Ar. ramidus* some 4.4 million years ago, as indicated by its retention of tree climbing features such as a pelvis that supported large climbing muscles, flexible wrists that allowed walking on all fours along the top of branches, and an opposable big toe that allowed it to grasp the branches with its feet: **'*Ar. ramidus* preferred a woodland-to-forest habitat rather than open grasslands'** (Tim D. White et al, '*Ardipithecus ramidus* and the Paleobiology of Early Hominids', *Science*, 2009, Vol.326, No.5949). In fact, the wealth of surrounding evidence from the *Ar. ramidus* fossil site in Ethiopia allowed the paleoanthropologist Andrew Hill to remark that **'There's so much good data here that people aren't going to be able to question whether early hominins were living in woodlands'** (Ann Gibbons, 'Habitat for Humanity', *Science*, 2009, Vol.326, No.5949), and fellow researcher Giday WoldeGabriel to state that *Ar. ramidus* lived **'in an environment that was humid and cooler than it is today, containing habitats ranging from woodland to forest patches'** (Giday WoldeGabriel et al, 'The Geological, Isotopic, Botanical, Invertebrate, and Lower Vertebrate Surroundings of *Ardipithecus ramidus*', *Science*, 2009, Vol.326, No.5949). Indeed, this **'good**

data' associated with the 'Ardi' dig has meant that paleobiologists have been able to reconstruct *Ar. ramidus'* habitat to an extraordinary level of detail: **'Ardi lived on an ancient floodplain covered in sylvan woodlands, climbing among hackberry, fig, and palm trees, and coexisting with monkeys, kudu antelopes, and peafowl'** (Ann Gibbons, 'Breakthrough Of The Year: *Ardipithecus ramidus*', *Science*, 2009, Vol.326, No.5960) where **'doves and parrots flew overhead'** (Ann Gibbons, 'Habitat for Humanity', *Science*, 2009, Vol.326, No.5949). Combine this environment with our knowledge of *Ar. ramidus'* diet, which indicates *Ar. ramidus* **'was a generalized omnivore and frugivore** [fruit eater]**'** (Gen Suwa et al, 'Paleobiological Implications of the *Ardipithecus ramidus* Dentition', *Science*, 2009, Vol.326, No.5949), and our knowledge of existing ape behaviour, which indicates *Ar. ramidus* **'almost certainly slept and fed in trees'** (Craig Stanford, 'Chimpanzees and the Behavior of *Ardipithecus ramidus*', *Annual Review of Anthropology*, 2012, Vol.41), and a picture begins to emerge of the ideal nursery conditions that enabled love-indoctrination to develop.

[287]These ideal nursery conditions also refute the long-held nurturing-avoiding theory, espoused by E.O. Wilson amongst others, that upright walking supposedly developed when our ancestors moved out onto the savannah: **'*Ar. ramidus* did not live in the open savanna that was once envisioned to be the predominant habitat of the earliest hominids'** (Giday WoldeGabriel et al, 'The Geological, Isotopic, Botanical, Invertebrate, and Lower Vertebrate Surroundings of *Ardipithecus ramidus*', *Science*, 2009, Vol.326, No.5949). In fact, the evidence that bipedality developed in **'forest or wooded environments'** is now so conclusive that Hill was able to assert that **'Savannas had nothing to do with upright walking'** (Ann Gibbons, 'Habitat for Humanity', *Science*, 2009, Vol.326, No.5949). Yes, because the development of bipedality is closely associated with the love-indoctrination process it had to have occurred while our ancestors were inhabiting ideal nursery conditions, which clearly suggested an arboreal environment—as I maintained when I originally put forward the love-indoctrination process in 1983.

[288]These recent fossil discoveries also confirm the third requirement for love-indoctrination to occur: the presence and influence of <u>more maternal mothers</u>. Scientists are able to deduce a remarkable amount of information about the social behaviour of our ancestors from their fossils, and, as a result of this evidence, are now beginning to acknowledge that these species exhibited low levels of aggression toward one another, and that females were not only not dominated by males, but dictated mate choice by choosing non-aggressive, cooperative males to

reproduce with—hallmarks you would expect of a society highly focused on maternal nurturing of their infants.

289 The first striking evidence provided by the fossil record to support these deductions is that these early humans had small canine teeth: **'male canine size and prominence were dramatically reduced by ~6 to 4.4 Ma'** (Gen Suwa et al, 'Paleobiological Implications of the *Ardipithecus ramidus* Dentition', *Science*, 2009, Vol.326, No.5949). This is relevant because **'canines function as weapons in interindividual aggression in most anthropoid species'** (ibid), particularly in aggressive male-to-male sexual competition for mating opportunities, and so canines **'inform aspects of social structure and behavior'** (ibid), where small canines are **'indicative of minimal social aggression'** (Tim D. White et al, '*Ardipithecus ramidus* and the Paleobiology of Early Hominids', *Science*, 2009, Vol.326, No.5949).

290 Furthermore, comparisons of canine size in *Ar. ramidus* with current apes indicate that *Ar. ramidus* males **'retained virtually no anatomical correlates of male-to-male conflict'** (C. Owen Lovejoy, 'Reexamining Human Origins in Light of *Ardipithecus ramidus*', *Science*, 2009, Vol.326, No.5949), a situation that would apply to our earlier ancestors *Sahelanthropus* and *Orrorin* since they too had small canines. Given that the reality of the animal kingdom involves fierce competition between sexually reproducing individuals seeking to reproduce their genes, this reduction in aggressive male competition for mating opportunities is an *extremely* significant anomaly, as Lovejoy recognises: **'Loss of the projecting canine raises other vexing questions because this tooth is so fundamental to reproductive success in higher primates. What could cause males to forfeit their ability to aggressively compete with other males?'** (ibid). Traditional attempts to answer this **'vexing'** question have argued either that large canine teeth were made redundant when humans adopted hand-held weapons—the so-called 'weapons replacement' hypothesis; or that large, overlapping canines made eating certain foods difficult and, therefore, were selected against; or that large canines had to make way for the large teeth of the robust australopithecines. However, the fossil record now shows that canines were reduced well before the emergence of the australopithecines; and as mentioned, it also shows that *Ar. ramidus* **'was a generalized omnivore and frugivore** [fruit eater]**'** like baboons and many other species of current primates who *have* retained their large canines. And with regard to weapon use rendering large canines redundant, the fossil record now

shows that our ancestors had developed small canines at least as early as *Sahelanthropus*, which is millions of years before any fossil evidence of weapon or tool use—and even if those ancestors brandished weapons such as branches or bones that would not leave 'evidence', the argument still fails to explain why having weapons *and* large canines would not be an advantage in any contest. A 1992 paper described the confusion that has surrounded the evolution of human canine reduction: **'the issue of human canine evolution has continued to be controversial and apparently intractable'** (Leonard O. Greenfield, 'Origin of the human canine: A new solution to an old enigma', *American Journal of Physical Anthropology*, 1992, Vol.35, No.S15). And the new discoveries have only increased this confusion. But as we can now see, the answer to the **'vexing'** and **'apparently intractable'** question of **'what could cause males to forfeit their ability to aggressively compete with other males'** is the love-indoctrination process. As will be explained below, conscious self-selection of integrativeness—especially the female sexual or mate selection of less competitive, less aggressive, more integrative males—developed to assist, speed up and help maintain love-indoctrination's development of integration. Indeed, male competition for mating opportunities is so **'fundamental to reproductive success'** that only active sexual selection against it can account for its reduction, as is made clear in this quote: **'Canine reduction did not result from a relaxation of selection pressure for large canines, but rather a positive selection against them'** (Arthur Klages, 'Sahelanthropus tchadensis: An Examination of its Hominin Affinities and Possible Phylogenetic Placement', *Totem: The University of Western Ontario Journal of Anthropology*, 2008, Vol.16, No.1). Indeed, it is now so clear that canine reduction could only be caused by **'a positive selection against them'** that the importance of sexual selection is now being recognised by leading anthropologists such as Lovejoy, Gen Suwa, Berhane Asfaw, Tim White and others, who write, **'In modern monkeys and apes, the upper canine is important in male agonistic** [aggressive] **behavior, so its subdued shape in early hominids and *Ar. ramidus* suggests that sexual selection played a primary role in canine reduction. Thus, fundamental reproductive and social behavioral changes probably occurred in hominids long before they had enlarged brains and began to use stone tools'** (Gen Suwa et al, 'Paleobiological Implications of the *Ardipithecus ramidus* Dentition', *Science*, 2009, Vol.326, No.5949).

291 As these authors make clear, the reduction in the size of canines was such a remarkable achievement that it required **'fundamental reproductive**

and social behavioral changes' in which **'sexual selection played a primary role'**. These scientists are describing a society that switched from being patriarchal—dominated by male sexual selection with males aggressively competing for mating opportunities—to matriarchal, dictated by female sexual selection where females choose mates that are less aggressive. However, what these scientists don't explain is the *only* mechanism that could allow such a switch: love-indoctrination. This remarkable reversal where females are empowered, and males **'forfeit their ability to aggressively compete with other males'**, is discussed in more detail in chapter 5, however, it is sufficient to emphasise at this point that the fossil record is increasingly providing compelling evidence that female sexual selection was occurring very early in human history, at least as early as *Sahelanthropus* some 7 million years ago, and that it **'emerged in concert with habituation to bipedality'** (C. Owen Lovejoy, 'Reexamining Human Origins in Light of *Ardipithecus ramidus*', *Science*, 2009, Vol.326, No.5949), which again is in accord with love-indoctrination, all of which I first predicted in 1983.

292 Another significant factor revealed by the fossil record is the difference in the size between males and females, including their canines, a phenomena known as sexual size dimorphism. Since **'sexual size dimorphism is generally associated with sexual selection via agonistic male competition in nonhuman primates...if a species showed very strong size dimorphism, it probably was characterized by intense male mate competition'** (J. Michael Plavcan, 'Sexual Size Dimorphism, Canine Dimorphism, and Male-Male Competition in Primates', *Human Nature,* 2012, Vol.23, No.1). Conversely, scientists recognise that a low level of sexual size dimorphism is an indicator of a society in which males do not aggressively compete for mating opportunities. As mentioned, the fossil records for our human ancestors show that **'There is no evidence of substantial canine dimorphism in earlier hominins, including *Sahelanthropus*, *Ardipithecus*, and *Australopithecus anamensis*, or later hominins'** (ibid). In addition to this low level of canine dimorphism, *Ar. ramidus* exhibited low levels of body size dimorphism, which, in terms of behaviour, **'were probably the anatomical correlates of comparatively weak amounts of male-male competition, perhaps associated with...a tendency for male-female codominance as seen in *P. paniscus*** [bonobos]**'** (Gen Suwa et al, 'Paleobiological Implications of the *Ardipithecus ramidus* Dentition', *Science*, 2009, Vol.326, No.5949). As will be described shortly, the prevailing view about bonobos

is that rather than having achieved **'male-female codominance'**, they have, in fact, gone further and achieved female dominance, a matriarchy.

293 So the three requirements of the love-indoctrination process of bipedality, ideal nursery conditions and selection for more maternal mothers are now being dramatically confirmed by the fossil record. However, as I mentioned earlier and will elaborate on shortly, the problem with this nurturing, true explanation, and why its early permutations were discarded by the scientific establishment, is that it has been an unbearably confronting, exposing truth for our present human-condition-afflicted human race that has been so unable to adequately nurture our infants to the extent our instincts expect. This new evidence has left those scientists who continue to deny the importance of nurturing in our development in a predicament in which they are forced to ask the right questions even though they are **'vexing'**, but refuse to acknowledge the truthful answer, because until the human condition was explained nurturing was an off-limits subject. The following passage from Lovejoy exemplifies this predicament: **'Why did early hominids become the only primate to completely eliminate the sectorial canine complex** [large projecting canines that are continuously sharpened against a lower molar]**? Why did they become bipedal, a form of locomotion with virtually no measurable mechanical advantage? Why did body-size dimorphism increase in their likely descendants? These are now among the ultimate questions of human evolution'** ('Reexamining Human Origins in Light of *Ardipithecus ramidus*', *Science*, 2009, Vol.326, No.5949). Lovejoy further reduced these **'ultimate questions'** to this one, final sentence that admits the reality of a cooperative past: **'Even our species-defining cooperative mutualism can now be seen to extend well beyond the deepest Pliocene** [well beyond 5.3 million years ago]**'** (ibid). Yes, as stated at the outset of this chapter, the great outstanding mystery for biologists has been how could the cold, selfish, competitive, gene-based natural selection process have possibly created such warm, unconditionally selfless, cooperative, loving instincts in us humans? But to answer that question of questions required the explanation of the human condition that would finally make sense of *why* we haven't been able to adequately nurture our infants—because with that compassionate insight it at last becomes psychologically safe to admit that nurturing is what made us human, thus allowing these **'ultimate questions of human evolution'** to be answered.

Chapter 4:6 Bonobos provide living evidence of the love-indoctrination process

[294]While these recent fossil discoveries are providing exciting confirmation that our ape ancestors completed the development of the love-indoctrination process, of the living primate species, only *Pan paniscus*, or bonobos (or pygmy chimpanzees as they were once called because of their comparatively gracile bodies), have not only developed love-indoctrination, they appear to have come close to completing the love-indoctrination process to become a fully integrated Specie Individual; they are certainly by far the most cooperative/harmonious/gentle/loving/integrated of the non-human primates. It follows then, that although there is no suggestion that bonobos or chimpanzees are a living human ancestor, comparisons have been made between bonobos and our ancestors. For instance, the physical anthropologist Adrienne Zihlman first proposed in 1978 **'that, among living species, the pygmy chimpanzee (*P. paniscus*) offers us the best prototype of the prehominid** [pre-human] **ancestor'** (Adrienne L. Zihlman et al, 'Pygmy chimpanzee as a possible prototype for the common ancestor of humans, chimpanzees and gorillas', *Nature*, 1978, Vol.275, No.5682), using the then earliest known human ancestor, *Australopithecus*, to compare the two species' physical characteristics, including their bipedality, canine teeth and sexual size dimorphism. In 1996 Zihlman refined her assessment to include similarities with the (at the time) newly discovered *Ardipithecus*. In a further example, the primatologist Frans de Waal has noted the extraordinary similarity between our ape ancestor and bonobos, saying, **'The bonobo's body proportions—its long legs and narrow shoulders—seem to perfectly fit the descriptions of Ardi, as do its relatively small canines'** (*The Bonobo and the Atheist*, 2013, p.61 of 289). Yes, bonobos (who, along with their chimpanzee cousins, share 98.7 percent of their DNA with humans) are physically extremely similar to our fossil ancestors, but beyond the physical similarities, scientists are suggesting bonobo behaviour *also* corresponds with that of our ancestors. In addition to the view expressed above, that, like bonobos, *Ardipithecus* were not male dominated, Zihlman has suggested that **'the *Pan paniscus* model offers another way to view the social life of early hominids, given their sociability, lack of male dominance and the female-centric features of their society'** ('Reconstructions reconsidered: chimpanzee models and human evolution', *Great Ape Societies*, eds William C. McGrew et al, 1996, p.301 of 352).

A female bonobo rests in a day nest of compacted branches and leaves high in the forest canopy at the Kokolopori Bonobo Reserve, Democratic Republic of the Congo.

A group of bonobos at the Lola Ya Bonobo Sanctuary, Democratic Republic of the Congo

With her infant beside her, Kame shares provisioned sugar cane with Senta, a non-related juvenile, at the Wamba bonobo research station, Democratic Republic of the Congo, 1987.

[295]So given the exceptionally maternal, matriarchal, cooperatively behaved bonobo species have developed the love-indoctrination process, we should expect that they provide living evidence of the three elements previously identified as being required for that process to occur: bipedalism, ideal nursery conditions, and selection for more maternal mothers.

[296]In relation to bipedalism, research confirms that bonobos are extraordinarily bipedal; in fact, they are **'the most bipedal of all the extant** [living] **apes'** (Roberto Macchiarelli et al, 'Comparative analysis of the iliac trabecular architecture in extant and fossil primates by means of digital image processing techniques', *Hominoid Evolution and Climate Change in Europe: Vol.2*, eds Louis de Bonis et al, 2001, p.71 of 372). As was explained earlier, we can now account for this bipedality, along with bonobos' peaceful cooperative nature: the longer infancy is delayed, the more and longer infants had to be held, thus the greater need and, therefore, selection for arms-freed, upright walking.

Upright Bonobos. Note the well-developed breasts (similar to those of female humans) for suckling their young—another indication of how important a role nurturing is playing in bonobo society.

[297] As mentioned, ideal nursery conditions are the second requirement for establishing love-indoctrination—and the bonobos have certainly benefited from a comfortable environment in the food-rich, relatively predator-and-competitor-free, ideal nursery conditions of the rainforests south of the Congo River. (Bonobos, for example, don't have to compete with gorillas, who have a diet similar to that of the bonobo but do not live south of the Congo River.) These ideal nursery conditions, which have been so conducive to their successful development of the love-indoctrination process, and the resulting love-indoctrinated cooperativeness of the bonobos compared with that of their chimpanzee cousins who live north and east of the Congo River, is apparent in this quote: **'we may say that the pygmy chimpanzees historically have existed in a stable environment rich in sources of food. Pygmy chimpanzees appear conservative in their food habits and unlike common chimpanzees have developed a more cohesive social structure and elaborate inventory of sociosexual behavior... Prior to the Bantu (Mongo) agriculturists' invasion into the central Zaire basin, the pygmy chimpanzees may have led a carefree life in a comparatively stable environment'** (Takayoshi Kano & Mbangi Mulavwa, 'Feeding ecology of the pygmy chimpanzees (*Pan paniscus*)'; *The Pygmy Chimpanzee*, ed. Randall Susman, 1984, p.271 of 435). Indeed, it is an indication of how difficult it is to develop love-indoctrination that even the bonobos, living as they do in their ideal conditions, and who **'have developed a more cohesive social structure'** than their chimpanzee relatives, still find it necessary to employ sex as an appeasement device to help subside residual tension and aggression between individuals; this is the **'elaborate inventory of sociosexual behavior'** referred to in this quote. I should mention here that bonobos also use a rich and constant array of vocalisations to assist the developing **'cohesive social structure'** of the group. Indeed, **'Bonobos are the most vocal of the great apes'** ('What is a Bonobo?', Bonobo Conservation Initiative, accessed Feb. 2014; see <www.wtmsources.com/119>), and **'are excitable creatures who frequently "comment" on minor events around them through high-pitched peeps and barks. Even if most of these vocalizations are noticeable only at close range, one definitely hears more vocal exchange in a group of bonobos than in a group of chimpanzees'** (Frans de Waal & Frans Lanting, *Bonobo: The Forgotten Ape*, 1997, p.10 of 210). But just how extraordinarily **'cohesive'** or integrated the bonobo species *has* become through their development of the love-indoctrination process is apparent in the following quote from the primatologist Barbara Fruth, who has spent

many years studying bonobos in their natural habitat: **'up to 100 bonobos at a time from several groups spend their night together. That would not be possible with chimpanzees because there would be brutal fighting between rival groups'** (Paul Raffaele, 'Bonobos: The apes who make love, not war', Last Tribes on Earth.com, 2003; see <www.wtmsources.com/143>). So there is relatively little conflict between individual bonobos or even between groups of bonobos, which is another indication that this species is well on its way to integrating its members into the Specie Individual—a feat that all evidence indicates is what our ape ancestors succeeded in achieving.

© Takeshi Furuichi

A group of bonobos in the forest of the Congo

[298]But while a situation conducive to the development of integration can be reached of there being sufficient food, shelter and territory, from a genetic point of view there can never usually be enough of that other key resource, mates—simply because the more successfully an individual can breed, the more their genes can carry on and multiply. Since females are limited in how often they can reproduce due to pregnancy—and, in the case of mammals, lactation—it is the males who have the opportunity to breed continuously, and so it is the competition for mating opportunities amongst males that is the most difficult form of genetic selfishness to overcome. Whilst bonobos haven't been able to fully develop love-indoctrination and thus integration, as evidenced by the fact that they have to use sex to quell residual tension, they have been able to develop it sufficiently to bring to an end this most difficult of all forms of selfish competitiveness, male competition for mating opportunities, as the fossil evidence now suggests our ape ancestors did. In fact, bonobos have been *so* successful at reining in male competition for mating opportunities that within their society there has been what could be described as a gender-role reversal, with bonobo females forming alliances and dominating social groups, both of which are distinctly male behaviours in chimpanzee, gorilla, orangutan and other non-human primate societies. Bonobo society is matriarchal, female-dominated, controlled and led. Further, in bonobo society, the entire focus of the social group does appear to be on the maternal, or female, role of nurturing infants, as this observation by America's leading ape-language researcher, the biologist and psychologist Sue Savage-Rumbaugh evidences: **'Bonobo life is centered around the offspring. Unlike what happens among common chimps, all members of the bonobo social group help with infant care and share food with infants. If you are a bonobo infant, you can do no wrong...Bonobo females and their infants form the core of the group'** (Sue Savage-Rumbaugh & writer Roger Lewin, *Kanzi: The Ape at the Brink of the Human Mind*, 1994, p.108 of 299). It is just such circumstances one would expect in a society that exhibits the third element necessary for love-indoctrination, selection for more maternal mothers.

[299]The following photographs of bonobos with infants reveal something of just how exceptionally nurturing bonobo females are—and even males, because the bottom right photo is of a male lovingly playing with an infant. Bonobos clearly have the environmental comfort and the freedom from fighting and tension in their world needed to develop the ability to love their infants.

Bonobos nurturing their infants

[300]The 2011 French documentary *Bonobos*, which was directed by Alain Tixier, contains marvellous footage evidencing the ability of bonobos to nurture their infants and the wondrous effects such nurturing has on their offspring. A short segment from the documentary showing the tenderness of the bonobo mothers and the absolute joy and zest for life of their infants can be seen in this YouTube clip at <www.wtmsources.com/107>. If you can't understand the French commentary, the story is about a young bonobo called Beny who was sold as a pet after his mother was killed by poachers. Fortunately, he was rescued by Claudine André, who took him to her wonderful bonobo sanctuary, Lola Ya Bonobo in the Democratic

Republic of the Congo, and later released him back into the Congo forest. While the documentary's commentary is superficial, there are two very revealing comments in the short accompanying film that discusses the making of the documentary. The first comment was made by Tixier, who said that **'The choice to do a film about bonobos was because they're surely the most fascinating animals on the planet. They're the closest animals to man. They're the only animals capable of creating the same "gaze" as a human. When you look at a bonobo you're taken aback because you can see behind the eyes it's not just curiosity, it's understanding. We see human beings in the eyes of the bonobo.'** The second was provided by the film's animal advisor, Patrick Bleuzen, who remarked that **'Once I got hit on the head with a branch that had a bonobo on it. I sat down and the bonobo noticed I was in a difficult situation and came and took me by the hand and moved my hair back, like they do. So they live on compassion, and that's really interesting to experience.'**

Photograph by Finlay MacKay for *TIME*

The human-like gaze of the bonobo Kanzi

Chapter 4:7 The role of nurturing in our development has been an unbearably confronting truth

301 Before continuing, it needs to be re-emphasised just how unbearably confronting this discussion about the importance of nurturing in the development of humans has been without the compassionate understanding of the human condition. The tragic reality has been that ever since the terrible battle broke out between our original all-loving instinctive self or soul and our newer conscious mind and our present immensely upset angry, egocentric and alienated human condition emerged (as was described in chapter 2), no child has been able to be given anything like the psychosis-and-neurosis-free, pure love that all infants were given back in this nurturing phase in our primate past. And unable, until now, to explain *why* we humans have become so psychologically upset and thus unable to adequately love our infants, such talk about the role that nurturing has played in the maturation of our species—and continues to play in our individual upbringing today, since we are all born still instinctively expecting to receive such pure, unconditional love—*has* been unbearable. So yes, it *is* only now that we can explain *why* we humans have become psychologically embattled and unable to adequately nurture our children that we can safely admit the crucial role that nurturing has played in human development. (Much more will be said in chapter 5 about how unconfrontable this nurturing explanation of human origins has been and, as a result, how human-condition-avoiding mechanistic science has denied the truth of the nurturing origins of our unconditionally selfless moral nature.)

302 While it has been unbearable and thus unconfrontable, the truth, nevertheless, is that nurturing was the all-important influence in the maturation of our species and remains the all-important influence in the maturation of our individual lives. The female gender created humanity, and, while under the duress of the upset state of the human condition it has rarely been possible to adequately nurture our offspring, the importance of nurturing in producing a secure, sound adult remains paramount. The archetypal image of the Madonna and child that is such a feature of Christian mythology—which I depicted in the drawing that appears at the beginning of this chapter—*is* all-meaningful because for Christ to have been such a sound, unresigned, denial-free-thinking person he must have had an exceptionally nurturing mother. So when the author

Olive Schreiner wrote the following passage she was, in fact, articulating the agonising reality for virtually all mothers who suffer under the duress of the human condition: **'They say women have one great and noble work left them, and they do it ill...*We* bear the world and *we* make it. The souls of little children are marvellously delicate and tender things, and keep for ever the shadow that first falls on them, and that is the mother's or at best a woman's. There was never a great man who had not a great mother—it is hardly an exaggeration. The first six years of our life make us; all that is added later is veneer'** (*The Story of an African Farm*, 1883, p.193 of 300). Yes, as this commentator in a story about the failure of many mothers to adequately nurture their offspring summarised: **'For a lot of women the only really important anchor in their lives is motherhood. If they fail in a primary role they feel should come naturally it is devastating for them'** ('The Deserted Mothers' Club', *The Weekend Australian Magazine*, 30 Nov. 2013). But, as will be explained in chapter 7:16D, the role that men had to take up of championing the search for knowledge when the fully conscious mind emerged was so psychologically crippling that they *also* contributed greatly to the corruption of the souls of each new generation of humans—so the unbearable 'guilt' from not being able to adequately nurture our children has not been confined to women, as this quote from the best-selling author of books for and about children, John Marsden, makes clear: **'The biggest crime you can commit in our society is to be a failure as a parent and people would rather admit to being an axe murderer than being a bad father or mother'** ('A Single Mum's Guide to Raising Boys', *Sunday Life*, *The Sun-Herald*, 7 Jul. 2002).

Chapter 4:8 The emergence of consciousness assisted the love-indoctrination process by allowing the sexual selection of integrativeness

303 To return now to the description of the biological origins of our moral instincts.

304 Given selfish competition for mating opportunities, particularly amongst males, is such a powerful force, the gender-role reversal apparent in bonobo society—which, as described, scientists are now beginning to confirm also occurred in the society of our ape ancestors—was an absolutely extraordinary achievement. In fact, this reversal from patriarchy to matriarchy is *so* extraordinary an achievement—especially the *speed* of its development, occurring over only some 1 million years since their split with chimpanzees, which is very fast in evolutionary terms—that

it must have been facilitated by some special factor. And indeed it was—it was helped along by the emergence of the most powerful tool of all for developing the order of matter on Earth: the conscious mind.

305 As mentioned earlier (in ch. 4:4), the reality is that developing love-indoctrination to the point where love or unconditional selflessness becomes instinctive is a very precarious process, akin to trying to swim upstream to an island in a fast flowing river—any difficulty or breakdown in the nurturing process and you are invariably 'swept back downstream' once more to the old competitive, selfish, each-for-his-own, opportunistic, 'animal condition' situation. So while love-indoctrination could allow unconditionally selfless behaviour to emerge in our ape ancestors, it was a very difficult, and also a very slow, process to both get underway and maintain. What the situation needed was a mechanism to assist, speed up and help maintain love-indoctrination's development of integration—assistance that came from the emergence of a conscious mind that enabled the conscious self-selection of integrativeness, especially the female sexual or mate selection of less competitive and aggressive, more integrative males with whom to mate. (As the title of his 1871 book, *The Descent of Man and Selection in Relation to Sex*, intimates, Charles Darwin actually suggested the role mate selection could have played in our human development, although he didn't understand its significance in the context of the love-indoctrination process.) Again, in 1983 I predicted that there must have been female sexual selection against male aggression in our ape ancestors as part of the love-indoctrination process—a prediction that the aforementioned reports of reduced canine size now confirm.

306 While the explanation for why humans became conscious while other animals haven't is the subject of chapter 6, an extremely brief account of it should be given here to support the forthcoming explanation as to how this conscious self-selection of integrativeness developed. In short, it was the love-indoctrination process that enabled consciousness to emerge, for while a mother's nurturing of her infants enabled unconditionally selfless behaviour to develop and, over time, become instinctual, this training in unconditional selflessness produced a *further* accidental by-product: it produced brains trained to think selflessly and thus truthfully and thus effectively and thus become 'conscious' of the relationship of events that occur through time. Other species that can't develop love-indoctrination and thus unconditional selflessness can't think truthfully and thus effectively because unconditional selflessness, which they are unable

to develop an orientation to, is the truthful theme or meaning of existence. As has been pointed out in each of the first three chapters in this book, you can't hope to think truthfully and thus effectively if you're lying. Species whose behaviour is governed by genetic selfishness have emerging minds that are, in effect, dishonestly orientated; their minds are alienated from the truth—they won't, in fact, allow selflessness-recognising, truthful and thus effective thinking—which means they can never make sense of experience and thus never become conscious. What all this means is that the human mind has been alienated from the truth *twice* in its history: firstly, in our pre-love-indoctrinated past when, like all other animals (except bonobos, who have almost completed the process of love-indoctrination), our brains were blocked from thinking truthfully; and, secondly, in our present state, where our minds have been alienated from the truth as a result of the human condition. Again, I emphasise that this is an extremely brief account of how love-indoctrination liberated consciousness, the much more complete explanation of which appears in chapter 6.

307 (Incidentally, people wonder how we can know that other species, like dolphins and elephants, aren't fully conscious like we humans are. As just explained, developing consciousness depends on overcoming the competitive, selfish 'animal condition' and becoming orientated to selflessness, so if you are still preoccupied with selfish, competitive dominance, as other animals are, you can't become fully conscious. Indeed, as all good animal trainers know, the key to managing and training non-human animal species, of both sexes, is to recognise that their great preoccupation is in achieving dominance whenever possible. Humans' deepest preoccupation, however, is with giving and receiving selfless love, a preoccupation we mistakenly project onto other animals, especially our pets, resulting in all the problems we have in managing them effectively. Animals are still victims of the 'animal condition', and so controlling them requires asserting dominance. If other animal species *had* achieved full consciousness they would not still be stranded in a world dictated by the rules of a selfish, competitive dominance hierarchy; they would be as preoccupied by giving and receiving love as we humans fundamentally are.)

308 And so it was an emerging conscious intellect that the love-indoctrination process had liberated that then began to support the development of selflessness and thus help maintain and speed up that all-precious process—because consciousness made it possible to recognise

the importance of selflessness and, having realised that, begin to actively select for it. Unconditional selflessness or love *is* the theme of existence, so to not be able to recognise that, as species that haven't been able to develop love-indoctrination and become conscious haven't been able to do, means they are, in effect, 'locked out' of the ideal state—that being the essential agony of the 'animal condition'. We can, therefore, appreciate how immensely relieving it must be to be liberated from that condition. For our ape ancestors to effectively be set free from the tyranny of the selfish gene-based natural selection process must have felt like heaven had opened up! Love could finally be indulged in! They had finally escaped the stupor of the animal condition and the opportunity and desire to then develop love, select for gentleness and kindness, would have been immense. (As will be explained in chapter 8, a similar opportunity to be transformed by being released into a world of love appears for humans now that we can finally escape the agony of the human condition!) It follows then that bonobos, who, as mentioned, have been able to rein in divisive aggressive behaviour, bring it under control, and develop love and achieve the extraordinary transition from patriarchy to matriarchy, have largely escaped the stupor of the animal condition and as a result are the most conscious, the most intelligent of all animals next to humans—as these quotes evidence:

'**Everything seems to indicate that Chim** [a bonobo] **was extremely intelligent. His surprising alertness and interest in things about him bore fruit in action, for he was constantly imitating the acts of his human companions and testing all objects. He rapidly profited by his experiences...Never have I seen man or beast take greater satisfaction in showing off than did little Chim. The contrast in intellectual qualities between him and his female companion** [a chimpanzee] **may briefly, if not entirely adequately, be described by the term "opposites"'** [p.248 of 278] **... Prince Chim seems to have been an intellectual genius. His remarkable alertness and quickness to learn were associated with a cheerful and happy disposition which made him the favorite of all** [p.255] **...Chim also was even-tempered and good-natured, always ready for a romp; he seldom resented by word or deed unintentional rough handling or mishap. Never was he known to exhibit jealousy...** [By contrast] **Panzee** [the chimpanzee] **could not be trusted in critical situations. Her resentment and anger were readily aroused and she was quick to give them expression with hands and teeth** [p.246]' (Robert M. Yerkes, *Almost Human*, 1925).

309 The pre-eminent ape language researcher Sue Savage-Rumbaugh reinforced this view when she wrote that '**Individuals who have had first hand interactive experience with both *Pan troglodytes*** [**chimpanzees**] **and *Pan***

paniscus **(Yerkes and Learned, 1925; Tratz and Heck, 1954) have been left with the distinct impression that pygmy chimpanzees** [bonobos] **are considerably more intelligent and more sociable than *Pan troglodytes*'** and that **'Each individual who has worked with both species in our** [ape language research] **lab is repeatedly surprised by their** [bonobos'] **communicative behavior and their comprehension of complex social contexts that are vastly different from anything seen among *Pan troglodytes*'** ('*Pan paniscus* and *Pan troglodytes*: Contrasts in Preverbal Communicative Competence', *The Pygmy Chimpanzee*, ed. Randall Susman, 1984, pp.396, 411-412 of 435). The bonobo that Savage-Rumbaugh has worked most with is Kanzi, and you can see in the photo of Kanzi that was included a few paragraphs back something of this **'intelligent' 'comprehension'** in his eyes. As the filmmaker Alain Tixier said about the intelligent gaze of bonobos, **'They're the only animals capable of creating the same "gaze" as a human. When you look at a bonobo you're taken aback because you can see behind the eyes it's not just curiosity, it's understanding. We see human beings in the eyes of the bonobo.'**

[310]The following extract from an article titled 'The Bonobo: "Newest" apes are teaching us about ourselves' also illustrates just how extraordinarily aware, cooperative, empathetic and intelligent bonobos are: **'Barbara Bell…a keeper/trainer for the Milwaukee County Zoo…works daily with the largest group of bonobos…in North America…"It's like being with 9 two and a half year olds all day," she** [Bell] **says. "They're extremely intelligent…They understand a couple of hundred words," she says. "They listen very attentively. And they'll often eavesdrop. If I'm discussing with the staff which bonobos (to) separate into smaller groups, if they like the plan, they'll line up in the order they just heard discussed. If they don't like the plan, they'll just line up the way they want." "They also love to tease me a lot," she says. "Like during training, if I were to ask for their left foot, they'll give me their right, and laugh and laugh and laugh. But what really blows me away is their ability to understand a situation entirely." For example, Kitty, the eldest female, is completely blind and hard of hearing. Sometimes she gets lost and confused. "They'll just pick her up and take her to where she needs to go," says Bell. "That's pretty amazing. Adults demonstrate tremendous compassion for each other"'** (*Chicago Tribune*, 11 Jun. 1998). More recently, the bonobo researcher Vanessa Woods described bonobos as **'the most intelligent of all the great apes'** ('Bonobos – our better nature', blogs.discovery.com, 21 Jun. 2010; see <www.wtmsources.com/134>).

[311]I should say that while some scientists suggest that chimpanzees are as intelligent as bonobos, there is an undeniable freedom in the mind of

a bonobo that is apparent in their capacity to be interested in the world around them, and in their empathy, compassion and even humour, as evidenced by the quotes above. While it *is* true that chimpanzees show a mental dexterity, it is a limited, opportunistic, self-interested mental focus (similar to a very primitive version of the alienated, deadened, self-centred, narrowly-focused minds of humans today), not the broad, free, open, curious, aware, all-sensitive, loving, truly thoughtful, conscious mind that bonobos have. It does have to be remembered that mechanistic science doesn't even have an interpretation of the word 'love', so we can't expect it to be capable of being interested in, or of acknowledging, the kind of open, curious, aware, all-sensitive and loving conscious mind that bonobos have. Indeed, the extraordinarily cooperative, unconditionally loving and truly aware characteristics of bonobos would motivate human-condition-avoiding, alienated mechanistic scientists to find a way to demote them at every opportunity, just as they found a way to demote the work of primatologist Dian Fossey, Sue Savage-Rumbaugh and my own work—something that will be described in chapter 5.

312 In terms then of being able to consciously favour more selfless and integrative behaviour in order to achieve greater overall selflessness and, with it, integration, the fastest, most effective way of doing so would be through selecting mates who are more selfless because that way you are eliminating selfishness at the fundamental level of your species' genetic make-up—which, as mentioned in pars 289-291, is what scientists studying the fossil record are now inferring took place: **'sexual selection played a primary role in canine reduction'**. Further, since it is the males who are most preoccupied with competing for mating opportunities, it is the females who are in the best position to implement this selection of less aggressive individuals with whom to mate. Significantly, primatologists—despite not recognising the process of love-indoctrination (and, therefore, the different degrees of success various primate species have had in developing it), or that love-indoctrination liberated consciousness that in turn allowed self-selection of integrativeness, especially through mate selection—*have* verified this female-driven selection of cooperative integrativeness amongst primates (the underlinings in this and some subsequent quotes are my emphasis): **'Male** [baboon] **newcomers also were generally the most dominant while long-term residents were the most subordinate, the most easily cowed. Yet in winning the receptive females and special foods, the subordinate, <u>unaggressive veterans got more than their fair share, the newcomers next to nothing. Socially</u>**

inept and often aggressive, newcomers made a poor job of initiating friendships' (Shirley Strum, *National Geographic*, Nov. 1987); and **'The high frequencies of intersexual association, grooming, and food sharing together with the low level of male-female aggression in pygmy chimpanzees** [bonobos] **may be a factor in male reproductive strategies. Tutin (1980) has demonstrated that a high degree of reproductive success for male common chimpanzees was correlated with male-female affiliative behaviours** [again, 'affiliative' being an evasive, denial-complying, mechanistic term meaning friendly/cohesive/social/loving/integrative]**. These included males spending more time with estrous females, grooming them, and sharing food with them'** (Alison & Noel Badrian, 'Social Organization of *Pan paniscus* in the Lomako Forest, Zaire'; *The Pygmy Chimpanzee*, ed. Randall Susman, 1984, p.343 of 435).

[313]While the observations made in the popular documentary series *Orangutan Island* (about the rehabilitation of a group of juvenile orphaned orangutans on a protected island in Borneo) probably can't be considered a product of rigorous scientific research, they do provide revealing footage and interesting commentary about orangutan behaviour. In one episode, an orangutan named Daisy, who is the dominant young female in the group (in the series she is described as 'Sheriff Daisy'), is seen strongly reprimanding another young female, Nadi, who, according to the narrator, repeatedly behaves selfishly: **'As usual, Daisy is keeping a watchful eye on all the action and she spies someone who is not playing fair—it's repeat offender Nadi who is refusing to share** [a jackfruit]**...Daisy decides it's her duty to step in...Daisy is refusing to allow Nadi anywhere near the eating platform because Nadi's been upsetting the order in this peaceful community...Daisy is making Nadi pay for her behaviour, so to avoid starving Nadi has no choice but to leave'** (*Animal Planet*, episode 'House of Cards', 2007). Whether or not it is an accurate interpretation of events, the footage appears to fully support the commentary—a female is seen to be maintaining **'the order in this peaceful community'**. (Further illustrations of strong-willed, female primates insisting on integrative behaviour will be provided shortly in ch. 4:11.)

Chapter 4:9 Sexual selection for integrativeness explains neoteny

[314]Yes, since it is the males who are the most preoccupied with competing for mating opportunities, the females must have been the first to select for selfless, cooperative integrativeness by favouring integrative rather than competitive and aggressive mates—and it was this process of the

conscious self-selection of integrativeness, especially the sexual selection of less aggressive males with whom to mate, that greatly helped love-indoctrination subdue the males' divisive competitiveness. Moreover, by seeking out less aggressive, more integrative mates, the females were, in essence, selecting those who have been most love-indoctrinated. This raises the next point to be explained, which is that the most love-indoctrinated and thus most integrative individuals will be those who have experienced a long infancy and exceptional nurturing and are closer to their memory of their love-indoctrinated infancy; specifically, those that are younger. The older individuals became, the more their infancy training in love wore off. In the case of <u>our ape ancestors, they began to recognise that the younger an individual, the more integrative he or she was likely to be</u> and, as a result, began to idolise, foster, favour and select for youthfulness because of its association with cooperative integration. The effect, over many generations, was to retard physical development so that adults became more infant-like in their appearance—which explains how we first came to regard <u>neotenous</u> (infant-like) features, like large eyes, dome forehead, snub nose and hairless skin, as beautiful and attractive.

315 Consider, for instance, the neotenous or infant-like large eyes of seal pups and frogs, and the large eye spot markings together with the soft, typically infant-like, moppish ears of giant pandas—they are what make these animals so 'appealing'. The following drawing of a panda depicted without its trademark spotted eyes and round ears, and with pricked ears and small eyes instead, shows just how quickly it loses its 'cute' appeal.

'Would we care if they weren't so cute? White out the black eye spots and give the ears points, and the panda loses much of its appeal'
Good Weekend, The Sydney Morning Herald, 23 Sep. 1989

[316]To indicate the effects of the love-indoctrination, mate selection neotenising process, I have compiled the following series of photographs of infant and adult non-human primates, specifically bonobos and chimpanzees.

[317]Firstly, the following three photographs, of an adult chimpanzee, an infant chimpanzee and an adult bonobo, show the similarity between the infant chimpanzee and the adult bonobo, indicating, as stated, the effects of the love-indoctrination, mate selection neotenising process.

Photographs by: Taronga Zoo; Michael Rayner; and Animals Animals

Adult chimpanzee; infant chimpanzee; and an adult bonobo

[318]In addition to their remarkably neotenous physical appearance, there is also a marked variance in features between individual bonobos (which is also apparent in the very different facial features of the two bonobos pictured standing upright earlier, after par. 296), suggesting the species is undergoing rapid change. This in turn suggests that the bonobo species has hit upon some opportunity that facilitates a rapid development, which evidence indicates is the ability to develop integration through love-indoctrination and mate selection.

[319]The photographs below of an infant and adult chimpanzee also show the greater resemblance humans have to the infant, again illustrating the effect of neoteny in human development.

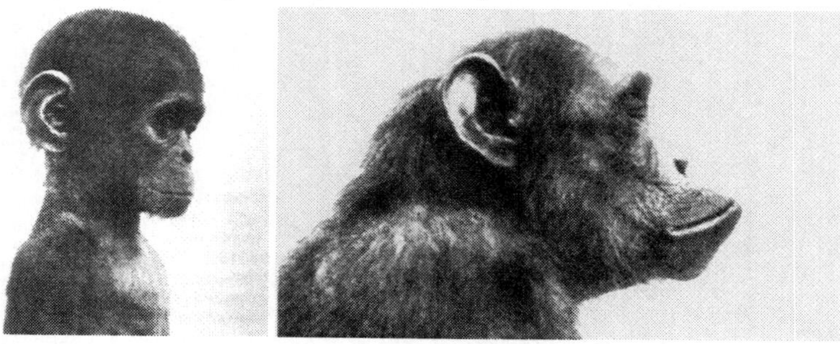

Infant and adult chimpanzee from *The Mismeasure of Man*, Stephen Jay Gould, 1981

[320] And finally, the following photograph of a chimpanzee foetus at seven months shows body hair on the scalp, eyebrows, borders of the eye lids, lips and chin—precisely those places where hair is predominantly retained in adult humans today, again illustrating the effect of neoteny or pedomorphosis in human development. ('Pedomorphosis' comes from the Greek *pais*, meaning 'child', and *morphosis*, meaning 'shaping'.) Clearly, humans are an extremely neotenised—love-indoctrinated—ape. Interestingly, a report on the fossilised remains of our 4.4-million-year-old ancestor *Ar. ramidus* describes it as having a **'short face and weak prognathism** [projecting jaw] **compared with the common chimpanzee'** (Gen Suwa et al, 'The *Ardipithecus ramidus* Skull and Its Implications for Hominid Origins', *Science*, 2009, Vol.326, No.5949), probably indicating that this physical retardation or neotenisation was well underway by this point in our ancestral history.

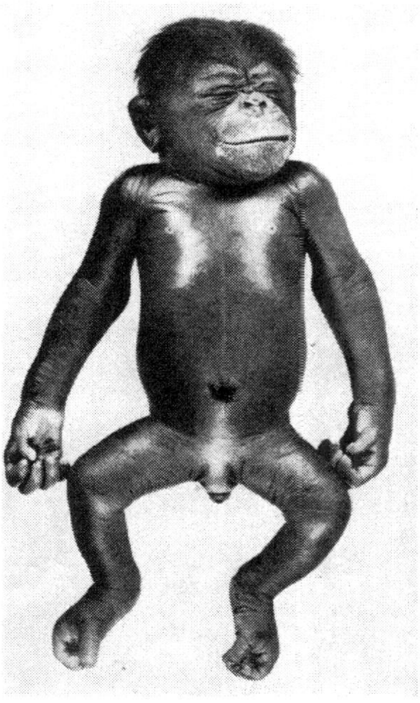

Chimpanzee foetus at seven months from *The Life of Primates*, Adolph H. Schultz, 1969

[321] So, we humans did learn to recognise that the older individuals became, the more their infancy training in love wore off and, therefore, the younger an individual, the more integrative he or she would likely be, and it was this selection for youthfulness that had the effect of retarding our development so that we became more infant-like in our appearance as adults. As stated, this was how we came to regard neotenous features—large eyes, dome forehead, snub nose and hairless skin—as attractive.

Chapter 4:10 The importance of nurturing in bonobo society

322 As an indication of how bonobos have been able to develop the nurturing love-indoctrination process more than chimpanzees, bonobo females have, on average, one offspring every five to six years and provide better maternal care than chimpanzees. Bonobos are born small, develop more slowly than other ape species, and stay in a state of infancy and total dependence for a comparatively long period of time—being weaned at about five years of age and remaining dependent on their mothers up until between the ages of seven and nine. Chimpanzees are weaned at about four years of age and remain dependent up to the age of six (on average). So infancy in bonobos can last up to five years, and in the case of humans it *does* last five years. Biologists have noted that an individual's lifespan is, amongst other factors, **'related to the postnatal development rate'** (Richard Cutler, 'Evolution of longevity in primates', *Journal of Human Evolution*, 1976, Vol.5, No.2), and so it is possible to deduce that the selection for a longer infancy period had the side effect of lengthening all the stages of maturation, which may explain how we humans acquired our comparatively long life span.

323 The primatologist Takayoshi Kano's long-running study of bonobos in their natural habitat in the Democratic Republic of the Congo has observed this prolonged infancy in practise, reporting that **'The long dependence of the son may be caused by the slow growth of the bonobo infant, which seems slower than in the chimpanzee. For example, even after one year of age, bonobo infants do not walk or climb much, and are very slow. The mothers keep them near. They start to play with others at about one and a half years, which is much later than in the chimpanzee. During this period, mothers are very attentive...Female juveniles gradually loosen their tie with the mother and travel further away from her than do her sons'** (Frans de Waal & Frans Lanting, *Bonobo: The Forgotten Ape*, 1997, p.60 of 210). As alluded to in this passage, the bond between mother and son is of particular significance in bonobo society where the son will maintain his connection with his mother for life and depend upon her for his social standing within the group. For example, the son of the society's dominant female, the strong matriarch who maintains social order, will rise in the ranks of the group, presumably to ensure the establishment and perpetuation of unaggressive, non-competitive, cooperative male characteristics, both learnt and genetic, within the group. Again, historically, it is the male primates

who have been particularly divisive in their aggressive competition to win mating opportunities and, therefore, the gender most needing of love-indoctrination—as this quote makes clear: **'Patient observation over many years convinced** [Takayoshi] **Kano that male bonobos bonded with their mothers for life. That contrasts with chimpanzee males who rarely have close contact with their mothers after they grow up, instead joining other males in never-ending tussles for dominance'** (Paul Raffaele, 'Bonobos: The apes who make love, not war', Last Tribes on Earth.com, 2003; see <www.wtmsources.com/143>).

324 The following quote (part of which has already been referred to), from Sue Savage-Rumbaugh and the writer Roger Lewin's book, *Kanzi: The Ape at the Brink of the Human Mind*, provides this further insight into bonobo society and its emphasis on nurturing: **'Bonobo life is centered around the offspring. Unlike what happens among common chimps, all members of the bonobo social group help with infant care and share food with infants. If you are a bonobo infant, you can do no wrong. This high regard for infants gives bonobo females a status that is not shared by common chimpanzee females, who must bear the burden of child care all alone. Bonobo females and their infants form the core of the group, with males invited in to the extent that they are cooperative and helpful. High-status males are those that are accepted by the females, and male aggression directed toward females is rare even though males are considerably stronger'** (1994, p.108 of 299).

325 *Also* **'Unlike what happens among common chimps'** is the fact that bonobos regularly share their food, and while chimpanzees restrict their plant-food intake to mainly fruit, bonobos eat leaves and plant pith as well as fruit, a diet more like that of gorillas. While bonobos have been known to capture and eat small game, including small monkeys, to supplement their diet with protein, they are not known to routinely hunt down and eat large animals such as colobus monkeys, like chimpanzees do, with **'hunting behavior** [by bonobos] **very rare'** (Tetsuya Sakamaki quoted by David Quammen, 'The Left Bank Ape', *National Geographic*, Mar. 2013). (In chapter 5:6 it is explained how mechanistic science has had to try to argue that bonobos are not gentle and peaceful because their cooperative behaviour serves as an unbearably exposing and confronting reminder of our now immensely angry, egocentric and alienated, unloving and unloved lives. For example, you will recall how, in par. 103, E.O. Wilson brazenly attempted to paint bonobos as indistinguishable from chimpanzees by stating that, like chimpanzees, bonobos **'do not share'**, and that bonobos **'hunt in coordinated packs in the same manner as chimpanzees'**. However, as I pointed out, even

the most basic research shows that Wilson had to fudge his evidence to support his lie—but this is how desperate resigned humans have been to find some excuse for our species' divisive behaviour.) Also, while physical violence is customary amongst chimpanzees it is rare among bonobos where, although the males are stronger, male aggression has been tamed and, unlike other great apes but like our ancestors, there is actually little difference in sexual size dimorphism between the male and female of the species. And like our ancestors *Sahelanthropus*, *Orrorin* and *Ardipithecus*, bonobos have reduced canine teeth, another indication they are less aggressive. The practice of infanticide, while not uncommon amongst chimpanzees, also appears to be non-existent within bonobo societies where even orphan bonobos are cared for by the group. In chimpanzee society orphans are occasionally adopted by a female but are not especially cared for by the group. Social groups of bonobos also have much greater stability than social groups of chimpanzees, with bonobos periodically coming together in large, harmonious, stable groups of up to 120 individuals—as Barbara Fruth's observation, included in par. 297, indicated: **'up to 100 bonobos at a time from several groups spend their night together. That would not be possible with chimpanzees because there would be brutal fighting between rival groups.'** The Japanese primatologist Genichi Idani similarly reported that **'In societies of African great apes other than bonobos, only antagonistic** [aggressive] **relationships have been reported between unit-groups...**[**Bonobo**] **neighborhood relationships...allow peaceful coexistence of two unit-groups** [and] **provide the first nonhuman primate community model which is close to...human society'** ('Relations between Unit-Groups of Bonobos at Wamba, Zaire: Encounters and Temporary Fusions', *African Study Monographs*, 1990, Vol.11, No.3). So, bonobos are much gentler than their chimpanzee cousins; they are relatively placid, peaceful and egalitarian, exhibiting a remarkable sensitivity to others—and not just towards their own kind, as will be evidenced shortly with an account of how a bonobo cared for a stunned bird.

Chapter 4:11 The importance of strong-willed females in developing integration

326 The following extract from the 1995 *National Geographic* documentary *The New Chimpanzees* provides a good example of the important role a strong matriarchy plays in the maintenance of cooperative, integrative behaviour. To quote from the narration: **'An impressively stern** [bonobo]

female enters and snaps a young sapling. Once she picks herself up she does something [that would be] **entirely surprising for a female chimp, she displays** [the female is shown assertively dragging the sapling through the group]**, and the males give her sway** [a male is shown cowering out of her way]**. For this is the confident stride of the group's leader, its alpha female, whom** [Takayoshi] **Kano has named Harloo.'**

[327] While this example is specific to bonobos, those who have studied primates will typically tell you of the presence in their study group of an extraordinarily self-assured, secure-in-self and strong-willed female. This is because all primates are trying to develop the nurturing of integrativeness but only our ancestors and the bonobos have had the right conditions to actually achieve it. During a trip to Kenya in 1992, my partner Annie Williams and I were invited to visit the anthropologist Shirley Strum's 'Pumphouse Gang' study troop of baboons, which had been made famous through numerous articles in *National Geographic* magazine. During our visit I noticed that Strum kept on her desk the skull of a baboon named Peggy. Displaying a skull is a bit macabre but Strum said she did so in memory of Peggy who was an extraordinarily confident, strong-willed, authoritative, charismatic individual who successfully led the 'Pumphouse Gang' for many years. As Strum has written: **'She** [Peggy] **was the highest-ranking female in the troop, and her presence often turned the tide in favor of the animal she sponsored. While every adult male outranked her by sheer size and physical strength, she exerted considerable social pressure on each member of the troop. Her family also outranked all the others...another reason for the contentment in this particular family was Peggy's personality. She was a strong, calm, social animal, self-assured yet not pushy, forceful yet not tyrannical'** (*Almost Human: a journey into the world of baboons*, 1987, pp.38-39 of 294).

From the author's 1992 visit to Africa. L to R: the author with Dr Shirley Strum's Chololo Ranch Baboon Project study group in Kenya; with Dr Susanne Abildgaard and young chimpanzee at Jane Goodall's Chimpanzee Rehabilitation Centre in Burundi; and with the Susa Mountain Gorilla study group in Rwanda.

328 In her famous studies of gorillas, which she documented in her 1983 book, *Gorillas in the Mist*, the primatologist Dian Fossey described the strength of character that had to be developed to curtail male aggression, and the centred, security of self needed to be an effective, love-indoctrinating mother, when she wrote about how **'Old Goat'** was **'an exemplary parent'** and that, as a result, her son **'Tiger'** was **'a contented and well adjusted individual'**. While gorillas have not been able to develop love-indoctrination to the same degree as bonobos—seemingly because they have lacked the necessary ideal nursery conditions—denial-free, honest studies of their behaviour, notably Fossey's, have revealed the strong relationship between nurturing and integrativeness that is the love-indoctrination process. The following extracts from *Gorillas in the Mist* reveal more about Old Goat's nurturing of Tiger, as well as the extreme fragility of the love-indoctrination process and how any disruption to it would result in a regression back to the competitive, each-for-his-own, opportunistic, divisive, 'animal condition' existence: **'Like human mothers, gorilla mothers show a great variation in the treatment of their offspring. The contrasts were particularly marked between** [the gorilla mothers] **Old Goat and Flossie. Flossie was very casual in the handling, grooming, and support of both of her infants, whereas Old Goat was an exemplary parent'** (p.174 of 282). The effect of Old Goat's **'exemplary parenting'** of Tiger is apparent in the following extract: **'Like Digit, Tiger also was taking his place in Group 4's growing cohesiveness. By the age of five, Tiger was surrounded by playmates his own age, a loving mother, and a protective group leader. He was a contented and well-adjusted individual whose zest for living was almost contagious for the other animals of his group. His sense of well-being was often expressed by a characteristic facial "grimace"'** (pp.186-187). The **'growing cohesiveness'** (developing integration) brought about by **'loving mother**[s]**, and a protective group leader'** is love-indoctrination.

329 Incidentally, with regard to the **'protective group leader**[s]**'**, namely the male silverback gorillas, their large size is not only due to having to compete for dominance but also reflects that while bonobos depend on the safety of trees for a secure, threat-free environment, gorillas evidently selected for physical size and great strength, particularly in the males, to protect their groups from external, predatory threats—as the anthropologist Adolph H. Schultz noted, the adult male gorilla **'is a remarkably peaceful creature, using its incredible strength merely in self-defence'** (*The Life of Primates*, 1969, p.34 of 281).

[330]Fossey's account of the love-indoctrinated Tiger later in life illustrates how nurtured love is critical in producing and maintaining the integrated group. In it she describes how the secure, integrative, loving Tiger tried to maintain integration or love in the presence of an aggressive, divisive gorilla after the group's integrative silverback leader, Uncle Bert, was shot by poachers: **'The newly orphaned Kweli, deprived of his mother, Macho, and his father, Uncle Bert, and bearing a bullet wound himself, came to rely only on Tiger for grooming the wound, cuddling, and sharing warmth in nightly nests. Wearing concerned facial expressions, Tiger stayed near the three-year-old, responding to his cries with comforting belch vocalizations. As Group 4's new young leader, Tiger regulated the animals' feeding and travel pace whenever Kweli fell behind. Despondency alone seemed to pose the most critical threat to Kweli's survival during August 1978. Beetsme...was a significant menace to what remained of Group 4's solidarity. The immigrant, approximately two years older than Tiger and finding himself the oldest male within the group led by a younger animal, quickly developed an unruly desire to dominate. Although still sexually immature, Beetsme took advantage of his age and size to begin severely tormenting old Flossie three days after Uncle Bert's death. Beetsme's aggression was particularly threatening to Uncle Bert's last offspring, Frito** [son of Flossie]. **By killing Frito, Beetsme would be destroying an infant sired by a competitor, and Flossie would again become fertile. Neither young Tiger nor the aging female was any match against Beetsme. Twenty-two days after Uncle Bert's killing, Beetsme succeeded in killing fifty-four-day-old Frito even with the unfailing efforts of Tiger and the other Group 4 members to defend the mother and infant...Frito's death provided more evidence, however indirect, of the devastation poachers create by killing the leader of a gorilla group. Two days after Frito's death Flossie was observed soliciting copulations from Beetsme, not for sexual or even reproductive reasons—she had not yet returned to cyclicity and Beetsme still was sexually immature. Undoubtedly her invitations were conciliatory measures aimed at reducing his continuing physical harassment. I found myself strongly disliking Beetsme as I watched his discord destroy what remained of all that Uncle Bert had succeeded in creating and defending over the past ten years...I also became increasingly concerned about Kweli, who had been, only a few months previously, Group 4's most vivacious and frolicsome infant. The three-year-old's lethargy and depression were increasing daily even though Tiger tried to be both mother and father to the orphan. Three months following his gunshot wound and the loss of both parents, Kweli gave up the will to survive...It was difficult to think of Beetsme as an integral member of**

Group 4 because of his continual abuse of the others in futile efforts to establish domination, particularly over the indomitable Tiger...Tiger helped maintain cohesiveness by "mothering" Titus and subduing Beetsme's rowdiness. Because of Tiger's influence and the immaturity of all three males, they remained together' (*Gorillas in the Mist*, pp.218-221).

331 It is clear from this account how very easily any disruption to the love-indoctrination process can cause a regression to the competitive, opportunistic, each-for-his-own, pre-love-indoctrination, 'animal condition' situation—and thus how utterly incredible it is that the bonobos have been able to overcome the agony of the 'animal condition' and be as free as they are from the tyranny of the selfish gene-based natural selection process.

Chapter 4:12 Descriptions of bonobos provide an extraordinary insight into what life for our human ancestors was like before the emergence of the human condition

332 Indeed, the following selection of quotes provides further insight into how extraordinarily integratively orientated bonobos have become and, through that stunning evidence, just how wonderful our species' time in the innocent, cooperatively behaved, loving, 'Garden of Eden', fully integrated, 'heavenly' 'Golden Age' must have been. I doubt you will find a better clue to our glorious past, and now future, than what you are about to read.

333 Firstly, the following extract, which has already been referred to, demonstrates how extraordinarily sensitive, cooperative, loving and intelligent bonobos are, as well as how few exist in captivity: **'Barbara Bell...a keeper/trainer for the Milwaukee County Zoo...works daily with the largest group of bonobos...in North America..."It's like being with 9 two and a half year olds all day," she** [Bell] **says. "They're extremely intelligent... They understand a couple of hundred words," she says. "They listen very attentively. And they'll often eavesdrop. If I'm discussing with the staff which bonobos (to) separate into smaller groups, if they like the plan, they'll line up in the order they just heard discussed. If they don't like the plan, they'll just line up the way they want." "They also love to tease me a lot," she says. "Like during training, if I were to ask for their left foot, they'll give me their right, and laugh and laugh and laugh. But what really blows me away is**

their ability to understand a situation entirely." For example, Kitty, the eldest female, is completely blind and hard of hearing. Sometimes she gets lost and confused. "They'll just pick her up and take her to where she needs to go," says Bell. "That's pretty amazing. Adults demonstrate tremendous compassion for each other".'

[334] The intensity of personal relationships within bonobo society is also beautifully apparent in the following anecdote from the aforementioned book, *Kanzi: The Ape at the Brink of the Human Mind*, in which Sue Savage-Rumbaugh recounts the extreme elation and affection shown by her famous bonobo research subject, the young adult male Kanzi, when reunited with his mother, Matata, after a number of months apart (a picture of Matata and Kanzi appears at the beginning of this chapter): **'I sat down with him** [Kanzi] **and told him there was a *surprise* in the colony room. He began to vocalize in the way he does when expecting a favored food—"eeeh....eeeh....eeeh." I said, *No food surprise. Matata surprise; Matata in colony room.* He looked stunned, stared at me intently, and then ran to the colony room door, gesturing urgently for me to open it. When mother and son saw each other, they emitted earsplitting shrieks of excitement and joy and rushed to the wire that separated them. They both pushed their hands through the wire, to touch the other as best they could. Witnessing this display of emotion, I hadn't the heart to keep them apart any longer, and opened the connecting door. Kanzi leapt into Matata's arms, and they screamed and hugged for fully five minutes, and then stepped back to gaze at each other in happiness. They then played like children, laughing all the time as only bonobos can'** (pp.143-144).

[335] And, as referred to earlier, such displays of sensitivity are not limited to their own kind, as this description from *Bonobo: The Forgotten Ape*, by the primatologist Frans de Waal and photographer Frans Lanting, reveals: **'Betty Walsh, a seasoned animal caretaker, observed the following incident involving a seven-year-old female bonobo named Kuni at Twycross Zoo in England. One day, Kuni captured a starling. Out of fear that she might molest the stunned bird, which appeared undamaged, the keeper urged the ape to let it go. Perhaps because of this encouragement, Kuni took the bird outside and gently set it onto its feet, the right way up, where it stayed, looking petrified. When it didn't move, Kuni threw it a little way, but it just fluttered. Not satisfied, Kuni picked up the starling with one hand and climbed to the highest point of the highest tree...She then carefully unfolded its wings**

and spread them wide open, one wing in each hand, before throwing the bird as hard as she could [to free it]**'** (1997, p.156 of 210). The situation reminds one of the compassion extended to Patrick Bleuzen, the animal advisor on the aforementioned French documentary, *Bonobos*, who was comforted by a bonobo after he was hit on the head by a falling branch during production.

336 This next account (which is very reminiscent of the earlier mentioned orangutan, Sheriff Daisy, and her maintenance of cooperative order) reveals something of bonobos' strong moral sense: **'When a maintenance man was working on the heating and cooling system at the Iowa Primate Learning Sanctuary, Kanzi the bonobo didn't like it. So what did the bonobo do? He went to his computer and used his words to tell a researcher to put on a monkey suit and chase after the man. Julie Gilmore, a veterinarian at the sanctuary, said bonobos like to know if everyone is getting along, who is in charge and whether everyone is following the rules. Gilmore said that when the worker came, Kanzi was convinced he wasn't following the rules. She said the worker was luckily a good sport about the whole thing and allowed the monkey suit chasing to commence. The bonobo was so happy that the man was chased away, he watched a video of the action about 100 times. That's just one example of how the bonobos at the sanctuary communicate with the researchers…** [At the research sanctuary] **Sue Savage-Rumbaugh…taught Kanzi language using a chart of symbols…She was able to train Kanzi to know what word each symbol represented. Kanzi, who is now 33, now understands thousands of words and grammar'** ('Talking to another species', *The Des Moines Register*, 21 Nov. 2013). It's worth noting that the big difference between Sheriff Daisy and Kanzi is that Daisy was a juvenile orangutan. Because of the scarcity of food where orangutans live in the rainforests of Borneo and Sumatra they are forced to live competitively as adults; in effect, they are thrown out of the integrative, loving state they are able to experience when they are young. Kanzi, however, is a fully grown, 33-year-old adult and still expects to be living in an integrative, cooperative, loving state. Such is the ideal love-indoctrinating nursery conditions of the world from which bonobos come.

Chapter 4:13 Milwaukee County Zoo's fabulous group of bonobos

The author (left) and Professor Harry Prosen
at the Milwaukee County Zoo bonobo exhibit

337 On 3 February 2014 I had the incredible good fortune to spend a day with one of the largest and most socially authentic captive breeding groups of bonobos in the world at the Milwaukee County Zoo in Wisconsin in the USA. Visiting the bonobos in their native habitat in the heart of the Congo basin is almost impossible because of the region's remoteness and impenetrable jungle, as well as the horrific civil war and economic collapse that is ravaging the Democratic Republic of the Congo. And even those who *do* manage to travel to where the bonobos are found only very occasionally get to glimpse them. So being able to spend hours right beside such a large group of these animals who are living virtually as we humans once did during our original pre-conscious, pre-human-condition-afflicted, innocent, loving, harmonious, cooperative state was so wonderful and precious it was almost beyond description. And since those who work with such a group on a regular basis surely know more about the lives of bonobos than anyone else, how lucky was I to be accompanied on my visit by one of those people, and someone who I

strongly believe must be the most gifted and accomplished psychiatrist in the world today—Professor Harry Prosen, who provided the Introduction to this book. Harry arranged my visit to the Zoo, whose staff most generously, and for the first time ever, brought almost the whole group into the display area for us to see—a group that numbered 16 of their 23 bonobos (including all the females and infants), which, incidentally, is a significant portion of the world's captive population of only some 250 bonobos (in 2014).

338 A lifetime of successfully treating humans led Harry to identify what he refers to as **'an empathy deficit'**—which is an acceptable but, in truth, evasive, denial-complying, mechanistic scientific term for 'lack of love during upbringing'—as being the main cause of humans' psychological problems. Indeed, Harry has spent a lifetime studying the effects of empathy/love, and the lack thereof, in not only humans but other species, including bonobos, who he has recognised as the most empathetic/loving of all non-human species. Three examples illustrate just how honest Harry is able to be about the need for empathy/love/unconditional selflessness in the lives of highly social mammals and how psychologically devastating the effects are of not receiving it—and thus how truly effective a psychiatrist he is. In the first example, Harry was asked to assist, via a video link, a girl in a mental institution who was completely uncommunicative, virtually mute. After Harry talked to her she finally mumbled the word 'Mary', to which Harry intuitively replied, 'Mary Magdalene?' When the girl seemed to respond, Harry added, 'Wasn't she a prostitute?', which reached into the girl's core psychosis and from there she was able to open up and heal. In another example, Harry was asked to help in a situation at a zoo where an elephant had died, leaving both its elephant companion and human keeper catatonic with grief. Harry intuitively moved the catatonic keeper in front of the catatonic elephant, which prompted the elephant to put its trunk on the shoulder of the keeper, and that sharing of feeling, of empathy, of compassion, between the two freed them both from their catatonic states. And in a case that was reported around the world, the Milwaukee County Zoo sought Harry's assistance with an adult bonobo named Brian, whose **'infancy had diverged from the normal pattern for bonobos in which mother and baby are in constant contact'** (Jo Sandin, *Bonobos: Encounters in Empathy*, 2007, p.48 of 109). Harry ascertained that Brian's extremely distressed behaviour, including induced vomiting and

self-mutilation, **'rose from desperate, but futile, attempts to calm himself in the midst of extreme anxiety'**, and **'prescribed surrounding Brian with a world that was absolutely predictable'**—a calming **'therapeutic environment'** (ibid) that finally allowed Brian's healing to begin. Little wonder Harry is psychiatric consultant to the Bonobo Species Preservation Society. And little wonder he was so enthralled when, in 2004, he read my love-indoctrination explanation for humans' and bonobos' moral nature in a copy of the *Human Condition Documentary Proposal* (<www.worldtransformation.com/doco>) that we had sent to scientists around the world, especially those working with bonobos. Indeed, Harry was so excited by the proposal, in which I outlined all the explanations of the human condition, of Integrative Meaning and of the nurturing origins of our moral nature and conscious mind that are presented in this book, he wrote to us saying, **'The Proposal is never out of my brief case. I read it over and over and it is I think one of the most astonishing and outstanding things of our time. It is a gift and I hope you don't give an inch to your detractors'** (WTM records, 17 Jun. 2005). In fact, Harry became so supportive of my work and so concerned about the attacks that were being made on it that, as he described in his Introduction to this book, he came to Australia in 2007 to give evidence in support of my work in a defamation trial. Harry has certainly become a precious friend on this most difficult of tasks of delivering understanding of the human condition to the world. Indeed, Harry's love and support for my work makes me wish I had been able to share these explanations with the anthropologist Ashley Montagu (1905-1999) because they too would have made complete sense of all that he was so honestly able to recognise about the importance of love in human life—as demonstrated by his amazingly truthful paper, 'A Scientist Looks at Love', which is referred to in par. 369. Yes, since Harry has often said, **'All the great theories that I have encountered in my lifetime of studies of psychiatry can be accounted for under your explanations of human origins and behaviour'**, I'm sure the very rare honesty of Montagu's thinking would have meant that he would have also become a very valuable ally in trying to deliver these liberating but at the same time confronting explanations to the world.

339 This issue of the confronting nature of my explanations of the origin and nature of human behaviour, and resulting extreme resistance to it that led to the vicious attacks upon it and the biggest defamation case in Australia's history, became all too apparent to Harry when he

tried valiantly to have my nurturing explanation that so fully accounts for human and bonobo moral behaviour referred to in Jo Sandin's 2007 book about the bonobos at the Milwaukee County Zoo, *Bonobos: Encounters in Empathy* (which was quoted from above), which had been commissioned by the Zoological Society of Milwaukee. In chapter 5 of her book Sandin did manage to include oblique references to the love-indoctrination explanation, such as **'there is evidence from the work of evolutionary biologists to suggest that it was the distant ancestors of bonobos that brought empathy into the evolutionary process'** and **'certain life experiences are necessary for a bonobo to develop into an empathetic and wise adult'**, and **'the nurturing care Lody received (and remembered) constituted a key element in the bonobo's maturing into an empathetic and wise male'** (p.52), but despite all the respect Harry has at the Zoo for being such an effective psychiatric consultant for all their social animals, and despite wanting to, she was unable to include the following comment that Harry asked be added at the end of chapter 5 of her book: **'In summary, to the fascinating and indeed fundamental question for biologists of how the extraordinary empathy and even altruism we are observing amongst bonobos developed, our observations point to nurturing, maternalism and associated matriarchy as key influences. Certainly our observations do appear to be confirming of the nurturing explanation for empathy and true altruism that was first put forward by the American philosopher John Fiske in 1874** [Fiske's work will be described in chapter 5:3] **and, more recently, by Australian biologist Jeremy Griffith in his various books, in particular in Part 2 of his 2004 *Human Condition Documentary Proposal.*'** My work was mentioned in the concluding chapter of Sandin's book, but the key reference to **'the nurturing explanation for empathy and altruism'** wasn't included. I have to emphasise that failure to acknowledge the nurturing explanation for humans' and bonobos' moral nature is consistent with the practice of human-condition-avoiding, denial-complying mechanistic science the world over. The next chapter in this book (5) will describe at length the immense resistance to the nurturing explanation, as well as the mountain of nurturing-avoiding, dishonest explanations for the moral nature of humans and bonobos that have been put forward in its place. For example, it will be mentioned there how in 2005 I submitted an abstract of a paper titled 'Nurturing as the Prime Mover in Primate Development and Human Origins' for presentation at the International Primatological Society's (IPS) 2006 Congress in Uganda (the submission and subsequent correspondence can be viewed at <www.worldtransformation.

com/ips-2006-congress>), only to be rejected on the grounds that **'Both reviewers felt this abstract presents no data nor a testable hypothesis'**. Despite arguing that my nurturing, love-indoctrination explanation for humans' moral instincts **'contains a great deal of supportive evidence in the form of many summaries of data-supported studies of bonobos and other primates by leading primatologists'**, and **'is an entirely testable, validatable hypothesis, as the evidence just described about bonobos shows'**, and submitting this protest to the President and 38 members of the IPS Congress Committee, the rejection was upheld! Humans have certainly not wanted to face the truth of the importance of nurturing in their lives; again, they **'would rather admit to being an axe murderer than being a bad father or mother'**!

The author (left) with Trish Khan, Jan Rafert and Professor Harry Prosen

340 To return, however, to my visit to the wonderfully cared for bonobos at the Milwaukee County Zoo, which, next to a visit my partner Annie and I made to the Tiva River in Africa, which I describe in par. 658, is the highlight of my life. Having thought and written about bonobos for so long it was an incredibly exciting moment for me to finally meet them—and what I saw exceeded my expectations about their cooperative, integrative and intelligent nature. The Zoo's curator of primates and small mammals, Jan Rafert, and primate area supervisor, Trish Khan, were unaware of the nurturing, love-indoctrination explanation for bonobos' moral behaviour, suggesting that bonobos' secure, food rich environment was what allowed for selection for cooperation, with Jan saying these ecological conditions allowed bonobos to **'not only take care of their own family but**

take care of each other'; and Trish suggesting that **'there is just no competition to produce offspring because** [female bonobos] **are copulating throughout their entire cycle...that's the big theory, isn't it'**. These suggestions are, in effect, aspects of the Social Ecological Model that is discussed and dismantled in chapter 5:8. Jan and Trish did, however, confirm to me all the bonobo behaviours that I have already described: their cooperative care for each other; their nurturing love; their sensitivity; how at being reunited after years of separation they are, as Jan described it, **'beside themselves with glee'**; how, according to Trish, **'their intelligent, understanding, ingenious minds so tests the wits of the keepers that the keepers are often exhausted by the end of the day'**; their constant vocal communication, including specific vocalisations for the different keepers; their tendency towards bipedalism; their use of sex to create and cement bonds and relieve tension; their robust sense of humour and laughter; the strong, forceful personalities of the group's dominant females; the lifelong bond between mother and son; the absence of the territorial wars that occur between groups of chimpanzees (because bonobos tolerate other groups of bonobos—they are, in essence, all one large group, a Specie Individual); and, above all for me, the confirmation of their moral sense, their constant maintenance of social order. For example, Jan emphasised that when young males within bonobo society begin to become competitive and aggressive like adolescent males of other species **'The group will tolerate it for only so long before everyone comes down on it very forcefully'** to prevent such behaviour developing. Yes, as was stated about Kanzi's situation, **'bonobos like to know if everyone is...following the** [social] **rules'** that achieve and maintain integration.

341 With all other highly social species that I have observed, such as meerkats, I have had the feeling that their groups are still more a collection of individuals, but that wasn't the case with the bonobos. With them it was as though they were all part of one organism, all deeply aware of and in tune with each other. While there were outbursts of anxiety and occasional tensions and even fights amongst them—because bonobos haven't as yet completed the love-indoctrination process that enables the fully cooperative, utterly harmonious, completely integrated state to develop—there was a high degree of harmony in the group; to such an extent, in fact, that there was an overall togetherness, a real peacefulness and tranquility, a unity, a security, an each-knew-all-about-and-trusted-and-supported-and-loved-everyone-else-feeling between them. It was

like they took each other for granted, just as we take our arms or legs or ears for granted because they are simply part of our whole being. It was like they all felt they belonged and were part of something bigger; it wasn't like they had to trust that this was the case, it was that they *knew* it was the case. As is very apparent in some of the marvellous footage WTM Founding Member James Press took of the bonobos for inclusion in Harry's video introduction to this book that appears on our WTM website (see <www.worldtransformation.com/prosen-intro>), when one of them looked at you with the extraordinary awareness and thoughtfulness that their facial expression exhibits, it was as though that individual was fully connected by an invisible cable or link to all the other members of their group behind them and around them; that at that moment they were the looking-out-at-the-world component of the whole group. Yes, they all moved about together like one big roly-poly organism that would suddenly materialise in front of you, and then, just as suddenly, vanish all together into another corner of their enclosure. I derived a deep, calming, reassuring, and even happy, life-as-it-should-be, feeling from being with them, which was quite amazing. To use some of the extraordinarily honest words from Sir Laurens van der Post that were included in par. 80, I felt part of a time when **'All on earth and in the universe were still members and family of the early race seeking comfort and warmth through the long, cold night before the dawning of individual consciousness in a togetherness which still gnaws like an unappeasable homesickness at the base of the human heart.'**

342 It is just such a tragedy that the phenomenal integration and emerging conscious awareness that exists amongst bonobos has not been able to be admitted. There was no signage on their enclosure indicating their incredible cooperative, loving, nurturing nature and amazing presence of mind. They were just another variety of animal for the public to see. No one was told that what you are seeing here is a species living on the threshold of the completely integrated, heavenly, Garden-of-Eden state of **'togetherness'** we humans once lived in. What was on display was one of the most amazing and special sights that can be seen on planet Earth, yet no one was being made aware of it. In fact, as I discuss in chapter 5, I think the nurturing, integrative, near-conscious nature of bonobos is so confronting that there is a very real inclination to try to dismiss them as nothing special, emphasise that they do have fights, etc, etc—and even block them out of our awareness, which *has* occurred to the extent that

they have been described as **'the forgotten ape'** (Frans de Waal & Frans Lanting, *Bonobos: The Forgotten Ape*, 1997).

343 As I said, those who work on a regular basis with this large, socially authentic group of bonobos at the Milwaukee County Zoo surely know as much or even more about bonobo life than anyone else in the world, and since Sandin's book *Bonobos: Encounters in Empathy* wonderfully documents the lives and interactions over many years of this large group of bonobos (as supplied by their keepers at the Milwaukee County Zoo and Harry) it has to be one of the most informative books on bonobo behaviour. Available through Amazon, this book evidences all the interactions and behaviours of bonobos that have been mentioned in this chapter. It contains, for example, references to bonobos' **'Natural nurturing patterns** [that]**...has yielded mother-reared youngsters as socially robust as they are physically healthy'** (p.43), and recounts how **'Bonobo babies born at the Milwaukee County Zoo immediately cling to their mothers. They are carried constantly. Infants fostered here are greeted with enthusiasm and cuddled by males and females'** (p.66). I should mention that it also describes **'individuals whose life history** [of missing out on nurtured love] **has left them deficient in bonobo social graces...so-called problem bonobos'** (p.26), and how the group enforces **'the community's code of conduct'** (p.28) on such **'rule-breaking'** individuals (p.29), which is all *extremely* confronting for the human race today where almost everyone's **'life history** [of not being adequately nurtured in their infancy and childhood] **has left them deficient'**, **'problem'** humans. As will be described at length in chapter 5, it is little wonder there has been resistance to acknowledgment of the importance of nurturing in the lives of both bonobos and humans.

344 The following paragraph, from Sandin's book, in particular reveals the whole love-indoctrination process at work: **'We see Maringa, for almost two decades the undisputed empress of the group, in full diva mode, throwing tantrums and slyly peeking to note their effect. We glimpse the intricate etiquette that orders relationships among group members. We witness the matriarch and her "sisters" in power administering swift and painful rebukes in response to infractions of the colony's behavior code. We view Maringa's own slide from power. Here we observe the philosopher-king Lody, long the dominant male, ruling by empathy and wisdom, wielding power but always within the context of a matriarchal society. We watch younger males challenge his pre-eminence as he acknowledges his gradual physical decline. We experience tender moments between mothers and children, mischievous testing of the patience of tribal**

elders and compassionate interventions on behalf of weaker bonobos' (p.60). The **'behavior code'** among bonobos is a code that is dedicated to maintaining cooperative order, integration. While bonobos have not yet completed the love-indoctrination process, the **'code of conduct'** or **'rules'** or **'intricate etiquette'** represents the final stages of the behavioural management needed to develop the fully integrated, all-loving and all-sensitive state that became instinctive in our distant human forebears.

345 With regard to the **'compassionate interventions on behalf of weaker bonobos'**, this is but one of many illustrations from the book: **'Dr. Prosen has been particularly impressed by the attention given Brian by Lody, the group's dominant male. When Brian seemed too panicky to move, Lody often would take his hand and walk him into a different area such as the playroom or the outdoor yard. It was in Lody's company that Brian first sat down and ate with a group, a "very big deal for him," recorded in animal-management notes August 30, 1999. Regularly, Lody would postpone his own meals to sit with and comfort Brian when the younger bonobo was having difficulties...In Lody, Dr. Prosen said, the presence of weakness seems to activate his own compassionate strength'** (pp.49-50). Also, **'Harry Prosen...and keeper Barbara Bell credit Lody's wise leadership with the high level of emotional health of Milwaukee's bonobos and points to numerous instances of empathetic behaviour'** (p 14), and **'His** [Lody's] **days of mourning after the death of Kidogo constitute one of the most poignant chapters of the community's life'** (p.14).

346 I know that all of nature is under dire threat from humans' now horrifically psychologically upset state, but to lose bonobos—who are categorised by the International Union for Conservation of Nature, the world's main authority on the conservation status of species, as a species **'facing a very high risk of extinction in the wild'**—is an unbearable prospect. The very latest reports indicate that the bushmeat trade (where wild animals including bonobos are killed for meat) has now penetrated into even the most remote areas of the Congo basin where the bonobos live. Now that the human condition has been explained and we can at last admit that the bonobos are living in this delicate nursery state of nurturing love and, as a result, are highly sensitive, loving and aware, killing them off is a terrible, terrible case of 'the massacre of the innocents'. It is *truly* unbearable to think about—not to mention what we lose in terms of the astonishing and irreplaceable evidence they provide of our origins. So please God, let these ameliorating, peace-bringing understandings of our human lives get out to the world as soon as possible. (I should reiterate here that when I talk

of God, I am not invoking an abstract, metaphysical, supernatural being, I am referring to the all-pervading integrative power in our world.)

347 This concern for bonobos in their home in the Congo raises another aspect of the absolutely wonderful work that is being carried out by those associated with Milwaukee County Zoo. Dr Gay Reinartz, the director of the Bonobo & Congo Biodiversity Initiative (BCBI) for the Zoological Society of Milwaukee and coordinator of the Bonobo Species Survival Plan for the Association of Zoos and Aquariums, has since 1997 been travelling at least twice a year to the remote and dangerous Congo, even during its horrific civil war, to help protect the bonobos. What started out as a research project to discover whether Salonga National Park, which forms the heart of the bonobos' range, harbours a self-sustaining bonobo population and what habitats they prefer, has evolved into a multi-faceted conservation program that is **'the center of what is now an international effort to protect bonobos in the heart of their natural home'** (*Bonobos: Encounters in Empathy*, p.75). In 2011, the BCBI program received the international conservation award from the Association of Zoos and Aquariums, the highest recognition for an international conservation program.

348 So, bonobos provide an extraordinary insight into what life for our human ancestors was like, graphically illustrating the loving, cooperative, nurturing environment our ape ancestors must have been blessed with to complete the love-indoctrination process and develop the unconditionally selfless moral instincts we have within us. The fossil evidence indicates these ideal 'nursery' conditions occurred somewhere in Africa, just exactly where is not yet clear, but, as we intuitively recognise, Africa *was* **'the cradle of mankind'**.

Chapter 4:14 'A golden race...formed on Earth'

349 In all, love-indoctrination was an incredibly fortuitous development. It gave us humans our unconditionally selfless, integrative-behaviour-orientated, Specie-Individual, humanity-creating, **'awe'**-inspiring, **'distinct'** from other **'animals'**, moral instinctive self or soul. And it gave us our fabulous conscious mind, a development that meant we were free from the stupor of the animal condition, but did, of course, then lead to our horrifically upset state of the human condition—*but*, with understanding of the human condition now found, we can, as will be explained in chapter 8, finally be transformed to an existence that is free of *both*

conditions. Love-indoctrination also freed our hands to hold tools and carry out innumerable tasks, create all the things our conscious mind was capable of envisaging and ultimately undertake the experiments in knowledge that led to these insights; a fully conscious mind in a whale or a dog would be frustrated by its inability to implement its understandings. Further, love-indoctrination is what likely gave us humans the relatively long life span that has been so instrumental in the accumulation of knowledge. If we only lived to 30, which is considered a long life in the animal kingdom, instead of the 70 plus years we do (in the case of dogs, for example, no matter how much you look after one, it will still only live for 10-15 years), we would likely not have had sufficient time to properly assimilate and manage in our minds all the difficult nuances of the human condition, or to learn about and add to all the understandings humanity was accumulating.

350 In conclusion, nurturing was the main influence or prime mover in the development of humans—*not* upright walking, or tool use, or language, or mastery of fire, or migration from the forest to the savannah, or any one of the other explanations that evasive, denial-complying, mechanistic biologists have been putting forward in the mountain of dishonest books that have been published on human origins. The human race had an immensely happy, all-sensitive-and-all-loving, thrilled-and-enthralled-with-all-of-life, zest-for-living-drenched, ecstatic infancy and childhood, which, as will be described in chapter 7:2, lasted from some 12 million years ago right up to some 2 million years ago when our conscious mind became fully developed and with it the upset state of the human condition emerged—NOT the brutish, barbaric, backward, primitive, savage, bestial, demonic, competitive, aggressive, warring, sombre, morose, unhappy drudge of an existence that our species' early existence has for so long been portrayed in documentaries. When Buddhist scripture anticipated what a human-condition-resolved-and-thus-human-condition-free state would be like—that **'Human beings are then without any blemishes, moral offences are unknown among them, and they are full of zest and joy'** (Maitreyavyakarana; *Buddhist Scriptures*, tr. Edward Conze, 1959, pp.238-242)—it was also describing what life was like in the human-condition-free state that existed in our past when we had become free of the animal condition but had not yet encountered the upset state of the human condition.

351 Yes, the recognition in all our mythologies and in the work of our most profound thinkers of a wonderful, all-loving, innocent past for the

human race wasn't some without-any-factual-base, romantic, fanciful dream of some impossible, unrealistic, idyllic, utopian existence, as mechanistic science has tried to dismiss it as, but a *completely real* time in our species' distant past that recent fossil evidence is now confirming, and that bonobos provide ample living evidence of. Indeed, we can now finally reconcile our scientific knowledge with all the truths contained in our mythologies and religions—for example, we can finally appreciate that what the eighth century BC Greek poet Hesiod wrote in his poem *Theogony* about the 'Golden Age' in our species' past equates perfectly with the life of the bonobos that has just been described: **'When gods alike and mortals rose to birth / A golden race the immortals formed on earth...Like gods they lived, with calm untroubled mind / Free from the toils and anguish of our kind / Nor e'er decrepit age misshaped their frame...Strangers to ill, their lives in feasts flowed by...Dying they sank in sleep, nor seemed to die / Theirs was each good; the life-sustaining soil / Yielded its copious fruits, unbribed by toil / They with abundant goods 'midst quiet lands / All willing shared the gathering of their hands.'**

352 As will be described at some length next in chapter 5, there is, therefore, an immense amount of dishonest biological thinking that now has to be dismantled and redressed. In particular, it will be described how mechanistic science's reluctance to recognise the compelling evidence that bonobos provide for the nurturing origins of our moral nature has meant that little research has been done on bonobos—so little, in fact, that they have been, as mentioned, referred to as **'the forgotten ape'**. And as for the role nurturing is presently playing in bonobo society and did play in the emergence of humanity, chapter 5 will also describe how mechanistic science has denied such a confronting truth by maintaining that maternalism was nothing more than mothers protecting their dependent infants and providing training in life skills. And as for our upset behaviour, mechanistic science has simply blamed it on genes (or 'nature', as the contrived genetic excuse has been called in the 'nature versus nurture' debate) because, as the child psychologist Oliver James points out, **'Believing in genes removes any possibility of "blame" falling on parents'** (*They F*** You Up: How to Survive Family Life*, 2002, p.13 of 370). Other scapegoat excuses include 'chemical causes', as this dialogue from the 1989 film *Parenthood* illustrates: Counsellor: **'He's a very bright, very aware, extremely tense little boy who is only likely to get tenser in adolescence. He needs some special attention.'** Karen: **'It's because he was our first. I think we were very**

tense when Kevin was little. I mean, if he got a scratch, we were hysterical. By the third kid, you know, you let him juggle knives.' Counsellor: **'On the other hand, Kevin may have been like this in the womb. Recent studies indicate that <u>these things are all chemical</u>.'** Gil: '[points at Karen] **She smoked grass.'** Karen: **'Gil! I never smoked when I was pregnant...Will you give me a break?'** Gil: **'But maybe it affected your chromosomes.'** Counsellor intervening: **'You should not look on the fact that Kevin will be going to a special school as any kind of failure on your part.'** Gil: **'Right, I'll blame the dog.'**

353 Thankfully, the human race is now in a position to end all this lying and acknowledge the crucial role that nurturing has played in human origins, and in our own lives—which means all those dishonest, denial-complying, reverse-of-the-truth, mechanistic explanations, papers, books and documentaries that have been put forward to supposedly explain human nature and our origins will need to be re-written and re-filmed to present the honest story about our human journey. Yes, we are now able to admit that Rousseau *was* right when he said that **'nothing is more gentle than man in his primitive state'**. It was *not* our distant ancestors who were without the ability to be gentle and loving, as we have so long been taught, rather it is us fully conscious, immensely psychologically upset, angry, egocentric and alienated, soul-destroyed, human-condition-afflicted modern humans who have *lost* that ability.

354 At last the artificially relieving but, in truth, absolutely awful dishonest, alienated, cold, psychotic and neurotic darkness that Plato so accurately portrayed us as living in with his cave analogy can disappear from human existence, leaving in its place a world of warm sunlight once more, a world of alienation-free truthfulness, happiness and togetherness, a world **'full of zest and joy'**. All that remains now is for the human race to make the right choice between continuing on its habituated path of denial that will lead to terminal alienation and the extinction of the human race, or choosing the path of truth that has now opened up that leads to our species' transformation. The dangerous impediments to this critical juncture and the awe-inspiring potentials that lie beyond will be revealed in the final four chapters of this book.

355 (A more complete description of the origin of humans' unconditionally selfless, altruistic, moral instinctive self or soul can be found in *Freedom Book 1* at <www.worldtransformation.com/freedom-origin-of-morality>.)

Chapter 5

End Play for The Human Race

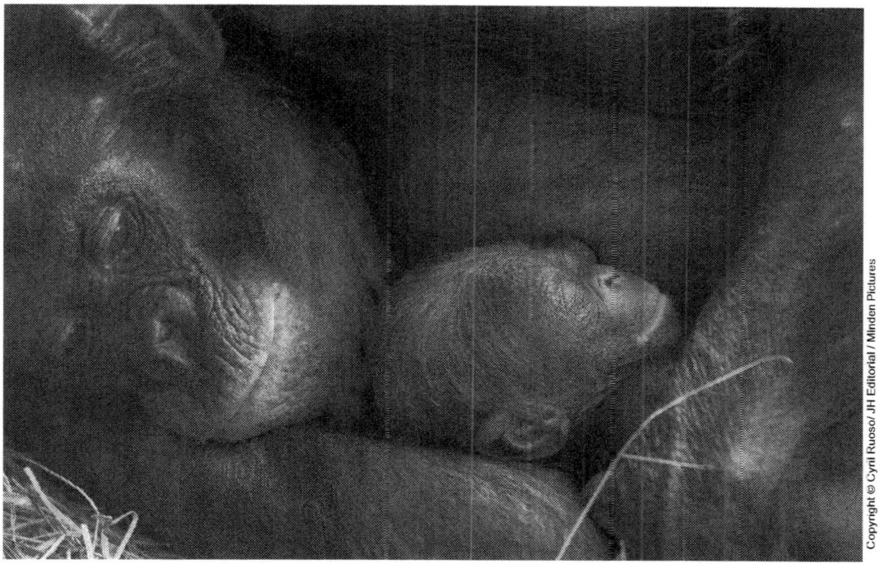

Bonobo mother with her infant

Chapter 5:1 Summary

[356] **This fifth chapter reveals how science, the designated vehicle for human enquiry, has evaded the unbearable truth of the importance of nurturing in the development of our species, and in our own lives, by inventing theories about our origins that are now so dangerously dishonest they threaten the human race with extinction.**

Chapter 5:2 The danger of denial becoming so entrenched that it locks humanity onto the path to terminal alienation

[357] As explained in chapters 1 and 3, the truth about the origin of our corrupted human condition and of the integrative meaning or purpose of existence were two extremely obvious truths that, until we found the compassionate explanation and understanding for both, were so unbearably exposing, condemning and confronting that we, the human race, had

no choice but to live in complete denial of them. For exactly the same reason, the truth of the importance of nurturing in both the maturation of our species and in the maturation of our own lives—how nurturing created our moral instincts—that was just described in chapter 4 has been another extremely obvious truth that has had to be denied because it too was unbearably condemning of our present unloved and unloving, human-condition-afflicted lives. And as we will see in chapter 6, the nature of consciousness is a further obvious truth that has also been so unbearably condemning that it too has been fiercely denied, to the extent that our conscious mind—nature's greatest creation—has been deemed an inexplicable mystery.

358 In light of all this unavoidable denial, it is not surprising that these four great outstanding mysteries in science—of the explanation of the human condition, of the meaning of our existence, of the origin of our unconditionally selfless moral instincts, and of why we humans became conscious when other animals haven't—hadn't been solved. If you can't confront the human condition you are in no position to find the explanation for it; and if you can't admit to Integrative Meaning you obviously can't explain what the meaning of our lives is; and if you can't admit to the importance of nurturing you are in no position to explain the origin of humans' unconditionally selfless moral nature; and if you can't confront the truth of the nature of consciousness you are in no position to explain how it arose and what has prevented its development in other species. It is *not* lack of mental cleverness or Einstein-like genius that made all the great missing explanations in science that are presented in this book impossible to find but alienation, the lack of innocent, soul-guided soundness and security of self. The truths and insights revealed in this book are, in fact, obvious and easy to find—as long as you are not living in a resigned, dishonest, alienated, human-condition-denying state. Cleverness had nothing to do with the finding of all these great breakthroughs in science—what *was* required was simple, truthful, soul-guided sound thinking. As Berdyaev recognised in the quote of his that was included in par. 128, **'the moral reformation of life…**[depends on] **a revelation of a clear conscience, unclouded by social conventions** [the most entrenched and insidious of which is denial]'. That exceptionally honest writer, Simone Weil, saw the problem clearly when she wrote that **'A new type of sanctity…a fresh Spring…a new revelation of the universe and of human destiny…**[requires] **the exposure of a large portion of truth and beauty hitherto**

concealed under a thick layer of dust [denial]**. More genius is needed than was needed by Archimedes to invent machines and physics. A new saintliness... The world needs saints who have genius, just as a plague-stricken town needs doctors'** ('Last thoughts', 1942; *The Simone Weil Reader*, ed. G. Panichas, 1977, p.114 of 529). There is certainly truth in the novelist Aldous Huxley's observation that **'We** [resigned humans] **don't know because we don't want to know'** (*Ends and Means*, 1937, p.270).

359 Yes, the recurring theme running through this book is that the truth *cannot* be found from an evasive and thus dishonest position. Living in denial of the human condition and of any truths that bring it into focus means your search for truth and understanding is only ever going to end in a completely lost state of confused madness. For instance, as we saw in chapter 1:8, mechanistic science initially tried to deny the existence of the human condition by dishonestly claiming that our selfish and competitive behaviour was a natural consequence of a supposed 'survival of the fittest' natural selection process, and that our selfless moral nature was actually a selfish strategy to help relatives or kin reproduce our genes. But that extreme dishonesty was both contrary to and offensive of the fact that we humans do have unconditionally selfless, loving, moral instincts, and so, in response to that backlash, a much cleverer way was concocted to deny our species' extreme sickness, its psychosis (as has been mentioned, the word psychosis literally means 'soul-illness'). This contrivance was E.O. Wilson's Multilevel Selection theory for eusociality, which maintains there is no psychosis involved in the human condition, but that we simply have some unconditionally selfless instincts that exist alongside competitive, 'survival of the fittest' selfish ones, and that our human condition is the product of a conflict between those two instinctive states! Yes, this theory renders the human condition inconsequential, virtually benign—with the consequence of this extremely sophisticated denial being that the human race now faces terminal alienation; madness. Indeed, if this extreme dishonesty on the part of mechanistic science had not been exposed, as it is throughout this presentation, then the stage *was* set for the extinction of our species.

360 In the case of humanity's denial of Integrative Meaning, it was explained in chapter 3:3 that the way the upset human race eventually found to eliminate *that* truth was by making the integrative theme or purpose of existence out to be a supernatural deity we termed 'God', a subject or realm that supposedly had no place in science—an outrageous

display of dishonesty that has led to all manner of blind, purpose-*less*, ultimately unaccountable, ridiculous interpretations of the natural selection process. And—as we will see in chapter 6—a similar fate has also befallen the study of consciousness, which has been so characterised by denial and dishonesty that it too has resulted in the human race becoming lost in a completely bewildered state of intellectual confusion about the all-important issues of what consciousness actually is and how it emerged. *And*, lastly, in the case of that other key issue that is so critical to our understanding of ourselves, namely the origins of our moral nature, what will be revealed throughout this chapter is that the denial of the importance of nurturing in the maturation of our species and in the maturation of our own lives has seen the 140 years that have passed since the American philosopher John Fiske first presented the nurturing explanation for our moral nature *squandered* through the development of extremely dishonest, alienated, mad biological theories—a tragic journey that has now culminated in the immensely dangerous Social Ecological Model, and its latest incarnation, the Self-Domestication Hypothesis, to 'explain' our species' altruistic instincts.

[361]In short, what has just been outlined illustrates how, even though mechanistic science had no choice other than to practise denial of any truths that brought the issue of the human condition into focus until we found the true explanation of the human condition, such denial was an extremely dangerous practice. Yes, what has been described illustrates that, as necessary as they have been, the longer denials are practised the more refined they become, so that in the end, after many decades of development, they inevitably become so sophisticated, so cleverly refined and so entrenched they effectively lock humanity onto a path to terminal alienation—to total derangement, death and extinction.

Chapter 5:3 The nurturing origins of our moral soul is an obvious truth

[362]Like the truths of the origin of our corrupted human-condition-afflicted state and of Integrative Meaning, the importance of nurturing in both the maturation of our own lives and in the maturation of our species—that nurturing created our moral instincts—*is* an obvious truth. While we have had to live in denial of it, we all intuitively know that a mother's love is crucial to the creation of a well-adjusted human

and that we are *all* born with an instinctive expectation of receiving unconditionally selfless love from our mother. And it's *also* obvious that such powerful instincts to nurture with love and be nurtured with love can't have come from nowhere. To be so strong in us they must have played a significant role in our species' development. So yes, if we weren't living in denial of the importance of nurturing in human life, it wouldn't be hard to work out that our unconditionally selfless moral instincts were borne out of the mother-infant relationship. While, as pointed out in par. 82, Charles Darwin was not a denial-free thinker who could confront the issue of the human condition, he was nevertheless an exceptionally honest, truthful and thus effective thinker who could see that our **'social instinct seems'** to be **'developed'** from **'parental' 'affections'**, writing in his 1871 book, *The Descent of Man*, that **'The feeling of pleasure from society is probably an extension of the parental or filial** [family] **affections, since the social instinct seems to be developed by the young remaining for a long time with their parents; and this extension may be attributed in part to habit, but chiefly to natural selection'** (ch.4). And although the renowned anthropologist Richard Leakey wasn't thinking honestly when he wrote that the **'bond between mother and infant'** occurs so that her infant can have a period of **'prolonged learning'** about its **'environment'** (the dishonesty of this thinking will be explained shortly in par. 381), he *was* thinking truthfully when he then emphatically asserted that **'the basis of all primate social groups is the bond between mother and infant. That bond constitutes the social unit out of which all higher orders of society are constructed'** (Richard Leakey & Roger Lewin, *Origins*, 1977, p.61 of 264). In his acclaimed television series and accompanying book that was dedicated to explaining **'the ascent of man'**, the great science historian Jacob Bronowski also recognised that the **'real vision of the human being'**—of an unconditionally selfless, loving, sound, integratively behaved individual, which Christ so exemplified—is a direct product of the nurturing that takes place between a mother and her child, saying that **'But, far more deeply, it** [a sound mind] **depends on the long preparation of human childhood... The real vision of the human being is the child wonder, the Virgin and Child, the Holy Family'** (*The Ascent of Man*, 1973, pp.424-425 of 448). It's true, the **'family'**, especially the **'Virgin** [uncorrupted, soul-intact, innocent, psychologically sound and secure mother] **and Child'**, *is* **'Holy'**; **'that bond'** *does* lie at the heart of what makes us, and made us as a species, truly human—namely loving and cooperative.

363 Yes, the crucial role played by the nurturing, loving **'bond between mother and infant'** in **'the ascent of man'** *is* a truth we are all intuitively aware of. That awareness is apparent when, for example, Africa is described as **'the cradle of mankind'**—Africa *is* where humanity was *nurtured* into existence. Similarly, when the anthropologist Loren Eisely wrote that **'Man is born of love and exists by reason of a love more continuous than in any other form of life'** (*An Evolutionist Looks at Modern Man*, c.1959), he was recognising the truth that humanity *was* **'born of love'**.

364 In fact, the nurturing explanation for our extraordinary unconditionally selfless, all-loving, social, moral instinctive self or soul is so obvious that only three years after Darwin tentatively ascribed the origin of our **'social instinct'** to **'parental' 'affections'**, it was put forward as a developed theory by the aforementioned John Fiske in his 1874 book, *Outlines of Cosmic Philosophy: based on the Doctrine of Evolution*. Indeed, if we were to prioritise the information we humans need to truly understand our world and our place in it, the first item would be to explain the origins of the variety of life, which is what Darwin did with his idea of natural selection. And the second would have to be to explain the origins of the particular variety of life that is the cooperative organism we call humanity, which is what Fiske did with his nurturing explanation for our species' original instinctive unconditionally selfless, loving, moral, social sense. *However*, while both these fundamentally important insights *were* made available to us way back in the mid-1800s, the only one I was taught at school and university was Darwin's idea of natural selection. It wasn't until 2004 when I chanced upon a comment about Fiske that I learnt of his remarkable contribution—which was many years after I had worked out that nurturing was the obvious explanation for our moral instincts. (As I have already pointed out, I first presented the nurturing, love-indoctrination explanation that is described in chapter 4 to the scientific community in 1983 in a submission to *Nature* magazine. I have submitted it elsewhere many times since, but to no avail, with each submission either rejected or ignored—something I will talk further about in ch. 5:12.) The trail I followed in 2004 that led me to Fiske began with a reference in a 1992 paper by the linguist Robin Allott to **'human love evolv**[ing]**'** from the **'mother/infant bond'** ('Evolutionary Aspects of Love and Empathy', *Journal of Social and Evolutionary Systems*, Vol.15, No.4). Following the source provided for this concept led to the work of the scientist and evangelist Henry Drummond—via the historian Dorothy Ross, who had

written that **'To Darwin's principle of natural selection by means of the struggle for survival, he** [Drummond] **added another principle that he considered far more important—"the Struggle for the Life of Others," or "altruistic Love," which developed in the course of evolution from the necessities of maternity'** (*G. Stanley Hall: The Psychologist as Prophet*, 1972, p.262 of 482). Going then to the 1894 writings of Drummond, I found that in responding to the question as to why humans have such long infancies, Drummond had written that **'The question has been answered for us by Mr. John Fiske, and the world here owes to him one of the most beautiful contributions ever made to the Evolution of Man. We know what this delay means ethically—it was necessary for moral training that the human child should have the longest possible time by its Mother's side'** (*The Ascent of Man*, 1894, ch. 'The Evolution of a Mother'). Progressing then to Fiske, in 1874 he had written that **'Throughout the animal kingdom the period of infancy is correlated with feelings of parental affection...The prolongation** [of infancy] **must...have been gradual, and the same increase of intelligence to which it was due must also have prolonged the correlative parental feelings, by associating them more and more with anticipations and memories. The concluding phases of this long change may be witnessed in the course of civilization. Our parental affections now endure through life...I believe we have now reached a... satisfactory explanation of...Sociality'** (*Outlines of Cosmic Philosophy: based on the Doctrine of Evolution*, Vol.IV, Part II, ch. 'Genesis of Man, Morally'). The **'prolongation'** of **'infancy'** was not, however, **'due'** to the **'increase of intelligence'**, rather, as will be explained in chapter 6:3, the prolonged infancy and its nurturing of selflessness *liberated* the fully conscious, intelligent mind, which only developed strongly *after* the love-indoctrination process was well established; this is evidenced by the emergence of the large brain that appears in the fossil record of our ancestors *after* the ape stage. Nevertheless, nurturing, **'parental affection'** was put forward by Fiske as a developed theory to explain **'human lov**[ing]**'**, **'altruistic'**, **'ethical'**, **'moral'**, **'social'** instincts way back in 1874.

365 Since Drummond has provided us with an absolutely wonderful description of Fiske's nurturing explanation for our moral nature, I will present a condensation of his account: **'The...pinnacle of the temple of Nature...is...The Mammalia, THE MOTHERS...**[It is] **That care for others, from which the Mammalia take their name...All elementary animals are orphans... they waken to isolation, to apathy, to the attentions only of those who seek their doom. But as we draw nearer the apex of the animal kingdom, the spectacle of a protective Maternity looms into view...**[the] **love of offspring...That early world,**

therefore, for millions and millions of years was a bleak and loveless world. It was a world without children and a world without Mothers. It is good to realize how heartless Nature was till these arrived...the ethical effect...of this early arrangement was nil...There was no time to love, no opportunity to love, and no object to love...Now, before Maternal Love can be evolved out of this first care...Nature must...cause fewer young to be produced at a birth...make them helpless, so that for a time they must dwell with her...And...she...dwell with them...In this [Mammal] **child...infancy reaches its last perfection...On the physiological side, the name of this impelling power is lactation; on the ethical side, it is Love...Millions of millions of Mothers had lived in the world before this, but the higher affections were unborn. Tenderness, gentleness, unselfishness, love, care, self-sacrifice—these as yet were not, or were only in the bud...To create Motherhood and all that enshrines...required a human child...The only thing that remains now is...that they** [human mother and child] **shall both be kept in that school as long as it is possible to...give affection time to grow...No animal except Man was permitted to have his education** [in love] **thus prolonged...We know what this delay means ethically—it was necessary for moral training that the human child should have the longest possible time by its Mother's side... A sheep knows its lamb only while it is a lamb. The affection in these cases, fierce enough while it lasts, is soon forgotten, and the traces it left in the brain are obliterated before they have furrowed into habit** [Note here recognition that the training in love wears off with age, which, as was explained in chapter 4:9, is why there was the selection for neotenous youth in the love-indoctrination process]**...To her [the human mother] alone was given a curriculum prolonged enough to let her graduate in the school of the affections...unselfishness has scored; its child has proved itself fitter to survive than the child of Selfishness... A few score more of centuries, a few more millions of Mothers, and the germs of Patience, Carefulness, Tenderness, Sympathy, and Self-Sacrifice will have rooted themselves in Humanity...However short the earliest infancies, however feeble the sparks they fanned, however long heredity took to gather fuel enough for a steady flame, it is certain that once this fire began to warm the cold hearth of Nature and give humanity a heart, the most stupendous task of the past was accomplished...**[And here Drummond quotes Fiske] **"From of old we have heard the monition, 'Except ye be as babes ye cannot enter the kingdom of Heaven'; the latest science now shows us—though in a very different sense of the words—that unless we had been as babes, the ethical phenomena which give all its significance to the phrase 'Kingdom of Heaven' would have been non-existent for us. Without the circumstances of Infancy...we should never have comprehended the meaning**

of such phrases as 'self-sacrifice' or 'devotion''' ' (*The Ascent of Man*, 1894, ch. 'The Evolution of a Mother').

366The outstanding question is, *why* did Fiske's fundamentally important explanation for the origins of our moral instincts that created **'humanity'—'one of the most beautiful contributions ever made to the Evolution of Man'**—virtually vanish from scientific discourse; *why* weren't we taught the nurturing explanation for our altruistic moral nature at school, or when I and others studied biology at university; *why* did I have to work the idea out myself? *Why* was this **'altruistic' 'principle'** that was **'considered far more important'** than the **'principle of natural selection'**, and which Fiske explained was able to be developed in our forebears by **'the necessities of maternity'**, allowed to *so* disappear from biological discourse that in the 140 years that have elapsed since Fiske presented his explanation a veritable mountain of books have been published presenting all manner of unaccountable, dishonest theories for the origins of our species' extraordinary moral nature? *Why*, when we had the truth, has there been such a colossal amount of tragically misguided effort that, as we will see, has now resulted in the dangerously dishonest, misleading Social Ecological/Self-Domestication explanations for our moral soul? And *why*, in turn, has *my* nurturing, love-indoctrination explanation for our moral soul—and, indeed, *all* my work—not just been rejected and ignored, but (as I will document in ch. 5:12) *so* ruthlessly attacked that I was made a pariah, and those helping disseminate these insights ostracised?

Chapter 5:4 The problem has been that the nurturing origin of our moral soul has been devastatingly, unbearably, excruciatingly condemning

367As mentioned in chapter 4:7, the answer to all these '*whys*' is that the nurturing explanation for our moral soul has been devastatingly, unbearably, excruciatingly condemning of humans' present inability to nurture children with the real, unconditional love that their instincts expect. Indeed, in his aforementioned paper, Allctt noted that when the nurturing explanation for our moral instincts was put forward by Fiske, and supported by a few others, it **'attracted a good deal of opprobrium [abuse]'**. But as I also pointed out in chapter 4:7, since the upset state of the human condition emerged some 2 million years ago, *no* child has been able to be given the amount of love its instincts expect, and unable, until now, to explain the human condition and thus provide the explanation for why this provision of love has been so compromised, the human race has had no choice but to deny the role nurturing has played in the development of humanity and in the maturation of our own individual lives. The great difficulty we have admitting the importance of nurturing in human development is evident in the comments that were referred to in chapter 4:7, that **'The biggest crime you can commit in our society is to be a failure as a parent and people would rather admit to being an axe murderer than being a bad father or mother'**; and that **'For a lot of women the only really important anchor in their lives is motherhood. If they fail in a primary role they feel should come naturally it is devastating for them.'**

368 In my 40 years of constant thinking and writing about the human condition I have assembled a collection of the rare occasions when the human-condition-avoiding, denial-practising, dishonest world we live in momentarily dropped its guard and let some truth through. The following two quotes are cases in point—they are the most honest admissions I have come across of both the importance of nurturing in human life and the now dire inability of mothers to adequately nurture their children due to the corrupting effects of our species' heroic search for self-understanding. Firstly, consider this quote from the author Olive Schreiner that also featured in chapter 4:7: **'They say women have one great and noble work left them, and they do it ill...*We* bear the world, and *we* make it. The souls of little children are marvellously delicate and tender things, and keep for ever the shadow that first falls on them, and that is the mother's or at best a woman's. There was never a great man who had not a great mother—it is hardly an exaggeration. The first six years of our life make us; all that is added later is veneer...The mightiest and noblest of human work is given to us, and we do it ill.'**

369 Then there is this powerful extract from the anthropologist Ashley Montagu's extraordinarily honest 1970 paper, 'A Scientist Looks at Love': **'love is, without question, the most important experience in the life of a human being...One of the most frequently used words in our vocabulary...** [yet] ***love* is something about which most of us are still extremely vague...There is a widespread belief that a newborn baby is a selfish, disorganized wild creature who would grow into a violently intractable savage if it were not properly disciplined.** [However,] **The newborn baby is organized in an extraordinarily sensitive manner...He does not want discipline...he wants love. He behaves as if he expected to be loved, and when his expectation is thwarted, he reacts in**

a grievously disappointed manner. There is now good evidence which leads us to believe that not only does a baby want to be loved, but also that it wants to love; all its drives are orientated in the direction of receiving and giving love. If it doesn't receive love it is unable to give it—as a child or as an adult. From the moment of birth the baby needs the reciprocal exchange of love with its mother...It has, I believe, been universally acknowledged that the mother-infant relationship perhaps more than any other defines the very essence of love...survival alone is not enough—human beings need and should receive much more...We now know that babies which are physically well nurtured may nevertheless waste away and die unless they are also loved. We also know that the only remedy for those babies on the verge of dying is love...The infant can suffer no greater loss than deprivation of the mother's love. There is an old Eastern proverb which explains that since God could not be everywhere he created mothers...maternal rejection may be seen as the "causative factor in...every individual case of neurosis or behavior problem in children."...Endowed at birth with the need to develop as a loving, harmonic human being, the child learns to love by being loved...To love one's neighbor as oneself requires first that one must be able to love oneself, and the only way to learn that art is by having been adequately loved during the first six years of one's life. As Freud pointed out, this is the period during which the foundations of the personality are either well and truly laid—or not. If one doesn't love oneself one cannot love others. To make loving order in the world we must first have had loving order made in ourselves...Nothing in the world can be more important or as significant...love is demonstrable, it is sacrificial, it is self-abnegative [self-denying]**. It puts the other always first. It is not a cold or calculated altruism, but a deep complete involvement with another. Love is unconditional...Love is the principal developer of one's capacity for being human, the chief stimulus for the development of social competence, and the only thing on earth that can produce that sense of belongingness and relatedness to the world of humanity which is the best achievement of the healthy human being...Scientists are discovering...that to live as if to live and love were one is the only way of life for human beings, because, indeed, this is the way of life which the innate nature of man demands. We are discovering that the highest ideals of man spring from man's own nature...and that the highest of these innately based ideals is the one that must enliven and inform all his other ideals, namely, *love*...Contemporary scientists working in this field are giving a scientific foundation or validation to the Sermon on the Mount and to the Golden Rule: to do unto others as you would have them do unto you, to love your neighbor as yourself...In an age**

in which a great deal of unloving love masquerades as the genuine article, in which there is a massive lack of love behind the show of love, in which millions have literally been unloved to death, it is very necessary to understand what love really means. We have left the study of love to the last, but now that we can begin to understand its importance for humanity, we can see that this is the area in which the men of religion, the educators, the physicians, and the scientists can join hands in the common endeavor of putting man back upon the road of his evolutionary destiny from which he has gone so far astray—the road which leads to health and happiness for all humanity, peace and goodwill unto all the earth' (*The Phi Delta Kappan*, Vol.51, No.9).

370 But while these quotes *are* incredibly honest, with understanding of the psychologically upset state of the human condition finally found (the explanation of which was presented in chapter 2), we can at last explain what is, in fact, fundamentally wrong with what Schreiner and Montagu have said, which is that rather than loving our infants being **'the mightiest and noblest of human work'**, and of there being **'nothing in the world…more important'** than being loved, the incursion of the human condition saw a **'mightie**[r]**'** and **'more important'** task assigned to humans, which was to persevere with humanity's corrupting, love-destroying search for knowledge until we found the understanding that would finally liberate the human race from that condition and allow the practice of nurturing to once more regain its place as the **'mightiest and noblest of human work'**.

371 So there has been a *very* good reason for why humans **'have literally been unloved to death'**, but until we could compassionately explain that reason we had no choice but to leave **'the study of love to the last'**. It is *only* now that we can explain the human condition, explain that humanity has had to be preoccupied with its soul-corrupting, love-destroying, anger-egocentricity-and-alienation-producing heroic search for knowledge, that we can explain *why* we have been so alienated as parents that we have been unable to give our offspring anything like the alienation-free, sound, secure, unconditional love needed to create **'The real vision of the human being'** of the sound child, the **'child wonder'**. Yes, it is *only* now that we can afford to admit that the playwright Samuel Beckett was only slightly exaggerating the brevity today of a truly loved, soulful, happy, innocent, secure, sane, human-condition-free life when he wrote, **'They give birth astride of a grave, the light gleams an instant, then it's night once more'** (*Waiting for Godot*, 1955), or that the

psychiatrist R.D. Laing was right when he wrote that **'To adapt to this world the child abdicates its ecstasy'** (*The Politics of Experience* and *The Bird of Paradise*, 1967, p.118 of 156).

372 Such is the enormous paradox of the human condition: we humans appeared to be horribly bad but are, in fact, heroically wonderful—but until that reconciling biological understanding was found we had no choice but to be prepared to create and live in a world that was devoid of truth—and love! As the song *The Impossible Dream* described this predicament, we had to be prepared **'to march into hell for a heavenly cause'**. So what is really needed to balance Schreiner's and Montagu's honest but unfairly condemning revelations about how inadequate parents have been in their ability to provide their children with the unconditional love their instincts expect, and, as a result, how hurt and alienated children have been, is a presentation emphasising just how incredibly, amazingly, extraordinarily heroic *all* parents have been to have even had children while they were living under the horrific duress of the human condition. Yes, a balancing presentation was needed, which this book now supplies, about how the human race has had to live with 2 million years of unjust condemnation—about how every day humans have had to get out of bed and face a world that, in effect, hated them, that considered them horrible mistakes, blights on Earth, defiling, bad, awful, even evil, sinful creatures, when, as explained in chapter 2, humans are nothing less than the heroes of the whole story of life on Earth!!

373 It follows then that there has been a justifiable reason for each of the '*whys*' listed earlier—*except* for the last one of why I and those advocating my work have been so thoroughly persecuted for presenting the nurturing explanation for our moral instincts. As emphasised, it was *only* when we could explain the human condition and thus finally understand our inability to nurture our offspring that it would be safe to admit the critically important understanding of our species' nurtured origins and, as Montagu said, put **'man back upon the road of his evolutionary destiny from which he has gone so far astray'** and restore **'health and happiness for all humanity'**, and since it is precisely that explanation of the human condition that I have presented, inclusive of that nurturing explanation, there is *NO* justification for the rejection, ostracism and persecution that I have been subjected to. Quite the reverse, in fact—such a response represents the very height of irresponsibility and an abuse of science's mandate to support endeavours that seek to understand and ameliorate

the plight of man. The seriousness of this ill-treatment will be revisited shortly (in ch. 5:12).

374 From the perspective, however, of mechanistic science, this need to deny the importance of nurturing in our human origins until we could explain the human condition has meant that biologists had to find some way of supporting this denial. And, as we are now going to see, *this* denial—like that which led to the corruption of Darwin's idea of natural selection into a 'survival of the fittest' process, through to the development of E.O. Wilson's Multilevel Selection theory for eusociality—was achieved through the invention of the Social Intelligence Hypothesis. (Obviously, the entire need for denial should have been eradicated 30 years ago when I presented the nurturing explanation in accompaniment with the explanation of the human condition, but that, again, is an issue I will return to in ch. 5:12.)

Chapter 5:5 To deny the importance of nurturing, the Social Intelligence Hypothesis was invented

375 Since humans are primates, the obvious area of research that has the most potential to shed light on our origins is the field of primatology, but it is in this most enlightening of fields that some of the most dishonest thinking about the origins of our species' moral sense has been taking place. Despite John Fiske having presented the nurturing explanation for our moral nature way back in 1874, the great majority of primatologists have been so fearful of the truth of nurturing that they have persevered along the habituated path of denial, serving up completely dishonest interpretations of primate behaviour. This denial is particularly palpable if we compare their dishonest studies with the work of the rare few honest primatologists who have dared to recognise the role that nurturing plays in primate society.

376 For instance, the obviousness—if you're not practising denial—of the nurturing, love-indoctrination process, and how extremely confronting a truth it is, is apparent in Dian Fossey's study of gorillas. As described in pars 328-330, Fossey was an extraordinarily strong-willed woman for whom the universal practice of denial in mechanistic science held no sway. Few, if any, however, have been able to cope with the honesty of Fossey's studies, and, as a result, she has been misrepresented as merely a fanatical gorilla conservationist—such as in the 1988 film of her life,

Gorillas in the Mist. A read, however, of her wonderfully insightful treatise on gorilla behaviour—the 1983 book *Gorillas in the Mist* upon which the film was unfaithfully based—shows just how courageous a scientist Fossey was. She watched the lives of troops of gorillas over many generations and gave a denial-free, honest account of what she saw, which was the whole love-indoctrination process at work. Fearlessly, she wrote that **'Like human mothers, gorilla mothers show a great variation in the treatment of their offspring...Flossie was very casual in the handling, grooming, and support of both of her infants, whereas Old Goat was an exemplary parent'**. Old Goat's offspring, the **'exemplary parent**[ed]**' 'Tiger', 'was taking his place in Group 4's growing cohesiveness. By the age of five, Tiger was surrounded by playmates his own age, a loving mother, and a protective group leader. He was a contented and well-adjusted individual whose zest for living was almost contagious for the other animals of his group...**[However,] **The immigrant...menace...Beetsme...developed an unruly desire to dominate...I found myself strongly disliking Beetsme as I watched his discord destroy...**[the group's] **cohesiveness'** On reading this, one can appreciate why the whole nurturing, love-indoctrination process has been so determinedly denied—Old Goat was an **'exemplary parent'** who created a **'well-adjusted'** offspring with a wonderful **'zest for living'**, while the **'menac**[ing]**'**, **'unruly'**, **'discord'**-creating, **'cohesiveness'-'destroy**[ing]**'**, non-**'loving'**, and by inference unloved, Beetsme was **'dislik**[able]**'**; the implication for humans being that if you don't give your child love you're a bad person, and, as has been stated, humans **'would rather admit to being an axe murderer than being a bad father or mother'**.

[377] In his 1989 book, *Peacemaking Among Primates*, the primatologist Frans de Waal recounts a meeting that was held between an unnamed psychiatrist and Harry F. Harlow, a psychologist who, in the 1950s, studied the extremely damaging effect isolation and touch deprivation had on rhesus monkey infants; their discussion shows just how unbearable and confronting both the concept and the importance of nurturing love has been for those studying primates, and just how fearless Fossey was, by comparison, in her honesty: **'For some scientists it was hard to accept that monkeys may have feelings. In** [the 1979 book] ***The Human Model...***[authors Harry F.] **Harlow and** [Clara E.] **Mears describe the following strained meeting: "Harlow used the term 'love', at which the psychiatrist present countered with the word 'proximity'. Harlow then shifted to the word 'affection', with the psychiatrist again countering with 'proximity'. Harlow started to simmer, but relented when he realized that the closest the psychiatrist had probably ever come to love was**

proximity'" (pp.13-14 of 294). Yes, humans' present human-condition-afflicted, unloved and unloving lives—where the **'closest'** the immensely upset human race has **'probably ever come to love'** is **'proximity'**—has meant that our ability to even acknowledge the existence of unconditional love, which is what nurturing essentially is, and the ramifications of not receiving it, has been nigh impossible. Since this *is* the end play time predicted in the Bible when **'the love of most will grow cold'** (Matt. 24:12), it is no wonder mechanistic science practises the *extreme* dishonesty that Allot described in his aforementioned paper, when he reported that **'Love has been described as a taboo subject, not serious, not appropriate for scientific study'**; this despite, as Montagu acknowledged, **'love'** being **'One of the most frequently used words in our vocabulary'**!

378 It follows then that the bonobos—whose extraordinarily maternal, nurturing, loving treatment of their infants and the resulting remarkable integrative, loving behaviour they exhibit as adults was described at length in chapter 4—have been *extremely* exposing, confronting and condemning of the unloved and unloving human race. As I mentioned in par. 298, the biologist, psychologist and bonobo authority Sue Savage-Rumbaugh has bravely admitted that nurturing is the focus of bonobo society; like Fossey, she has let the truth out of the bag that the cooperative behaviour of bonobos *is* a product of the infant-focused, nurturing of love, love-indoctrination process—writing, with the assistance of her co-author, the writer Roger Lewin, that **'Bonobo life is centered around the offspring. Unlike what happens among common chimps, all members of the bonobo social group help with infant care and share food with infants. If you are a bonobo infant, you can do no wrong...Bonobo females and their infants form the core of the group.'** But *also* like Fossey, Savage-Rumbaugh's honesty appears to have made her the target of the human-condition-avoiding, nurturing-denying, mechanistic scientific establishment, because in 2012 a campaign was launched to discredit and marginalise her ground-breaking work with bonobos. I have been told by a scientist whom I respect and who knows Savage-Rumbaugh personally that she *has* become somewhat erratic, but if that is the case I strongly suspect the genesis of such instability would be years of unfair and undermining criticism from the mechanistic establishment.

379 So the question now is, how was denial of the obvious role nurturing plays in the lives of more developed/integrated/social mammals, especially primates, and most especially bonobos—which the studies of

Fossey and Savage-Rumbaugh bear witness to, and which provide such powerful evidence for how we acquired our cooperative, unconditionally selfless, moral instincts—achieved? Clearly, such a denial wasn't going to be easy, but looking an obvious truth in the face and finding a way to deny it—such as finding a way to deny the extremely obvious truths of Integrative Meaning and of our corrupted human condition—is something we humans are masters at!

380 So yes, how did mechanistic science manage to look the obvious truth of the importance of nurturing in the face and deny it? There have been two ways. The first was to portray maternalism as nothing more than a mother providing her dependent offspring with food and protection. As was explained in chapter 4:4, the truth is that mothers' maternal instincts to nourish and protect their offspring *did* provide the base from which the love-indoctrination process was able to develop, however, in love-indoctrination, maternalism became about much more than a mother looking after her infant—it became a case of actively *loving* that infant. It is not of little significance that we speak of 'motherly love', not 'motherly protection'. The problem, however, with this method of denying the nurturing, loving significance of maternalism is that in the case of the extremely exposing-of-the-truth, infant-focused, maternal bonobo society, their environment has historically been food-rich and competitor-and-predator-free, so it doesn't make sense to argue that their exceptionally maternal behaviour has been driven by the need to either source food or provide protection.

381 The second method used to deny the significance of nurturing in bonobo life, and, by inference, its significance in the lives of our ape ancestors, is the one Richard Leakey referred to in his and Roger Lewin's 1977 book, *Origins*—that the extended infancies in primates is due to the infant's need for **'prolonged learning'** about their **'environment'**. A more complete rendition of this alleged explanation for the need for the nurturing that we see in the society of more developed mammals, particularly in bonobos and humans, is that **'The more sophisticated species also exhibit longer infant and juvenile stages, which are probably related to the time required for their more advanced mental development and their integration into complex social systems'** (*Encyclopedia Britannica*, accessed Feb. 2013; see <www.wtmsources.com/136>). As will be explained at some length in chapter 6:3 (in pars 498-507), this so-called 'Social Intelligence Hypothesis' (sometimes referred to as the 'Machiavellian Intelligence Hypothesis'), and its more

sophisticated version, the Ecological Dominance-Social Competition (EDSC) Model, essentially maintain that the long mother-infant association is needed to ensure the infant learns the skills necessary to manage **'complex social'** situations, and that it was this need that also led to **'more advanced mental development'**, ultimately the fully conscious, intelligent mind in humans.

382 As will be described in chapter 6:3 when the truthful explanation of the origins of consciousness is presented, even human-condition-avoiding mechanistic scientists have raised serious concerns over the viability of both the Social/Machiavellian Intelligence Hypothesis (S/MIH) and the EDSC Model, but from the human-condition-confronting, truthful view of biology some very obvious flaws with both models can be pointed out immediately—particularly in regard to their core argument that our fully conscious intelligent mind emerged as a result of having to learn to manage complex social situations, a process that supposedly required and explained the long infancy.

383 Social problem-solving is, of course, an obvious benefit of being conscious, but *all* activities that animals have to manage would benefit enormously from conscious intelligence—from the ability to reason how cause and effect are related, to understand change, to make sense of experience, to be insightful—so it is completely illogical to suggest that it wasn't until the need arose to manage complex social situations that consciousness developed. No, any sensible analysis of how and when consciousness emerged must be based on the question of 'What has *prevented* its development in other animals?' A lack of social situations doesn't explain why the fully conscious mind hasn't appeared in non-human species because there was ample need for a conscious mind prior to the appearance of complex social situations.

384 No, the only accountable explanation for the emergence of the fully conscious mind in humans and for what is blocking its emergence in other species is the nurturing, love-indoctrination explanation—which, to recap very briefly, states that the nurturing of selflessness *liberated* the fully conscious, intelligent mind from the block that exists in non-human species' minds against thinking selflessly and thus truthfully and thus effectively. (Again, this will all be fully explained in chapter 6:3.)

385 Another obvious flaw with both the S/MIH and the EDSC Model is that if it wasn't for the psychologically upset state of the human condition there would be no need to learn, and become intelligent enough to

master, the art of managing **'complex social systems'**. Through the process of love-indoctrination, we humans became *so* instinctively integrated that there was no disharmony/conflict/discord/**'complex**[ity]' to have to manage. Prior to the emergence of the human condition some 2 million years ago our species lived instinctively as one organism. What did the Greek poet Hesiod say? **'Like gods they lived, with calm untroubled mind, free from the toils and anguish of our kind…They with abundant goods 'midst quiet lands, all willing shared the gathering of their hands.'** The pre-human-condition-afflicted, integrated, cooperative, social, Specie Individual state, such as that which largely exists in bonobo society today, simply wasn't a situation that called for infants to be skilled in social management techniques.

386 And the even more blatant flaw with the two arguments that have been used to dismiss the important role nurturing played in the lives of our forebears, and continues to play in the development/integration/socialisation of other mammals, especially other primates, and most especially bonobos, is that they suggest nurturing is nothing more than a mother providing her dependent infant with food and protection, and that the long mother-infant association is needed only to allow the time to impart social skills. But, as stated earlier, humans can't have developed such powerful instincts to nurture our offspring with love, or such powerful expectations of receiving unconditional love as children, if nurturing hadn't played a fundamental role in the history of our species' development—as I said, such instincts and expectations do not appear out of thin air. The response to this statement, however, from advocates of the S/MIH and the EDSC Model would be that mothers simply don't have powerful instincts to nurture their offspring with love, and children simply don't have expectations of being loved; rather mothers have powerful *instincts to teach* their offspring how to manage **'integration into complex social systems'**, and children have *instinctual expectations of being taught* such skills. In other words, it is not a case of instincts to love and be loved 'coming out of thin air', but a case of such instincts never existing in the first place. But that is absurd; indeed, it is offensively dishonest, because everyone does intuitively know that what Schreiner and Montagu wrote about infants' need for love is true. Yes, what an infant needs from its parents—and from its mother in particular—*is* unconditional love, *not* training in the management of complex social situations! Certainly, when the need for denial is critical any excuse will do, and the art of denial is

to then stick to that excuse like glue, but that does *not* mean we are so unaware we are practising denial that we are unable to recognise and admit the truth when that denial is no longer needed—which, with the human condition now explained, it no longer is.

387 We all do actually know that to achieve the **'loving order in the world'** that Montagu recognised, the **'cold'**, **'bleak and loveless'** (as Drummond described it), ruthlessly selfish, competitive, must-reproduce-your-genes 'animal condition' had to be overcome, and the only means by which that could be achieved was through the mother-infant, nurturing-of-love situation: the love-indoctrination process. As was emphasised in chapter 4:4, the problem is that love-indoctrination is an extremely difficult process to develop and maintain to the point where the fierce competition to reproduce your genes is contained and integration achieved. Only our ape ancestors managed to develop love-indoctrination to the point where competition amongst males especially was contained and unconditional love and integration developed, something the fossil record is now confirming (as described in chapter 4:5). Bonobos are well on their way, but all the other relatively developed/integrated/social mammals are still battling to develop love-indoctrination to the point where it has overcome selfish competition amongst males to reproduce their genes. But they *are* trying to do so; they *are* trying to indoctrinate their infants with love to the degree their circumstances allow. Reports from anyone who has worked with the relatively developed/integrated/social mammals and who is not under the control of the thought police—the truth-denying, mechanistic scientific constabulary—give accounts of the development of love through nurturing, such as this of the nurturing, loving behaviour of elephants: **'After years of research and scientific observation it has been shown that elephant's social structure and familial bonds are similar, if not deeper, than the bonds developed among** [present immensely psychologically upset] **human beings. There are deeply stirring accounts, by such scientists as Joyce Poole, Cynthia Moss and Dr. Dame Daphne Sheldrick, of elephants weeping and expressing grief at the loss of their calves...and other herd members. There are recorded behaviors of near spiritual proportion...Calves frequently die of heartbreak from the loss of their mothers and abuse by human beings...There are also great displays of affection and mutual respect rarely viewed in the social structure of humans. Joyce Poole, internationally known expert on elephants, states, "I have never seen (wild) calves 'disciplined'. Protected, comforted, cooed over, reassured and rescued,**

yes, but punished, no. Elephants are raised in an incredibly positive and loving environment"' ('The Heart of Africa', accessed Feb. 2013; see <www.wtmsources.com/102>). Note that Poole's comment that elephant calves are never **'disciplined'** echoes Savage-Rumbaugh's observation that bonobo infants **'can do no wrong'**. Revealingly, like elephant orphans, orphaned **'Gorillas and bonobos... just die. They see their mothers killed and they give up'** (Vanessa Woods, *Bonobo Handshake*, 2010, p.67 of 278)—because, as stated above, they suffer **'heartbreak'**; their emotional desire for, their instinctive expectation of receiving, and their attachment to, a loving true (Integrative-Meaning-compliant) world is so great they literally cannot survive without it. In par. 334, a quote from Savage-Rumbaugh was included that described the rapturous joy expressed by the bonobos Matata and her adopted son, Kanzi, at their *loving* reunion. When the Friends of Bonobos charity designed their fundraising **'A Bonobo Mother's Love T-shirt'** to have a picture of a mother bonobo cradling her infant, they weren't 'anthropomorphising' or inappropriately humanising bonobos, as the mechanistic thought police would argue, they were unwittingly conveying the simple truth. Bonobo mothers aren't merely giving infants training in how to manage **'integration into complex social systems'**, that is an absurd suggestion—they are giving them **'love'**. But, again, when the need for denial is desperate, any excuse will do. The *truth* is that in the relatively developed/integrated/ social species of mammals, nurturing *has* moved beyond the primitive, pre-love-indoctrination, 'must nourish and protect' maternal situation to the 'must love' maternal situation—they *are* attempting to develop love-indoctrination.

Chapter 5:6 Dismissing maternal love as training to manage complex social situations still left the extraordinarily cooperative lives of bonobos, and of our ape ancestors, to somehow be explained

388 While both the dishonest S/MIH and the EDSC Model have been relied upon to dismiss the mother-infant bond as being nothing more than a mother nourishing and protecting her offspring, and training them in the art of managing complex social situations, a big problem remained: how to account for the remarkable cooperative behaviour of bonobos, and our own unconditionally selfless moral instincts? So the question now is, what nurturing-of-love-denying 'explanation' did

human-condition-avoiding mechanistic scientists come up with to 'solve' this problem?

389 The answer is that mechanistic scientists initially tried to portray the competitive aggression and violence that can be found in all ape species (except bonobos) as evidence of what our ape ancestors were supposedly like. But when it was found that bonobos didn't fit this model, they attempted to ignore the anomaly they represented altogether. And *then*, when these scientists could no longer ignore the extraordinary integration that is so apparent in bonobo society, they conceded that bonobos are cooperative but found a way to explain how they became cooperative that did not invoke, or credit in any way, nurturing. (At this point, it should be stated that, just as our fear of the human condition and resulting denial of it has been so great that we, the upset human race, have hardly been aware that we are living in denial of it—recall in chapter 1:2 how Resignation needed to be described to re-connect us to the real horror of the human condition—our fear of the truth of the importance of nurturing in human development and resulting denial of it has also become *so* developed and entrenched in us that we are hardly aware that we are practising it. It is almost instinctive in us now to avoid the significance of nurturing in human history, as though it's a rule we live by but with only a subliminal awareness that we are abiding by it.)

390 In regard to the initial strategy stated above—of relating our aggressive behaviour to that of apes—as was explained in chapter 1:8, ever since Darwin presented his idea of natural selection, humans have been misrepresenting it as a 'survival of the fittest' process, and using that misinterpretation to support the reverse-of-the-truth lie that, just like other animals, we humans have competitive, selfish and aggressive instincts that our intellect has to heroically control. Chimpanzees appeared to support this lie—they were obviously human-like, and so were used as a model for our ancestors, with anthropologists such as Raymond Dart and Robert Ardrey pointing to chimpanzees' intense and violent male competition, rape, infanticide and inter-group warfare as indicative of the behavioural heritage of our ancestors. Dart argued for the **'predatory transition from ape to man'** ('The Predatory Transition from Ape to Man', *International Anthropological and Linguistic Review*, 1953, Vol.1, No.4), while Ardrey was even more emphatic, saying that **'Man had emerged from the anthropoid background for one reason only: because he was a killer'** (*African*

Genesis, 1961, p.29 of 380). More recently, in 1999, a leading anthropologist and the author of *Demonic Males*, Richard Wrangham, put forward the so-called Chimpanzee Violence Hypothesis, which claimed that **'selection has favored a hunt-and-kill propensity in chimpanzees and humans, and that coalitional killing has a long history in the evolution of both species'** ('Evolution of Coalitionary Killing', *Yearbook of Physical Anthropology*, 1999, Vol.42). Obviously these theories were immensely popular because the idea that our instincts are wildly aggressive made our intellect's supposed role as mediator seem all the more heroic. While it all amounted to a reverse-of-the-truth lie (because, as explained in chapter 1, our instincts are loving while our intellect is the offending, divisive influence), it was, nevertheless, a very human-condition-relieving thesis.

391 Incidentally, even though we do now have fossil evidence supporting the love-indoctrination explanation for our moral instincts, without the living evidence that the bonobos provide, it would be very difficult to prove the nurturing of love explanation for our unconditionally loving moral nature. Thank goodness for bonobos!

392 Yes, with their peaceful and gentle society, the bonobos exposed this initial strategy for the lie it was, but in doing so exposed themselves to the wrath of mechanistic science as an unbearably exposing and confronting reminder of our now immensely angry, egocentric and alienated, unloving and unloved lives. So, as stated, mechanistic science's strategy to deal with *this* problem was to simply ignore the anomaly that bonobos represented. Indeed, this strategy was so successful that the first in-depth study of bonobos, which only occurred in 1954, was **'ignored and forgotten by the scientific community'** (Frans de Waal & Frans Lanting, *Bonobo: The Forgotten Ape*, 1997, p.11 of 210) because it dared to describe them as **'an extraordinarily sensitive, gentle creature, far removed from the demoniacal primitive force of the adult chimpanzee'** (E. Tratz & H. Heck, 'Der africkanische Anthropoide "Bonobo": Eine neue Menschenaffengattung', *Säugetierkundliche Mitteilungen*, Vol.2). In fact, it was this ongoing denial that led de Waal and the photographer Frans Lanting to title their 1997 collaboration, *Bonobo: The Forgotten Ape*. This book, which acknowledged rather than ignored the extraordinary sensitivity of bonobos, was, not surprisingly, vilified: **'De Waal's bonobo research came under sustained attack'** (*The Future of Bonobos: An Animal Akin to Ourselves*, Alicia Patterson Foundation, 2002; see <www.wtmsources.com/122>) from anthropologists such as Craig Stanford who wrote in a paper that **'It is clear that much of the research on these two intensively studied**

apes [in the case of bonobos, I would argue 'superficially' studied] **remains fraught with untested assumptions'** and that **'reported differences have been inflated'** ('The Social Behavior of Chimpanzees and Bonobos: Empirical Evidence and Shifting Assumptions', *Current Anthropology*, 1998, Vol.39, No.4). Responding to Stanford's criticisms, de Waal insightfully wrote that **'Two strategies have emerged to keep bonobos at a distance so as to preserve chimpanzee-based scenarios of human evolution, which traditionally emphasize warfare, hunting, tool use, and male dominance. The first strategy is to describe the bonobo as an interesting but specialized anomaly that can be safely ignored as a possible model of the last common ancestor (see Wrangham and Peterson 1996). The second strategy, adopted by Stanford, is to minimize the differences between the two *Pan* species: if bonobos behave, by and large, like chimpanzees, there is no reason to question the latter species' prominence as a model'** (ibid). (Another illustration of the use of this second strategy was provided in par. 103, when it was documented how E.O. Wilson attempted to suggest that bonobos and chimpanzees were indistinguishable because both **'do not share'** and both **'hunt in coordinated packs'**. However, as was demonstrated there, even the most basic research shows that Wilson had to fudge the evidence to support his lie.)

393 The problem that emerged with these dismissive strategies was that modern technology has increasingly made the bonobos more accessible, and their extraordinarily integrative behaviour, that is so different to chimpanzees, almost impossible to ignore. *Bonobos*, the French documentary that was referred to in par. 300, is a case in point.

394 The fact is, with their extraordinarily loving behaviour, bonobos have represented an ever-growing thorn in the side of a mechanistic scientific fraternity that desperately wanted to avoid their significance. And since bonobos couldn't be ignored forever, something had to be done to at least minimise their confronting presence. What happened was that while their **'extraordinarily sensitive, gentle'**, peaceful, cooperative behaviour could not very well be denied (and, in any case, there was, as will be explained shortly in ch. 5:9, a growing desire among ideal-behaviour-emphasising-but-human-condition-avoiding left-wing biologists to be able to emphasise cooperativeness and gentleness), it was hoped that at least a way could be found to explain why bonobos were cooperative in a manner that still avoided acknowledging the unbearable significance of their remarkable nurturing, maternal behaviour. And the way that was found was through a theory known as the 'Social Ecological Model'

that sought to explain social behaviour in terms of ecological factors that influence social interactions.

[395] Before describing this Social Ecological Model it should be documented how mechanistic science has dismissed the fossilised evidence of our species' cooperative past by employing a strategy of evasion almost identical to that which has been applied to the bonobos.

Chapter 5:7 Fossil evidence of our species' cooperative past has also been dismissed and then ignored by mechanistic science

[396] Yes, just as the evidence provided by bonobos of our species' cooperative past has been denied by mechanistic science—first by claiming that our ancestors behaved competitively and aggressively 'like most ape species', and, when bonobos clearly didn't fit that model, by simply ignoring the anomaly altogether—the fossilised evidence of our cooperative past has also been dismissed as irrelevant and, when that strategy became untenable, it too was simply ignored.

[397] For instance, it was stated in chapter 4:5 that the fossil record reveals that our ape ancestors had small canine teeth. Although it was explained there that the only accountable explanation for the reduction in canine size in our ape ancestors is that it was caused by female sexual selection against male mating aggression—that small canines are **'indicative of minimal social aggression'**—mechanistic scientists initially tried to avoid the implications these small canines raised by maintaining that they were only a recent development, and that the fossil record would inevitably provide evidence of an aggressive heritage For example, in 1915 it was written, **'That we should discover such a race [a human race in which the canine teeth were pointed, projecting, and shaped as in anthropoid apes], sooner or later, has been an article of faith in the anthropologist's creed ever since Darwin's time'** (Sir Arthur Keith, *The Antiquity of Man*, p.459 of 519). However, as subsequent fossil discoveries pushed back, by millions of years, the age at which our ape ancestors still had small canines it became increasingly difficult to maintain this **'article of faith'** of an aggressive **'human race'**. As noted in chapter 4:5, other attempts to account for our ancestors' canine reduction that did not cite a reduction in aggression have also been rendered untenable by these ongoing discoveries, such as by what we now know of their diet.

[398]So, through the recent discoveries of *Ardipithecus*, *Orrorin*, and the 7-million-year-old *Sahelanthropus*, and the conclusive evidence they provide of our species' cooperative heritage, it appears that human-condition-avoiding, mechanistic scientists have been cornered into simply ignoring this evidence. For example, Richard Wrangham published the Chimpanzee Violence Hypothesis in 1999, and yet failed to mention the discovery, only 7 years earlier, of *Ardipithecus*—a find that had confirmed the existence of small canines at least 4.4 million years ago. Similarly, E.O. Wilson's attempt in his 2012 book, *The Social Conquest of Earth*, to portray war as a **'universal and eternal'** presence in human history (a claim that was repudiated in chapter 1:10) makes no mention of the fact that small canines characterise our ape ancestors, despite Wilson discussing both *Australopithecus* and *Ardipithecus* in his book. In 2009 the popular science magazine *Scientific American* (a subsidiary of the leading journal *Nature*) failed to make a single reference to a suite of papers—which had been some 15 years in the making—on *Ardipithecus ramidus* that had been published in a 2009 special edition of *Science*, **'the most extensive special issue of *Science* since Apollo 11'** (Tim D. White, Letters, *Scientific American*, 2010, Vol.302, No.1). As Tim White, team leader of the *Ardipithecus* researchers, asked with justifiable perplexity, **'How and why did the *Scientific American* editorial miss that story?'** (ibid). It is as if the fossil evidence of a cooperative past has become too strong to refute but the implications too daunting to acknowledge, rendering the majority of mechanistic scientists speechless. While a handful of scientists have broken the silence since the *Ardipithecus* discoveries were published, predictably most have done so to deny their cooperative implications on the only grounds left, which is by arguing that *Ardipithecus*, *Orrorin* and *Sahelanthropus* are not part of the human lineage, despite all the evidence indicating that they are.

[399]Yes, mechanistic science's strategies for dealing with the fossil record's evidence of a cooperative past have followed a pattern similar to that which was employed to deal with the bonobos—first denounce the evidence as irrelevant on the false basis that our past is aggressive, and, when that position becomes untenable, simply ignore the evidence.

[400]A further point of significance, and one that was raised in par. 294, is that the fossil record, particularly the 1992 discovery of *Ardipithecus*, clearly shows the physical and behavioural similarities between our pre-australopithecine ape ancestors and bonobos. But if bonobos themselves

needed to be ignored, it follows that any evidence that links our ancestors to them will be treated the same way—as indeed it has. As Frans de Waal explains in his 2013 book, *The Bonobo and the Atheist*, **'a scientist on the Ardi** [*Ardipithecus*] **team...could think only of chimps as a comparison...**[despite] **The bonobo's body proportions—its long legs and narrow shoulders—seem**[ing] **to perfectly fit the descriptions of Ardi, as do its relatively small canines. Why was the bonobo overlooked?'** (pp.60-61 of 289).

Chapter 5:8 The Social Ecological Model

[401]To return now to mechanistic science's determination to avoid the unbearable significance of bonobos' remarkable nurturing, maternal behaviour through the development of the theory known as the Social Ecological Model (SEM).

[402]In essence, the SEM holds that the abundance of food available to the bonobos was what gave rise to their extraordinary harmony and co-operation. The model maintains that a plentiful supply of food meant that groups of bonobos no longer had to split up to feed, which gave females the opportunity to socialise more and, in time, form coalitions to protect themselves against aggressive male competition for mating opportunities, thus forcing males to adopt more cooperative strategies. To quote from a description of the SEM in a 2001 paper by the anthropologists Richard Wrangham and David Pilbeam: **'The absence of gorillas made high-quality foliage more available for proto-bonobos than for chimpanzees. As a result, proto-bonobos experienced a reduced intensity of scramble competition compared to chimpanzees. Reduced scramble competition allowed more stable parties, which then made several forms of aggression more dangerous and costly, and less beneficial, to the aggressors. This change in the economics of violence led through various social consequences to female-female alliances, concealed ovulation, and reduced individual vulnerability to gang attacks. All these favored a reduction in the propensity for male aggressiveness'** ('African apes as time machines', *All Apes Great and Small: Volume 1: African Apes*, eds Birutė Galdikas et al, 2002, p.12 of 316). So, under the dictates of the SEM, bonobos were able to become, to quote Wrangham again, **'a species biologically committed to the moral aspects of what, ironically, we like to call "humanity": respect for others, personal restraint, and turning aside from violence as a solution to conflicting** [social] **interests'** (*Demonic Males*, 1997, p.230 of 350). Wrangham is also reported to have referred to **'those loving bonobos'** ('Chimps and Chumps', *National Review*, 1999, Vol.51, No.18).

403 So this model claimed to explain bonobos' extraordinary **'respect for others'**, **'personal restraint'**, **'loving'**, **'moral'**, cooperative nature without a single mention of the importance of nurturing! Its appeal, however, can be judged by its adoption, in one form or another, by almost all the leading bonobo researchers (including Alison and Noel Badrian, Ben Blount, Christophe Boesch, Barbara Fruth, Brian Hare, Gottfried Hohmann, Takayoshi Kano, Suehisa Kuroda, Toshisada Nishida, Amy Parish, David Pilbeam, Barbara Smuts, Randall Susman, Frans de Waal, Frances White and Richard Wrangham), and by the fact that it remains virtually uncontested since being put forward in the early 1980s. While not all these scientists necessarily agree with Wrangham that bonobos behave in a **'loving'** way—with some maintaining bonobos are capable only of reciprocal selflessness—all acknowledge that bonobos are extraordinarily cooperative and all use the SEM to explain that cooperation.

404 But while the SEM serves the purpose of denying the significance of nurturing in bonobo society, the arguments it uses to do so—namely that stable parties allow females to form coalitions to counter male aggression, and that those coalitions are successful in dominating males—are, in fact, seriously flawed.

405 Firstly, with regard to bonobo females being able to form coalitions to protect themselves against male aggression, the SEM argues that the opportunity to spend more time together—which, in the case of bonobos, they say was made possible by an abundance of food—allowed bonobos to form **'stable parties'** that in time allowed the formation of **'female-female alliances'** to protect themselves against **'male aggressiveness'** arising from competition for mating opportunities. However, while it is not uncommon amongst mammals for situations to occur where **'stable parties'** are able to and do form, it is rare for such **'stable parties'** to lead to **'female-female alliances'** that have the power to prevent **'male aggressiveness'**—which is surprising because such an alliance would certainly seem beneficial since male aggression does come at a cost to females, as these quotes indicate: **'Sexual harassment has significant negative consequences for females even when it doesn't end up in coerced sex; it can increase stress and interfere with normal activities, reducing health, fitness and longevity. Some forms of harassment may even culminate in injury or death'** (Linda Mealey, *Sex Differences: Developmental and Evolutionary Strategies*, 2000, p.130 of 480); and **'Females in many mammalian species experience both sexual aggression and infanticide by males'** (Barbara Smuts & Robert Smuts, 'Male Aggression and Sexual Coercion of Females

in Nonhuman Primates and Other Mammals: Evidence and Theoretical Implications', *Advances in the Study of Behavior*, 1993, Vol.22). If **'stable parties'** were all that were required to form **'female-female alliances'** that could stop **'male aggressiveness'** then such alliances would be common amongst mammals, but they aren't. In short, the formation of such alliances *do* occur within primate societies (especially among bonobos), but if the SEM is correct it should *also* be regularly occurring in the societies of non-primate mammals, but it isn't. To quote the anthropologists Barbara and Robert Smuts: **'the use of female coalitions to thwart aggressive males appears to be rare in other mammals compared with nonhuman primates'** (ibid).

406 The only conclusion for why the strategy of forging **'female-female alliances'** to prevent **'male aggressiveness'** is rare outside primate species is that the strategy is not genetically successful—in other words, it is not in the interest of the female's chances of reproducing her genes, otherwise natural selection would have 'discovered' it and the practice would be a common one amongst all mammals. As such, there has to be some other factor aside from **'stable parties'** enabling females to form these types of coalitions, and the regularity with which these alliances appear within primates compared with non-primate mammals suggests it is something unique to primates. But while the SEM cannot account for this peculiarity, the love-indoctrination explanation certainly does. As was explained in chapter 4:4, primates, whose arms are semi-freed from walking and are thus able to hold a helpless infant, are uniquely placed to develop love-indoctrination should conditions be favourable—and, as we've established, the bonobos have the most favourable environmental conditions amongst the primates and, as a result, *have* been able to develop the most love-indoctrinated integration.

407 Yes, it would appear that it is only amongst primate species, where more nurturing is able to be selected for and both males and females are able to be indoctrinated with unconditional selfless love, that the **'economics of violence'** changes enough for it to be possible to form **'female-female alliances'** to help rein in **'male aggressiveness'**.

408 Secondly, in attempting to explain how male aggression could be quelled by **'female-female alliances'**, the SEM relies on the superficially persuasive but biologically flawed argument that biologists have desperately been resorting to more and more (especially left-wing biologists, who are wanting to promote cooperation and gentleness), which is that cooperation is more advantageous than competition and can therefore

be selected for. This reliance is evident in Wrangham's argument that **'female alliances'** made male **'aggression more dangerous and costly, and less beneficial, to the aggressors'** and that **'turning aside from violence'** was **'a solution to conflicting** [social] **interests'**. While it may appear persuasive to suggest that males could stop competing for mating opportunities once female coalitions formed—because aggression became more **'dangerous and costly, and less beneficial'** than cooperation, or simply because **'turning aside from violence'** is a **'solution to conflicting** [social] **interests'**—the biological reality is that, without love-indoctrination, genetic selfishness will see males *continue* to aggressively compete for any and all mating opportunities. As mentioned in par. 290, this fact of life was pointed out by C. Owen Lovejoy when, in discussing our ape ancestors, he wrote that **'Loss of the projecting canine raises other vexing questions because this tooth is so fundamental to reproductive success in higher primates. What could cause males to forfeit their ability to aggressively compete with other males?'** The same **'vexing'** question arises in regard to bonobos, something that was pointed out by the primatologist Gottfried Hohmann when he observed that **'The** [bonobo] **males, the physically superior animals, do not dominate the females, the inferior animals?...It is not only different from chimpanzees but it violates the rules of social ecology'** ('Swingers: Bonobos are celebrated as peace-loving, matriarchal, and sexually liberated. Are they?', *The New Yorker*, 30 Jul. 2007). Frans de Waal made a similar observation when he wrote about bonobos that **'dominance by the "weaker" sex constitutes a huge violation of every biologist's expectations'** (*Bonobo: The Forgotten Ape*, 1997, p.76 of 210).

409 This biologically flawed argument that cooperation can be selected for by natural selection because it is more advantageous than competition is, in fact, the same biologically flawed argument that E.O. Wilson relied upon in his Multilevel Selection theory for eusociality. Recall in chapter 1:10 how Wilson argued that a group composed of selfless members who consider the welfare of the group above that of their own will be more cooperative and thus successful when competing against groups who have selfish, non-cooperative members, and, as such, competition between groups can lead to the selection of unconditionally selfless traits. Again, while it is superficially persuasive to suggest that a group with cooperative members will defeat a group with competitive members, the genetic reality is that whenever an unconditionally selfless, altruistic trait appears those that are selfish will take advantage of it, thus negating the establishment of a group of cooperators in the first

place. Any selflessness that *might* arise through group selection will be constantly exploited by individual selfishness from *within* the group; as the biologist Jerry Coyne pointed out, **'group selection for altruism would be unlikely to override the tendency of each group to quickly lose its altruism through natural selection favoring cheaters** [selfish individuals]**'** ('Can Darwinism improve Binghamton?', *The New York Times*, 9 Sep. 2011).

[410] The flawed reasoning of this second aspect of the SEM can be seen even more clearly in this description of it by the anthropologist Brian Hare: **'there are some species that outcompeted others by becoming nicer...** [because] **it's very costly to be on top. Often in primate hierarchies, you don't stay on top very long. Everyone is gunning for you. You're getting in a lot of fights. If you don't have to do that, it's better for everybody'** ('Why Some Wild Animals Are Becoming Nicer', *Wired*, 7 Feb. 2012). While it is superficially persuasive to argue that **'it's better for everybody'** if **'you don't have to'** **'fight'** to stay **'on top'**, the idea **'violates'** the fundamental fact about the gene-based natural selection process, which is that you do have to be constantly **'gunning'** to **'stay on top'** because if you are not others will be and you won't reproduce your genes. That's the reality of the 'animal condition': fierce competition exists between sexually reproducing individuals seeking to reproduce their genes.

[411] It should be mentioned that advocates of the SEM argue that the bonobos' practices of **'concealed ovulation'** (where females do not give signs of their fertile period) and of making sex continually available have contributed to the reduction of male competition for mating opportunities, *however*, such practices can *only* develop once competition for mating opportunities has begun to subside because while ever competition to mate remains intense, if a female makes mating available to every male, and/or doesn't advertise she is ready for fertilisation, she can't ensure she mates with the strongest, most virile male and, therefore, that her offspring will be successful competitors. Making sex continually available and concealing ovulation could *assist* love-indoctrination, but not initiate the development of love/unconditional selflessness. No, these practices do not *lead* to a reduction in competition, rather they are a *result* of male competition having largely *been* contained, which, as explained, **'selection against aggression'** can't achieve.

[412] So even though it is seductive to run the argument that cooperation can be selected for by natural selection because it is more advantageous than competition, it is so blatantly biologically incorrect that for biologists

to employ it means they are deliberately lying—resorting to a desperate form of bluff—but such has been the extent of the need to avoid the nurturing explanation for bonobos' unconditionally selfless behaviour, and, by extrapolation, for humans' **'moral'** natures. So it is not surprising that this now insatiable need to account for our moral instincts in a non-confronting way has led to the development of a somewhat more refined form of the SEM.

Chapter 5:9 A brief history of left-wing dishonest mechanistic biology

413 Before introducing this new, 'improved' version of the SEM, it should be mentioned that this seductive but patently untrue argument that cooperation is more advantageous than competition and can therefore be selected for has a long history of use by cooperation-not-competition-supporting, selflessness-not-selfishness-emphasising, human-journey-to-find-knowledge-opposing left-wing biologists (the pseudo idealistic philosophy of the left-wing was briefly explained in chapter 2:9 and will be more fully dealt with in chapters 7:15 and 7:16). For example, in 1880, the zoologist Karl Kessler said that **'the progressive development of the animal kingdom…is favoured much more by mutual support than by mutual struggle'** (Address titled *On the law of mutual aid* to the St Petersburg Society of Naturalists, Jan. 1880). Even up to the 1960s this so-called 'naive' misrepresentation of natural selection as being socialistic rather than individualistic was still occurring, with, for instance, the behaviourist Konrad Lorenz writing frequently of *behaviour* having **'a species-preserving function'** (there are many mentions of this phrase in his 1963 book, *On Aggression*). As was explained in chapter 3, the development of species *is* part of the integrative process, but the *behaviour* of a species is characterised by extremely selfish competition between its sexually reproducing members. In fact, it was biologist George Williams' exasperation with this misrepresentation of natural selection as not being a selfish process that motivated him to write his famous 1966 book, *Adaptation and Natural Selection*—a publication that laid the foundations for the selfishness-justifying, right-wing theory of Sociobiology/Evolutionary Psychology. When, however, Williams' theory was dishonestly used to misrepresent our cooperative, moral instincts as being nothing more than a product of kin-selection-based selfish reciprocity (only assisting those who share your genes in order to propagate

your own), left-wing biologists initially tried to maintain that we do have unconditionally selfless moral instincts by arguing that they are derived from by-products of natural selection—what the biologists Stephen Jay Gould and Richard Lewontin described in 1979 as **'a lot of [building] cranes'** acting in conjunction with **'natural selection'** ('Darwin Fundamentalism', *The New York Review of Books*, 12 Jun. 1997, in which Gould elaborated upon his and Lewontin's by-products or 'Spandrels' theory). However, when Gould and Lewontin were unable to specify what the particular 'by-products'/'spandrels'/'cranes' were that achieved this feat of creating our genuinely moral instincts (the by-product was nurturing but they couldn't confront and admit that truth), the left-wing was left with nowhere to go but back to the now highly discredited 'cooperation is more advantageous than competition and can therefore be selected for', group-selection-type argument. In 1994, despite the situation where **'group selection has been regarded as an anathema by nearly all evolutionary biologists'** (Richard Lewontin, 'Survival of the Nicest?', *The New York Review of Books*, 22 Oct. 1998), the biologist David Sloan (D.S.) Wilson desperately tried to **'re-introduce group selection...as an antidote to the rampant individualism we see in the human behavioral sciences'** (David Sloan Wilson & Elliot Sober, 'Re-Introducing Group Selection to the Human Behavioral Sciences', *Behavioral and Brain Sciences*, 1994, Vol.17, No.4). In fact, it was D.S. Wilson's theory of Multilevel Selection that argued that natural selection operated at the group level as well as the individual level that (as described in chapter 1:10) E.O. Wilson commandeered to re-assert the right-wing emphasis on selfishness by claiming that even though multilevel selection supposedly confirmed we have unconditionally selfless instincts derived, he said, from warring between groups, it still allowed for the existence within us of selfish instincts derived from individual-level selection. But since the idea that our moral instincts were derived from warring between groups was unpalatable to the cooperation-not-competition-supporting, selflessness-not-selfishness-emphasising left-wing, they came up with an alternative, non-warring group-level explanation, which was that our moral instincts arose from having to cooperate to defend ourselves against predators. This argument was first put forward by the anthropologist Robert Sussman in his 2005 book, *Man the Hunted: Primates, Predators, and Human Evolution*. In summarising his theory, he said, **'Our intelligence, cooperation and many other features we have as modern humans developed from our attempts to out-smart the predator'** (presentation at the American Association for the Advancement of Science's Annual Meeting, 19 Feb. 2006). While the threat of predators would have encouraged cooperation, the biological reality is that it was not going to *make* us social; it was not going to overcome the fundamental problem of genetic selfishness, which is that wherever selflessness develops it is going to be subverted by selfish opportunists. The fact is, species have been living with the threat of predators since life first emerged and it has never been able to bring about full integration. What we see instead is the eventual development of dominance hierarchy as a means to try to contain the rampant selfish competition and opportunism. The whole reason E.O. Wilson put forward the argument that warring *between* groups was the full integration/eusociality threshold breaker was because there had to be an *extreme* need for cooperation if selfish opportunism was going to be defeated; there had to be a situation of conflict where groups of cooperators would defeat groups of non-cooperators. Arguing that groups of cooperators survived the threat of predators better than groups of non-cooperators doesn't create anything like the same selection pressure as actual conflict between groups. Indeed, this **'defence against predators'** argument overlooks the basic reason Wilson put forward the supposedly plausible, but actually still flawed, between-group warfare argument, and, apparently sensing this deficiency, what left-wing biologists *then* did was attempt to bolster it by adding the old 'by-products/many cranes/matrix of influences' illusion that Gould and Lewontin had first used. In a desperate attempt to somehow find a cooperation-emphasising biological account of human origins, the left-wing were now being forced to throw everything into the pot—group selection *and* a multitude of vague 'influences'! This is evident in the 2011 book *Origins of Altruism and Cooperation* where, for example, Robert Sussman and Robert Cloninger wrote in the Foreword that **'Research in a great diversity of scientific disciplines is revealing that there are many biological and behavioral mechanisms that humans and nonhuman primates use to reinforce pro-social or cooperative behavior. For example, there are specific neurobiological and hormonal mechanisms that support social behavior. There are also psychological, psychiatric, and cultural mechanisms'** (viii of 439). Yes, it was being alleged that a matrix of **'many biological and behavioral mechanisms'** created **'pro-social or cooperative behavior'**, but the question remains, how exactly did it do it? The *illusion* is that the origin of our moral instincts has been explained when it hasn't—but, again, in the desperation to counter the right-wing's selfishness-emphasising doctrine such extreme illusion was deemed necessary!

414 (As was mentioned in chapter 1 when the human-condition-avoiding, fundamentally dishonest right-wing biological theories for human origins were first presented, a much more complete description of all these right-wing and left-wing dishonest biological theories about human origins can be found in *Freedom Book 1* at <www.worldtransformation.com/freedom-the-denials-in-biology>.)

415 Clearly, avoiding the nurturing explanation for our moral instincts meant that neither the right-wing nor the left-wing were ever going to provide an accountable explanation of our moral instincts. While the selfishness-emphasising-but-still-human-condition-avoiding right-wing at least had the advantage of being able to truthfully emphasise that natural selection is a selfish process to support their position, the selflessness-emphasising-but-human-condition-avoiding left-wing had to rely on the patently dishonest 'cooperation beats competition' group-selection-type argument. And since both camps were, in any case, avoiding the entire issue of the human condition itself and, therefore, denying Integrative Meaning, it wasn't possible for either to explain what was explained in chapter 3, which is that even though the gene-based natural selection process is dedicated to developing the order of matter it couldn't, outside of the love-indoctrination scenario, select for the self-eliminating, unconditionally selfless traits that would allow full integration. And unable to access this human-condition-confronting reconciling explanation of the paradoxical nature of the gene-based natural selection process, these two positions became more and more desperate, with the right-wing position, which stressed the fact that genes are selfish, producing the selfishness-justifying theories of Social Darwinism ⟶ Sociobiology/Evolutionary Psychology ⟶ and Multilevel Selection that were described in chapter 1; and the idealism-stressing left-wing position, which attempted to emphasise the greater truth that natural selection is an integrative process, embarking on the journey just outlined, where group selection was put forward as the explanation for the origin of our moral nature, then, when group selection was dismissed as unsound biology, a multitude of influences was proposed, and then, when that failed, group selection was once again brought back into play, and then, when *that* inevitably fell short, a completely desperate combination of both group selection and a matrix of influences was resorted to! Of course, without the reconciling explanation as to why the right-wing and the left-wing emerged in the first place, both were

bound to become sillier and sillier, and, in the end, completely mad—and, as is being described, that is what happened: the right-wing ended up developing the extremely mad and dangerous theory of Multilevel Selection, while the left-wing ended up developing the extremely mad and dangerous theory of the SEM. (I might mention that I have written more about just how farcical science has become in *Freedom Book 1* at <www.worldtransformation.com/freedom-science-a-farce>. Based around the shocking revelations in the 2010 documentary, *Secrets of the Tribe* (directed by José Padilha), I describe the equally dishonest and thus never can be reconciled and thus now totally polarised right-wing and left-wing anthropological theories that are reviewed in the documentary and which seek to account for the behaviour of the relatively innocent Yanomamö Indians of the Amazon. This farcical situation that so characterises science now, where no one cares about the pursuit of truth-based understanding anymore, only about imposing their own twisted philosophy on the world, was perfectly captured at the end of the documentary when it played George and Ira Gershwin's well-known song *Let's Call the Whole Thing Off*, which features the lyrics, **'Things have come to a pretty pass…It looks as if we two will never be one…You like potato and I like potahto…Let's call the whole thing off'** (1937). Yes, science is so dishonest and thus inept now it may as well **'call the whole thing off'**, abandon its task of finding understanding of human behaviour! I might also mention that right-wing anthropologists' portrayal of so-called 'primitive' races as not being relatively innocent, but fierce and aggressive—as described in the program in relation to the Yanomami—will be exposed for the lie that it is later in pars 685-691.)

416 Significantly, in relation to the two arguments employed by the SEM to explain bonobo behaviour—that stable parties allowed bonobo females to form coalitions to counter male aggression for mating opportunities, and that those coalitions are successful in dominating males and eliminating aggression—in 2009 the leading architect of the SEM, Richard Wrangham, admitted that **'The circumstances in which females are able to form effective alliances among each other, and the frequency and effectiveness of this strategy, remain important** [unexplained] **problems for detailed examination in bonobos, chimpanzees, and other primates'** (*Sexual Coercion in Primates and Humans*, 2009, p.464 of 504). So the whole basis of the SEM, of the **'circumstances in which females are able to form effective alliances among each other'**, and the **'effectiveness of this strategy'** in stopping male aggression,

is being undermined by its leading architect and proponent! This is somewhat like E.O. Wilson conceding there were serious problems with his theory of Sociobiology when he moved on to develop his Multilevel Selection theory for eusociality! But where else could nurturing-avoiding biologists go in their efforts to explain bonobos? Nowhere—so, despite its **'important problems'**, support of the SEM continued. What will now be described is how nurturing-avoiding, mechanistic biologists tried to make the SEM more accountable of bonobos', and our ape ancestors', extraordinarily integrative, moral behaviour.

Chapter 5:10 The Self-Domestication Hypothesis

417 In terms of providing a nurturing-avoiding, human-condition-escaping explanation for bonobo behaviour, the problem with the SEM is that it only offers a supposed explanation for bonobos' *lack* of aggression, and so still falls well short of being able to provide a supposed explanation for bonobos' extraordinary **'personal restraint'**, **'respect for others'**, **'loving'**, **'moral'**, cooperative, harmonious, gentle state. And there is a *big* difference between not being aggressive and being loving. Given this shortfall, nurturing-avoiding, mechanistic biologists clearly needed to come up with a more sophisticated version of the SEM, one that could supposedly account for the bonobos' extraordinary cooperative, gentle, peaceful, loving nature. This supposed solution was provided in 2012, with the presentation of the so-called Self-Domestication Hypothesis (SDH) by anthropologists Brian Hare, Victoria Wobber and Richard Wrangham (one of the originators of the SEM) in a paper titled 'The self-domestication hypothesis: evolution of bonobo psychology is due to selection against aggression' (*Animal Behaviour,* 2012, Vol.83, No.3).

418 The first point to note is the use up front in the title of this paper of the idea that the **'evolution of bonobo psychology is due to selection against aggression'**, as if being able to **'select...against aggression'** is a normal, acceptable biological principle, a fait accompli, when it isn't. As emphasised, genes are selfish; outside of the love-indoctrination situation they don't allow for **'selection against aggression'** between sexually reproducing individuals. What is being put forward is the superficially persuasive but biologically flawed 'cooperation is more advantageous than competition and can therefore be selected for' argument that the SEM relies on, but to be putting it up front in their title is an outrageous

bluff, a desperate deception, an all-out effort to create the illusion that **'selection against aggression'** is sound, acceptable biology.

419 Another point that should be made before looking at the soundness or otherwise of the SDH is that by concluding their 2012 paper with the following statement, its proponents suggest that it not only explains bonobo cooperation but potentially human morality as well: **'The self-domestication hypothesis is therefore a potentially powerful tool for understanding the processes by which selection shapes both psychological and other seemingly unrelated traits, including those in humans.'** (Incidentally, the **'psychological' 'traits'** they refer to are *behaviours*, such as tolerance and playfulness, *not a psychosis*—so like the Multilevel Selection theory for eusociality, the SDH does not address the psychology of the human condition, rather it is just another desperate attempt to deny it.)

420 Which brings us to the accountability of the SDH. Supposedly inspired by research into domestic dogs and the experiments of the Russian scientist Dmitri Belyaev in domesticating silver foxes for the fur industry, the SDH proposes that **'selection against aggression'** inadvertently involves selection for youthfulness or juvenileness, so that adults in subsequent generations end up retaining **'pro-social'** juvenile behavioural traits such as **'increased tolerance'**, **'increased adult play'**, a **'decrease in predatory motivation'**, and **'decreased xenophobia** [fear of outsiders]**'**. (Hare, Wobber and Wrangham describe these traits as **'correlated by-products'** of the original SEM-derived **'selection against aggression'** process.) This 'juvenilisation' is known as the **'domestication syndrome'** because it is thought to account for changes between wild animals and their domestic descendants, such as changes between wolves and dogs.

421 Essentially, when the proponents of the SDH say that **'In addition to showing less severe forms of aggression compared to chimpanzees, bonobos show differences...that appear analogous to the domestication syndrome'** (ibid), what they are claiming is that the increase in **'pro-social'** behaviour that characterises the **'domestication syndrome'** bridges the gap between the mere **'lack of aggression'** that the SEM could only hope to account for, and bonobos' extraordinary **'personal restraint'**, **'respect for others'**, **'loving'**, **'moral'**, cooperative, harmonious, gentle behaviour. However, juvenile **'pro-social'** behaviour *does not* replace or override the selfish genetic need to aggressively compete for the fundamental biological needs of food, shelter, territory and mates. In fact, species that have been domesticated, like dogs and foxes, still aggressively compete for food, territory and mating

opportunities, something that is almost entirely absent in bonobo behaviour. So the SDH's claim to bridge the gap and explain bonobos' selfless, loving behaviour is simply another giant bluff. The truth is that without the involvement of love-indoctrination to first establish unconditionally selfless love in the system, retarding stages of maturation alone can't create a state of unconditionally selfless love.

422 Certainly, we humans have domesticated dogs and even silver foxes by selecting for tamer and more social juvenile characteristics, the effect of which has been to retard the development of these animals so that they retain the tamer, more tolerant and more **'pro-social'** behaviour of juveniles into adulthood—with the juvenile physical characteristics of floppy ears, more neotenous faces, etc, also carrying through into adulthood. However, while retarding development does bring the tamer, more tolerant, more **'pro-social'** characteristics of the juvenile stages into adulthood, it doesn't free the genes from their need to be selfish, and so doesn't eliminate selfish competition and aggression—*only* love-indoctrination can do that. Juvenileness is a form of more 'friendly' and tolerant socialness but it isn't a selfless state. In fact, as stated, dogs and foxes who have been 'puppyfied' still aggressively compete for the resources of food, territory and mating opportunities, behaviour that lies in stark contrast to bonobos' selfless and loving behaviour. Being more friendly, prepared to mingle and socialise—like domesticated dogs—is an improvement on the SEM's reduced aggression theory for bonobo behaviour, but the truth is it still falls well short of being able to account for bonobos' unconditionally loving behaviour.

423 A brief summary of the love-indoctrination process that was described in chapter 4 may help clarify this failure of both the SEM and the SDH.

424 The love-indoctrination process states that by selecting for longer infancies (which primates, with their arms semi-freed from having to walk on all fours, were able to do because they could hold a helpless infant), and for more maternal mothers, all within an ideal nursery environment of ample food and few predators, an infant's brain is able to be inscribed or indoctrinated with unconditionally selfless love, thus allowing it to grow up to behave selflessly. An accidental, but fortuitous, side-effect of this indoctrination or training of a mind in selfless, truthful, effective thinking, however, was the emergence of consciousness, for once liberated, the conscious mind could then support the development of selflessness by consciously favouring (especially in mate selection)

more selfless individuals, thus greatly speeding up the development of selflessness. Since the training in selfless love tended to wear off with age, selection for selflessness became, to a degree, a selection for youthfulness, resulting in more youthful, neotenous characteristics in adults. Both the SEM and the SDH, in effect, describe this process *but without* the key element of the involvement of the nurturing of unconditional selflessness; they omit the whole process of love-indoctrination—a glaring omission that created two problems. First, it necessitated the development of the flawed, dishonest **'selection against aggression'** argument to attempt to explain how selfless cooperation could emerge without love-indoctrination. And, second, since this **'selection against aggression'** could not, in fact, create unconditionally selfless love, only a supposed reduction in aggression, it could *only* ever lead to less aggressive, tamer, more tolerant, more social characteristics in adults. The point is, unconditionally selfless love is *not* produced by the situation espoused by either the SEM or the SDH, whereas love-indoctrination *does* produce love, which *can* then be actively selected for. So neither the SEM nor the SDH explain the extraordinary **'personal restraint'**, **'respect for others'**, **'loving'**, **'moral'**, unconditionally selfless, cooperative, harmonious, gentle behaviour we see in bonobos and in our own moral instincts. The delaying of the onset of adult competitiveness and aggression that the SEM and the SDH describe does *not* produce an unconditionally loving individual; to produce that you have to be selecting for individuals that have been nurtured with love, but that is a process neither the SEM or the SDH recognise.

425 To suggest that selecting for juvenileness can lead to less aggressive juvenile **'psychology'** being carried through to adulthood to the point of eliminating selfishness was simply a deception—another bluff—by the proponents of the SDH. In interviews Hare has conducted about the SDH we can see him trying to bridge this gap between the extraordinarily cooperative, gentle, peaceful, unconditionally selfless, **'loving'** behaviour that bonobos display, and what the SDH is able to supposedly explain, namely **'pro-social'** traits such as tameness and **'increased tolerance'**, when he describes bonobos as **'peaceful'** ('Why Bonobos Don't Kill Each Other', *The New York Times*, 5 Jul. 2010) and **'kind'** creatures (Brian Hare speech at Poptech 2010; see <www.wtmsources.com/139>) who **'absolutely are upset if there is any hint of aggression in the group'** ('Bonobos–Making Love Not War', *Catalyst*, ABC, 20 Sep. 2007), and who find **'joy in working with others'** ('Dogged', *Smithsonian*, Oct. 2007)—as

if those traits and emotions are what his hypothesis is able to explain the origins of.

[426]This attempt by Hare to bridge the gap between the extraordinarily cooperative, selfless behaviour that bonobos display, and what the SDH is allegedly able to explain, is very similar to Wrangham's earlier claim that the SDH is able to account for bonobos' extraordinary (and these are his words) **'personal restraint'**, **'respect for others'**, **'loving'**, **'moral'** behaviour. Furthermore, it was also described earlier how, in their 2012 paper, Hare, Wobber and Wrangham said their **'self-domestication hypothesis is...a potentially powerful tool for understanding the...psychological...traits...in humans'**—**'a potentially powerful tool for understanding'** the origin of our unconditionally selfless moral nature no less! Hare has also proposed that **'bonobos display...what might be thought of as our better angels'** ('"Hippie chimp" genome may shed light on our dark side', *Science* on NBCNews.com, 13 Jun. 2012; see <www.wtmsources.com/141>), which again is our unconditionally selfless moral nature!

[427]The truth is, domestication or juvenilisation cannot *create* this type of behaviour without there having been love-indoctrination, it can only stymie the growth of adult types of behaviour, and so Hare and Wrangham are having to exaggerate its effect to account for bonobos' love and gentleness. Yes, it is only as *part* of the nurturing, love-indoctrination process that juvenilisation can produce real **'loving'**, **'moral'**, **'peaceful'**, **'kind'**, **'joy'** in cooperation, **'personal restraint'**, and **'respect for others'**, and an abhorrence **'of aggression'**.

[428]All my publications have included a description of the love-indoctrination, mate selection process, with an account of humans' domestication of dogs appearing in my 1988 book, *Free: The End Of The Human Condition* (see <www.worldtransformation.com/free-love-indoctrination>), and a description of humans' domestication of both dogs and foxes appearing in the 2009 edition of my book *The Great Exodus* (see <www.worldtransformation.com/exodus-mate-selection>). (I should note that Wrangham was sent *Free* in 1988, and over 2005-2006 all three SDH authors were sent another of my publications, *The Human Condition Documentary Proposal*, which also contains a description of the love-indoctrination, mate selection process (see <www.worldtransformation.com/doco-maternalism>)). The reason I referred to how **'domesticated dogs are derived from their common ancestral wild type by neoteny—retarding development at some juvenile stage'** (*Free: The End Of The Human Condition*, p.142 of 223) was because the domestication of dogs and foxes does dramatically illustrate some of the aspects involved in the

love-indoctrination, mate selection process, particularly how powerfully effective conscious selection can be in producing a change (it **'Explains [the] speed of human development'** (ibid. p.142)), and how the development of stages of maturation is retarded by selecting for youthfulness (it **'is a marvellous illustration of the development of neoteny'** (ibid. p.141)). However, I explained that **'self-selection'** (as I termed the process that bonobos and our ape ancestors employed to assist in the development of unconditionally selfless behaviour) differs to the selection we employed to domesticate dogs and foxes in that without love-indoctrination to create the unconditionally selfless love that could then be selected for, **'self-domestication'** (or, again, as I originally termed the process in all my books, **'self-selection'**) can *only* achieve tamer, more tolerant and more social characteristics in adults, *not* unconditionally selfless love. As I emphasised in *Free*, **'On their own genes could not develop selflessness but once there was love-indoctrination [they could]'** (p.47).

[429]An illustration of the difference between the effects of love-indoctrination and the effects of domestication put forward by the SDH, which is merely selecting against aggression, can be seen in the work of the famous 'dog whisperer' Cesar Millan. Millan is forever informing dog-owners that the mistake they are making in trying to control their dogs is that they are attempting to love them into behaving less aggressively when what they have to do to achieve control and reduce aggression is impose dominance. Millan is, in effect, recognising that domesticated dogs haven't overcome the 'animal condition' of selfishly having to ensure their genes reproduce, which is why they are still highly competitive for food, shelter, territory and a mate—a competitiveness that can only be partially overcome through the imposition of a dominance hierarchy, where each individual accepts its position in a hierarchy that is determined according to the competitive strengths of the various individuals involved.

[430]I might mention that in 2013 Hare published a bestselling book titled *The Genius of Dogs*, in which he again exaggerated the effects of domestication, this time in regard to dogs. Hare writes that in contrast to wolf packs where the leaders are the dominant breeding pair, **'the leader of a feral dog pack is the dog who has the most friends'** (B. Hare & V. Woods, 2013, p.174 of 367). This statement is supposedly supported by a report titled 'Effect of affiliative and agonistic relationships on leadership behaviour in free-ranging dogs', which found that **'formal dominance in free-ranging dogs may be a more consistent predictor of leadership than agonistic dominance'**

(Roberto Bonanni et al, *Animal Behaviour*, 2010, Vol.79, No.5). Although this finding distinguishes between **'formal dominance'** (in which the subordinate animal signals its submission during a greeting ceremony), and **'agonistic dominance'** (in which the subordinate animal signals its submission following a fight), it still makes very clear that feral dogs are highly competitive and operate within a **'dominance'** hierarchy.

431 In summary, there is a quantum difference between the claimed reduction in aggression that both the SEM and the SDH can supposedly produce and the very real love we see in bonobos. Neither the SEM or the SDH *begin* to offer an accountable explanation of that species' extremely loving behaviour, whereas the nurturing, love-indoctrination explanation *fully* accounts for it. Although genes are a tool for developing order, they are limited in the sense that they can't normally develop unconditional selflessness, which means that genetics is a selfish, cold, loveless process; it is not going to produce bonobos' warm, gentle, cooperative, loving behaviour—unless the love-indoctrination path is taken, for it alone has the power to superimpose love on an essentially selfish system. As Drummond said of nurturing love, it was only **'once this fire began to warm the cold hearth of Nature and give humanity a heart, the most stupendous task of the past was accomplished'**. In contrast, the SEM and the SDH are desperate and hopelessly flawed attempts to explain bonobo behaviour and the origins of our moral nature without admitting the critical role of nurturing.

432 (A more complete description of the SEM and the SDH and their limitations can be found in *Freedom Book 1* at <www.worldtransformation.com/freedom-social-ecological-model>.)

433 Having now analysed the mechanisms of both the SEM and the SDH, we now need to describe the immense danger they present to the human race.

Chapter 5:11 <u>End play for the human race</u>

434 As stated at the beginning of this chapter, the great danger of the practice of denial is that, in the end, it becomes so entrenched and sophisticated that it locks humanity onto a path to terminal alienation; to total madness and extinction. The development of the denial of the truth that nurturing created humanity, firstly in the form of the SEM, and in its most recent and most sophisticated incarnation, the SDH, dramatically illustrates this great danger.

435 The bonobos offer the most powerful evidence of the nurturing origins of our unconditionally selfless moral soul, but the SEM and, to an even greater extent, the SDH attempt to not only deny that evidence but bury it under seductive yet totally false explanations for their gentle, loving, cooperative nature. Misappropriating aspects of the truth about bonobos, such as their extraordinarily loving cooperation, their neoteny, and even burgeoning intelligence (which the SDH does using a version of the Social/Machiavellian Intelligence Hypothesis, the dishonesty of which was briefly explained in ch. 5:5), and using that to evidence the SDH, is a very sophisticated way of giving credibility to the lie that nurturing had no role to play in the development of our moral soul. But to bury such evidence of the origins of our unconditionally selfless moral soul that created the cooperative, integrative state that is humanity, is to threaten the human race with permanent estrangement from the truth about our all-loving true self or soul, which is the truth we need if we are to properly understand and, by so doing, heal our psychologically alienated condition. Burial of the truth about our soul stands in the way of us ever gaining an honest, ameliorating understanding of ourselves—of our origins, our present condition and future potential.

436 Further, to deny the importance of nurturing is to deny the importance of the main activity we need to practise if we are to produce humans who are sound and secure in self. Indeed, it is *only* through the nurturing of our offspring that the human race can hope to become healthy, psychosis-and-neurosis-free humans once again—to, as Montagu said, put **'man back upon the road of his evolutionary destiny from which he has gone so far astray'** and restore **'health and happiness for all humanity'**.

437 Yes, if we refuse to admit the critical role nurturing has played in the emergence of humanity and, as a direct result of that heritage, in the sound upbringing of humans today, then levels of alienation will only increase and terminal alienation *will* soon destroy the human race. Like E.O. Wilson's use of the Multilevel Selection theory to deny the true nature of the human condition, the SEM's and the SDH's denial of the importance of nurturing is such an alienating lie that it *will* lead the human race into a state of irretrievable madness.

438 Indeed, as if the danger just described wasn't enough, the even greater threat posed by the SEM/SDH is that it is now being developed into *another* denial of the true nature of the human condition that is even more seductive and thus dangerous than Wilson's because it appears to

take into account, albeit in a totally distorting manner, the most crucial evidence we have about human origins from our closest relatives, the bonobos! As the lead author of the SDH, Brian Hare, has said, **'They** [bonobos] **have done something in their evolution that even humans can't do. They don't have the dark side we do...If we only studied chimps, we'd get a skewed view of human evolution'**, and **'bonobos display...what might be thought of as our better angels'** ('"Hippie chimp" genome may shed light on our dark side', *Science* on NBCNews.com, 13 Jun. 2012; see <www.wtmsources.com/141>). So, according to Hare, chimpanzee-like instincts in humans account for our **'dark side'**, while the instincts allegedly accounted for by the SEM/SDH, as demonstrated by the bonobos, gave rise to our goodness, our unconditionally selfless moral instincts, our **'better angels'**. This argument presents a model for our 'good and evil'-afflicted human condition that is similar to that provided by Wilson's Multilevel Selection theory for eusociality, except that our 'good' instincts are supposedly derived from factors espoused by the SEM/SDH, rather than cooperation forged through warring with other groups, as Wilson suggested. As was emphasised when the Multilevel Selection 'explanation' for the human condition was presented and exposed, we humans *do* have unconditionally selfless, loving, 'good' instincts but they were derived from the nurturing love-indoctrination process that bonobos are developing; and we *do* practise divisive, 'bad' behaviour, but, again, that is derived not from genetic competitiveness but from a psychosis that emerged when our conscious mind was liberated by the love-indoctrination process, and, once liberated, rapidly developed to challenge our *completely* loving (not partially loving and partially selfish) 'good' instincts for the role of managing our lives. So, what Hare, Wobber and Wrangham are doing is precisely what Wilson was doing with his Multilevel Selection theory, which was to bring the human condition to our attention, but only to trivialise it—to subvert the truth about the all-important issue of our psychologically troubled condition!

439 This is all overwhelming; there are so many dangers associated with the SEM/SDH that it hardly bears thinking about. With E.O. Wilson's all-out atomic bomb attack on the truth about the human condition, and now Hare, Wobber and Wrangham's no-holds-barred asteroid attack on the best evidence we have for the origins of our moral soul, and also on the truth about the human condition, the honesty and resulting insight needed to save the human race is being buried at the bottom of the deepest, darkest ocean trench that has been filled with reinforced concrete for

good measure. It's truth-hating behaviour of the highest order. Basically, mechanistic science has become completely deranged—and since science is the facility charged with delivering us from the human condition, this state of affairs represents end play for the human race!

440 Without the exposé being presented in this book all hope for humanity would be lost. But this exposé still has to be recognised by the scientific establishment—a recognition that, as will now be described, has so far not occurred; indeed, for 30 years now all it has received from the scientific establishment is obscenely irresponsible, stone-wall resistance.

Chapter 5:12 *The* great obscenity

441 I and the 50 Founding Members of the World Transformation Movement (the organisation I established in 1983 to research and promote understanding of the human condition because mechanistic science wouldn't approach this all-important subject) have been trying to interest the scientific community in the world-saving insights into the human condition that are being presented in this book for a full 30 years now, but, save for a handful of supportive scientists and some positive responses from other eminent scientists and thinkers, it has so far all been to no avail. All our submissions have been either ignored or rejected by the scientific establishment. In fact, the situation is much, much worse than this, for instead of attracting interest, debate and support, I and the World Transformation Movement (WTM) have been so ferociously attacked by the two biggest, left-wing (dogmatic, pseudo idealistic, 'let's pretend there's no human condition and the world should just be ideal', dishonest) media organisations in Australia, namely our national public broadcaster, the Australian Broadcasting Corporation (ABC), and Fairfax Media, that I was made a pariah and the WTM completely marginalised. We endured this situation until, after 15 years of emotionally exhausting, and, for such a small group, financially taxing, defamation actions taken by us (which, we've been told by legal experts, involved the biggest defamation case in Australia's history), we finally managed to right the extremely serious wrong. In 2010, after an initial court judgment found that my work is of **'such a poor standard that it has no support at all from the scientific community'**(!), three judges in the New South Wales Court of Appeal unanimously found that the earlier lower court ruling did **'not adequately consider' 'the nature and scale of its subject matter'**, in particular

'that the work was a grand narrative explanation from a holistic approach, involving teleological elements', and that other important submissions **'were not adequately considered by the primary judge'**, including that the work can make **'those who take the trouble to grapple with it uncomfortable'** because it **'involves reflections on subject-matter including the purpose of human existence which may, of its nature, cause an adverse reaction as it touches upon issues which some would regard as threatening to their ideals, values or even world views'**! The earlier court ruling hopefully gives the reader some idea of the scale of the persecution that my work has had to endure, and also some idea of the immense relief that the final ruling brought to those supporting these understandings—in particular to myself; to my partner Annie Williams, whose loving support enabled me to cope with all the years of persecution (in particular it was Annie's care that enabled me to recover from 10 years of the debilitating illness of Chronic Fatigue Syndrome [CFS] that I developed in 1999 after endless years of fighting the persecution); to my brother Simon, who started the base group of people supporting these insights into the human condition; to my very close friend Tim Macartney-Snape, whose high standing in the Australian community (Tim is a twice-honoured Order of Australia recipient) had been jeopardised because of his courageous, unwavering support of my work; and to all the other Founding Members of the WTM who have also had to endure horrible persecution, ostracism and suffering. (I might mention that with regard to my CFS, apparently Charles Darwin also suffered from it, presumably because he too was under enormous pressure for having his revolutionary ideas **'greeted with violent and malicious criticism'** (*The Origin of Species*, 1968 Penguin edn, title page). Indeed, there has been some talk of renaming CFS the 'Charles Darwin Syndrome' (Roger Burns, *Chronic Fatigue Syndrome Changing the Name*, Sep. 1996; see <www.wtmsources.com/123>).)

[442] Our monumental struggle against persecution—which has been so great that we have had to forgo having children to ensure we have sufficient resources of time, energy and funds to effectively resist the persecution and maintain our efforts to ensure these ideas lead to the transformation of the human race—is documented at <www.worldtransformation.com/persecution>. After all that we have been put through, the fact that our little, but mighty, band of brothers and sisters supporting these human-race-liberating insights is still standing and able to mount this further assault on the citadel of denial/lying (mechanistic science) to try to crack it open and free the human race, *is* an absolute miracle.

443 Yes, this denial—this ignoring, rejecting and persecuting—has been completely and utterly unjustified; in fact, it has been *totally* irresponsible, *obscenely* irresponsible. *Certainly* all the world-saving insights that appear in this book, insights that I have been putting forward for 30 years, bring the historically unbearably confronting issue of the human condition into stark focus, *but all are presented within the framework of the compassionate explanation of the human condition, which means the need for denial of them has been removed*. As emphasised from the beginning of this book, the human race has always lived in hope, faith and trust that one day the redeeming explanation of our psychologically distressed and insecure human condition would be found, and the most fundamental reason of all for freedom of expression to be maintained in the world is to keep the door open to that possibility. Science especially, having been charged with that greatest of responsibilities of finding these human-condition-liberating insights, must remain open to human-condition-confronting-and-explaining insights (remembering, of course, that E.O. Wilson's claimed explanation of the human condition was *not* actually concerned with confronting and analysing the human condition, merely with avoiding it). *Yes, science holds the ultimate responsibility to consider, not ignore, or worse, persecute, scientific analysis of the human condition!* What has transpired has been obscenely irresponsible.

444 So you can imagine our shock when, in 2012, we learnt that the three leading anthropologists responsible for putting forward the SDH, Wrangham, Hare and Wobber—all of whom, as has been mentioned, were made aware many years ago of my love-indoctrination explanation for the origins of our and the bonobos' moral instincts—made no acknowledgment or even mention of my synthesis in their 2012 paper, despite acknowledging the work of many other researchers in a detailed section on **'evolutionary explanations for reduced aggressiveness in bonobos relative to that in chimpanzees'**. Worse, it would appear that since they were each informed of my synthesis (in the case of Wrangham, on four separate occasions, the earliest being in 1988), what they have done is take virtually all the elements from my synthesis—such as the bonobos' ability to throw light on our origins, and specifically the origins of our morality; that their social groups are much more stable than those of chimpanzees; the role of females in taming male aggression; the liberation of

consciousness; the role of self-selection; the neotenising, juvenilisation process; the use of the domestication of dogs and foxes as an illustration of the neotenising, juvenilisation process; the significance of ideal ecological conditions; the use of sex as a device to reduce tension; the reduced dimorphism between the sexes; the reliance of males on their mothers for social standing; the lack of aggression between groups of bonobos; the lack of routine hunting by bonobos, etc, etc—and, leaving out anything to do with nurturing, presented it as **'A new hypothesis'** (Ed Yong, 'Tame Theory: Did Bonobos Domesticate Themselves? A new hypothesis holds that natural selection produced the chimpanzee's nicer cousin in much the same way that humans bred dogs from wolves', *Scientific American*, 25 Jan. 2012). While it is *extremely* irresponsible to ignore and reject world-saving insights into the human condition, it is so, so much worse to actually take those insights and wantonly subvert or misappropriate the truth they contain. If that is indeed the case, and we believe there is no other plausible interpretation, then that is the very greatest of crimes against humanity.

445 What must be clear to the reader overall is that the obscenely irresponsible treatment of these world-saving insights simply must stop. The obscenity of the situation was made *dramatically* clear when:

- in December 2012 it was announced that an American billionaire had pledged $200 million to Columbia University's **'accomplished scholars whose collective mission is both greater understanding of the human condition and the discovery of new cures for human suffering'** (*The Educated Observer*, Winter 2013); and,
- in January 2013 in Europe the European Commission **'announced that it would launch the flagship Human Brain Project with a 2013 budget of €54 million (US$69 million), and contribute to its projected billion-euro funding over the next ten years'** ('Neuroscience: Solving the brain', *Nature*, 2013, Vol.499, No.7458) with the goal of providing **'a new understanding of the human brain and its diseases'** (Press Release, The Human Brain Project, 28 Jan. 2013; see <www.wtmsources.com/111>) to **'offer solutions to tackling conditions such as depression'** ('Scientists to simulate human brain inside a supercomputer', *CNN International Edition*, 12 Oct. 2012; see <www.wtmsources.com/129>); and,
- in April 2013 the President of the United States, Barack Obama, announced a **'Brain Initiative'**, giving **'$100 million initial funding'** (*The Sydney Morning Herald*, 4 Apr. 2013) to mechanistic science, in **'a research effort**

expected to eventually cost perhaps ten times that amount' ('Neuroscience: Solving the brain', *Nature*, 2013, Vol.499, No.7458), to also find **'the underlying causes of... neurological and psychiatric conditions'** afflicting humans in order to **'develop effective ways of helping people suffering from these devastating conditions'** (US National Institutes of Health, accessed Apr. 2013; see <www.wtmsources.com/114>); and,

- in April 2013 *BBC News Business* reported that **'Lord Rees, the Astronomer Royal and former president of the Royal Society, is backing plans for** [Cambridge University to open] **a Centre for the Study of Existential Risk** [meaning risk to our existence]. **"This is the first century in the world's history when the biggest threat is from humanity," says Lord Rees'** ('How are humans going to become extinct?', accessed Jul. 2013; see <www.wtmsources.com/120>). The article then referred to Oxford University's Future of Humanity Institute that was established in 2005, which is **'looking at big-picture questions for human civilization...**[and] **change...**[that] **might transform the human condition'** (Future of Humanity Institute, University of Oxford, accessed Jul. 2013; see <www.wtmsources.com/118>), quoting its Director and advisor to the Centre for the Study of Existential Risk, Nick Bostrom: **'There is a bottleneck in human history. The human condition is going to change. It could be that we end in a catastrophe or that we are transformed by taking much greater control over our biology'**!

446 *Finally*, *ultimately*, *at last*, the human race is attempting to address **'the human condition'**!! Yes, the truth is that the underlying, core problem involved in *all* of the **'big-picture questions'** and the **'Existential Risk'** of **'catastrophe'** is our **'neurological and psychiatric'**-troubled **'human brain'**—*that's* what had to be addressed. The **'biggest threat is from humanity'** itself, our species' inability to **'discover' 'the underlying causes of'** our psychologically distressed lives and, by so doing, end this **'bottleneck in human history'** and achieve the way-overdue **'cures for human suffering'**, which will **'transform' 'our biology'**. Yes, it *has* been a case of self-discovery or self-destruction—understand the human condition or all is lost! BUT, the mechanistic scientific paradigm, including all of its universities, has proved completely incapable of adopting a denial-free approach and truthfully addressing the human condition, and yet now, when the world is absolutely on its knees and desperate for the neurosis-and-psychosis-addressing-and-healing real insight into the human condition, that same mechanistic paradigm is being given millions and millions and millions

and millions of dollars to turn around and do what it has already shown it can't, but what I and the WTM have already done! And, I might add, we have done this, 40 years of work in all, without *any* outside financial support—from academic institutions or from public or private benefactors. We have funded the whole effort from our own self-sufficient initiatives, efforts, and contributions. Indeed, as has been mentioned, rather than receiving any encouragement or financial support from the world at large, we have been *attacked* by the establishment and had to generate the wherewithal ourselves to fight and defeat that enormously powerful institution-backed attack! For example, the principal media attack against us was carried out by the Australian Broadcasting Corporation, a body which receives well over a billion dollars annually in government funding. We have never been invited to participate in any major scientific or public forums, or been given a wage or grant by an academic institution to investigate any of the subjects that we have so effectively studied; quite the reverse—we've been brutally ostracised and forced to sustain and pay our own way to carry on our work and to fight the massive forces opposing it. Talk about David and Goliath! And it is not as though these huge financial initiatives that are now, in desperation, being made to specifically investigate the human condition can be oblivious to our existence and work because, for one thing, the WTM has the domain name **'humancondition.com'** and any planned attempt, such as these, to supposedly properly address the human condition would, one would assume, have to be aware of our existence. So the mechanistic paradigm is saying it is finally going to bring **'greater understanding of the human condition'**, but has not changed tack at all and acknowledged those who have done just that! The hypocrisy and obscenity of what is going on is *absolutely astronomical*.

447 In regard to how utterly dishonest and completely self-defeating the human-condition-*avoiding*, reductionist, mechanistic paradigm's attempt is to solve the mental problems that now threaten to overwhelm humanity, I should re-include the reference Professor Prosen made in this book's Introduction to an opinion piece by Benjamin Y. Fong of the University of Chicago that was published in *The New York Times* in August 2013. Fong wrote that **'The real trouble with the Brain Initiative is…the instrumental approach…**[such **biological reduction** is] **intent on uncovering the organic "cause"…of mental problems…rather than looking into psychosocial factors…By humbly claiming ignorance about the "causes" of mental problems…neuroscientists**

unconsciously repress all that we know about the alienating, unequal, and dissatisfying world in which we live and the harmful effects it has on the psyche, thus unwittingly foreclosing' the ability to **'alleviate mental disorder'**.

448 Yes, Fong's right, as he says here and elsewhere in his article, mechanistic science's **'synthetic'** focus on the **'organic'** rather than the **'psychological'** nature of mental problems can only end in denying humans **'the possibility of self-transformation'**. As Professor Prosen recognised in his Introduction, it is *only* the *psychosis-addressing, human-condition-confronting* position that I have taken that could hope to find, and now *has* found, the reconciling and human-race-transforming understanding of the human condition—and yet it is this approach that has been treated as heretical, an anathema and a threat by the mechanistic scientific establishment!!

449 A very brief summary of the 30 years' worth of submissions to the scientific establishment of these world-saving insights into the human condition, into the origins of our moral nature and conscious mind, and into the integrative meaning of existence is provided below; alternatively, the full presentation can be accessed at <www.worldtransformation.com/full-history-of-rejection>. (Incidentally, this link provides details of the presentations of the whole synthesis, including the nurturing explanation for our moral soul, that have been made to Wrangham, Hare and Wobber.) These 30 years' worth of submissions represent an *extraordinarily* long and determined yet futile effort—in fact, the ill-treatment they document is horrific, truly obscene—to have all these fully accountable insights properly considered by the scientific establishment and conveyed to the wider world. As has been emphasised, in the case of the origins of our moral instincts, there has been 140 years of dishonest biological thinking on this issue, mountains of books written and oceans of wasted effort since Fiske first solved the problem with the nurturing explanation—and 30 years now since I presented the nurturing answer to the problem in accompaniment with the explanation of the human condition, which is 30 years of terrible human suffering and acts of atrocity that have occurred on Earth from a lack of self-understanding in humans that should never have happened!!!

- In 1983 I wrote to Sir David Attenborough and Professor Stephen Jay Gould, presenting these insights, but received no real response from either.

- In 1983 I personally submitted an 8,000 word summary of this synthesis (<www.worldtransformation.com/nature>) to the then editor of *Nature* magazine, Sir John Maddox, and the then Features Editor of *New Scientist* magazine, Colin Tudge—both of whom rejected it.
- In 1988, 800 copies of my book *Free: The End Of The Human Condition* (<www.worldtransformation.com/free>) with first-rate publicity packages were sent to every relevant scientist and journal in the world for review, but despite a few significant commendations there was no real response.
- In 1989 Professor John Wren-Lewis personally presented *Free* to 10 science journals including *Nature*, *New Scientist* and *Endeavour*, none of which responded.
- In 1989 a booklet titled *Reconciliation* that contained a summary of the synthesis was circulated to 600 scientists, scientific journals and other relevant parties—to little response.
- In 1991 over 1,000 copies of my book *Beyond The Human Condition* (<www.worldtransformation.com/beyond>) with first-rate publicity packages were circulated to scientists, journals, universities, and relevant institutions and media—to little response from the scientific establishment overall.
- In 1992 *Beyond* was launched at the National Museum of Kenya and over 70 copies were given and sent to eminent scientists and influential people—however, no lasting support eventuated.
- In 1992 Professor Wren-Lewis published a paper in which he plagiarised my work, claiming the insights were his own! Redress was achieved and Wren-Lewis ceased his involvement (including a directorship) with our project.
- In 1993, 76 first-rate publicity packages with copies of *Beyond* were sent to all the leading literary agents and publishers in the world, but all declined to represent or publish the book.
- In 1995 two highly defamatory publications—an Australian Broadcasting Corporation (ABC) television program, and a full page *The Sydney Morning Herald* newspaper article—were made about me, my work and its supporters (<www.worldtransformation.com/persecution>).

- In 2002, 70 copies of my book *A Species In Denial* (<www.worldtransformation.com/asid>) were sent to the leading literary agents in the world, and another 70 copies were sent to the major international publishers, but all declined to represent or publish the book.
- In 2003 a further 800 copies of *A Species In Denial* with first-rate publicity packages were circulated to scientists, journals, universities, and relevant institutions and media. Despite a foreword by Professor Charles Birch, a commendation from Professor John Morton, and becoming a bestseller in Australia and New Zealand, the scientific community all but failed to respond.
- Between 2004 and 2006, 2,500 copies of a documentary proposal (<www.worldtransformation.com/doco>) on the human condition (in which I outlined all the main biological explanations that are presented here in *IS IT TO BE*), were sent to scientists, scientific publications and organisations. While the proposal received over 100 commendations from leading scientists and thinkers, and engaged Professor Harry Prosen's ongoing support of these explanations, it did not produce any substantial, long-term interest from the scientific community.
- In 2005 my proposal to present 'The Citadel Of The Darwinian Revolution—The Biology Of Our Human Condition—At Last Explained', at the Annual Meeting of the American Association for the Advancement of Science (AAAS) titled *Grand Challenges, Great Opportunities*, was rejected by the Program Committee.
- In 2005 an abstract of my paper 'Nurturing as the Prime Mover in Primate Development and Human Origins' (<www.worldtransformation.com/ips-2006-congress>) was submitted for presentation at the International Primatological Society's (IPS) 2006 Congress in Uganda, but was rejected.
- In 2006 journalist Jo Sandin was unable to include reference to my nurturing explanation for human and bonobo moral behaviour in her 2007 book about the bonobos at the Milwaukee County Zoo, *Bonobos: Encounters in Empathy*.
- In 2006 my book *The Great Exodus: From the horror and darkness of the human condition* (<www.worldtransformation.com/exodus>) was pub-

lished online, and sent to many relevant scientists—to no response from the scientific community.

- In 2008 the World Transformation Movement began presenting its online Introductory Videos—however, as yet there has been little response from the scientific community.
- In 2009 my book *Freedom* (<www.worldtransformation.com/freedom>) was published online, but has not yet generated any real response from the scientific community.
- In 2011 my book *The Book of Real Answers to Everything!* (<www.worldtransformation.com/real-answers>) was published online, but has so far failed to attract significant support from the scientific community.
- In 2012 Wrangham, Hare and Wobber published the paper 'The self-domestication hypothesis: evolution of bonobo psychology is due to selection against aggression', making no acknowledgment or even mention of my love-indoctrination synthesis, despite each being previously informed of it (in the case of Wrangham on four separate occasions)—even appearing to subvert and misappropriate virtually all the elements of my synthesis, while leaving out anything to do with nurturing.

---

450 History certainly teems with examples of those who have been persecuted for telling the truth, for daring to defy the great denial blanketing our planet, Christ's crucifixion being the most famous, but we now have another to match that ultimate example, which is what has been done to me; what's that rhetorical question in the Bible: **'was there ever a prophet** [a denial-free thinker] **your fathers did not persecute'** (Acts 7:52). HOWEVER, when all the truth about human existence is accompanied by the greater, dignifying, compassionate, redeeming and rehabilitating full truth of the explanation of the human condition, *that denial is no longer justified*. And, of course, that denial is so habituated—our fear of the truth has soaked into our bones—but, nevertheless, that door to the possibility of understanding the human condition *must be kept open*. There *has* to be sufficient strength of character, soundness, left on Earth for the full truth to be recognised and supported when it emerges, which—as has been fully evidenced—it now has, because if there isn't then all is lost. So the

question remains: *is it to be terminal alienation or truth and transformation for the human race?*

451 Since the great achievement of life on Earth has been the development of consciousness, it follows that for those charged with the responsibility of overseeing its maturity to enlightenment and sanity—namely the scientific establishment—to abuse that responsibility is the greatest of obscenities; worse, it is a spear through the very heart of all of life and meaning on Earth. It is the meanest, most bitter, most selfish, most bloody-minded, most hateful of the truth, most unnecessary behaviour this planet has ever witnessed.

452 The remaining three chapters in this book explain the emergence of consciousness in humans, the story of humanity's journey from ignorance to enlightenment, and, finally, how the transformation of the human race takes place, should the path to transformation be chosen over the path to terminal alienation. In regard to the next chapter's explanation of how consciousness emerged, unlike the mysteries of the human condition, of our meaning, and the origin of our moral soul, our ability to understand consciousness depended on recognising so many historically denied truths that mechanistic science can't be as easily criticised for ignoring its discovery as it can be for ignoring the other key insights—which is why the main damnation of science for its ill-treatment of these world-saving insights appears here in chapter 5.

Chapter 5:13 Nurturing now becomes a priority

453 Before progressing to those final chapters it needs to be emphasised that having found the compassionate biological understanding of the upsetting battle that humans have had to wage against the ignorance of our instinctive self as to the fundamental goodness of our conscious self, our species' preoccupation with that battle can at last end and focus return to the nurturing of our children. Further, now that we have found that *very* good reason as to *why* parents have been unable to nurture their children with anything like the unconditional love children instinctively expect and need, no longer is it necessary to deny the importance of nurturing both in the maturation of our species and in our own lives. Yes, with the arrival of understanding the terrible guilt that parents have felt for not being able to adequately nurture their offspring departs. Parents can now

understand that their inability to adequately love their offspring is *not* their fault, but the product of the ever-accumulating levels of upset in the human race as a whole that resulted from our species' heroic 2-million-year search for knowledge, specifically for self-knowledge, understanding of why we conscious humans are good and not bad. *So yes, we can finally now admit the importance of nurturing and turn our attention back to the nurturing of our offspring—in fact, it now becomes a matter of great urgency that we do both these things.*

454 As stated in par. 436, the nurturing of children is what is required to produce adults who are sound and secure in self. Ultimately, it is *only* through the nurturing of our offspring that the human race can hope to become healthy—namely psychosis-and-neurosis-free—once again; to, as Ashley Montagu said, put **'man back upon the road of his evolutionary destiny from which he has gone so far astray'** and restore **'health and happiness for all humanity'**. As will be explained in chapter 8, with understanding of the human condition found all humans can *immediately* be free of the agony of the human condition by taking up the Transformed Way of Living (which involves adopting a mental attitude where you leave that agony behind as dealt with), but to produce humans who are *completely* free of psychosis and neurosis and not needing to take up the Transformed Way of Living, they need to be nurtured with unconditional love in their infancy and childhood—a process that will take some generations to achieve while all the accumulated upset within our species subsides.

455 The initial difficulty then for virtually all parents is that since they are so psychologically upset, especially so alienated from their natural, loving, instinctive true self or soul as a result of humanity's 2-million-year upsetting search for knowledge, it is simply not possible for them to give their children, or indeed anyone for that matter, unconditional love. HOWEVER, while virtually all parents will have to accept that they will not be able to give their offspring unconditional love, the opportunity now to actually understand the very good reason for *why* they are unable to do so will bring so much calming, reassuring relief that nurturing won't be nearly as fraught with guilt and difficult as it has been. So again, while it will take a number of generations before the majority of parents will be able to generate real, natural, unconditional love, the ability now to understand our species' upset state, and the realisation now that each and every person no longer has to continue

with the upsetting battle to prove their worth (which, as will be explained in chapter 8, is what makes the fabulously exciting new Transformed Way of Living possible), is going to bring *enormous* relief to all men and women, which will help them immensely with the whole activity of nurturing. Indeed, the overall relief and excitement alone that will now emerge in the human race from the realisation that we have finally ended our 2-million-year struggle against ignorance will mean that children are going to find themselves in, if not an unconditionally loving world, then at least an anxiety-free, truthful, compassionate, happy one. And significantly, since the adult world can finally break the silence of their denial and tell children the truth about their horrifically upset condition, that honesty will also make an immense difference to the psychological wellbeing of children.

456 Basically, finding understanding of the human condition makes possible a *whole new world* that is relieved of the horrible frustration and agony of that condition; the terrible siege state where humans couldn't understand themselves can at last be lifted, and, with its lifting, life can return to every aspect of human existence, including to the impossible situation parents have been in trying to nurture their children. Yet again, we see how critical it is that science recognises these denial-free understandings of the human condition so that the public at large is made aware of them. The ignoring, and, when pushed, attack and repression, of these answers MUST stop, and they must instead be supported—because only then can this great transition for the human race begin. Yes, this end to humanity's great, heroic but horrifically upsetting battle to find knowledge, ultimately self-knowledge, understanding of the human condition, together with this change from having to live in denial of the importance of nurturing and of our extremely corrupted condition, changes the whole direction of the human race. Rather than plunging towards ever greater levels of upset, especially towards terminal levels of denial/alienation, suddenly humanity can be heading towards a world free of the human condition.

457 (Much more will be said in chapters 7:16B, 7:16C and 7:16D about the devastating consequences of parents' inability to nurture their infants while under the duress of the human condition, and how parents are to cope with the truth about the importance of nurturing.)

Chapter 6

What is Consciousness, and Why, How and When Did Humans' Unique Conscious Mind Emerge?

Chapter 6:1 <u>Summary</u>

458 Earlier in this book about the plight of humankind under the duress of the human condition, four great outstanding questions were identified as having to be solved for peace to finally come to our species' troubled existence. They were to explain the human condition itself (the psychosis-addressing-and-solving explanation for which was presented in chapter 2); to explain the meaning of our lives (which was done in chapter 3); to explain how we humans acquired

our extraordinary unconditionally selfless moral sense (the answer to which was presented in chapter 4); and, finally, to explain how humans became conscious when other animals haven't, the answer to which will now be presented in this chapter.

Chapter 6:2 What is consciousness?

459 To address the question of the origins of humans' conscious mind it is obviously first necessary to explain what consciousness actually is, but as with the truths of our corrupted human condition, of Integrative Meaning and of the nurturing origins of our moral sense, the truth about the nature of consciousness has been so unbearably condemning of our corrupted condition that we humans have had no choice but to deny it until the human condition was truthfully explained, which it now has been.

460 The fact is, the subject of consciousness is *so* closely correlated with the unbearably depressing subject of the human condition that 'consciousness' has become synonymous with—indeed, code for—the problem of the human condition. The science writer Roger Lewin articulated the problem precisely when he described the great difficulty humans have had of trying to **'illuminate the phenomena of consciousness'** as **'a tough challenge...perhaps the toughest of all'** (*Complexity*, 1993, p.153 of 208), using the philosopher René Descartes' own disturbed reaction when he tried to **'contemplate consciousness'** to illustrate the nature and extent of the difficulty: **'So serious are the doubts into which I have been thrown...that I can neither put them out of my mind nor see any way of resolving them. It feels as if I have fallen unexpectedly into a deep whirlpool which tumbles me around so that I can neither stand on the bottom nor swim up to the top'** (p.154).

461 So yes, while anyone who has searched the term 'consciousness' on the internet will have found it to be a subject cloaked in mystery and confusion, there has actually been a *very* good reason for this, and it's *not* because consciousness is an impenetrably complex subject, as we may think—it is because it raises the unbearable issue of the human condition and, as a result, has been *deliberately* left in an obscure, cryptic state. Resigned humans simply haven't wanted to know what consciousness is.

462 There have been two powerful reasons why the subject of consciousness raised the unbearable issue of the human condition and why, therefore, examination of it led to such a fearful, all-our-moorings-taken-

from-under-us, **'deep whirlpool'** of terrible depression for upset humans—and why, it follows, the subject has been deemed a 'no-go zone'.

[463] The first reason is that even beginning to *vaguely* contemplate the nature of our human situation invariably led us to the initial obvious question of 'What makes humans unique?', with the answer clearly being that we are conscious—but thinking about *that* marked the start of a very slippery, dangerous slope because it quickly led to the depressing accusation: 'Well, if we are such a clever species why do we treat each other and our planet so appallingly?' Yes, trying to think about consciousness meant trying to fathom what—when we humans are the only fully conscious, reasoning, intelligent, extraordinarily clever, 'can-get-a-man-on-the-Moon' animal—is so intelligent and clever about being so competitive, selfish and aggressive, in fact, *so* ruthlessly competitive, brutal and even murderous, that human life has become all but unbearable and we have nearly destroyed our own planet?!

[464] The prospect that our intelligent, rational, insightful, aware, understanding human mind has, it seems, unintelligently, irrationally, thoughtlessly, indifferently and stupidly almost destroyed the whole planet we live on and brought human existence to a state of unbearably lonely, alienated, egocentricity-crazed, aggressive dysfunctionality has been an *extremely* confronting matter to think about. While our fully conscious, intelligent mind *is* surely the culminating achievement of the grand experiment in nature that we call life, it also appeared to be to blame for all the devastation and human suffering in the world! Rather than being wonderful, our conscious mind appeared to be the plague of the planet! *That* is how **'serious are the doubts'** that thinking about consciousness produced within us!

[465] *And*, our inability—until now—to understand this dichotomy of being the most clever and brilliant but also the most apparently destructive, evil and stupid force to have ever appeared on Earth has meant that we humans have understandably been *extremely* insecure and defensive about, and thus hesitant to explore the inner workings of, our supposedly wonderful conscious mind.

[466] The second reason why the subject of consciousness has been so confronting is because thinking about the nature of consciousness quickly brought us into contact with the truth of the selflessness-dependent, loving, integrative meaning of existence, the unbearably depressing implications of which were described in chapter 3. The explanation

of what consciousness actually is will reveal the problem because, as we will see, while there is a simple and obvious explanation for consciousness itself (like there is for the human condition, the meaning of our existence and the origin of our moral soul), that explanation has had very confronting implications.

467 As briefly described in chapter 2:3, humans can be distinguished from other animals by the fact that we are fully conscious; that is, sufficiently able to understand and thus manage the relationship between cause and effect to wrest management of our lives from our instincts—and even to reflect upon our existence, specifically the issue of our immensely upset human condition that wresting management from our instincts brought about. This consciousness is a product of the nerve-based learning system's ability to remember, for it is memory that allows understanding of cause and effect to develop.

468 To elaborate, nerves were originally developed as connections for the coordination of movement in multicellular animals. An incidental by-product of the development of nerves was that of memory. While the actual mechanism by which nerves are able to store impressions is not yet fully understood, we do know it involves chemical processes. What is important, however, is that nerves do have the capacity for memory because once you have the ability to remember past events, you are then in a position to compare them with current events and identify regularly occurring experiences. This knowledge of, or insight into, what has commonly occurred in the past enables the mind to predict what is likely to occur in the future and to adjust behaviour accordingly. Thus, the nerve-based learning system (unlike the gene-based learning system) can *associate* information, *reason* how experiences are related, learn to *understand* and become CONSCIOUS of the relationship of events that occur through time.

469 In the brain, nerve information recordings of experiences (memories) are examined for their relationship with each other. To understand how the brain makes these comparisons, think of the brain as a vast network of nerve pathways onto which incoming experiences are recorded or inscribed, each on a particular path within that network. Where different experiences share the same information, their pathways overlap. For example, long before we understood the force of gravity we had learnt that if we let go of a weighted object it would invariably fall to the ground—falling to the ground was a shared characteristic. The value of recording

information as a pathway in a network is that it allows related aspects of experience to be physically related. In fact, the area in our brain where information is related is called the 'association cortex'. Where parts of an experience are the same they share the same pathway, and where they differ their pathways differ or diverge. All the nerve cells in the brain are interconnected, so with sufficient input of experiences onto a nerve network of sufficient size, similarities or consistencies in experience show up as well-used pathways, pathways that have become highways. (In terms of illustrating the brain's capacity to associate information, it has been found that in the convolutions of our brain's cerebral cortex there are about 20 billion nerve cells with 10 times that number of interconnecting dendrites which, if laid end to end, would stretch at least from Earth to the Moon and back.)

470 An 'idea' represents the moment information is associated in the brain. Incoming information could reinforce a highway, slightly modify it or add an association (an idea) between two highways, dramatically simplifying that particular network of developing consistencies to create a new and simpler interpretation of that information. For example, the most important relationship between different types of fruit is their edibility. Elsewhere, the brain has recognised that the main relationship connecting experiences associated with living organisms is that they appear to try to stay alive, at least for a period of time. Suddenly it 'sees' or deduces ('tumbles' to the idea or association or abstraction) a possible connection between eating and staying alive which, with further experience and thought, becomes reinforced as 'seemingly' correct. 'Eating' is now channelled onto the 'staying alive' highway. It is at this point of information association that the issue of the human condition appears, for any subsequent thought would try to deduce the significance of 'staying alive' and, beyond that, the meaning of life and the relevance of selfishness and selflessness. Ultimately the brain would arrive at the unbearable truth of Integrative Meaning, which, as explained in chapter 3, upset humans have not wanted to face because it confronted us with our *lack* of integrativeness, with our divisiveness.

471 The process of forgetting also plays a part in forging the relationship between experiences. Since duration of nerve memory is related to use, our strongest memories will be of those highways, those experiences that have the greatest relativity. Our experiences not only become related or associated in the brain, they also become concentrated because the brain

gradually forgets or discards inconsistencies or irregularities between experiences. Forgetting serves to cleanse the network of less consistently occurring information, preventing it from becoming cluttered with meaningless (non-insightful) information.

[472]Our language development took the same path as the development of understanding. Commonly occurring arrangements of matter and commonly occurring events were identified (became clear or stood out). Eventually all the main objects and events became identified and, as language emerged, named. For example, we named those regularly occurring arrangements of matter with wings 'birds' and what they did we termed 'flying'.

[473]Once insights into the nature of change are put into effect, the self-modified behaviour starts to provide feedback, refining the insights further. Predictions are compared with outcomes, leading all the way to the deduction of the meaning of all experience, which, as emphasised, is to order or integrate matter.

[474]Consciousness, therefore, is the ability to understand the relationship of events sufficiently well to effectively manage and manipulate those events; it represents the point at which the confusion of incoming information clears, starts to fit together or make sense and the mind becomes effective, a master of change. Chimpanzees, for example, demonstrate consciousness when they effectively reason that by placing boxes one on top of the other they can create a stack that they can then climb upon to reach a banana tied to the roof of their cage. It should be pointed out, however, that it is one thing to be able to stack boxes to reach bananas—to manage immediate events—but quite another to effectively manage events over the long term, to be secure managers of the world. In fact, as will be explained in chapter 7, bonobos (and, to a lesser degree, the other great apes) demonstrate the same degree of consciousness that humans exhibit during infancy, the stage when, between the ages of two and three, we develop sufficient consciousness to recognise that we are at the centre of the changing array of experiences around us. It is when we become aware of the concept of 'I' or self and discover conscious free will, the power to manage events. This is followed by the childhood stage, during which we revel in this free will, 'play' or experiment with it, while adolescence is when we encounter both the sobering responsibility of free will and the agonising identity crisis brought about by the dilemma of the human condition, of whether

or not we are meaningful beings. Adulthood, it follows, is the mature stage that humanity is now in a position to enter, for we finally have the understanding of the human condition that allows us to end our species' insecure state of adolescence forever.

475 In summary, 'insight' was the term given to the nerve highways, the correlation our brain made of the consistencies or regularities it found between events through time. Once humans could deduce these insights—these laws governing events in time past—we were in a position to predict or anticipate the likely turn of events. We could learn to understand—be conscious or aware of—what happened through time. Our intellect could understand the design inherent in, and the process behind, changing information; it could *learn* the predictable regularities or common features in experience, and thus ultimately distil the integrative theme or meaning or purpose or direction of existence.

476 So, yes, consciousness has been an immensely difficult subject for humans to investigate—not because of the practical difficulties involved in determining how our brain works, but because *resigned humans didn't want to know how it worked*. While we couldn't explain our divisive, psychologically upset state of the human condition we had to avoid admitting too clearly how the brain functions because admitting that information could be associated and simplified—admitting to insight—was only a short step away from realising the *ultimate* insight of Integrative Meaning—a truth that immediately confronted us with our own inconsistency with that meaning.

477 As emphasised in chapter 3, to admit to Integrative Meaning meant having to face the fact that our competitive and aggressive behaviour is seemingly totally at odds with the integrative direction of life, no less. The development and maintenance of the order of matter requires that the parts of developing wholes cooperate *not* compete. *Integrative* Meaning confronts us squarely with our *divisive* human condition. Better then to evade the existence of purpose in the first place by avoiding the possibility that information could be associated, refined and simplified. It is the same reason we sidestepped the term 'genetic refinement' for the process of the genetic refinement of the integration of matter on Earth, preferring instead the much vaguer term, 'genetics'. We had to evade the possibility of the refinement of information in *all* its forms because, again, admitting that information could be simplified or refined was admitting to an ultimate refinement or law, confronting us with our inconsistency

with that law, namely with the law of Integrative Meaning, against which we previously had no defence.

478 In fact, not only have we avoided the idea of meaningfulness, we have also shied away from any deep, meaningful thinking that might *lead* to confrontation with Integrative Meaning. But while ensuring deeper insights remained elusive may have saved us from exposure, in the process we buried the truth, becoming, as a result, *extremely* superficial in our thinking. In short, we became masters of *not* thinking—deeply-alienated-from-the-truth beings.

479 Demonstrating our masterful evasion of the nature of consciousness we used words like 'conscious', 'intelligent', 'understanding', 'reason' and 'insight' regularly without ever actually identifying what we are conscious of, intelligent about, understanding, reasoning or having an insight into, which is how events or experiences are related. The conventional obscure, evasive definition of intelligence is **'the ability to think abstractly'**. The other imprecise, obscure, evasive phrase used whenever we wanted to refer to the uniqueness of our intelligence without actually saying what our conscious, understanding, insightful intelligence is, was to say that 'We are the species that is able to reflect upon itself.' So to name the area of the brain that associates and simplifies information as the 'association cortex' was, in fact, a slip of our evasive guard. Of course, when we weren't 'on our guard' against exposure few would deny that information can be associated, simplified and meaning found. In fact, most of us would say we do it every day of our lives—if we didn't, we wouldn't have a word for 'insight'. That is the amazing aspect about our denial of anything that brings the dilemma of the human condition into focus: it is not unusual for us humans to accept an idea up to a point, but as soon as it starts to lead to a confronting conclusion, pretend it doesn't exist—and do so without batting an eyelid.

480 To illustrate how we avoided acknowledging the fundamental ability of the brain to associate and reduce information to essentials (and thus be forced to deduce the integrative meaning or theme or purpose in experience), consider the following case from my files of a *Newsweek* magazine cover story (7 Feb. 1987). While the title and subject of the nine-page article raised the crucial question of 'How the brain works', the author referred to the association capability of the brain in such an unnecessarily complicated way it was effectively buried: **'Productive thought requires not just the rules of logic but a wealth of experience and background**

information, plus the ability to generalise and interpret new experiences using that information.' The **'ability to generalise and interpret'** is the ability to associate information, but the meaning is all but lost in the sentence.

481 In case it is thought this convoluted description may have been due to poor expression rather than deliberate evasion on the part of the author, it should be pointed out that apart from a mention of **'chunking or grouping of similar memories together'** and one unavoidable mention of the **'association cortex'**, the article contains no other reference to the brain's fundamental ability to associate information. The entire nine-page piece, on how our brain works, hangs on this one inept sentence. If you are not intending to be evasive it is not difficult to clearly describe the mind's ability to associate information, as has just been done.

482 More recently, *National Geographic* magazine's February 2014 edition carried a feature story on **'how the human mind really works'**, but again the article didn't go anywhere near the issue of **'how the human mind <u>really</u> works'**. Titled **'Secrets of the Brain – New technologies are shedding light on biology's greatest unsolved mystery: how the brain really works'**, the article referred to **'President Barack Obama's 'BRAIN Initiative' 'last spring'**, and mentioned **'disorders such as schizophrenia, autism and...depression'**, however, there was not even an *attempt* in this article to address and explain the all-significant feature of **'the human mind'** of consciousness; in fact, there was no mention of 'consciousness', and certainly not of the dilemma of the human condition associated with being a conscious species. The article was completely mechanistic, totally evasive of any even oblique reference to the crippling psychosis and neurosis involved in the **'disorders'** now plaguing our conscious **'human mind'**. The denial being practised in the 1987 *Newsweek* story is chicken feed compared to the level of denial in this 2014 story. Benjamin Fong was certainly right when, as mentioned in par. 447, he predicted that **'The real trouble with the Brain Initiative'** would be its **'synthetic'** focus on the **'organic'** rather than the **'psychological'** nature of mental problems.

483 I should emphasise that our evasion and denial is often obviously false and yet we believed it, because we had to. For instance, in the case of Integrative Meaning, we are surrounded by examples of integration everywhere we look—every object, enduring or otherwise, is a hierarchy of ordered matter, testament to the development of order of matter—and yet we have denied it, just as mechanistic science couldn't even provide a definition for two of humanity's most commonly used and important

words/concepts: 'love' and 'soul'. The hypocrisy inherent in denial is palpable yet understandable.

484 The most honest and yet still somewhat intellectually evasive description I have found in the denial-complying mechanistic paradigm of what gives rise to consciousness is one that appears in Wikipedia: **'Memory...allows sensory information to be evaluated in the context of previous experience...and...allows information to be integrated over time so that it can generate a stable representation of the world'** (Wikipedia, accessed Apr. 2013; see <www.wtmsources.com/128>). Yes, consciousness *is* the ability to **'generate a stable representation of the world'**, the ability to generate a representation of the world that is accurate and can therefore be effectively used to base anticipations of future events upon.

485 So it's not difficult to explain consciousness if you are not needing to avoid what it is. To look at someone with a track record of thinking truthfully, it was mentioned in par. 81 that the naturalist Eugène Marais wrote these amazingly insightful words about the human condition: **'The highest primate, man, is born an instinctive animal. All its behavior for a long period after its birth is dominated by the instinctive mentality...it has no memory, no conception of cause and effect, no consciousness...As the...individual memory slowly emerges, the instinctive soul becomes just as slowly submerged... For a time it is almost as though there were a struggle between the two.'** Notice the ease with which Marais describes consciousness—**'memory'** makes possible **'conception of cause and effect'** which gives rise to **'consciousness'**. It really is as obvious and straight forward as that! Indeed, when, as cited in par. 140, the writer Aldous Huxley commented that **'Non-rational creatures do not look before or after'**, he was simply stating the obvious truth that to be **'rational'** is to be able to **'look before or after'**; to have the capability to make sense of cause and effect. Which raises the question: 'Why on earth was this not explained to us by science, why weren't we taught this at school?' But as has now been explained, the issue of our conscious intelligent mind prompted the unbearable self-realisation, 'Well, if I'm so cleverly insightful why can't I manage my life in a way that is not so mean and indifferent to others; indeed, why, if I am such a brilliantly intelligent person, am I such a destructively selfish, angry, egocentric, competitive and aggressive monster?' And as was emphasised, the other problem was that explaining the nature of consciousness very quickly brought us into contact with the unbearably depressing truth of Integrative Meaning. An appropriate definition of 'consciousness' is 'the ability to make sense of

experience', but applying such a definition immediately highlights the problem with the issue of consciousness, because, due to the depressing implications, we humans haven't wanted to 'make sense of experience' because we have not wanted to recognise the truth of Integrative Meaning. So, to ask people to look into the issue of consciousness was to expect them to confront the issue of their own less-than-ideal human condition, which is why 'consciousness' has become a code word for the issue of the human condition.

[486] We can, therefore, appreciate why, when Descartes tried to **'contemplate consciousness'**, it caused him such fearful depression that he said, **'It feels as if I have fallen unexpectedly into a deep whirlpool which tumbles me around so that I can neither stand on the bottom nor swim up to the top.'**

[487] The great psychiatrist R.D. Laing was someone who acknowledged both the importance of the issue of consciousness (the human condition), and how truly difficult a **'realm'** it has been for upset humans to study when he wrote: (the underlining is my emphasis) **'The requirement of the present, the failure of the past, is the same: to provide a thoroughly self-conscious and self-critical human account of man...Our alienation goes to the roots. The realization of this is the essential springboard for any serious reflection on any aspect of present inter-human life** [pp.11-12 of 156] **...We respect the voyager, the explorer, the climber, the space man. It makes far more sense to me as a valid project—indeed, as a desperately urgently required project for our time—to explore the inner space and time of consciousness. Perhaps this is one of the few things that still make sense in our historical context. We are so out of touch with this realm** [so in denial of the issue of the human condition] **that many people can now argue seriously that it does not exist. It is very small wonder that it is perilous indeed to explore such a lost realm** [p.105]' (*The Politics of Experience* and *The Bird of Paradise*, 1967).

[488] Clearly, the need to find a way to avoid confronting what consciousness entailed was immense, and the initial obvious excuse that mechanistic science devised to do so was to simply assert that the phenomenon of the human mind is just too extraordinary for science to be able to explain; that, as a mechanistic scientist might say, 'While science might expect to be able to explain the physical mechanisms operating in our brain it is beyond our present powers of explanation to account for how a collection of neurons could produce the subjective, uniquely personal experience of the human mind.' Viewed superficially, it *is* incredible that the mental state, where we have our own personal feelings and particular sense of self,

can arise from a bunch of nerves, but really it is simply a consequence of there having been sufficient development of nerves' ability to understand cause and effect. In the same light, it *is* amazing that 92 elements operating under the law of Negative Entropy could give rise to the amazing variety and complexity of life that we see around us, or that the simple combustion engine could give rise to the amazing phenomenon that is a Ferrari sports car, but they have. We can try to argue that knowing how the combustion engine works doesn't explain the mystique that surrounds a Ferrari, but, in fact, it does. Consciousness *is* amazing but it *is* simply a result of memory which allows us to understand cause and effect. So while the claim that it was beyond our current powers to explain the phenomenon of consciousness no doubt helped upset humans avoid the issue of the human condition, the argument was in truth nothing more than intellectual bluff.

489 Given his aversion to the subject, it is not surprising that the whole body-mind dichotomy excuse for not being able to explain consciousness appears to have begun with Descartes when, in 1641, he wrote that the mind was **'entirely and truly distinct from my body'** (*Meditations on First Philosophy*, Meditation VI, Part 9). This idea of consciousness as a sort of **'ghost in the machine'** (Gilbert Ryle, *The Concept of Mind,* 1949, p.15 of 334) of the brain, which could not be explained in physical terms, was embraced by scientists such as Thomas Huxley, who in 1868 wrote, **'how it is that any thing so remarkable as a state of consciousness comes about as the result of irritating nervous tissue, is just as unaccountable as the appearance of the Djin when Aladdin rubbed his lamp'** (T.H. Huxley & W.J. Youmans, *The Elements of Physiology and Hygiene: A text-book for Educational Institutions*, 1868, p.178 of 420). By the 1990s this excuse that the state of consciousness cannot be accounted for had become so established it was actually referred to as the 'hard problem' of consciousness, as distinct from the 'easy problem' of explaining the neurological activities of the brain. The philosopher who coined these terms, David Chalmers, wrote that **'It is widely agreed that experience arises from a physical basis, but we have no good explanation of why and how it so arises. Why should physical processing give rise to a rich inner life at all?** ('Facing up to the Problem of Consciousness', *Journal of Consciousness Studies*, 1995, Vol.2, No.3); and, **'How does the water of the brain turn into the wine of consciousness?...How is it that all of this matter adds up to something as complex, as interesting and as unique as consciousness?'** ('David Chalmers: The Conscious Mind', *Thinking Allowed with Jeffrey Mishlove*, TV series and DVD collection, #H230). In contrasting the 'easy problem' with the supposed 'hard

problem' of explaining our subjective mind, the cognitive scientist Steven Pinker wrote that **'The Hard Problem...is why it feels like something to have a conscious process going on in one's head—why there is first-person, subjective experience...The problem is hard because no one knows what a solution might look like or even whether it is a genuine scientific problem in the first place... To appreciate the hardness of the Hard Problem, consider how you could ever know whether you see colours the same way that I do. Sure, you and I both call grass green, but perhaps you see grass as having the color that I would describe, if I were in your shoes, as purple'** ('The Brain: The Mystery of Consciousness', *TIME*, 29 Jan. 2007). In truth, it is intellectual nonsense to argue that we can't reach a shared understanding of the nature of our world. The philosopher John Searle spoke honestly about consciousness when he wrote that **'Brain processes cause consciousness but the consciousness they cause is not some extra substance or entity. It is just a higher level feature of the whole system'** ('The Problem of Consciousness', *Social Research*, 1993, Vol.60, No.1).

490 The cognitive scientist Daniel Dennett was approaching the truth about the real problem surrounding consciousness when he wrote that **'With consciousness, however, we are still in a terrible muddle. Consciousness stands alone today as a topic that often leaves even the most sophisticated thinkers tongue-tied and confused. And, as with all the earlier mysteries, there are many who insist—and hope—that there will never be a demystification of consciousness'** (*Consciousness Explained*, 1991, p.22 of 528). Yes, the psychologically upset and insecure human race hasn't wanted to know what consciousness is, because it didn't want to confront the human condition! What was required for humans to tolerate the **'demystification of consciousness'** was for the human condition to be compassionately explained, at which point the contrived **'muddle'** could be resolved; as Pinker wrote, **'I admit that the theory** [that consciousness is beyond our powers of explanation] **could be demolished when an unborn genius—a Darwin or Einstein of consciousness—comes up with a flabbergasting new idea that suddenly makes it all clear to us. Whatever the solutions to the Easy and Hard problems turn out to be, few scientists doubt that they will locate consciousness in the activity of the brain'** ('The Brain: The Mystery of Consciousness', *TIME*, 29 Jan. 2007). Yes, when my professor of biology at Sydney University, the late Charles Birch, observed that **'Biology right now awaits its Einstein in the realm of consciousness studies'** (*Ockham's Razor*, ABC Radio National, 16 Apr. 1997), he too was referring to the need for a biologist to find understanding of the human condition. But as I explained in par. 358, it wasn't **'genius'**, extreme cleverness, a high IQ, that was required to

'suddenly make…it all clear' and explain how consciousness arose from the basic **'activity of the brain'**, namely memory, it was simply a sound, denial-free, honest approach to thinking about the human condition. Indeed, as will now be pointed out, focusing on the need for a high IQ to solve problems was just another expression of the insecurity surrounding the whole issue of consciousness—just another way to laud our intelligence and gloss over its corrupting effects.

491 The response taken by adolescents when faced with the unbearable issue of the human condition perfectly illustrates the response virtually all humans have had to confronting the nature of consciousness—so, if the reader will bear with me, I will now briefly describe just how determinedly adolescents blocked out the issue of the human condition.

492 When <u>the concept of Resignation was explained in chapter 1:2</u> (pars 43-52) it was described how when upset humans were around 14 or 15 years of age they each tried to face down the issue of the imperfection in the world of humans around them and in their own behaviour—the issue of the human condition—only to find it a suicidally depressing exercise. As Carl Jung said, **'When it** [our **'shadow'**, the negative aspects of ourselves] **appears…it is quite within the bounds of possibility for a man to recognize the relative evil of his nature, but it is a rare and shattering experience for him to gaze into the face of absolute evil.'** To avoid subjecting themselves to that **'shattering experience'** ever again, adolescents were forced to resign themselves to never revisiting the issue of the human condition. In fact, every moment thereafter was spent carefully avoiding any encounter with the issue, and denying any truths that brought it into focus. <u>What was not described in chapter 1:2 was the strategy a resigned person adopted *after* Resignation</u>. While living in denial of their corrupted condition brought some relief, resigned humans still needed to find ways to feel good about themselves. Yes, not only did they have to find ways to avoid and deny the issue of the human condition, they also had to find ways to convince themselves they were the opposite of flawed and corrupted and, it would seem, **'the face of absolute evil'**. So, unable to refute the negative view of themselves with compassionate understanding, they became focused on emphasising and developing whatever positive views of themselves they could find or create. In practice this meant they became preoccupied competing for power, fame, fortune and glory. Winning a football match, or behaving in some way that brought praise—basically carrying out any activity that would relieve the insecurity of their condition—became all

that mattered. Once the full extent of the imperfection of their corrupted condition became apparent at the time of Resignation, the need for reinforcement to counter that uncertainty about their worth became extreme, desperately so for those who were more insecure. So after Resignation, the 'ego', which the dictionary defines as **'the conscious thinking self'** (*Concise Oxford Dictionary*, 5th edn, 1964), became extremely selfishly focused or centred on the need for reinforcement and relief from criticism. Thus, in addition to practising complete denial of any confronting truths, and becoming very angry towards any criticism, extreme ego-centricity was one of the main characteristics to come out of Resignation. Basically after Resignation, the angry, egocentric and alienated responses to the dilemma of the human condition that had emerged in infancy and which had been gradually developing throughout childhood and early adolescence suddenly greatly amplified. Indeed, the extremely self-worth-embattled, heroic mantra of the resigned person became one of, as the adages go, **'Give me liberty** [from criticism] **or give me death'**, **'No retreat, no surrender'**, **'Death before dishonour'**—in essence, 'I don't believe the criticism is deserved, and in any case it's too unbearable to accept, so I will *never* tolerate any insinuation that I am a bad person'! (A more complete description of the process of Resignation can be found in *Freedom Book 1* at <www.worldtransformation.com/freedom-resignation>.)

493 We can visualise the situation faced by resigning adolescents by imagining them in a room, with one end containing all manner of depressing truths about the world of humans around them and, most depressing of all, about their own lives, and the other end the few positive thoughts and views about the world and about themselves. Well, not surprisingly, resigned humans chose to live entirely at the end of the room where they could be surrounded by the few positives they could identify or develop—especially positive reinforcements for themselves gained through achieving power, fame, fortune and glory. And not only did they stay jammed right up against the wall at that positive end, as removed as humanly possible from any negative truths, they made sure they stood with their nose flat against the wall so they couldn't even *see* the other side of the room. That is how narrow and limited the existence of a resigned human has been. The upset, resigned mind has been fixated on a few positives about themselves while blocking out a whole 'room'—in fact, a whole universe of subjects and thoughts and awarenesses. In this light, we can appreciate the accuracy of Plato's analogy of humans

living imprisoned **'a long way underground'** in a cave where all they could see were shadowy illusions of the real world.

494 This 'jammed-down-one-end-of-the-room' analogy of how the upset, resigned human race has had to live in denial of the whole issue of the human condition also applies to the aspect of the human condition that involves our species' apparently destructive conscious intelligence. Unless we were exceptionally free of upset behaviour, exceptionally nurtured with unconditional real love in our infancy and childhood and thus exceptionally secure in self and thus able to avoid Resignation, thinking truthfully about what consciousness is was only ever going to raise the question of why aren't we cognisant of change and insightful enough to behave in an ideal way? To avoid raising those depressing questions resigned humans had to avoid any truthful analysis of what consciousness actually is. And not only that, they needed to shift their focus entirely away from the issue of what consciousness entails onto any positives they could find about their intellect. The resigned mind had to focus on telling themselves that their intellect is a brilliant talent, capable, for instance, of inventing a machine to send man to the Moon and back. To cope with the whole issue of the goodness or otherwise of their intellect they maximised all the positive aspects about their conscious intellect, which is their cleverness, and minimised all the negative aspects, which is their state of alienation—namely the extent of their soundness or lack thereof—with the best method of doing so to simply not think about and acknowledge what the nature of conscious intelligence really is.

495 And so the resigned world started to regard, or more specifically, measure, people according to their level of mental dexterity, with the creation of IQ tests and societies whose membership was restricted to the most clever, such as Mensa. The resigned world only allowed people into university who had high IQs and could pass exams that tested for a person's intellectual brilliance, never for their soulful soundness. It tested children for their ability to remember endless streams of ridiculously superficial and irrelevant facts such as Queen Isabella the 5th married King Arnold the 12th in 1522 and together they fought The War Of The Itchy Armpits in 1591, or something like that; never asking the *real* questions of why there were kings and queens and poor people—selfishness, inequality and indifference to others—and why humans fought and killed each other in wars. And it created game shows that glorified those with the

best memories for mundane, superficial facts or for successfully spelling words or completing mathematical sums. It was an extremely escapist, evasive *intellectual* world, not a sound, soulful *instinctual* world. The emphasis was entirely on intellectual brilliance, not on soulful soundness. It never measured people for how alienated they were, or for the speed at which their minds could block out confronting truth, or the speed at which they could override their instinctive moral sense and exploit others, or for how mentally insecure and thus egocentrically self-preoccupied and thus indifferent to others they were. It only stressed how smart you were, never how corrupted and destructive your intellect was. And it wasn't as though the resigned world didn't know who was soul-corrupted, upset and alienated and who was innocent; after all, to ignore, deny, repress and, in the extreme, persecute to the point of, in the case of Christ, *crucifying* honest, truthful innocent soundness because it found it too confronting, it had to first be able to recognise it. It would have been as easy—indeed, probably much easier—to design exams that tested a person's level of alienation or soundness or soulfulness quotient, their SQ, than it was to design exams that tested their intelligence quotient, or IQ.

496 Yes, the truth is, our intelligence has been an *extremely* insecure and defensive entity. It has been an instrument of dishonest denial, not of honest thoughtfulness. It has been preoccupied with escapist, superficial, alienated, **'fake'**, **'phony'**, sophisticated intellectualism, not with confronting, penetrating, truthful, thoughtful instinctualism. Such has been the human condition. Thank goodness then that it has at last been explained and the horrifically debilitating practice of denial can end.

Chapter 6:3 Why, how and when did consciousness emerge in humans?

497 As stated at the beginning of this chapter, our self-adjusting conscious mind is a 'fabulous' phenomenon, the culminating achievement of the grand experiment in nature that we call life. Unlike the gene-based natural selection way of processing information, where different arrangements of matter with their different properties are compared for their integrative potential, the conscious mind is able to make these comparisons with the information abstracted from its source. The gene-based natural selection process *is* a form of thinking, of comparing the integrative potential of different arrangements of matter which can be thought

of as different 'ideas'—but that 'thinking' or 'learning' or 'refinement', that experimentation, that 'selection' of one 'idea' over another, has to be carried out in practise. The conscious mind, however, has the incredible capacity to be able to do this 'thinking', this comparing of 'ideas', without having to carry them out in practise. Yes, for all the problems it has given rise to, the power of the conscious mind to separate information from its source *is* fabulous, it *has to be* nature's greatest invention.

[498]The obvious question then is how did our wondrously powerful conscious mind emerge from the gene-based natural selection process? To solve problems it's first necessary to ask the right questions, and in the case of consciousness, the obvious initial question is 'Why did we humans develop consciousness when other animals haven't?' Since consciousness occurs at a certain point in the development of a mind's efficiency in associating information, and since conscious intelligence—the ability to reason how cause and effect are related, to understand change, to be insightful—would obviously be a very great asset for any animal to acquire, one would assume that fully developed conscious intelligence would have been actively selected for as soon as animals were able to develop a reasonably elaborate central nervous system, and would have thus appeared in many species—and yet it hasn't. As was described in chapter 5:5, despite this being an obvious assumption, the conventional mechanistic explanation for the emergence of conscious intelligence in humans, and its absence in other animals, is that it occurred as a result of the need to manage complex social situations—for example, in *The Social Conquest of Earth*, E.O. Wilson writes that in order **'to feel empathy for others, to measure the emotions of friends and enemy alike, to judge the intentions of all of them, and to plan a strategy for personal social interactions...the human brain became...highly intelligent'** (2012, p.17 of 330). Termed the Social Intelligence Hypothesis, and occasionally the Machiavellian Intelligence Hypothesis, this theory was first formally put forward by the psychologist Nicholas Humphrey in 1976: **'In broad terms, the MIH was originally developed to explain the special intelligence attributed to monkeys and apes (Humphrey 1976) as adaptations for dealing with the distinctive complexities of their social lives, such as volatile social alliances. The term "Machiavellian" was used by Byrne & Whiten (1988) to capture the central concept of adaptive social manoeuvring within groups made up of companions subject to similar pressures to be socially smart, and the spiralling selection pressures this implies'** (Andrew Whiten & Carel van Schaik,

'The evolution of animal "cultures" and social intelligence', *Philosophical Transactions of The Royal Society B*, 2007, Vol.362, No.1480).

499 The first point to make is one that was explained in chapter 4 and re-emphasised in chapter 5 (pars 381-385)—that if it wasn't for the psychologically upset state of the human condition there would be no need to learn, and become intelligent enough to master, the art of managing **'social' 'complexities'**. Through the process of love-indoctrination, we humans became *so* instinctively integrated that there was no disharmony/conflict/discord/**'complex**[ity]**'** to have to manage. Before the emergence of the human condition some 2 million years ago our species lived instinctively as one organism—as the Greek poet Hesiod said, **'Like gods they lived, with calm untroubled mind, free from the toils and anguish of our kind...They with abundant goods 'midst quiet lands, all willing shared the gathering of their hands.'** The pre-human-condition-afflicted, integrated, cooperative, social, Specie Individual state that the fossil record of our ancestors now evidences, and which largely exists in bonobo society today, simply wasn't a situation that involved **'social' 'complexities'**. No, the bonobos' harmonious society, and humanity, was borne out of a genuinely altruistic, unconditionally selfless, all-loving, empathetic-towards-others, non-devious, opposite-of-**'smart'**-cunning-**'manoeuvring'**-**'Machiavellian'** situation.

500 However, coming back to the obvious question of 'Why did humans develop conscious intelligence when other animals haven't?', as I pointed out in par. 383, while the ability to solve social problems *is* an obvious benefit of having a conscious mind, *all* activities that animals have to undertake would benefit enormously from being able to understand cause and effect, so it is completely illogical to suggest that it wasn't until the need to manage extremely complex social situations that consciousness developed. No, any sensible analysis of the question of the emergence of consciousness must be based on the question of 'What has *prevented* its development in other animals?' Consciousness is such a powerful asset for an animal to acquire that something must have blocked its selection in other species. The lack of social situations doesn't explain why the fully conscious mind hasn't appeared in non-human species because there was ample need for a conscious intelligent mind prior to the appearance of complex social situations.

501 It should also be pointed out that while **'Most contemporary workers in animal cognition, particularly those working with primates, are enthusiastic about the social intelligence hypothesis'** (Kay Holekamp, 'Questioning the social

intelligence hypothesis', *Trends in Cognitive Sciences*, 2007, Vol.11, No.2), the hypothesis has at least two flaws that have been recognised by human-condition avoiding, mechanistic scientists. Firstly, research has shown that there actually *isn't* a correlation between more complex social groups and greater social learning, which you *would* expect if the reason for the long mother-infant association was to teach the skills needed to live in complex social groups. For instance, **'several phenomena have been identified...for which the social intelligence hypothesis cannot account...Evidence of this sort has also accumulated in the literature on primate social intelligence. For example, a comparative analysis found that innovation, tool use and the incidence of social learning in primates co-vary across species and that the frequency of occurrence of social learning is not correlated with group size among primate species'** (ibid). The **'comparative analysis'** in question says in regard to primates, that **'social group size and social learning frequency are not correlated'** (Simon Reader & Louis Lefebvre, 'Social Learning and Sociality', *Behavioral and Brain Sciences*, 2001, Vol.24, No.2).

502 Secondly, evidence that having to manage more complex social situations has *not* led to greater intelligence is also being furnished through mechanistic science's studies of relative brain size (which is widely regarded as an indicator of greater intelligence), with research showing that highly social species such as meerkats and hyenas have not developed a larger brain in proportion to body size beyond that of less social species: **'no association exists between sociality and encephalization** [brain size in proportion to body size] **across Carnivora** [which include meerkats and hyenas] **and that support for sociality as a causal agent of encephalization increase disappears for this clade** [group]**'** (John Finarelli & John Flynn, 'Brain-size evolution and sociality in Carnivora', *Proceedings of the National Academy of Sciences*, 2009, Vol.106, No.23).

503 But rather than accepting these fundamental flaws and abandoning the S/MIH, a more sophisticated version of it was actually put forward in 1989 by the biologist Richard Alexander. This refinement, which became known as the 'Ecological Dominance–Social Competition' (EDSC) model, holds that a species must first somehow overcome or dominate its environment *before* the S/MIH can apply, arguing that **'as our hominin ancestors became increasingly able to master the traditional "hostile forces of nature," selective pressures resulting from competition among conspecifics** [companions] **became increasingly important, particularly in regard to social competencies. Given the precondition of competition among kin–and reciprocity–based coalitions (shared with chimpanzees), an autocatalytic** [self-fuelling] **social arms race was initiated,**

which eventually resulted in the unusual collection of traits characteristic of the human species, such as…an extraordinary collection of cognitive abilities' (M.V. Flinn et al, 'Ecological dominance, social competition, and coalitionary arms races: Why humans evolved extraordinary intelligence', *Evolution and Human Behavior*, 2005, Vol.26). While it does not answer the obvious question of why *all* activities that animals have to manage wouldn't profit from a conscious intellect, this prerequisite of **'ecological dominance'** is meant to explain why other species in complex social situations haven't developed intelligence—apparently because they hadn't dominated their environment first.

504 The concept of **'ecological dominance'** put forward by the EDSC model describes a supposed mastery of the environment achieved, it says, through developments such as tool use, projectile weapons and controlled use of fire that **'roughly coincided with the appearance of *Homo erectus*, 1.8 mya** [million years ago]**'** (ibid). While it is true that conscious intelligence was increasing rapidly during the reign of *Homo erectus*, as will be revealed in chapter 7:2, conscious intelligence first emerged in our ape ancestor, prior to the emergence of the australopithecines (who appeared some 4 million years ago) and so well before *Homo*; and further, the dramatic increase in intelligence evident in *Homo erectus* was *not* a result of **'ecological dominance'** but, as described throughout this book, due to *psychological* factors—as has been the continual increase in conscious intelligence throughout the genus *Homo*. (It is important to differentiate here that the **'ecological dominance'** described by the EDSC model is *not* the same ecologically dominant/beneficial situation our ape ancestors found themselves in when living in the 'ideal nursery conditions' that allowed love-indoctrination to develop, which, as was briefly described in chapter 4:8 and as is about to be described in more detail, *did* have the side effect of liberating conscious intelligence.)

505 This flaw in the EDSC model's theory that conscious intelligence did not arise until the emergence of *Homo erectus* because until that time we hadn't dominated our environment is also evidenced by the fossil record, which shows that brain size dramatically increased in humans *before* they became **'ecologically dominant'**. Yes, the model's argument that **'ecological dominance should arise prior to or along with increases in brain size'** (ibid) is a flaw that even mechanistic, human-condition-avoiding scientists recognise, as this study points out: **'a great deal of encephalization** [brain size relative to body size] **occurred before humans were dominant… The EQ** [encephalization quotient] **of the first instance of *Homo, Homo habilis*,**

had already doubled relative to our nearest relatives today, chimpanzees (*Pan troglodytes*). These hominins were still largely foragers, scavengers (not yet organized hunters), and prey for more powerful predators' (R.D. Horan et al, *A Paleoeconomic Theory of Encephalization*, selected paper presented at the annual meetings of the American Economic Association, San Francisco, Jan. 2009). No, it certainly doesn't make sense that developments such as tool use, projectile weapons and the controlled use of fire enabled us to become intelligent because such advancements would have *required* conscious intelligence to both invent and subsequently manage. Something must have enabled consciousness to emerge, which *then* allowed us to become clever enough to invent and manage these early technologies.

506 There are several other theories for how humans developed consciousness, including models based around the need to solve problems resulting from climate change, sexual selection for consciousness as an indicator of fitness, and the discovery of cooking, however, without being able to acknowledge the human condition, they—as with the S/MIH/EDSC models—are unable to explain in a fully accountable way why consciousness is *not* a powerful, fitness-assisting asset for an animal to have in all situations (not just social situations).

507 I should note here that there has been an attempt, using what's called the Expensive Brain Hypothesis, to explain why consciousness has not been widely developed. This theory first equates brain size with intelligence and then claims the reason intelligence has not evolved is because **'The high proportion of energy necessarily allocated to brain tissue may therefore constrain the response of natural selection to the beneficial impact of increased brain size on an animal's survival and/or reproductive success'** (Karin Isler & Carel van Schaik, 'Metabolic costs of brain size evolution', *Biology Letters*, 2006, Vol.2, No.4). The fact is, natural selection has the ability to select for an asset if the benefits outweigh the costs, and since consciousness is such an *extremely* valuable asset, the costs in energy of developing a conscious mind would not, you would expect, be great enough to inhibit its development—after all, natural selection has selected for assets that require an *immense* amount of energy, such as the males of many species being able to expend colossal amounts of energy during the mating season; consider, for instance, the peacock, which grows an extravagant new train each season, or antelopes engaging in a period of endless and ferocious rutting each year. Another point is that bonobos, who are verging on having developed full consciousness, don't have huge brains like we have.

As will be described in pars 540 and 547, what has driven *our* brain and conscious mind to become so developed is the difficulty of having to try to manage the extreme complexities of *our* human condition, as evidenced by the fact that the very large brain developed when the human condition developed, which was *after* the time we became conscious.

508 The fact is, other animal species have been able to develop all manner of extraordinary mental abilities, many superior to our own, but never full consciousness. (The reason we can know that other animals like dolphins and elephants haven't developed full consciousness was explained in par. 307.) For instance, in the United States, the nutcracker bird buries around 30,000 nuts throughout the summer months, each in a different location, but come winter and the cover of snow it can recall the location of 90 percent of them. The goby fish can memorise the topography of the tidal flats at high tide so that when the tide retreats it knows the exact location of the next pool to flip to when the one it is in evaporates. And then there is the male common canary, which has a specific part of its brain that expands dramatically every spring so it can learn new mating songs, only to shrink again once the need for it ends at the conclusion of the mating season (just as peacocks shed their tattered tail feathers at the end of the season). So again the question is, if other animals have been able to develop such extraordinary mental abilities, what's stopping them from developing *full* consciousness? As will now be explained, the simple and very obvious explanation—if you are not living in denial of the truth of selflessness-dependent, Integrative Meaning—is that genetics is such a selfish process that it will normally block the development of a selflessness-recognising, truthful, effective-thinking conscious mind.

509 To now explain more fully this obvious—if you are not living in denial—explanation for what has blocked the development of consciousness in almost all species. The explanation begins by re-stating what was pointed out in chapter 3:4, which was that one of the limitations of the gene-based learning system is that it normally can't select for unconditionally selfless, altruistic, self-sacrificing behaviour because altruistic traits tend to self-eliminate; they tend not to carry on and so normally can't become established in a species. The effect is that the gene-based learning system actively resists altruistic behaviour. For instance, whenever a female kangaroo comes into estrous, the males pursue her relentlessly. Despite both parties almost falling with fatigue,

the chase continues. It is easy to see how this behaviour developed: if a male relaxed his efforts he would lose his opportunity to reproduce. Self-interest is fostered by natural selection with the result that genetic selfishness has become an extremely strong force in animals. It is clear then that there would be no chance of a variety of kangaroo that considered others above itself developing. Unless, of course, they could develop love-indoctrination, but while a kangaroo can look after a joey in its pouch, the pouch is more an external womb, allowing little behavioural interaction between mother and infant. It is the selfless treatment—the active demonstration of love—that trains the infants in selflessness or love. Also, since grass, which they live on, is not very nutritious, kangaroos have to spend most of their time grazing, which leaves relatively little time for social interaction between mother and infant and thus limited training in love.

510 Genetic refinement normally acts against any inclination towards selfless behaviour because selflessness disadvantages the individual that practises it and advantages the recipients of the selfless treatment—such is the meaning of selflessness. Selflessness normally can't be reinforced by genetic refinement; indeed, it is emphatically resisted by it. It follows then that in terms of the development of consciousness, the gene-based learning or refinement system was, in effect, totally opposed to any altruistic, selfless thinking. In fact, genetic refinement developed *blocks* in the minds of animals to prevent the emergence of such thinking. And it is this block against truthful, selflessness-recognising-thinking in the minds of almost all animals that prevents them from becoming conscious of the true relationship or meaning of experience.

511 To explain more fully how these blocks against selflessness-recognising-thinking developed, an example of how genes resist self-destructive behaviour will be helpful. In what are termed 'visual cliff' experiments, newborn kittens are placed on a table and while they will venture towards the edge, they won't allow themselves to go beyond the edge and fall—a sheet of glass is actually placed over the table to prevent them from accidentally slipping off the edge, but the point is the glass is unnecessary because the kittens instinctively know not to travel beyond the table's edge. Presumably, this instinctive orientation against doing so evolved because any cat that did venture too close to a precipice invariably fell to its death, leaving only those that happened to have an instinctive block against such self-destructive practices. Natural selection

or genetic refinement develops blocks in the mind against behaviour that doesn't tend to lead to the reproduction of the genes of the individuals who practise that behaviour.

512 And so just as surely as cats were eventually selected for their instinctive block against self-destruction, most animals have been selected with an instinctive block against selfless thinking because such thinking also tends not to lead to the reproduction of the genes of the individuals who think that way. The effect of this block was to prevent the developing intellect from thinking truthfully and thus effectively.

513 As pointed out when Integrative Meaning was explained in chapter 3, selflessness or love is the theme of existence, the essence of integration, the meaning of life. While the upset, alienated human race has learnt to live in denial of this truth of the selfless, loving, integrative meaning of existence, it is, in fact, an extremely obvious truth and one that is deduced very quickly if you are able to think honestly about the world. We are surrounded by integration; every object we look at is a hierarchy of ordered matter, witness to the development of order. It follows then that if you aren't able to recognise and thus appreciate the significance of selfless Integrative Meaning you are not in a position to begin to think straight and thus effectively; you can't begin to make sense of experience. All your thinking is coming off a false base and is therefore derailed from the outset from making sense of experience. As was mentioned in par. 115, the philosopher Arthur Schopenhauer wrote that **'The discovery of truth is prevented most effectively...by prejudice, which...stands in the path of truth and is then like a contrary wind driving a ship away from land'**. You can't think effectively with lies in your head, especially with such important lies as denial of Integrative Meaning. Your mind is, in effect, stalled at a very superficial level of intelligence with little ability to understand the relationship of events occurring around you.

514 To elaborate, any animal able to associate information to the degree necessary to realise the importance of behaving selflessly towards others would have been at a distinct disadvantage in terms of its chances of successfully reproducing its genes. It follows then that those animals that *don't* recognise the importance of selflessness are genetically advantaged, which means that eventually a mental block would have been 'naturally selected' to prevent the emergence of the ability to make sense of experience, to prevent the emergence of consciousness. At this point in development, genetic refinement favoured individuals that were not

able to recognise the significance of selflessness, thus ensuring animals remained incognisant, unconscious of the true meaning of life.

515 Having denied the truth of Integrative Meaning and the importance of selflessness, it is not easy for the alienated human race to appreciate that conscious thought depends on the ability to acknowledge the significance of selflessness/love/Integrative Meaning. However, our own human-condition-induced mental block or alienation is, in fact, the perfect illustration of and parallel for this block in the minds of animals. Unable to think truthfully about the selfless, loving integrative theme of existence, all our thinking has been coming off a false base and, as a result, we too have been unable to think effectively. Alienation has rendered the human race almost stupid, incapable of deep, penetrating, meaningful thought.

Wallace & Gromit by Nick Park and Bob Baker of Aardman Animations

516 When it comes to thinking truthfully and thus soundly, humans are now almost as mentally incognisant as animals—a state of affairs that is parodied in the popular animated cartoon *Wallace & Gromit*. In the series, Wallace is a lonely, sad—alienated—human figure whose dog Gromit is very much on an intellectual par with him in his world. Both wear the same blank, stupefied expression as together they muddle their way through life's adventures!

517 Yes, as R.D. Laing was quoted as saying in par. 56, **'Our alienation goes to the roots'**, there is now **'fifty feet of solid concrete'** between us and our condemning instinctive self or soul. But to be alienated is to not know you

are alienated because if you did you wouldn't be alienated, you wouldn't be blocking out the truth—all of which means it will be difficult to accept that we humans are now 'almost as mentally incognisant as animals', so these further references to just how alienated we humans have become may help to reveal the extent of our alienation.

[518]Plato, that extraordinary denial-free thinking prophet, wrote that **'when the soul** [our instinctive orientation to Integrative Meaning] **uses the instrumentality of the body** [uses the body's intellect with its preoccupation with denial of Integrative Meaning] **for any inquiry...it is drawn away by the body into the realm of the variable, and loses its way and becomes confused and dizzy, as though it were fuddled** [drunk]**...But when it investigates by itself** [free of intellectual denial]**, it passes into the realm of the pure and everlasting and immortal and changeless, and being of a kindred nature, when it is once independent and free from interference, consorts with it always and strays no longer, but remains, in that realm of the absolute** [Integrative Meaning]**, constant and invariable'** (*Phaedo*, c.360 BC; tr. H. Tredennick). He also wrote that the **'capacity** [of **a mind...to see clearly**] **is innate in each man's mind** [we are born with an instinctive orientation to Integrative Meaning], **and that the faculty by which he learns is like an eye which cannot be turned from darkness** [the state of living in denial] **to light** [the denial-free truth] **unless the whole body is turned; in the same way the mind as a whole must be turned away from the world of change until it can bear to look straight at reality, and at the brightest of all realities which is what we call the Good** [Integrative Meaning or God]**'** (c.360 BC; *Plato: The Republic*, tr. H.D.P. Lee, 1955, p.283 of 405). Yes, the human race *has* to stop living in denial of Integrative Meaning, **'the Good'**, if we are to begin to think effectively—otherwise we will stay in the situation where our mind **'loses its way and becomes confused and dizzy, as though it were fuddled** [drunk/stupid/alienated]'.

[519]While our **'capacity' 'to see clearly'** is, as Plato said, **'innate'**, denial and its alienating effects came about through our own corrupting search for knowledge and through our encounter with the already upset, human-condition-afflicted, corrupt world. And just as this search for knowledge and our encounter with the upset world began at birth and continued throughout our lives, the extent of our insecurity about our corrupted state and associated block-out or alienation *also* increased throughout our lives, until eventually we were walking around, in effect, free of criticism but totally in the dark in terms of our access to truth and meaning. So it follows that it was when humans were very young that

they were most able to think truthfully. The famous psychoanalyst Sigmund Freud recognised this when he said, **'What a distressing contrast there is between the radiant intelligence of the child and the feeble mentality of the average adult'** (*The Freud Reader*, ed. P. Gay, 1995, p.715). Christ also recognised the mental integrity of the young when he said, **'you have hidden these things from the wise and learned, and revealed them to little children'** (Bible, Matt. 11:25). Albert Einstein famously echoed these sentiments when he noted that **'every child is born a genius'**, while the philosopher Richard Buckminster Fuller acknowledged that **'There is no such thing as genius, some children are just less damaged than others'** (NASA Speech, 1966), and that **'All children are born geniuses. 9999 out of every 10,000 are swiftly, inadvertently de-geniused by grown-ups'** (Mario M. Montessori Jr, Paula Polk Lillard & Richard Buckminster Fuller, *Education for Human Development: Understanding Montessori*, 1987, Foreword). R.D. Laing also observed that **'Each child is a new beginning, a potential prophet** [denial-free, truthful, effective thinker]**'** (*The Politics of Experience* and *The Bird of Paradise*, 1967, p.26 of 156) and pointed out that **'Children are not yet fools, but** [by our treatment of them] **we shall turn them into imbeciles like ourselves, with high I.Q.'s if possible'** (ibid. p.49). Many exceptionally creative people have also made statements to the effect that genius is the ability to think like a child. For example, one of the most accomplished artists of all time, Pablo Picasso, famously said about his struggle to paint well that **'It's taken me a lifetime to learn to paint like a child.'** The poet Charles Baudelaire similarly wrote that **'genius is no more than childhood recaptured at will'** (*The Painter of Modern Life*, 1863). And just like the already mentioned quotes about our alienated condition by Samuel Beckett that **'They give birth astride of a grave, the light gleams an instant, then it's night once more'**, and R.D. Laing that **'To adapt to this world the child abdicates its ecstasy'**, the artist Francis Bacon said **'the shadow of dead meat is cast as soon as we are born'** (*The Australian*, 15 Jun. 2009, reprinted from *The New Republic*). Similarly, in *House of Cards*, a 1993 film based on a screenplay by Michael Lessic, one of the characters makes the following intuitive comment about how sensitive and vulnerable innocent children have been to the horror of the alienated world of adults: **'I used to watch Michael** [a character in the film] **about two hours after he was born and I thought that at that moment he knew all of the secrets of the universe and every second that was passing he was forgetting them** [he was having to live in denial of them]**.'** The poet Percy Bysshe Shelley was another who decried the fate of adult humans when he wrote that **'Our boat** [our being] **is asleep on Serchio's stream, its sails are folded like thoughts in**

a dream' (*The Boat on the Serchio*, 1821). The Bible also offers this account of our estranged, resigned adult condition: **'This people's heart has become calloused** [alienated]**; they hardly hear with their ears, and they have closed their eyes'** (Isa. 6:10 footnote). And in his incredibly honest poem *Intimations of Immortality from Recollections of Early Childhood*, William Wordsworth provided this description of how quickly we humans become alienated from our all-sensitive and truthful, innocent instinctive self or soul: **'There was a time when meadow, grove, and streams / The earth, and every common sight / To me did seem / Apparelled in celestial light / The glory and the freshness of a dream / It is not now as it hath been of yore / Turn wheresoe'er I may / By night or day / The things which I have seen I now can see no more // The Rainbow comes and goes / And lovely is the Rose / The Moon doth with delight / Look round her when the heavens are bare / Waters on a starry night / Are beautiful and fair / The sunshine is a glorious birth / But yet I know, where'er I go / That there hath past away a glory from the earth // ...something that is gone / ...Whither is fled the visionary gleam? / Where is it now, the glory and the dream? // Our birth is but a sleep and a forgetting / The Soul that rises with us, our life's Star / Hath had elsewhere its setting / And cometh from afar / Not in entire forgetfulness / And not in utter nakedness / But trailing clouds of glory do we come / From God, who is our home / Heaven lies about us in our infancy! / Shades of the prison-house begin to close / Upon the growing Boy / ...And by the vision splendid / Is on his way attended / At length the Man perceives it die away / And fade into the light of common day / ...Forget the glories he hath known / And that imperial palace whence he came'** (1807).

520 Indeed, we have become *so* alienated that as adults we will readily intellectually focus on a safely sectioned-off area of inquiry or activity—such as solving a maths equation, or mastering a computer problem, or debating whether God has been destroyed by the big bang theory of the origins of the universe, or ordering our wardrobe, or polishing our car, or making a cake, or even sending man to the Moon—and yet we won't go beyond those safety limits and risk encountering *anything* to do with the issue of 'self', the depressing subject of the human condition.

521 This extraordinary, indeed mad, situation was well summarised by General Omar Bradley when, as mentioned in par. 118, he said, **'The world has achieved brilliance...without conscience. Ours is a world of nuclear giants and ethical infants.'** Yes, we humans will apply all our vigour to protesting an environmental cause or the rights of an indigenous race or the demand for peace, or any one of a number of other politically

correct causes, but we will *not* look at the nightmare of angst in ourselves, the *real* devastation and issue of our own condition and, beyond that, the human condition that needs to be addressed if we are to bring about a caring, equitable and peaceful world—because the fact is, no matter how much we try to restrain and conceal our upset eventually our world *will* become an expression of ourselves and thus as devastated as we are. To fix the world we have to first fix ourselves. As will be made very clear in chapter 7:16O, the truth is that the main function of politically correct causes has been to allow upset humans to feel that they are doing good when they are actually *avoiding* what is required to make a positive difference—namely confronting the issue of the human condition. Human life has been preoccupied with maintaining the many delusions and false ways of making us feel good about ourselves and with all manner of escapisms from reality rather than with meaningful thinking and progressive actions as we claim it is. In short, the human condition is *the* all-important issue that had to be looked at to free ourselves from our condition and truly bring peace to our world and yet it is the one issue we have refused to look at. As Carl Jung recognised, **'Man everywhere is dangerously unaware of himself. We really know nothing about the nature of man, and unless we hurry to get to know ourselves we are in dangerous trouble'** (Laurens van der Post, *Jung and the Story of Our Time*, 1976, p.239 of 275). R.D. Laing also recognised this when he wrote that **'The requirement of the present, the failure of the past, is the same: to provide a thoroughly self-conscious and self-critical human account of man'** (*The Politics of Experience* and *The Bird of Paradise*, 1967, p.11 of 156). Yes, the human condition *is* the elephant in our living rooms that we pretend not to see, the all-important issue that we assiduously practise denying.

522 The point is, the human mind has been alienated from the truth *twice* in its history: once when we were like other animals, instinctively blocked from recognising the truth of selflessness, and then again in our species' current state, during which we have become insecure about our divisive condition with no choice but to live in Plato's dark cave of denial of the significance of the selfless, loving integrative meaning of existence.

523 But while humans have gradually retreated from consciousness into virtual unconsciousness because of our insecurity about our non-ideal, soul-corrupted, 'fallen', human-condition-afflicted state, we were, to our knowledge, the first species to become fully conscious. So, the next

question is, how were our ape ancestors able to overcome the block that exists in the minds of the great majority of animals and become capable of making sense of experience, become conscious? (As mentioned in par. 387, all animals are trying to develop love-indoctrination and to what degree they have been able to develop it will dictate to what degree they have been able to develop at least a rudimentary level of consciousness, but no other existing species has developed full consciousness like humans have, and bonobos almost have.)

524 Understanding how the nurturing love-indoctrination process was able to develop selfless, moral instincts in our ape ancestors (and to some degree in some other primates today) allows us to answer this crucial question, because the reason we were able to become fully conscious is that the nurturing of selfless instincts breached the block against thinking truthfully by superimposing a new, truthful, selflessness-recognising mind over the older, effectively dishonest, selfless-thinking-blocked one. Since our ape ancestors could develop an awareness of cooperative, selfless, loving meaning, they—and, by extension, humans—were able to develop truthful, sound, effective thinking and so acquired consciousness, the essential characteristic of mental infancy.

525 To use a comparative example, chimpanzees are currently in a relatively early stage of mental infancy—they have the conscious mental powers of approximately a two-year-old human and demonstrate rudimentary consciousness, in that they are beginning to relate information or reason effectively and make sufficient sense of experience to recognise that they are at the centre of the changing array of events they experience. Experiments have shown that they have an awareness of the concept of 'I' or self and, as mentioned earlier, are capable of reasoning how events are related sufficiently well to know that they can reach a banana tied to the roof of their cage by stacking and climbing upon boxes.

526 In the case of bonobos, evidence suggests that this species is the most intelligent or conscious next to humans—as is apparent in these quotes that were referred to in pars 308-310: **'Everything seems to indicate that** [Prince] **Chim** [a bonobo] **was extremely intelligent. His surprising alertness and interest in things about him bore fruit in action, for he was constantly imitating the acts of his human companions and testing all objects. He rapidly profited by his experiences...Never have I seen man or beast take greater satisfaction in showing off than did little Chim. The contrast in intellectual qualities between**

him and his female companion [a chimpanzee] **may briefly, if not entirely adequately, be described by the term "opposites"'** (Robert M. Yerkes, *Almost Human*, 1925, p.248 of 278). Sue Savage-Rumbaugh reinforced this view when she wrote that **'Individuals who have had first hand interactive experience with both *Pan troglodytes* [chimpanzees] and *Pan paniscus*** [bonobos or pygmy chimpanzees] **(Yerkes and Learned, 1925; Tratz and Heck, 1954) have been left with the distinct impression that pygmy chimpanzees are considerably more intelligent and more sociable than *Pan troglodytes*'** and that **'Each individual who has worked with both species in our lab is repeatedly surprised by their** [bonobos'] **communicative behavior and their comprehension of complex social contexts that are vastly different from anything seen among *Pan troglodytes*'** ('*Pan paniscus* and *Pan troglodytes*: Contrasts in Preverbal Communicative Competence', *The Pygmy Chimpanzee*, ed. Randall Susman, 1984, pp.396, 411-412 of 435). The following extract demonstrates how extraordinarily aware, cooperative, empathetic and intelligent bonobos are: **'Barbara Bell...a keeper/trainer for the Milwaukee County Zoo...works daily with the largest group of bonobos...in North America... "It's like being with 9 two and a half year olds all day," she** [Bell] **says. "They're extremely intelligent...They understand a couple of hundred words," she says. "They listen very attentively. And they'll often eavesdrop. If I'm discussing with the staff which bonobos (to) separate into smaller groups, if they like the plan, they'll line up in the order they just heard discussed. If they don't like the plan, they'll just line up the way they want." "They also love to tease me a lot," she says. "Like during training, if I were to ask for their left foot, they'll give me their right, and laugh and laugh and laugh. But what really blows me away is their ability to understand a situation entirely." For example, Kitty, the eldest female, is completely blind and hard of hearing. Sometimes she gets lost and confused. "They'll just pick her up and take her to where she needs to go," says Bell. "That's pretty amazing. Adults demonstrate tremendous compassion for each other"'** (*Chicago Tribune*, 11 Jun. 1998). More recently, the bonobo researcher Vanessa Woods described bonobos as **'the most intelligent of all the great apes'** ('Bonobos – our better nature', blogs.discovery.com, 21 Jun. 2010; see <www.wtmsources.com/134>).

527 Again, as was explained in par. 311, there are some scientists who suggest that chimpanzees are as intelligent as bonobos, but there is an undeniable freedom in the mind of a bonobo that is apparent in their capacity to be interested in the world around them, and in their empathy, compassion and even simple, childish humour, as evidenced by the quotes above. While it is true that chimpanzees show a mental

dexterity, it is a narrow, opportunistic, self-interested mental focus (similar to a very primitive version of the alienated, deadened minds of humans today), not the broad, free, open, curious, aware, all-sensitive, loving, truly thoughtful, conscious mind that bonobos have. It does have to be remembered that mechanistic science doesn't even have an interpretation of the word 'love', so we can't expect it to be capable of showing interest in or acknowledging the kind of open, curious, aware, all-sensitive and loving conscious mind that bonobos have. Indeed, the extraordinarily cooperative, unconditionally loving and truly aware characteristics of bonobos would motivate, and indeed (as we saw in chapter 5) *have* motivated, human-condition-avoiding, alienated mechanistic scientists to find a way to demote them at every opportunity, just like they found a way to demote my work and that of Dian Fossey and Sue Savage-Rumbaugh.

528 So how did the process of nurturing overcome the instinctive block against selfless thinking/behaviour? It makes sense that at the outset the brain of our ape ancestors was relatively small with a limited amount of association cortex, the brain matter in which information is associated. These brains had instinctive blocks preventing the mind from making deep meaningful/truthful/selflessness-recognising perceptions. At this stage, however, these small, inhibited brains were being trained in selflessness, so although there was not a great deal of unfilled cortex available, what was available was being inscribed with a truthful, effective network of information-associating pathways. The mind was being taught the truth and given the opportunity to think clearly, in spite of the existing instinctive blocks or 'lies'. While at first this truthful 'wiring' would not have been very significant due to the small size of the brain, it had the potential for much greater development. Further, as was explained in par. 274, with this selfless training of the brain occurring over many generations, the selfless 'wiring' in the brain would have gradually become instinctive or innate. Genes would inevitably follow and reinforce any development process—in this they were not selective. The difficulty lay in getting the development of unconditional selflessness to occur, for once it was regularly occurring it would naturally become instinctive over time, which it did—our instinctive moral soul, the 'voice' of which is our 'conscience', was formed. We are born with a brain that has instinctive orientations that incline us to behave unconditionally selflessly, and to expect to be treated in the same way.

529 Thus, the mind was trained or programmed or 'brain-washed' or 'indoctrinated' with the ability to think in spite of the blocks working against such training; it had, at last, been stimulated by the truth. Of course, it must be remembered that in this early stage of development the emphasis was on training in *love*, not on the liberation of the conscious ability to *think*, which was incidental to Negative Entropy's push for our ape ancestors to become an integrated group of multicellular animals; but, once fully liberated, the conscious mind takes on a life of its own, it is free to develop—and that development follows an inevitable path.

530 Yes, after assisting the love-indoctrination process by allowing for the conscious selection of less aggressive mates, the conscious mind progresses down its own particular path of development. As summarised earlier, and this will be elaborated upon in some detail in chapter 7, this path of progression kicks off in the infancy stage, during which the conscious mind is sufficiently aware of the relationship of events that occur through time to recognise that the individual doing the thinking is at the centre of the changing array of experiences around it. It is during infancy that the conscious individual becomes aware of the concept of 'I' or self, which is what bonobos and, to a lesser degree, the other great apes are capable of. Infancy is also when the individual discovers conscious free will, the power to manage events. Childhood is the stage during which the conscious individual revels in this free will, 'playing' or experimenting with it, while adolescence is the stage during which the conscious individual encounters both the sobering responsibility of free will and the agonising identity crisis brought about by the conflict with the already established instinctive orientations and the dilemma that results, which, in the case of humans, we call the human condition: the question of whether or not we are meaningful beings. Adulthood is when the conscious mind ends the insecure stage of adolescence by finding understanding of that corrupted condition, which is the mature, transformed state that humanity can, if it chooses correctly between terminal alienation and transformation, now enter.

531 As was pointed out at the end of chapter 5, this explanation of the origin of consciousness was always going to be an extremely difficult explanation for mechanistic science to appreciate because it depends on recognising so many truths that have historically been denied. To arrive at this explanation of how consciousness emerged and then developed into the highly intelligent brain we humans have today hinges on being able to

recognise the truth of Integrative Meaning and its theme of unconditional selflessness—and from there why animals would have developed blocks in their minds preventing selfless, truthful, effective thinking and thus consciousness—and from there how the nurtured training of selflessness in humans would have liberated truthful thinking and thus consciousness—and from there how the emergence of consciousness would have led to a terrible battle with our instinctive self—and from there how the psychological upset that resulted from the battle would have demanded a more developed, intelligent, bigger brain in order to find understanding of why we had become divisively behaved. Yes, that journey has involved so many unbearably confronting truths that it is not surprising that the human-condition-avoiding mechanistic paradigm has found it difficult to acknowledge. This journey from one unbearably confronting truth to another also reveals how if you are living in denial of truth you have no chance of making sense of our world and place in it—as evidenced by the mountain high pile of books that have been written about consciousness without ever managing to penetrate the subject. So, sadly, the President of the United States Barack Obama's 2 April 2013 **'Brain Initiative'**, which will see **'$100 million initial funding'** (*The Sydney Morning Herald*, 4 Apr. 2013) going to *human-condition-avoiding, denial-committed, mechanistic science* to find **'the underlying causes of...neurological and psychiatric conditions'** and **'develop effective ways of helping people suffering from these devastating conditions'** (US National Institutes of Health, accessed Apr. 2013; see <www.wtmsources.com/114>), namely the human condition (and as described in par. 445, there are very similar initiatives occurring in Europe), is so self-defeating it *is*, in effect, nothing more than an act of pure desperation.

532 In conclusion, it was the process of nurturing love-indoctrination that not only gave our species its instinctive orientation to behaving cooperatively—our moral soul—it *also* liberated consciousness in our forebears. As already pointed out in chapter 4, since nurturing is largely a female role and females controlled the selection of cooperative mates, it is true to say that the female gender created humanity. Throughout humanity's infancy and childhood, a period of time that, as will be explained in the next chapter, lasted from some 12 to 2 million years ago, nurturing played the most important role in the group; it was a matriarchal society in which males had to support this focus on nurturing and protect the group from external threats. However, as will also be described in the next chapter, humanity's matriarchal structure came to

an end when our ignorant, ideal-behaviour-demanding instinctive self began to criticise our conscious mind's search for knowledge and men, in their role as group protectors, had to take up the battle of resisting that naive, idealistic criticism from our instinctive self that was threatening to stop the maturation of humanity from ignorance to enlightenment. At this point, the patriarchal society that has existed for the last 2 million years came into being.

533 What now needs to be presented is a description of all the psychological stages that our conscious mind progressed through in that heroic journey from ignorance to enlightenment.

534 (A more complete description of the nature and origins of consciousness can be found in *Freedom Book 1* at <www.worldtransformation.com/freedom-consciousness>.)

Chapter 7

The Greatest, Most Heroic Story Ever Told: Humanity's Journey from Ignorance to Enlightenment

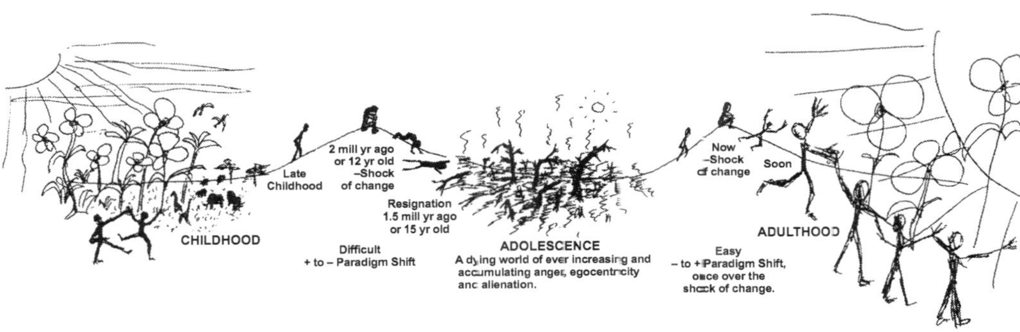

Drawing by Jeremy Griffith © 1996 Fedmex Pty Ltd

Chapter 7:1 Summary

[535] With the desperately needed answers to the great outstanding mysteries facing the human race of the human condition, of the meaning of our existence, and of the origins of our unconditionally selfless moral instincts and conscious mind, now presented in this book, the full story of life on Earth can at last be told. In particular, we can explain where all species appear on the 'ladder' of integration of matter, and why, thus far, it is only us, the human race, that has managed to progress to the 'rung' of integrating the members of our species into the Specie Individual. Indeed, it was that unique achievement that led to the emergence of our conscious mind that had to then search for knowledge, ultimately self-knowledge—the result of which was the breakdown of the integrated Specie Individual into the divisive, human-condition-afflicted state. But with the human condition now understood, humanity is finally in a position to *return* to the fully integrated Specie Individual situation once more, but this time in a knowing or understanding position—as the exceptionally denial-free, effective-thinking prophet Moses predicted, we can finally become '*like God* [the integrated state], *knowing*' (Bible, Gen. 3:5).

536 **What remains to be presented in this book is a description of the actual stages that our conscious mind progressed through from ignorance to enlightenment of our human condition, and, hopefully, with that understanding now found, to the transformation of the human race into the *'knowing'* or enlightened form of the integrated Specie Individual. Yes, we can now explain and describe the *entire psychological journey* our species has been on—and since that psychological journey is the *real* journey we fully conscious, human-condition-afflicted humans have been on, and since it has never before been able to be presented, this chapter provides the first ever true description of our species' development. While there have been libraries of books written about human origins, this is the first denial-free, truthful description of our species' emergence from our ape ancestors. And, since the main dishonest denial that has been maintained by mechanistic science was that our distant ancestors were controlled by competitive, selfish, brutish instincts that our conscious mind had to try to restrain, when, in fact, the true story is the absolute opposite of this account—that our ape ancestors were actually loving and gentle and that it was with the emergence of our conscious mind that we *became* competitive and selfish brutes—what is going to be presented is the completely new, never before told story of the human race!**

537 **Yes, what follows is the most amazing and epic journey of any species to have existed on Earth—and it's our own story, the story of the incredibly, phenomenally brave and heroic human race!**

Chapter 7:2 <u>The stages of humanity's maturation from ignorance to enlightenment</u>

538 We begin this most amazing story of humanity's journey from ignorance to enlightenment by meeting with our ancestors, not in person, but in the wonderfully fortuitous fossilised remains we have of them.

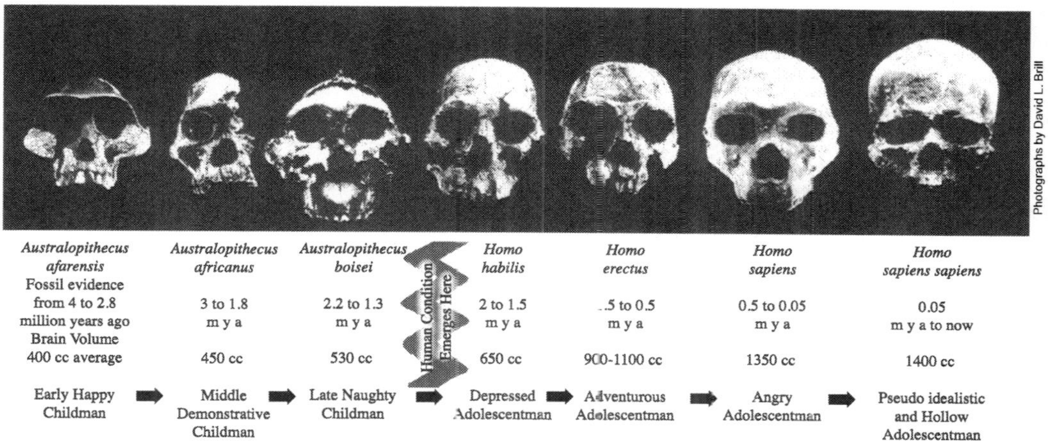

539 This sequence of fossil hominin (human ancestor) skulls, dating back to the beginning of the australopithecines, appeared in the November 1985 edition of *National Geographic* magazine. While anthropologists have since discovered more varieties of *Australopithecus* and *Homo* than those depicted here, these remain representative of the main varieties. The fossils that have been found to belong to our forebears who existed prior to the australopithecines are not included in this picture, but they too will be accounted for shortly.

540 In examining this sequence it is apparent that a sudden increase in the size of the brain case, and by inference the brain's volume, occurred around 2 million years ago. A larger brain case was needed to house a larger 'association cortex'. As explained in pars 469-475, the ability to 'associate' information is what made it possible to reason how experiences are related, learn to understand and become conscious of, or aware of, or intelligent about, the relationship between events that occur through time. It follows that the development of a *larger* association cortex meant that a greatly increased need for understanding had emerged, which we

are now able to explain would have resulted from the emergence of the dilemma of the human condition, where only self-understanding could bring an end to all the psychological upset that condition has produced and which has been so crippling of our species' development and that of our own lives. The inference we can take from this evidence is that the human condition became a full-blown problem some 2 million years ago with the emergence of *Homo*.

541 The descriptions I have provided below the picture of the various fossil hominid skulls document the stages humanity progressed through as this liberated consciousness developed. The names ascribed to each stage indicate parallels with our own human life-stages, because the stages that we, as conscious individuals, progress through are the same stages our human ancestors progressed through—'ontogeny recapitulates phylogeny': our individual consciousness necessarily charts the same course that our species' consciousness has taken as a whole. Eugène Marais recognised this when he wrote that **'The phyletic history of the primate soul can clearly be traced in the mental evolution of the human child'** (*The Soul of the Ape*, written between 1916-1936 and published posthumously in 1969, p.78 of 170). Wherever consciousness emerges it will first become self-aware (the 'Infancy' stage of the development of a conscious mind), then, inevitably, it will start to experiment with its power to effectively understand and thus manage change (the 'Childhood' stage), following which it will seek to understand the meaning behind all change ('Adolescence'), and from there it will obviously try to comply with that meaning ('Adulthood').

542 In the case, however, of consciousness developing within us individually and in our ancestors, that journey was significantly disrupted at adolescence by our search for the understanding of why we—individually and as a species—were not behaving in accordance with the integrative, cooperative meaning of existence; adolescence being the stage in the development of consciousness where the search for identity takes place, where the search for understanding of the meaning behind change, and the conscious organism's relationship to that meaning, occurs. In the case of our lives individually, and that of our species, the particular identity we needed to find understanding of was why we were behaving divisively when the meaning of existence is to behave cooperatively and lovingly. Until we could answer that question, our development, both individually

and collectively, was stalled (or what psychologists refer to as 'arrested') in an insecure adolescent state, unsure of our identity, particularly the reason why we haven't been ideally behaved—and preoccupied trying to validate our existence, prove that we are good and not bad, find *some* relief from the insecurity of the human condition. So, to mature from insecure adolescence to secure adulthood depended upon finding understanding of our divisive human condition.

543 To go over this important point again, since the human race as a whole has not—until now—been able to understand the dilemma of the human condition, humans haven't been able to properly enter adulthood. When stages of maturation aren't properly completed it doesn't mean subsequent growth stages don't take place, they do, but if a previous stage isn't properly fulfilled those subsequent stages are greatly compromised by the incomplete preceding stages. People do grow up, but in a state of arrested development. Without the explanation of the human condition, humans have been insecure, not properly developed—in fact, preoccupied justifying themselves and finding ways to escape feeling unworthy. As the actress Mae West once famously said about men, **'If you want to understand men just remember that they are still little boys searching for approval.'** It is only with understanding of the human condition now found that humans will be able to properly complete their adolescence and grow into secure adults, and the human race as a whole will be in a position to mature from insecure adolescence to secure adulthood. This is why descriptions such as 'The Angry Adulthood Stage of Humanity's Adolescence' appear in the following stages to describe our species' progression, for while we became adults in a physical sense, without the explanation of the human condition we were still psychologically stranded in adolescence.

544 In summary, infancy is 'I am', childhood is 'I can', adolescence is 'but who am I?', while adulthood is 'I know who I am.'

545 To apply these stages involved in the development of consciousness to the journey through which our species progressed, the term 'Infantman' pertains to the ape ancestor who first developed the nurturing training in selflessness that produced the fully cooperative state and, in doing so, liberated consciousness. As described in chapter 4:5, fossils from this period, which dates from 12 to 4 million years ago, are rare; however, scientists believe that our ape ancestors from this period

include *Sahelanthropus tchadensis* (who lived some 7 million years ago and is thought to be the first representative of the human line after we diverged from humans' and chimpanzees' last common ancestor); *Orrorin tugenensis* (who lived some 6 million years ago); and the two species of *Ardipithecus*: *kadabba* (who lived some 5.6 million years ago), and *ramidus* (who lived some 4.4 million years ago). The various stages of 'Childman', who, it follows, developed from 'Infantman', were, as just summarised under the picture of the skulls, the australopithecines who began to experiment with the power of conscious free will: 'Early Happy Childman' (*Australopithecus afarensis*), who developed into 'Middle Demonstrative Childman' (*Australopithecus africanus*), who then developed into 'Late Naughty Childman' (*Australopithecus boisei*). Again, at each stage greater experimentation in conscious self-management was taking place—from demonstrating the power of free will in mid-childhood, to beginning to challenge the instincts for the right to manage events in late childhood.

546 As has now been explained, this challenging of our instincts led to criticism from those instincts, which in turn upset our conscious mind. In late childhood this emerging upset expressed itself in the physical flailing out at the 'unjust world'—the **'naughty nines'**, as parents and teachers describe this stage. What occurs to bring an end to childhood at about the age of 12, and what would have occurred at the end of the childhood stage in our ancestors around 2 million years ago, is the realisation that physically protesting and flailing out doesn't achieve anything and that what we need to do instead is calm down, take stock of our situation, and try to understand what is causing us to be so upset; try to find the explanation for *why* we are being criticised. This significant change from being a frustrated, protesting, boisterous extrovert to a sobered, deeply thoughtful introvert signals the beginning of adolescence. Although unaware of the underlying reason, anthropologists have, nevertheless, recognised that a significant change did take place around 2 million years ago because it was at this juncture that they changed the name of the genus from *Australopithecus* to *Homo*; 'Childman', the australopithecines, became 'Adolescentman', *Homo*.

547 As stated, adolescence is the stage when the search for identity takes place, and the identity that 'Adolescentman', *Homo*, particularly sought to understand was their lack of ideality—the reason why

they were not ideally behaved. And once begun, this march of upset could *only* be brought to an end by finding sufficient knowledge to explain why the instincts' 'criticism' was undeserved. As such, there was an ever increasing need for mental cleverness to explain ourselves—specifically to find the liberating understanding of the human condition—hence the rapid increase in brain volume from 2 million years ago onwards.

548 As with our species' progression through the various stages of its childhood, when our species entered its adolescence we necessarily went through a similar progression of stages, starting with the early sobered Adolescentman stage (early *Homo habilis*), followed by the depressed Adolescentman stage (late *Homo habilis*), through to the adventurous Adolescentman stage (*Homo erectus*, who first left Africa), the embattled angry Adolescentman stage (*Homo sapiens*), and the pseudo idealistic and hollow Adolescentman stage that we currently occupy (*Homo sapiens sapiens*). But with the finding of understanding of why we have been divisively behaved now found, humans individually, and humanity collectively, can finally mature from this insecure adolescence to a secure adulthood: 'Adolescentman' *can* become transformed 'Adultman'. (Note, 'man' is an abbreviation for 'human' or 'humanity', however, the use of 'man' also denotes a recognition that while humanity's infancy and childhood was matriarchal or female-role-led because that was when nurturing of infants was all-important, with the emergence of the egocentric, male-role-led need to defy the instincts, search for knowledge and prove we humans are good and not bad, humanity's adolescence became patriarchal—a transition that will be more fully explained shortly (in pars 596-597). Since humanity's adulthood will be neither female or male led—because our species' maturation is complete—'Adultman' should more properly be described as 'Adulthuman'. By the same logic, since adolescence was *originally* female role-led, the description for that stage should be 'Adolescentwoman' rather than 'Adolescentman', but to avoid unnecessarily complicating the matter we will leave all as 'man' for now.)

549 We will now examine these various stages that a conscious mind has to progress through from infancy to childhood to adolescence and adulthood from the perspective of both the individual and humanity.

Chapter 7:3 INFANCY

The species: our early ape ancestors, including the ardipithecines — 12 to 4 million years ago

The individual: 0 to 3 years old

Photograph of Matata by Manny Rubio

Bonobos Matata and her adopted son, Kanzi

=

Drawing by Jeremy Griffith © 2006 Fedmex Pty Ltd

550 As explained in chapter 4, it was through the process of love-indoctrination, the exceptional nurturing of infants, that our ape ancestor was able to develop unconditionally selfless, fully altruistic, cooperative, moral, integrative, Specie Individual-forming behaviour—a process the bonobos are currently perfecting, as illustrated by the above photo of the bonobo Matata hugging her adopted son, Kanzi. Nurturing is what made us human, a truth the image of the Madonna and child has been the archetypal representation of. And, as was explained in chapter 6:3, with the development of selflessness consciousness was able to emerge, and with consciousness came all the stages of maturation that consciousness itself had to progress through—namely infancy, childhood, adolescence, and adulthood. (It should be mentioned that while the main stages of maturation in non-human species are generally described as infancy, then adolescence when individuals are on the threshold of sexual maturity, and finally the sexual maturity of adulthood, in the case of fully conscious humans our stages of maturation involve a *psychological* journey, so our stages of maturation (from infancy to childhood, adolescence and

adulthood) are fundamentally different to the infancy, adolescence and adulthood stages of other species.)

551 As stated, the first stage of infancy occurs when humans become sufficiently conscious, sufficiently aware of cause and effect to realise that 'I exist'; it is when we realise that we are at the centre of constantly changing experiences. This emerging consciousness and resulting self-awareness is apparent in bonobos who are the most intelligent of all non-human primates; indeed, some of that intelligence is apparent in Matata's expression. Yes, as was explained and evidenced in pars 308-311 and 526-527, bonobos are **'the most intelligent of all the great apes'**; and **'Everything seems to indicate that Chim** [a bonobo] **was extremely intelligent. His surprising alertness and interest in things about him bore fruit in action, for he was constantly imitating the acts of his human companions and testing all objects. He rapidly profited by his experiences...Never have I seen man or beast take greater satisfaction in showing off than did little Chim. The contrast in intellectual qualities between him and his female companion** [a chimpanzee] **may briefly, if not entirely adequately, be described by the term "opposites"'**; and, **'"It's like being with 9 two and a half year olds all day," she** [zoo keeper Barbara Bell] **says. "They're extremely intelligent...They understand a couple of hundred words," she says. "They listen very attentively. And they'll often eavesdrop. If I'm discussing with the staff which bonobos (to) separate into smaller groups, if they like the plan, they'll line up in the order they just heard discussed. If they don't like the plan, they'll just line up the way they want." "They also love to tease me a lot," she says. "Like during training, if I were to ask for their left foot, they'll give me their right, and laugh and laugh and laugh. But what really blows me away is their ability to understand a situation entirely."'**

552 As was explained in detail in chapter 4:5, recently published reports on the 'Infantman' fossils, *Sahelanthropus*, *Orrorin* and *Ardipithecus*, provide powerful validation of the love-indoctrination process, and also confirm that 'Infantman' closely resembled bonobos in many important respects. To briefly reiterate, there are, as has been emphasised, three main requirements for the nurturing-of-infants, love-indoctrination process to take place—upright walking, or bipedalism; ideal nursery conditions; and the presence and influence of more maternal mothers—and these recent fossil discoveries now show that at least as far back as 7 million years ago our ancestors were adapted to walking upright, as are bonobos, who walk upright more often than any other non-human primate. With regard to ideal nursery conditions, they also confirm that 'Infantman' lived in

forests and woodlands, not savannahs, and was semi-arboreal like the bonobos, who are the most arboreal of the African apes. Importantly, with regard to the need for more maternal mothers, it was also described in chapter 4:5 that the fossil record now strongly suggests that 'Infantman' lived in a cooperative society that was characterised by female sexual selection for less aggressive males, indicating a matriarchal society, which is consistent with a society where the raising of infants had become the central focus, as it is with bonobos. Based on these recent fossil finds, the anthropologist C. Owen Lovejoy reinforced both the reality and the duration of our cooperative past, writing that **'Even our species-defining cooperative mutualism can now be seen to extend well beyond the deepest Pliocene** [well beyond 5.3 million years ago]**'** ('Reexamining Human Origins in Light of *Ardipithecus ramidus*', *Science*, 2009, Vol.326, No.5949).

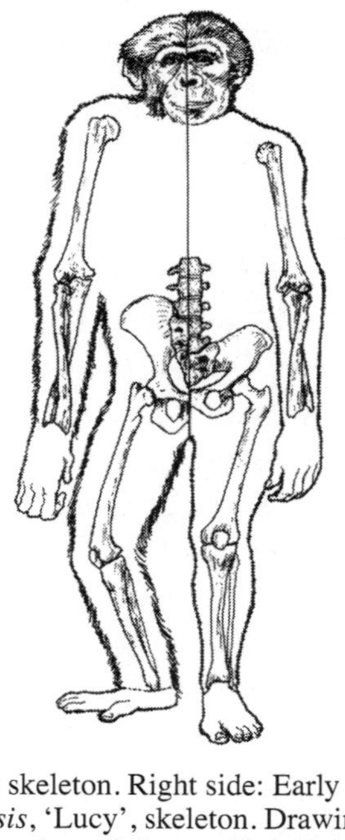

Left side: Bonobo skeleton. Right side: Early australopithecine, *Australopithecus afarensis*, 'Lucy', skeleton. Drawing by Adrienne Zihlman from 'Pygmy chimps, people, and the pundits', *New Scientist,* 15 Nov. 1984.

553 A comparative look at the physique of bonobos and the fossil evidence provided by 'Lucy', the 3.5-million-year-old 'Childman', *Australopithecus afarensis*, who was discovered in the Rift Valley of Africa in 1974 by a team headed by the paleoanthropologist Donald Johanson, is also informative, as Lucy's bone structure shows an amazing similarity to that

of a bonobo. While Lucy's skull indicates that she had a slightly larger brain than a bonobo, the two are very similar in stature and in the length of the lower limbs, and are fairly similar in overall body proportions. Lucy's pelvis shows that she walked fully upright. The pelvis of a bonobo, while not quite as adapted to bipedalism/upright walking as Lucy's, is significantly more adapted than that of a chimpanzee. Incidentally, *Ar. ramidus*, who it has already been argued is at a comparable stage of development to bonobos (recall Frans de Waal's observation that **'The bonobo's body proportions—its long legs and narrow shoulders—seem to perfectly fit the descriptions of Ardi, as do its relatively small canines'**), has a brain volume measuring between 300 and 350cc, which is the same size as a bonobo.

554 In summary, the fossil record is now providing clear evidence that love-indoctrination had been established in 'Infantman' at least as far back as 7 million years ago, and this time frame will no doubt increase as more fossils emerge. The fossils also evidence that in bonobos, the most conscious and intelligent of the great apes, we have a living comparison of our 'Infantman' ancestor on the verge of 'Childman'.

Chapter 7:4 CHILDHOOD

The species: the australopithecines — 4 to 1.3 million years ago
The individual: 4 to 11 years old

Drawing by Jeremy Griffith © 1996 Fedmex Pty Ltd

555 As stated, childhood is the stage when consciousness begins to experiment in self-adjustment and manage events to its own chosen ends. It comprises three stages: 'Early Happy, Innocent Childhood', 'Middle Demonstrative Childhood', and 'Late Naughty Childhood', with each stage involving a greater degree of experimentation in conscious self-management—from demonstrating the power of free will in mid-childhood, to beginning to challenge the instincts for the right to manage events in late childhood.

Chapter 7:5 Early Happy, Innocent Childhood

The species: the early australopithecines including *Australopithecus afarensis* — 4 to 2.8 million years ago
The individual: 4, 5 and 6 years old

556 'Early Happy, Innocent Childhood' encompasses the time when the intellect becomes sufficiently able to understand the relationship between cause and effect to start actively experimenting—'playing'—with the conscious power to self-manage and self-adjust. At this stage we are still, as it were, holding onto our mother's apron strings with one hand while carrying out short experiments in conscious self-management with the other; we are still depending on our instinctive nurtured orientations to love for the overall management of our life. In the case of humans today, it is the 'Look at me, Daddy, I can jump puddles' stage where reinforcing admiration from parents of the emerging conscious ability to manage events is so important.

557 As mentioned, 'Lucy', the 3.5-million-year-old *Australopithecus afarensis* whose skeleton so closely resembles that of a bonobo, is an example of 'Childman', or more specifically 'Early Happy, Innocent Childman'.

Chapter 7:6 Middle Demonstrative Childhood

The species: *Australopithecus africanus* — 3 to 1.8 million years ago
The individual: 7 and 8 years old

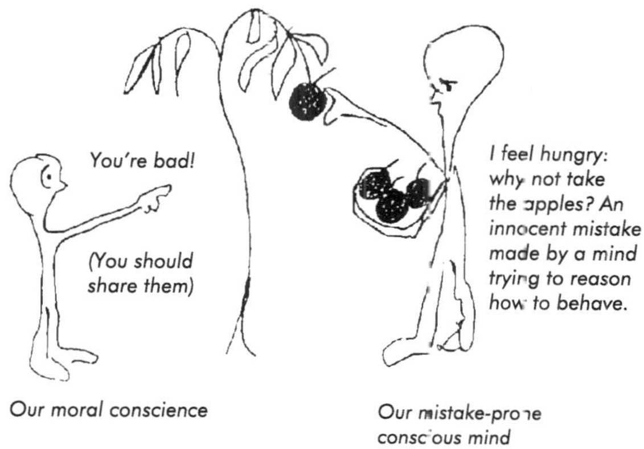

558 It is during the 'Middle Demonstrative Childman' stage that the intellect starts demonstrating the power of free will and experiences its first encounter with the frustrations of a conflict with the instincts, which is the human condition.

559 By mid-childhood the conscious mind is sufficiently able to make sufficient sense of experience to successfully manage and thus plan activities for not just minutes ahead, but for hours and even days—a development that empowers the individual to be both outwardly marvelling at, and demonstrative of, its intellectual power. It is at this stage of active self-management that the results of some experiments in self-adjustment begin to get the child into trouble. If we consider the behaviour of children today who have reached this stage, we can appreciate the kind of trouble that entails. Envisage a child sitting at a table with a birthday cake on it. Being new to this business of reasoning, the child thinks, innocently enough, 'Well, why shouldn't I take all the cake for myself', before duly doing so. While many mothers actually witness these grand mistakes of pure selfishness that young children make when they first attempt to manage their lives from a basis of understanding, they still have to be reasonably lucky to do so because, once done, the child generally doesn't make such a completely naive mistake again due to the criticism that experiment in self-management

attracts—from the child's instinctive moral conscience and from his conscious mind's awareness of the, in truth, very obvious integrative, selfless theme of existence—as well as from others present. (As was explained in chapter 2:5, this avalanche of criticism is the 'double and triple whammy' of condemnation that humans experienced when we searched for knowledge.) But despite the nasty shock from all the criticism and his desire to not make such a mistake again, the child, while unable to explain his actions, does feel that what he has done is not something bad, not something deserving of such criticism. In fact, by this stage in the child's mental development, he has become quite proud of the effort he's taken during his early happy innocent childhood stage to self-manage his life, successfully carrying out all kinds of tentative experiments in self-adjustment—drawing attention to his achievements with excited declarations like 'Look at me Daddy, I can jump puddles', and so on. So the child is only just discovering that this business of self-adjusting is not all fun and that 'playing' with the power of free will lead to some serious issues. Indeed, the frustrated feeling of being unjustly criticised for some of his experiments gives rise to the precursors of the defensive, retaliatory reactions of anger, egocentricity and alienation; some angry, aggressive nastiness creeps into the child's behaviour. Furthermore, in this situation of feeling unfairly criticised, it follows that any *positive* feedback or reinforcement begins to become highly sought-after, which is the beginning of egocentricity—the conscious thinking self or ego starts to become preoccupied trying to defend its worth, assert that it is good and not bad. At this point, the intellect also begins experimenting in ways to deny or deflect the unwarranted criticism, which, in this initial, unskilled-in-the-art-of-denial stage, takes the form of blatant lying: 'But Mum, Billy told me to do it', or 'But Mum, the cake accidentally fell in my lap.' These apparent misrepresentations weren't *actually* lies, rather they were inadequate attempts at explanation. Lacking the real excuse or explanation, it was at least *an* excuse, a contrived defence for the child's mistake. The child was evading the false implication that his behaviour was bad, in the sense that a 'lie' that said he wasn't bad was less of a 'lie' than a partial truth that said he was. Basically, the child has started to feel the first aggravations from the horror of the injustice of the human condition—and we can expect that exactly the same kind of mistakes in thinking and resulting frustrations with the ensuing criticism would have also occurred in the lives of our 'Middle Demonstrative Childman' ancestors, namely *Australopithecus africanus*. Experiments in thinking, such as 'There are some apples; why shouldn't I take them all for myself?' would have occurred, which would have resulted in criticism from their moral instincts and conscious mind's awareness of the integrative, selfless theme of existence—criticism that would have led to the beginnings of the psychologically upset behaviours of anger, egocentricity and alienation in *Aus. africanus*.

560 While children today have to—just as our *Aus. africanus* ancestors would have had to—negotiate this middle demonstrative stage where at times they behave in a way that is 'disobedient' of their moral instincts and 'defiant' of the integrative theme of existence, their conscious mind still doesn't know why it has been disobedient and defiant; it isn't able to understand and explain that it has become a conscious being. Also, now that they are capable of thought they can't stop thinking, so mistakes in self-management are going to continue to occur, as will the criticism those mistakes attract, and, it follows, the upset with that criticism. Yes, from demonstrating the power of free will, the child has started to feel the first aggravations from the horror of the injustice of the human condition. Of course, for our 'Middle Demonstrative Childman' *Aus. africanus* ancestors, love would still have very much been the dominant influence in their lives overall, which means any distress from upset would have been quickly healed with love. Furthermore, the defensive expressions of anger, egocentricity and denial would have been restricted to feelings and actions rather than expressed in words; in fact, there would not have been a strong call for language until the adolescent state emerged some 2 million years ago when the battle of the human condition developed and, with it, alienation, because it was only when we became variously alienated in self and thus variously alienated from each other that a strong need to try to justify and explain ourselves to one another arose. The anthropologist Richard Leakey's study of brain cases in fossil skulls for the imprint of Broca's area, the word-organising centre of the brain, evidences this development: **'*Homo* had a greater need than the australopithecines for a rudimentary language'** (*Origins*, 1977, p.205 of 264). Prior to the emergence of alienation we were all instinctively aware of and in sync with each other; apart from contact and warning calls, and expressions of excitement and joy, there was no need to develop a sophisticated, complex language.

561 While discussing the effects of alienation, it should be pointed out that infants would have stopped being quiet when mothers stopped being able to properly respond to their needs due to their alienated, soul-devastated condition. For instance, even the infants of relatively innocent, less alienated races of humans today, such as the Matabele of South Africa, the Australian Aborigine and the Bushmen of the Kalahari, rarely cry, as this quote about the Bushmen confirms: **'!Kung** [Bushmen]**... infants hardly ever cry'** (Dr Harvey Karp, *Cultures without Colic: Breastfeeding & Other Baby Lessons from the !Kung San*, accessed Apr. 2012; see <www.wtmsources.com/117>). Yes, motherese language developed as a way for alienated humans to try to pacify their distressed innocent infants.

Chapter 7:7 Late Naughty Childhood

The species: the robust australopithecines (*Aus. robustus* and *Aus. boisei*) — 2.2 to 1.3 million years ago
The individual: 9, 10 and 11 years old

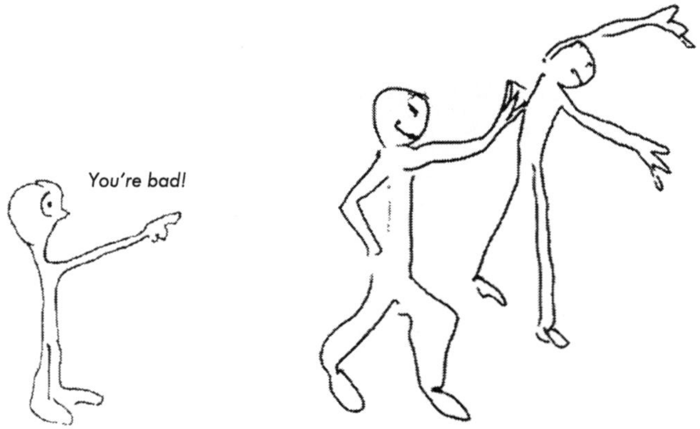

Our moral conscience

'Why shouldn't I push Johnny over in the playground?' –the frustrated, naughty, bullying final stage of childhood.

Drawing by Jeremy Griffith © 1991-2011 Fedmex Pty Ltd

562 The 'Late Naughty Childman' stage represents the time when the intellect naively lashes out at the increasing unjust criticism it is encountering as a result of its first tentative experiments in self-adjustment.

563 Since school teachers become very aware of the changing behaviour of the many children they have under their care, I asked a teacher to describe what she and her colleagues knew of the stages children and early adolescents go through. These are the main points from the response she collected: **'Six and seven-year-olds are considered to be very compliant, but by eight children are starting to test the waters and challenge the world a little.'** She continued, **'the eight-year-olds can be annoying and a little naughty'**, while **'nine and ten-year-olds can be hard to handle as they seem to hit a phase of recklessness'** and **'are considered naughty'**. She noted that **'Teachers love teaching 11 and 12-year-olds because it is during this stage that children become civilised'** but that **'Teachers consider students who are 14, 15 and 16 years old the most difficult to teach. The adolescents seem to be at complete odds with what is expected of them. Most teachers are terrified of these extremely uncooperative mid-teenage ages'** (WTM records, 1997). These insights support the explanations being given of the stages of maturation of consciousness through childhood—and confirm the agonising stage of Resignation that occurs in early adolescence, which was described in pars 43-52 and 492-493, and will be further explained within the context of humanity's development of consciousness shortly.

564 Of course, it was inevitable that as the human condition developed children were going to be increasingly influenced by the psychologically upset, angry, egocentric and alienated world in which they were raised, and it would become harder to differentiate what upset in them was a result of those external circumstances or from their own experiments in understanding. However, even in the original situation, where there was little or no upset in the world, we can expect that by the age of eight, children would justifiably be feeling resentful towards the 'criticism' emanating from their instinctive self of their tentative efforts to self-manage their life using understanding. And, unable to adequately cope with this 'criticism' with understanding of it, we can expect that each child would begin to retaliate against the criticism as the only form of defence available to them. The problem then, however, was that these early, relatively mild experiments in retaliation—of anger, selfishness and dishonest excuse-making in mid-childhood—had the alarming effect of greatly compounding the 'criticism' from the child's perfectly integratively orientated, moral instinctive self and from their awareness

of the integrative meaning of existence. From being mildly insecure we can expect the child to now feel guilty and that this drastic escalation in criticism and thus frustration would be a contributing factor to the turbulent, boisterous **'naughty nines'** that parents and teachers have labelled this stage. By the end of childhood, at the ages of 10 and 11, we can expect the resentment and frustration to be such that it would express itself in the form of taunting and bullying. The child would be belligerently lashing out at the unjust world: 'Why shouldn't I feel resentful and retaliate?', 'Why shouldn't I shove you around if I can, especially since I'm bigger and stronger?', 'Why can't I have my way?', and 'What's wrong with being selfish and aggressive anyway?'

565 In the situation that exists today, where the external upset is almost overwhelming, we can expect that almost all of the child's upset will have resulted from their encounter with external upset. The increasingly thoughtful child can see the whole horribly upset world and would be understandably totally bewildered and deeply troubled by it. Eight-year-olds will only be beginning to be consciously aware of the horror of the state of the world they have been born into, but by nine they will have become aware of that horror and be needing a lot of reassurance that 'Everything is going to be alright.' In fact, nine-year-olds can be so troubled by the imperfection of the world that they go through a process of trying not to accept that it is true. By 10, this despair about the state of the world reaches desperation levels with nightmares of distress for children. It is a very unhappy, lonely, anxious, needing-of-love time for them. So at 11 some enter a 'Peter Pan' stage where they decide they don't want to grow up; they decide they want to stay a child forever, surrounded by all the things they love, and not ever become part of the horror world they have discovered. It is no wonder **'Teachers love teaching 11 and 12-year-olds'** who have **'become civilised'**—they're essentially tame compared with the **'reckless'**, **'naughty'**, flailing-out-at-the-world **'nine and ten-year-olds'**.

566 The fossil record evidences the description that has been given of these stages of 'Childman'. The early *Australopithecus afarensis*, who have been described as occupying the early happy, prime-of-innocence stage, and the subsequent *Australopithecus africanus*, who have been described as being in the middle demonstrative childhood stage, are both finely built compared with the much more robustly

built *Australopithecus boisei* and associated *Australopithecus robustus*, both of whom have been described here as being in the 'Late Naughty Childman' stage. Anthropologists have even placed the more robust late australopithecines on a separate, dead-end branch to *Homo* (and recently some anthropologists have gone so far as to reclassify them as an entirely separate genus, *Paranthropus*), but that has to be impossible because for branching to occur there has to be deflecting influences—such as when Darwin's finches gradually became adapted to different food niches on the Galápagos Islands—and in our case there was only one major development going on and that was the *psychological* one. In a situation where there is only one all-dominant influence causing change there is no opportunity for divergence to develop, and from our species' infancy, we have been under the all-dominant influence of what was occurring in our heads, namely the development of consciousness and its psychological consequences. Any other influence was so secondary as to be ineffectual in causing our path to branch. So, contrary to what human-condition-avoiding, psychosis-denying mechanistic anthropologists have been telling us, there has been no branching off from either the australopithecines or *Homo*. But if the robust australopithecines weren't a separate branch, the question remains: why was there such a big physical disparity between them and the much more gracile or fine featured preceding *Aus. afarensis* and *Aus. africanus*, and the variety of early humans they gave rise to, the also much more gracile *Homo habilis?* The answer has to lie in the psychological differences between the much quieter, love-immersed *Aus. afarensis* and *Aus. africanus*, the extroverted, boisterous, bullying *Aus. robustus* and *A. boisei*, and the introverted, sobered, quiet early adolescent *H. habilis*, who will be described shortly.

567 To quickly elaborate on these size disparities, 'Late Naughty Childman', *Aus. robustus* and *Aus. boisei*, had comparatively large frames and skulls that were especially heavily built with very pronounced cranial and facial bone structures. Anthropologists recognise that these skull modifications came about to support the much stronger facial muscles that were required to work the heavy jaw and huge grinding teeth that characterise these late australopithecines. We know from such evidence as the wear patterns on their teeth that the australopithecines were vegetarian, but why did the later australopithecines need larger

grinding teeth? What dietary change occurred, and why? Being extroverted, increasingly naughty and roughly behaved, the late australopithecines were like older children today who would rather be out playing than eating, but such an extremely physically assertive and energetic lifestyle required fuel. Not being sufficiently conscious to attempt self-management, all other animal species exert only enough energy to secure their necessary food, space, shelter and a mate—they are conservative energy users—but, with their rough, energetic play, late childhood humans became the first non-conservative energy users on Earth. To 'eat and run', 'Late Naughty Childman' would have needed a readily available food source that they could eat quickly and, being vegetarian, they would have needed a lot of it because vegetables don't convert into as much energy as meat for instance, which was not to appear on humanity's dining table until upset developed. (While the australopithecines, in their naughtiness, would have been capable of being rough and possibly even cruel at times to other animals, they were not yet sufficiently upset with innocence to be killing innocent animals on a regular basis, which, as will be explained shortly in par. 606, is what so-called 'hunting' was really all about for upset humans and what finally led to meat-eating.) This increased energy requirement is confirmed by recent studies of dental wear patterns in robust australopithecines that indicate their **'diet included more C_4 [grasses and sedges] biomass than any other hominin...**[and its massive jaw, grinding teeth and necessary facial structure] **represents an adaptation for processing large quantities of low-quality vegetation'** (Thure Cerling et al, 'Diet of *Paranthropus boisei* in the early Pleistocene of East Africa', *Proceedings of the National Academy of Sciences*, 2011, Vol.108, No.23) that was **'nearly identical'** (ibid) to the diet of its predecessor *Aus. afarensis*. Interestingly, anthropologists had traditionally argued that *Aus. boisei*'s massive facial structure developed to allow **'hominins to colonize increasingly seasonal and open environments'** that offered **'a diet of nuts, seeds and hard fruit'** (ibid)—but, as it turns out, *Aus. boisei* was not eating anything different to *Aus. afarensis*, just greater quantities. As will shortly be described, in contrast to these late australopithecines, *H. habilis* was an entirely different individual. Introspective, deeply thoughtful and sobered, *H. habilis* were no longer interested in physically intimidating the world, and did not, therefore, need great quantities of energy and thus food, hence their reversion to a more gracile frame.

568 Another feature of the fossil record of early humans is the evidence it provides of the overlap between the different varieties. For example, in what is the biggest overlap by far, australopithecines were possibly still in existence up to 1.5 million years after *H. habilis* appeared. Again, this evidence has been used to argue that the late australopithecines branched away from the *Homo* line, but now that we can take into account what was happening psychologically, the overlap becomes understandable. Just as there are very early models of cars still around today, long after they have been superseded, so groups of early, less intelligent varieties of humans endured long after they had been superseded. Of course, the best example of such overlapping in the anthropological record is the existence today of remnants of our infant ape ancestors, specifically the non-human primates, and particularly bonobos, chimpanzees, gorillas and orangutans. Apes are not a branched development from the human line at all—they are on exactly the same development path, but at a much, much earlier stage. The reason that more recent species in the homo lineage such as *H. Habilis* do not currently exist, whereas the earlier equivalents of our ancestors such as bonobos, chimpanzees, gorillas and orangutans still do, is that once consciousness was liberated, subsequent psychological development became inevitable, which meant that any species with a liberated consciousness would inevitably develop into the subsequent stages that a conscious mind undergoes, and, as we have seen, this development became increasingly rapid. Even the bonobos, who have only recently chanced upon the ideal nursery conditions to allow love-indoctrination to develop the selfless, cooperative behaviour that liberates consciousness, are, it would seem, from their intelligence and their continually changing physiology, becoming more and more discrete from their chimpanzee cousins who are still stalled in early infancy, unable to develop love-indoctrination.

Chapter 7:8 <u>ADOLESCENCE</u>

The species: *Homo* — 2 million years ago to the present
The individual: 12 to 50 plus years old

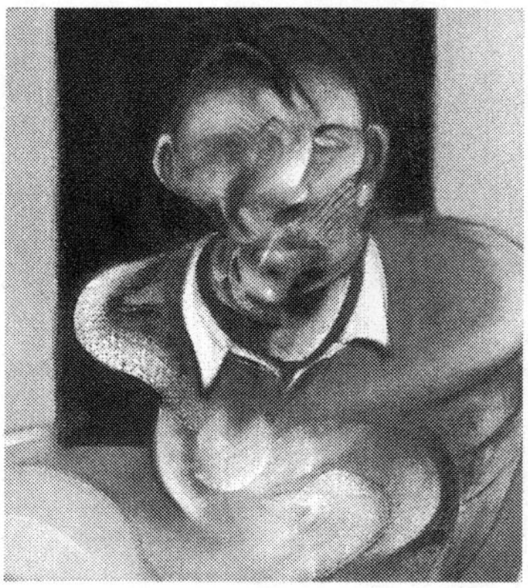

Detail of Francis Bacon's honest painting of the alienated human condition

=

Drawing by Jeremy Griffith © 1996 Fedmex Pty Ltd

Deluded, alienated 'Adolescentman'

569 As emphasised in chapters 1:2 and 7:2, adolescence encompasses the stage when we fully conscious humans search for our identity, for understanding of who we are—specifically for why we have not been ideally behaved. And since completing that upsetting search was to take some 2 million years, humanity's adolescence came to be characterised by ever-increasing levels of upset, which required the invention of more and more desperate means to try to contain that upset as well as find relief from it. Ultimately, during the final, agonising stages of this horror story, when upset reached such excruciating levels that we could no longer bear to confront and think about it, we came to a point where we had no choice but to resign ourselves to living in complete denial of the whole unbearably depressing issue of our corrupted condition. The result of this extremely dishonest block-out or denial was almost total separation from our all-sensitive and all-loving—but, at the same time, all-criticising—true self or soul. But when Resignation became an almost universal phenomenon amongst humans during the last 11,000 years it was indicative of the fact that the human race was fast approaching the psychologically dead state of terminal alienation—a state it *has* now virtually reached, and which all the apocalyptic-and–zombie-themed movies we are currently seeing in our cinemas are testament to. So, unfortunately,

the true story of our species' journey through the various stages of adolescence that now remains to be told is one that travels from soundness to insanity—keeping in mind that with the reconciling and rehabilitating understanding of the human condition now found, the human race can be (if science does its job and recognises this truthful explanation of the human condition) transformed back to life, back to a state of universal sanity, happiness and togetherness.

Chapter 7:9 Early Sobered Adolescentman

The species: the first half of *Homo habilis'* reign — 2 to 1.5 million years ago

The individual: 12 and 13 years old

570 The Early Sobered Adolescent Stage (of Humanity's Adolescence) signals the end of childhood and represents the time when we fully encounter the sobering imperfections of life under the duress of the human condition.

571 Throughout childhood we saw how the frustration with being criticised for searching for knowledge continued to increase until, in late childhood, the child's exasperation and resentment caused him to angrily lash out at the 'injustice of the world'. What happens at the end of childhood is that the child realises that physical retaliation doesn't make any difference and that the only possible way to solve the frustration is to find the reconciling understanding of *why* the criticism he is experiencing is not deserved. It is at this point that the child matures from a frustrated, extroverted protestor into a sobered, deeply thoughtful, introverted adolescent. Indeed, the *very* significant change that occurs in our psychological journey at this point, from the relatively human-condition-free state to

the very human-condition-aware state, is recognised by the fact that we separate those stages into childhood and adolescence—a demarcation that even our schooling system reflects by having children graduate from what is generally called primary school into secondary school at 12 to 13 years of age. As mentioned earlier, this critical juncture in our species' development was also recognised by anthropologists when they changed the name of the genus from *Australopithecus* (meaning 'southern ape') to *Homo* (meaning 'man'); 'Childman', the australopithecines, became 'Adolescentman', *Homo*.

Chapter 7:10 Depressed Adolescentman

The species: the second half of *Homo habilis'* reign — 2 to 1.5 million years ago

The individual: 14 to 21 years old

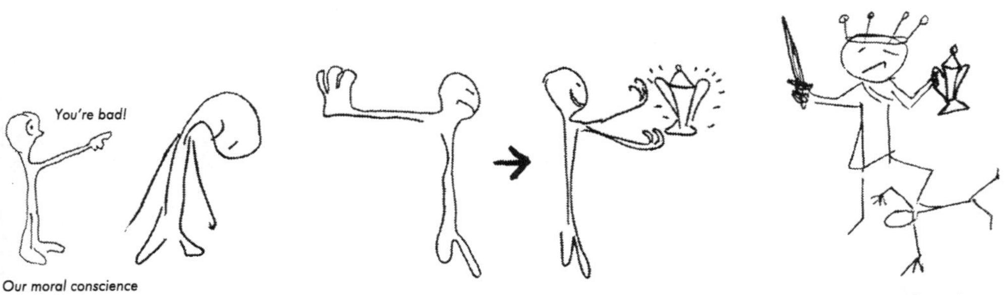

Approaching resignation *The moment of resignation* *The selfish, power-fame-fortune-and-glory seeking resigned adult*

572 The Depressed Adolescent Stage (of Humanity's Adolescence) represents the time when adolescents struggled with the depression caused by engaging the issue of the human condition. When the human race as a whole became excessively upset, as has happened during the reign of *H. sapiens sapiens*, the depression that resulted from confronting the issue of the imperfection of the world at large and within became so severe that most people in this adolescent stage had no choice but to resign to living in denial of the issue and any thinking that brought that issue into focus, which was almost all thinking—a process that resulted in the psychotic (psyche/soul repressed) and neurotic (neuron/mind repressed) alienated state. The agonising process of resigning to a life of living in denial of the human condition was described in pars 43-52, with this stage being the one that the school teachers described as **'the**

most difficult to teach. The adolescents seem to be at complete odds with what is expected of them. Most teachers are terrified of these extremely uncooperative mid-teenage ages.' The egocentric power, fame, fortune and glory-seeking strategy a resigned person adopted *after* Resignation to bring some relief to the insecurity of their situation was described in pars 492-493. It was explained there that not only did resigned humans find ways to avoid and deny the issue of the human condition, they also found ways to convince themselves they were the opposite of flawed and corrupted. It was described how, unable to refute the negative view of themselves with understanding, the resigned human could only counter the negatives by focusing on, emphasising and developing whatever positive view of both the world and themselves they could find. In particular, they became preoccupied finding ways to feel good about themselves by competing for power, fame, fortune and glory. If their team could win a football match, or if they could do something that brought praise—basically any activity that would relieve the insecurity of their condition—that was all that mattered. The more insecure, the more desperate was the need for reinforcement, and once the full extent of the imperfection of their corrupted condition became apparent at the time of Resignation, then the need for reinforcement to counter that uncertainty about their worth became extreme. So, after Resignation, the 'ego' or 'conscious thinking self' became extremely selfishly focused or centred on the need for reinforcement and relief from criticism. So, as well as practising complete denial of any confronting truths, and becoming very angry towards any criticism, extreme ego-centricity was the main outcome of Resignation. Basically, after Resignation, the angry, egocentric and alienated responses to the dilemma of the human condition that emerged in infancy and had been gradually developing throughout childhood and early adolescence suddenly became greatly amplified. Indeed, the extremely self-worth-embattled, heroic mantra of the resigned person became one of **'Give me liberty** [from criticism] **or give me death'**, **'No retreat, no surrender'**, **'Death before dishonour'**—in essence, 'I don't believe it is deserved, and anyway it's too unbearable to accept, so I will *never* tolerate any insinuation that I am a bad person'!

573 In the case of humanity's journey, Resignation would not have become the key feature it is in virtually all adult lives today during the time of *H. habilis*, or the *H. erectus* representatives of the next adventurous early adulthood stage of humanity's adolescence, or even the

subsequent *H. sapiens* representatives of the angry adulthood stage of humanity's adolescence. Initially, Resignation would have been rare—and, it follows, the further back in time, the rarer it would have been. This is because it is the upset from the lack of nurturing in infancy and early childhood that makes self-confrontation during the thoughtful early adolescent stage overwhelmingly depressing, and this lack of nurturing—basically, the alienated, detached-from-their-true-loving-soulful-self state of parents—was not a feature of human life until the final stages of humanity's adolescence. The upset from a developing mind's own efforts to self-adjust, while distressing and even depressing, was not sufficient to cause the mind to have to block out the truth of cooperative ideality. This is evidenced by the fact that even amongst *H. sapiens sapiens*, the 40-year-old-plus equivalent variety of humans living today, there have been adults who didn't resign, such as the prophets Abraham, Moses and Christ. Also, for adult members of the relatively innocent representative races of *H. sapiens sapiens*, such as the Bushmen of the Kalahari and the Australian Aborigines, to be as happy and full of the zest and enthusiasm for life and as generous, selfless and free in spirit as numbers of them are, or at least were when they were still living as hunter-foragers, means that they too must not have resigned. The English explorer and philosopher Bruce Chatwin acknowledged the unresigned soundness of Christ and also of the relatively innocent races when he wrote these extraordinarily honest words: **'There is no contradiction between the Theory of Evolution and belief in God** [Integrative Meaning] **and His Son** [the uncorrupted expression of our original instinctive orientation to Integrative Meaning] **on earth. If Christ were the perfect instinctual specimen—and we have every reason to believe He was—He must be the Son of God. By the same token, the First Man was also Christ'** (*What Am I Doing Here*, 1989, p.65 of 367). Yes, since the common dictionary definition of a 'prophet' is **'someone who speaks for God'** and Christ was an exceptionally uncorrupted expression of our original instinctive self or soul's orientation to Integrative Meaning, then Christ spoke for God; he was a prophet. In terms of being unresigned, Christ was certainly amongst **'the firstborn from among the dead** [resigned]**'** (Bible, Col. 1:18). I should mention that as a member of the Jewish race, Christ would have benefited from the genetic toughness of that race because it would have allowed his exceptionally well nurtured innocence to survive contact with the upset world where someone less genetically toughened may have not. As my headmaster at Geelong Grammar School, Sir James Darling (who in his full-page obituary in *The Australian* newspaper was described as **'a prophet in the true biblical sense'** (3 Nov. 1995)), said in one of his famous speeches about sensitive, innocent soundness not being enough for someone to be able to defy the alienated, dishonest world of denial and find the explanation of the human condition, **'he must be sensitive *and* tough'** (*The Education of a Civilized Man*, ed. Michael Persse, 1962, p.34 of 223)—which is why a member of the Bushmen race, which is a genetically relatively innocent race, could not have found the explanation of the human condition. (Note, later in par. 835 I cite more of Sir James Darling's speech, in which he spoke about needing to **'be sensitive *and* tough'**—and much more can be read about his incredibly visionary education program at Geelong Grammar School, where he deliberately set out to cultivate the innocent soundness needed to confront and solve the human condition, at <www.worldtransformation.com/darling>.) Sir Laurens van der Post also recognised that the Bushmen, although relatively innocent, did not have sufficient instinctive toughness to withstand the upset world when he described how **'mere contact with twentieth-century life seemed lethal to the Bushman. He was essentially so innocent and natural a person that he had only to come near us for a sort of radioactive fall-out from our unnatural world to produce a fatal leukaemia in his spirit'** (*The Heart of the Hunter*, 1961, p.111 of 233). A further difference between Christ and the more innocent so-called 'primitive' races such as the Bushmen is the greater level of self-restraint that accompanies toughness, which Christ, being a member of the more upset adapted Jewish race, would have also possessed. The story of how restraint has accompanied the rise in upset during the human journey will be described in chapter 7.

574 (Incidentally, understanding that Christ represents our **'instinctual'** self allows us to decipher the so-called 'Trinity' of main influences or forces on Earth that many religions recognise, with 'God the Father' being Integrative Meaning, while 'God the Son' and 'God the Holy Ghost or Spirit' are, respectively, the two great tools for developing integration or order, namely the gene-based and nerve-based learning systems that our instincts and conscious intellect represent.)

575 Ancient Greece must have also been home to quite a number of unresigned, denial-free, truthful, effective thinking so-called 'prophets' for that empire to have been so extraordinarily innovative, establishing as it did in that golden era so many of the foundation ideas for the Western world, across politics, philosophy, science, psychology, astronomy,

architecture and art. Certainly, the early Athenians Socrates and Plato were unresigned, denial-free thinking prophets; indeed, very early Athenian society must especially have been populated by relatively innocent people because they were sufficiently ego-free to both seek out uncorrupted, innocent shepherds to run Athens and, it follows, tolerate their authority. In fact, the prophet Muhammad observed **'that every prophet was a shepherd in his youth'** (Edward Rice, *Eastern Definitions*, 1978, p.260 of 433). It is the unnatural world of city living that is especially distressing to, and thus corrupting of, our original instinctive self or soul. The exceptional denial-free thinking prophet of our time, Sir Laurens van der Post, noted that during the turbulent period of Plato's time, Pericles, a close friend of Plato's stepfather, **'urged the Athenians therefore to go back to their ancient rule of choosing men who lived on and off the land and were reluctant to spend their lives in towns, and prepared to serve them purely out of sense of public duty and not like their present rulers who did so uniquely for personal power and advancement'** (Foreword to *Progress Without Loss of Soul*, Theodor Abt, 1983, p.xii of 389).

[576] As to the relatively alienation-free, natural, loving, nurturing ability of existing relatively innocent races, in Australian Aboriginal society **'All observers agree upon the extraordinary tenderness which parents display towards their children, and indeed, to all children whether of their own family and race or not'** (Ashley Montagu, *Coming into being among the Australian Aborigines*, 1974, p.345 of 426). In the case of the Bushmen of the Kalahari, **'Their love of children, both their own and that of other people, is one of the most noticeable things about the Bushman'** (*Tribes of the Kalahari Desert*, accessed Jun. 2010; see <www.wtmsources.com/103>). The **'Bushmen...mother carries her child with her at all times up to four years of age'** (Virginia Abernethy, *Population Pressure and Cultural Adjustment*, 2005, p.34 of 189). **'Children are breast-fed for up to 3½ years, and among the Bushmen lactation suppresses ovulation'** (G.N. Bailey, *Hunter-gatherer economy in prehistory: a European perspective*, 1983, p.114 of 247). And, as mentioned earlier, **'!Kung** [Bushmen]**...infants hardly ever cry.'** Similarly, a report on the studies of natural-living, more innocent and nurturing societies undertaken by the dentist and nutritionist Dr Weston Price offered these insights: **'For the next ten years** [during the 1930s], **he** [Dr Price] **travelled to various isolated parts of the earth, where the inhabitants had no contact with "civilisation" in order to study their health and physical development...Price took photograph after photograph of beautiful smiles, and noted that "healthy primitives" were invariably cheerful and optimistic. Such people were characterized by "splendid physical development". The women gave birth with ease. Their babies rarely**

cried and their children were energetic and hearty. Many others have reported a virtual absence of degenerative disease, particularly cancer, in isolated, so-called "primitive" groups' (Sally Fallon Morell, *Nasty, Brutish and Short?*, The Weston A. Price Foundation, accessed Apr. 2012; see <www wtmsources.com/110>).

577 Yes, the human race is not so instinctively adapted to upset now that humans are no longer capable of being innocent enough to avoid Resignation; with sufficient nurturing and shelter from upset behaviour an individual can avoid that path. The reality is that there has to be a great deal of upset in humans for that upset to become so unbearable that they have no choice but to pay the extremely high price of blocking out all access to their all-sensitive, enthralled-with-all-of-life, all-loving, soulful true self. The behaviour of a resigned person is essentially a form of death-like dissociation or autism; indeed, it matches perfectly the description a former president of the British Psychoanalytical Society, the psychiatrist and paediatrician D.W. Winnicott, gave for behaviour associated with autism: **'Autism is a highly sophisticated defence organization. What we see is *invulnerability*...The child carries round *the (lost) memory of unthinkable anxiety*, and the illness is a complex mental structure insuring against recurrence of the conditions of the unthinkable anxiety'** (*Thinking About Children*, 1996 posthumous publication of his writings, pp.220-221 of 343). Later in this description of the stages that the human race has progressed through under the duress of the human condition we will see that when upset became even more extreme how an even more dishonest, alienating, autism-equivalent psychological strategy than Resignation was invented to cope with the human condition. This was to take up born-again, pseudo idealism—an adaption that progressed from religion through to the extremely dishonest forms of pseudo idealism, from socialism, to the New Age Movement, to feminism, to environmentalism, to the politically correct movement and, ultimately, to totally truthless postmodern deconstructionism.

578 Since it requires a great deal of upset for Resignation to become unavoidable, we can expect that it has only become an *almost* universal phenomenon amongst adult humans from about 11,000 years ago when the advent of agriculture and the domestication of animals brought humans together in close proximity, the effect of which, as will be talked about in pars 671 and 724, was to rapidly spread and compound upset behaviour. The Biblical account of Noah's Ark is actually a metaphorical description of this time when Resignation 'flooded' the world and our soul and all its truths went under, 'drowned'—when our soul was

pushed into our subconscious, out of conscious awareness, and the highly competitive egocentric way of living became all-dominant. The only creatures to escape the horror of Resignation, to survive this 'drowning' of our soul, were the animals and the very few unresigned prophets, as symbolised in this case by Noah and his floating zoo. As it says in Genesis, **'Noah was a righteous man, blameless among the people of his time, and he walked with God** [he did not have to deny Integrative Meaning]**...God saw how corrupt the earth had become, for all the people on earth had corrupted their ways. So God said to Noah, "...make yourself an ark...I am going to bring floodwaters on the earth...Everything on earth will perish** [the soul and all the denial-free truths will perish when people resign to a life of denial]**. But I will establish my covenant with you** [but from here on prophets will have to preserve the truth of Integrative Meaning and all the other great truths that relate to it]**, and you will enter the ark...Go into the ark** [don't resign]**, you and your whole family, because I have found you righteous in this generation"'** (6:9, 12, 14, 17, 18; 7:1). And to think people have actually searched for the remnants of Noah's Ark, and tested ice cores from glaciers and sea beds in the Black Sea for evidence of a great flood event in the past, as if the story depicted an actual event rather than the metaphor/allegory it really is! Understandably, however—in a development that will be elaborated upon in ch. 7:16I—the more upset and thus insecure humans became, the less they could afford to confront the truths contained in religious scriptures and the more they needed to interpret their contents in literal and fundamentalist ways—'God is actually a person sitting in the clouds somewhere', 'Christ was actually physically resurrected from death', 'Christ's mother was actually a virgin', 'Abraham actually considered murdering his son', 'Judgment day actually heralds an afterlife in which some unlucky souls will be judged as evil and burnt in a fiery pit', etc, etc. But with the upset state of the human condition now defended, all religious metaphorical descriptions, parables and symbols—in fact, all mythology—can be safely explained and demystified, as is shown throughout this book and in the compilation of all my work, the book *Freedom*. Indeed, with understanding of the human condition found the Bible's *entire* contents can be fully demystified now—an exercise that will likely amaze the reader because they will discover that, unlike all the libraries of denial-complying books in the world, the Bible is a repository of extremely rare denial-free truth. In fact, as it turns out, the *whole* story of the human condition, bar its scientific explanation, is perfectly described in the Bible, albeit in the abstract, metaphysical and often metaphorical terms that denial-free, truthful-thinking prophets were limited to in those early pre-science times when the Bible was written. No wonder it is the world's bestselling and most widely distributed book, with 6 billion copies having been printed so far.

579 We now need to look at what happened during the second phase of the Depressed Adolescent Stage, in the years immediately *following* Resignation, as well as what happened in those same years, from 15 to 21 years of age, to those humans who *didn't have to resign*—a group that included virtually all our forebears who lived prior to the advent of *H. sapiens sapiens*.

580 Firstly, in looking at the situation of those who *did* resign to living a life of denial of the issue of the human condition and of any truths that brought the issue into focus, after resigning at about the age of 15 it normally took the individual another six years of procrastination to make sufficient mental adjustments to be able to embrace the new, post-resigned, extremely dishonest resigned way of living.

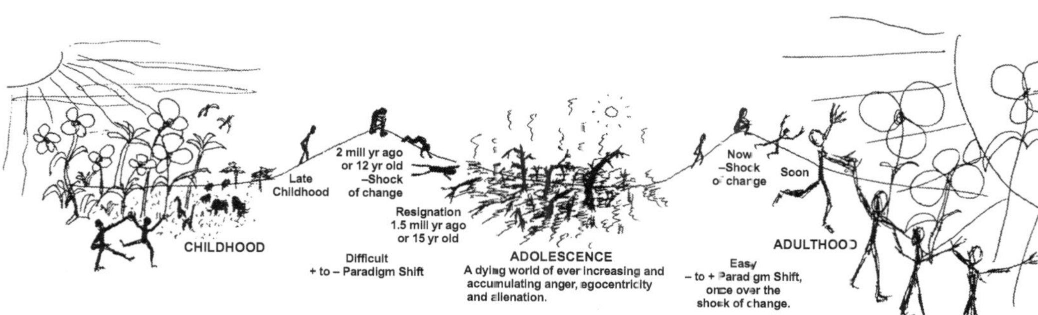

Humanity's Journey from Ignorance to Enlightenment

Drawing by Jeremy Griffith © 1996 Fedmex Pty Ltd

581 To best describe the situation presented by this sobered and then depressed period leading up to Resignation—and the six-year period of procrastination over having to take up a dishonest, soul-dead resigned life that followed—imagine we're all sitting on a ridge between two valleys. Behind us lies the valley of humanity's enchanted childhood, the 'Garden of Eden' in which everyone lived happily and extremely sensitively in a non-upset, cooperative, all-loving state. Before us, however, is a hell of smouldering wasteland of devastation and destruction, the wilderness of terrible upset and alienation. Of course, we didn't *want* to

go forward into that wasteland, but retreat was not an option. To leave all that happiness, laughter and togetherness behind was heartbreaking, but we had no choice but to turn our back on it; we couldn't throw away our conscious mind, we couldn't stop thinking, and, while we practised thinking, upset was an inescapable by-product that could only be brought to an end by finding understanding of our corrupted state—understanding that lay on the other side of that terrible wilderness of devastation, aloneness and alienation. Joe Darion's song *The Impossible Dream* offers up a beautiful description of how awesomely courageous humans have been undertaking our species' corrupting search for knowledge: **'To dream the impossible dream** [of one day, in the far future (which has now arrived), finding the liberating understanding of the human condition]**, to fight the unbeatable foe** [of our ignorant, ideal-behaviour-demanding instincts] **/ To bear the unbearable sorrow, to run where the brave dare not go / To right the unrightable wrong** [of being unjustly criticised]**, to love pure and chaste from afar / To try when your arms are too weary, to reach the unreachable star / This is my quest, to follow that star / No matter how hopeless, no matter how far / To fight for the right without question or pause / To be willing to march into hell for a heavenly cause / And I know if I will only be true, to this glorious quest / That my heart will lie peaceful and calm, when I'm laid to my rest / And the world will be better for this, that one man scorned and covered with scars / Still strove with his last ounce of courage, to reach the unreachable star.'**

[582] In short, there was no retreat for the resigned adolescent; like all the fully conscious humans who had gone before them, they *had to* find the courage to continue humanity's heroic search for knowledge. Procrastination got them nowhere, and so, more typically in recent times, after a few years' spent consuming lots of drugs and alcohol and partying long into the night to help them accept their fate, the adolescent had to 'get on with it' and take up the challenge of adulthood in a world where understanding of the human condition was yet to be found. In fact, it normally wasn't until they reached 21 that resigned adolescents finally managed to orientate themselves to their extremely compromised resigned life—an orientation that involved two main adjustments: firstly, they had to block out the negative reality that living so falsely and thus so dead in soul and intellect meant they would have to endure living a completely corrupted, **'phony'** and **'fake'** life; and secondly, they had to train their mind to block out all memory of the innocent childhood state and focus on whatever meagre positives they could find in the journey

ahead. I describe these positives as 'meagre' because the degree of happiness they provided was, in truth, of no comparison to the happiness the human race enjoyed while living in the magic state of our soul's true, all-loving and all-sensitive world.

583 The first tiny positive was the prospect of the adventure involved in trying to avoid, for as long as possible, the inevitable disaster of complete self-corruption. We may have been about to 'go under'—become totally corrupted—but at least we could hope to make a good fight of it. In fact, as will be described shortly, by the age of 21 young resigned adult men in particular could have so blocked out the truth of another ideal, soulful, integrative true world, and so adopted belief in a selfishness-justifying, competitive, survival-of-the-fittest meaning to life, that they deluded themselves that winning power, fame, fortune and glory would *genuinely* bring them validation, prove that they actually were good and not bad.

584 The second, in truth tiny, positive in the resigned existence was romance, the hope of 'falling in love', which (as will be described in par. 613) can be now understood as the hope of escaping reality through the dream of ideality that could be inspired by the neotenous image of innocence in women.

585 Although these two positives were tiny, resigned adolescents gradually built them up in their mind to the extent that they became everything. They had no choice but to mentally posture themselves and their resigned environment in such a way to be able to leave that ridge and take up humanity's journey to find liberating understanding of our species' upset, corrupted condition.

586 We now need to look at the journey into adulthood, and beyond, of those individuals who *didn't* resign—a group, as stated earlier, that included virtually all those forebears who lived prior to the emergence of *H. sapiens sapiens*.

587 Even for those who hadn't become so upset that they had to resign to a life of living in denial of the issue of the human condition, their lives still followed a parallel path to that of resigned humans, who were living with the delusion that by winning power, fame, fortune and glory they could genuinely validate themselves, prove that they were actually good and not bad. The reality of the resigned path, however, was that those on it were inevitably bound to discover that power, fame, fortune and glory couldn't bring them any real validation, but merely resulted in them becoming *more* upset, and thus *more* insecure about their meaning

and worth, and thus *more* dissatisfied. In contrast, while the unresigned were not living with the delusion that they could prove that they were champions and heroes of a competitive, survival-of-the-fittest, 'red-in-tooth-and-claw' world, they were, nevertheless, living with the naive illusion, the optimistic hope, that all the wrongness in the world could be righted; that they could make the world a better, more ideal place. The reality for the unresigned person was that the situation all around them, and even in themselves, only got worse as the upsetting search for knowledge continued. Everywhere humans were becoming more upset and thus cynical about the goodness of other humans, mean spirited and destructive. So while their disappointments and frustrations were coming off vastly different bases, both the resigned and the unresigned faced overwhelmingly difficult paths.

588 In the situation, however, where most of the population were resigned, as has been the case during the last 11,000 years of the reign of *H. sapiens sapiens*, the overwhelming problem for the unresigned person was that the resigned state of complete dishonesty and desperate competitiveness was a total mystery to them—because those who *were* resigned to living a life of extreme dishonesty and deluded competitiveness never admitted they were doing so. It was the 'silence' of the resigned state that was the most destructive of innocence, be that in children or unresigned adults. In effect, the resigned assumed that everyone else was also resigned and that it was, therefore, self-evident as to why they behaved so dishonestly, aggressively and competitively, but their behaviour was, in fact, a complete enigma to the unresigned. The idealistic innocence of the unresigned mind was so trusting and thus codependent to the resigned state that they were brutalised to the point where, almost invariably, their innocence was destroyed by the extreme dishonesty and defensiveness of the resigned state. The psychiatrist Wilhelm Reich wrote honestly about the effects of upset on innocence when he described how **'The living** [those relatively free of upset]**...is naively kindly...It assumes that the fellow human also follows the laws of the living and is kindly, helpful and giving. As long as there is the emotional plague** [the flood of upset in the world]**, this natural basic attitude, that of the healthy child or the primitive...**[or the unresigned adult, is subject to] **the greatest danger...For the plague individual also ascribes to his fellow beings the characteristics of his own thinking and acting. The kindly individual believes that all people are kindly and act accordingly. The plague individual believes that all people lie, swindle, steal and crave power. Clearly, then, the living is**

at a disadvantage and in danger' (*Listen, Little Man!* 1948, p.8 of 109). Sir Laurens van der Post described how the relatively innocent Bushmen race struggled to cope with upset when he wrote that '**mere contact with twentieth-century life seemed lethal to the Bushman. He was essentially so innocent and natural a person that he had only to come near us for a sort of radioactive fall-out from our unnatural world to produce a fatal leukaemia in his spirit'** (*The Heart of the Hunter*, 1961, p.111 of 233). In truth, this description also applies to the catastrophic effects the terminally alienated world today is having on many children, the result of which is the epidemic proportion of anxiety-related disorders like ADHD and autism that we are now seeing, a development I will describe in greater detail later (in chs 7:16B and 7:16C).

[589]The dilemma for the unresigned has been that no matter how much idealism, no matter how much selfless behaviour they threw at a problem, the bottom line truth was that *only* understanding of the human condition could stop the resigned from behaving the way they were behaving—simply because the resigned had no other way of coping in the meantime. We will see later in chapters 7:15 and 7:16 how various mechanisms, like religion, were developed to try to contain the dishonesty and devastation of the resigned way of living, but ultimately such measures were limited in their effectiveness. (I might mention here that I am able to describe the agonies of the unresigned life because that has been my personal situation—as evidenced by my ability to think freely about, and write at length about, the human condition, for if I was resigned to living in denial of the human condition I obviously wouldn't be able to confront the subject.)

[590]So while, in the final 15-to-21-year-old stage of adolescence, the resigned person procrastinated over having to take up such a dishonest, soul-less life, the unresigned person (or, in humanity's case, virtually all members of *H. habilis*, *H. erectus* and *H. sapiens*) had to adjust to the prospect of having their idealism disappointed, resisted and thwarted at every turn. For the unresigned, facing that valley of devastation was just as difficult in its own way as it was for the resigned, but, *like* the resigned, they had no choice but to accept that fate. And just as the resigned used those years between 15 and 21 to condition themselves to taking up the challenge of '**march**[ing] **into hell for a heavenly cause'**, so too did the unresigned. The 'adventure' for the unresigned was to see how much they could resist the corruption in the world, and if not change it then at least contain it. And the unresigned also used romance to inspire their horrifically

difficult—and as Resignation became all but universal—increasingly lonely undertaking. And so, after an initial period of mental adjustment, both the resigned and the unresigned began their 20s with a determination to make a difference, even if both realities were bound to become overwhelmed by the horror of life under the duress of the human condition.

591 The courage of all humans who have lived during humanity's heroic 2 million years in adolescence, during which time they had to face the inevitability of total self-corruption by the end of their lives, has been so immense it is, and possibly will be for all time, out of reach of true appreciation. Which makes it all the more tragic that, after such a stoic and courageous effort to overcome our human condition, all the deadening effects of living with that condition means that the human race now faces the very real prospect of becoming completely estranged/alienated from our soul's happy, loving and all-sensitive world—because if these understandings of the human condition continue to be resisted by the scientific establishment, humans will be perilously close to spending an eternity wandering mindlessly like waifs in the terrible wilderness of the darkness of denial and its resulting alienation, as it is so accurately described in the Bible: **'Today you** [Integrative Meaning/God] **are driving me from the land, and I will be hidden from your presence, I will be a restless wanderer on the earth'** (Gen. 4:14).

Chapter 7:11 Adventurous Adolescentman — includes the reconciling explanation of Men and Women

The species: *Homo erectus* — 1.5 to 0.5 million years ago
The individual: 21 to 30 years old

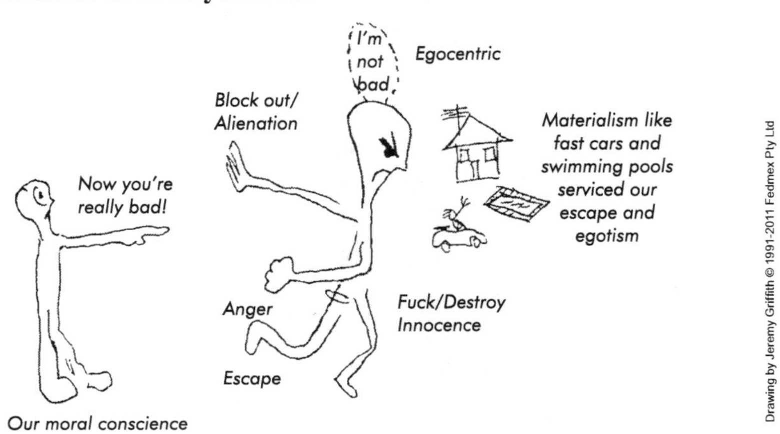

[592]The Adventurous Early Adulthood Stage (of Humanity's Adolescence) is the time we took up the battle to overthrow our idealistic instinctive self or soul's ignorance as to the fact of our conscious self's fundamental goodness.

[593]By 21 years of age, after about six years of blocking out the negatives and focusing only on the tiny positives available to them, resigned humans finally adjusted to life in Resignation. In fact, by 21 both resigned and unresigned young adults were able to arm themselves sufficiently well with a positive attitude to commit themselves to the battle that humanity as a whole was involved in of gradually, step by step, generation by generation, working towards one day accumulating sufficient knowledge to be able to explain and liberate the human race from the human condition. Indeed, by 21, young adults had made sufficient adjustments to be raring to go, with men in particular having become *so* focused on the positive of the adventure of attempting to make a good fight of the battle to validate themselves through winning power, fame, fortune and glory that they were cavalier and swashbuckling. Naive about just how quickly overwhelming the battle was going to become, both resigned and unresigned males had plenty of strength and resilience—plenty of 'rock-n-roll'. For their part, 21-year-old women had also become firmly focused on the few positives they had of the reinforcements they could receive from men for their physical beauty and of the satisfaction of being able to support men and nurture another generation of brave humans to carry on humanity's heroic struggle. (The different roles of men and women will be described more fully shortly.) Hence the significance of the long-held tradition in Western societies to hold a 'coming of age' party for offspring when they reached this milestone, at which they were typically given a 'key' symbolising that they were at last ready to leave home and 'face the world', and so with a big kiss from Mum and a slap on the back from Dad the young adult set off 'to see what life held for them'. Interestingly, the fact that young adults were considered sufficiently adapted to life under the duress of the human condition to be considered independent at 21 rather than at the round figure of 20 is an indication of how precisely all these stages with ages occurred, and also how uniform and powerful the effects of the human condition have been.

[594]So although Resignation was still far from universal, it was essentially during our 20s, in humanity's adolescence, that the human

race began refining all the techniques we needed to cope with living with the horror of the human condition. While we were in our teens we agonisingly adjusted to having accepted a life of living with upset, but in our 20s we took up the challenge of living out that life. Our forebear who lived in humanity's adventurous early adulthood stage was 'Adventurous Adolescentman', *H. erectus*, who existed between 1.5 million and 0.5 million years ago. Consistent with the description that has been given for this stage, fossil evidence has revealed that it was *H. erectus* who first adventured out from our ancestral home in Africa around 1.5 million years ago and initiated the spread of humans throughout the world.

595 While the many methods for coping with the human condition that were perfected during the 1-million-year reign of 'Adventurous Adolescentman' have been part of human life for so long now we tend to think of them as having always been part of our species' make-up, as part of our 'human nature', the fact of the matter is they all had to be invented. Yes, in truth, our behaviour underwent an *immense* transition—especially amongst those who became resigned and changed from living cooperatively, selflessly and lovingly, to living extremely competitively, selfishly and aggressively.

596 As has been mentioned, the dominant aspect of this transition that took place when humanity matured from childhood to adulthood was that <u>we went from being a matriarchal society to a patriarchal one</u>. Since it was this shift that had the most dramatic impact on the lives of men and women during this adventurous adolescence stage, I will now describe how it came about and its repercussions.

597 Throughout humanity's infancy and childhood our society was matriarchal or female-role led because the suckling of young and the nurturing of infants it led to provided the basis for the whole love-indoctrination process, and because females had to become sufficiently assertive to rein in the aggressive competition amongst males for mating opportunities. The honest-thinking philosopher Jean-Jacques Rousseau acknowledged women's nurturing role when he wrote: **'The first education is the most important, and this first education belongs incontestably to women; if the Author of nature had wanted it to belong to men, He would have given them milk with which to nurse the children'** (*On Education*, 1762, p.37 of 501). When Olive Schreiner wrote that **'They say women have one great and noble work left them...*We* bear the world and *we* make it. The souls of little children are marvellously delicate and tender things, and keep for ever the shadow that**

first falls on them, and that is the mother's or at best a woman's' (*The Story of an African Farm*, 1883, p.193 of 300), she was also acknowledging that it has been women's role to nurture infants. Thus, with women preoccupied nurturing the next generation it made sense that when our instinctive self threatened to stop our conscious mind's search for knowledge it was men who took on the role of resisting that threat. It also made sense for men to take up this role because historically men had always been the group protector, from marauding leopards, and so on, and since this resistance by our ignorant instinctive self of our mind's need to search for knowledge was a group threat, namely a threat to humanity, it especially made sense that men went out to meet it. Indeed, since the threat of ignorance that emerged 2 million years ago was *such* a serious threat to our species—if it prevailed we would never find understanding and never fulfil our responsibility to master intelligence—winning the battle against it became our species' priority, which is why our society became patriarchal or male-role led. Of course, this role differentiation where men took up the task of fighting against the ignorance of our instinctive self also made sense because nurturing depends on love, while fighting is an aggressive, non-loving behaviour; better then to leave women out of the upsetting battle to preserve as much upset-free, loving innocence as possible to nurture the next generation. The Bible acknowledges that human society changed from being a matriarchal, soul-centric, nurturing society to a patriarchal, egocentric, embattled society after the upsetting search for knowledge began when, after describing how Adam and Eve took the **'fruit' 'from the tree of...knowledge'** (Gen. 3:3, 2:17), Moses wrote, **'To the woman he** [God] **said...Your desire will be for your husband, and he will rule over you'** (3:16). Wives and children in virtually all cultures have not adopted their husband or father's surname because of some cultural coin toss, but because they were living in a patriarchal world.

598 So, in keeping with their already established roles as the group protectors, it was men who took up the loathsomely upsetting job of championing the ego over the ignorance of our original instinctive self, leaving women to their loving, nurturing role. The problem this role differentiation gave rise to, however, was that in not being responsible for or participating in the terrible battle to overthrow ignorance women were naive or unaware of the ramifications of fighting the battle, and, as a result, were unsympathetic towards both the battle and the frustrated anger and egocentricity it produced in men—a situation that placed

men in the awful predicament of being misunderstood and unjustly condemned by women. Women, not responsible for the fight against ignorance, and so not partaking in the battle itself, did not and could not be expected to understand what happened in the battle or the effect it had on men; they could understand the *search* for knowledge, but not what the battle involved, as this comment reveals: **'Shirley MacLaine can't find a man to love. The 48-year-old actress...**[said she] **longs for a "close and warm relationship" but hasn't met a suitable partner. "Most men I meet seem to be too involved in trying to be successful or making a lot of money," she said. "I feel sorry for all of them. Men have been so brainwashed into thinking they have to be so outrageously successful—to be winners—that life is very difficult for them. And it's terribly destructive, as far as I am concerned, when you are trying to get a serious relationship going"'** (Sydney's *Daily Mirror*, 14 Dec. 1982). The journalist, businesswoman and 2013 'Australian of the Year', Ita Buttrose, acknowledged women's comparative lack of ego when she said, **'That's the difference between men and women. Men have egos and we don't'** (*Enough Rope*, ABC, episode 47, 21 Jun. 2004). Yes, contrary to what feminists would have us believe, men and women are different—indeed, as the author Camille Paglia said herself to her feminist sisters, **'Wake up, men and women *are* different'** (*The Australian*, 4-5 Jul. 1992).

599 Life has certainly been **'very difficult'** for men, they have had an absolutely horrible job: they have had to be strong enough, in effect, to kill soul—to search for knowledge and determinedly defy our beautiful, cooperatively orientated, original instinctive self. And since that defiance resulted in becoming angry, egocentric and alienated, which are all divisive, un-Godly traits, men were, in effect, in violation of God, the integrative ideals of life! So from an initial state of upset, men had then to contend with a sense of guilt, which very greatly compounded their insecurity and frustrations and made them even more angry, egocentric and alienated. (As was explained in chapter 2:5, this avalanche of criticism is the 'double and triple whammy' of condemnation that humans experienced when we searched for knowledge.) How tough were men going to have to be to continue to do their job without receiving any respect or appreciation for why they were having to do what they were having to do! No wonder they have become so incredibly upset.

600 But what could men do? They couldn't explain themselves to women because they couldn't explain the human condition, and not able to explain the human condition, they weren't able to defend their

immensely upset, corrupted state. Indeed, instead of being able to explain that they were, in fact, the heroes of the whole story of life on Earth because they had to succeed (and have now succeeded) in championing the cause of nature's greatest invention, namely the conscious mind's battle to establish itself on Earth, men have had to endure being *completely* misunderstood and misrepresented as the villains of the piece. Yes, there was *very* significant meaning in the sparse, tightly written prose that the great American novelist Ernest Hemingway used to describe the stoic lives of men; just consider the titles alone of some of his books: *Death in the Afternoon*, *For Whom the Bell Tolls*, *Winner Take Nothing*, *To Have and Have Not*, *The Old Man and the Sea* and *Islands in the Stream*. The following comment about the feminist movement describes the immensely frustrating situation endured by men: **'One of the reasons that men have been so quiet for the past two decades, as the feminist movement has blossomed, is that we do not have the vocabulary or the concept to defend ourselves as men. We do not know how to define the virtues of being male, but virtues there are'** (Asa Baber, *Playboy*, Jul. 1983). The truth about men was certainly not as it appeared—they weren't selfish, world-destroying monsters; quite the reverse. So when the leading feminists Germaine Greer and Gloria Steinem said, respectively, **'As far as I'm concerned, men are the product of a damaged gene. They pretend to be normal but what they're doing sitting there with benign smiles on their faces is they're manufacturing sperm'** (*The Sydney Morning Herald*, 14 Nov. 1991), and **'A woman needs a man like a fish needs a bicycle'** (*TIME*, 18 Sep. 2000), they were making two of the most embarrassingly wrong statements in history, and yet, in her secret self, every woman has been making them for 2 million years! (Incidentally, Greer's assertion that all men are doing is **'manufacturing sperm'** reflects human-condition-avoiding, psychosis-denying mechanistic science's dishonest explanation for men's competitive nature, which is that they are competing to reproduce their genes.)

601 Yes, most unfortunately, women had little empathy with and respect for the effects of our species' battle. They tended to be soul-sympathetic, not ego-sympathetic. For example, while the embattled egos of men needed to build towering buildings symbolising their will and determination to defy and defeat the unjustly condemning world that surrounded them, **'If civilization had been left in female hands, we would still be living in grass huts'** (Camille Paglia, *Sexual Personae*, 1990). This famous statement from Paglia can be understood both literally and metaphorically, because the

fundamental situation was that if the soul (which women represented) had its way the intellect would never have been allowed to search for knowledge. The soul's ignorance *had to be* defied if knowledge and ultimately self-understanding was to be found. To give in to soul was to go nowhere, to remain in **'grass huts'**.

602 The gulf between men and women—acknowledged in the title of John Gray's bestselling 1992 book, **'Men are from Mars, Women are from Venus'**—is palpably clear in this conversation I once overheard: 'She: **You men are wholly monstrous, foreign bodies, in fact cancers on this planet.** He: **Yes, well, haven't you heard that women are so meaningless as to not even exist'**—adding, while walking away: **'The trouble with women is they come with a brain because their brain is missing a cog, they have no idea what's going on.'** This bitter exchange provides a true measure of the extent of 'the war between the sexes'. The fact is, women have not understood men at all, as this quote admits: **'Men are a knot, I'll never untie, around a box I long to peer into'** (Kate Llewellyn, poem 'Men', *The Sydney Morning Herald*, 20 Apr. 1996).

Men and Women, War and Peace by Michael Leunig,
Melbourne's *The Age* newspaper, 2 Sep. 1989

603 This cartoon by Michael Leunig titled *Men and Women, War and Peace* recognises the immense gulf that has existed between men and women's situation and their view of the world. Women have not appreciated the battle—they have not been able to empathise with what has been going on. In Leunig's cartoon, this lack of empathy is shown by the woman's tears of disappointment and sadness.

Women have not been, as it were, 'mainframed': as intuitively understanding of the battle as men—in the same way men have never been as mainframed to the role of nurturing as women intuitively are. While both men and women have had to live in denial of the battle of the human condition, men retained an awareness of the battle because they were having to fight it. Sir Laurens van der Post described this limitation of women in his 1976 book, *Jung and the Story of Our Time*, when he related a dream Carl Jung had about a blind woman named Salome. Sir Laurens wrote that **'Salome was young, beautiful *and* blind'**, explaining the symbolism of Salome's blindness with the following words, **'Salome was blind because the anima** [the soulful, more feminine side of humans] **is incapable of seeing'** (p.169 of 275). Regarding the caption on Leunig's cartoon, paradoxically *real* **'peace'** could only come by men winning their **'war'** against the ignorance of our instinctive self or soul.

604 It was because women lacked empathy with the battle taking place on Earth that they have had to work through men; as it states in the Bible, women were man's **'helper'** (Gen. 2:20). Soul helped the intellect; it did not lead it. Not being mainframed to the battle going on in the world has made it very difficult for women to create profound works of literature, music and art, as evidenced by almost all the quotes employed in this book's analysis of the human condition coming from males. Christ did not choose 12 men as disciples because of the cultural conditioning of his day, as some have claimed. Christ was never influenced by arbitrary tastes and attitudes. He was only influenced by the truth, and the truth is women have not been in the best position to mediate in a battle they were not directly engaged in, which is why in religions such as Christianity women have been excluded from the priesthood. Humanity's recent 2-million-year adolescent search for its identity, search for understanding of itself, particularly for the reason for its divisive nature, *has* been a patriarchal journey, just as humanity's infancy and childhood *was* a matriarchal stage.

605 The following passage from the philosopher Friedrich Nietzsche acknowledges all that has just been explained about women's greater ability to nurture; that women have had to work through men; and that women haven't been aware of, or mainframed to, the deeper battle that has been waging on Earth: **'Woman understands children better than a man...The man's happiness is: I will. The woman's happiness is: He will. "Behold, now the world has become perfect!"—thus thinks every woman when she**

obeys with all her love. And woman has to obey and find a depth for her surface. Woman's nature is surface, a changeable, stormy film upon shallow waters. But a man's nature is deep, its torrent roars in subterranean caves: woman senses its power but does not comprehend it' (*Thus Spoke Zarathustra: A Book for Everyone and No One*, 1892; tr. R.J. Hollingdale, 1961, p.92 of 342).

[606]What now needs to be described is that without the ability to explain the all-important role that men were playing, and thus defend themselves against women's lack of appreciation of that role—a situation that led to such unappreciative criticisms as those from Greer and Steinem—what men, in their anger, frustration and desperation, did is they turned on women and attacked their condemning innocence by violating it through sex, as in the 'fucking' or destroying of innocence. But before describing that extremely tragic development I first need to explain that women were not the *original* victims of men's upset. *That* unfortunate distinction went to animals because their innocence—in fact, the innocence of *all* of nature—also criticised men. Nature was a friend of our original instinctive self or soul because we grew up with it—humanity spent all its infancy and childhood alongside nature in the 'Garden of Eden' that was Africa—and so by association the natural world, especially the innocent world of animals, *also* criticised men. Yes, the hunting and killing of animals was the first great expression of men's upset anger and egocentricity. One of the first adaptions to living in the embattled angry, egocentric and alienated state, ultimately epitomised by Resignation, was that we changed from being a relatively peaceful vegetarian species to a ruthless hunter of animals. It has always been claimed that the hunting in the 'hunter-forager' lifestyle that characterised virtually the entire 2-million-year period of humanity's adolescence was primarily driven by the need for protein-rich food—a mechanistic denial-complying belief that has so far protected upset humans from the condemning truth of the extreme aggression involved in hunting. But, in fact, research shows that 80 percent of the food consumed by existing hunter-foragers, such as the Bushmen of the Kalahari, is supplied by the women's foraging (*Kalahari Hunter-Gatherers*, eds Richard B. Lee & Irven DeVore, 1976, p.115 of 408). So if providing food was not the reason, why did men hunt? This obvious question has led to the development of the 'show-off' hypothesis by some human-condition-avoiding, psychosis-denying, mechanistic scientists, such as Kristen Hawkes, which suggests that men developed hunting in order to demonstrate their worth as potential

mates and so increase their reproductive chances. However, the honest, denial-free answer is that hunting was men's earliest ego outlet: men attacked animals because their innocence, albeit unwittingly, unfairly criticised men's lack of innocence; it condemned their upset aggressive lives. Also, by attacking, killing and dominating animals, men were demonstrating their power, which was a perverse way of demonstrating their worth. If men could not rebut the accusation that they were bad, they could at least find some relief from that guilt by demonstrating their superiority over their accusers. The exhibition of power was a substitute for explanation. This 'sport' of attacking animals, which were once our species' closest friends, *was* one of the earliest expressions of our upset. One of the definitions given for 'sport' is **'the pastime of hunting, shooting, or fishing with reference to the pleasure achieved: "we had good sport today"'** (*Encylopedic World Dictionary*, 1971). The **'pleasure'** of hunting was of the perverse, sadistic, sick kind, of attacking animals for their innocence and its implied criticism of us.

Photo on the front page of *The Australian* (27 Oct. 2011) of a shooter posing with a scimitar-horned oryx, which he shot to death at a game ranch in the Northern Territory of Australia. The species is officially listed as extinct in the wild. Note the pride and happiness on the face of the shooter and his partner.

607 Anthropological evidence supports the notion that hunting is an aspect of fully conscious, upset 'Adolescentman', because it was during the time of *H. erectus* that the first signs of hunting appeared in the fossil record. All the anthropological evidence indicates 'Childman' was a vegetarian, but with big game hunting came meat eating, an adaption

that would have revolted our original instinctive self or soul since it involved *killing* and *eating* our soul's friends. Even today, the act of killing animals, or just *seeing* animals get slaughtered, produces feelings of deep revulsion within us. But we weren't to be put off and in time, as our increasingly upset and driven (to find ego relief) lifestyle developed, we became somewhat physically dependent on the high energy value of meat. (With respect to whether the lifestyle of 'Adolescentman' should be described as hunter-forager or forager-hunter, since the priority was the heroic search for knowledge and hunting provided men with some retaliatory relief from the criticism of the innocent world they had to live in, it should be described as hunter-forager.)

608 Returning to the overall situation that the human race was faced with, in particular the issue of how the different roles that men and women had taken up played out. As upset increased, so too did humanity's insecurity about being corrupted—as did, it follows, the need to combat that insecurity with whatever form of relieving reinforcement we humans could find. And since it was men who had to especially take on the responsibility of championing the conscious-thinking self or ego over the ignorance of our original instinctive self, it was men who particularly came to need reinforcement of their worth through the acquisition of power, fame, fortune and glory. In the soundtrack to the 1986 African musical *Ipi Tombi*, the female narrator says, **'The women had to do all the work because the men were so busy being big, strong and brave'** (Narration: Sesiya Hamba, *Drinking Song*, lyrics by Thandi Lephelile). This quote acknowledges just how preoccupied men eventually became in trying to prove their worth, in defeating the implication that they weren't worthy. In the end men became *so* insecure/ego-embattled that it really did become a case of **'Give me liberty or give me death'**, **'No retreat, no surrender'**, **'Death before dishonour'**, **'Death or glory'**—they wouldn't do anything that didn't bring them glory and adulation, which meant someone else (namely women) had to do all the menial work if it was going to get done. The following two photographs from Richard B. Lee and Irven DeVore's *Kalahari Hunter-Gatherers* (1976), of the relatively innocent Bushmen members of the present *H. sapiens sapiens* variety of humans, perfectly illustrate the situation. In the first image, women are shown gathering the aforementioned 80 percent of the food, in addition to nurturing the children—basically doing all the practical work—while the other image, titled *Telling the Hunt*, shows the men sitting around together

with their backs contemptuously shunning innocent nature's condemning presence as they boast about their heroic conquests over innocent animals. Clearly, the basic adaptions humans made to the human condition are well established in these relatively innocent Bushmen members of the present *H. sapiens sapiens* variety of humans; these could as easily be photos of women shopping and businessmen discussing a company takeover.

Women with infants digging roots

Telling the Hunt

Photographs by Marjorie Shostak; and Laurence Marshall

609 Of course, to properly context hunting, it has to be appreciated that the destruction of innocence has been going on at *all* levels. After all, in resigning, which became almost a universal phenomenon some 11,000 years ago, humans also destroyed the innocent soul in themselves by repressing it. Yes, our ignorant soul and any innocent representation of it, has been unbearably condemning, especially of men, which is why men, having already turned on and attacked their innocent animal friends, *then* turned on and attacked the relative innocence of their partners in life, women. Unable to explain their behaviour to women, men were left in an untenable situation: they couldn't just stand there and accept women's unjust criticism of their behaviour—they had to do *something* to defend themselves—but because women reproduced the species, men couldn't kill women the way they destroyed animals, and so instead men violated women's innocence or 'honour' through rape. Men perverted sex, as in 'fucking' or destroying, making it discrete from the act of procreation. What was being fucked, violated, destroyed, ruined, degraded or sullied was women's innocence. The feminist Andrea Dworkin recognised this

underlying truth when she wrote that **'All sex is abuse'** (*Intercourse*, 1987). And so in another tragic repercussion of the human condition, women's innocence was oppressed and, to a degree, they came to share men's upset.

610 The consequences of this horrific development—which will now be looked at—have obviously been immense. Prior to the perversion of sex, women weren't viewed as sex objects and so nudity had none of the problems of attracting lust and so there was no need to conceal our nakedness with clothes. To quote Moses in the Bible, when Adam and Eve took the fruit from the tree of knowledge—set out in search of understanding—**'the eyes of both of them were opened, and they realized that they were naked; so they sewed fig leaves together and made coverings for themselves'** (Gen. 3:7). Clothing was not *originally* designed to protect the body from cold as children have been evasively taught at school, but to restrain lust, to the extent that once we became extremely upset even the mere sight of a woman's ankle or face became dangerously exciting to men, which is why in some cultures women are completely shrouded and persecuted if any part of their body is revealed in public. It was a reverse-of-the-truth lie to say, as it is frequently argued, that this concealment of women was introduced out of 'respect for women'. The truth is, it was enforced because women were being *disrespected*.

611 In the case of the convention of marriage, this institution was invented as one way of containing the spread of upset. By confining sex to one life-long, monogamous relationship, the souls of the couple could gradually make contact and coexist in spite of the sexual destruction involved in their relationship. As stated in the Bible, in marriage **'a man will leave his father and mother and be united to his wife, and the two will become one flesh. So they are no longer two, but one'** (Mark 10:7, 8). Brief relationships, on the other hand, kept souls repressed and spread soul repression. However, the more upset, corrupted, insecure and alienated humans became, the more they needed sexual distraction and reinforcement through sexual conquest (in the case of men) and sex-object attention (a development in women that will be explained shortly), and thus the more difficult it became for both sexes to remain content in a single, monogamous relationship. The saying **'the first cut** [the first falling out of love] **is the deepest'** is an acknowledgment of the deep and total commitment humans make to their first love. It reveals

that the original, relatively innocent relationship between a man and a woman was monogamous. Since sex killed innocence, ideally (although impractical for the majority of the human race who had to ensure the continuation of the species) if we wanted to free our soul from the soul-destroying hurt sex caused it we needed to be celibate; as Christ explained it, some priests **'renounce marriage** [for] **the kingdom of heaven'** (Matt. 19:12).

612 BUT, while sex *was* an attack on innocence, an act of aggression, it was also one of the greatest distractions and releases of frustration and, on a nobler level, it became an inspirational act of love, an act of faith in and affection for men. A sublime partnership between men and women did develop, for when all the world disowned men for their unavoidable divisiveness, women, in effect, stayed with them, bringing them the only warmth, comfort and support they would know. As it says in Genesis, **'The Lord God said, "It is not good for the man to be alone. I will make a helper suitable for him"...Then the Lord God made a woman...and he brought her to the man'** (2:18, 22). Friedrich Nietzsche gave an honest description of the roles that developed for men and women in humanity's heroic journey to overthrow ignorance when he famously wrote that **'Man should be trained for war and woman for the recreation of the warrior: all else is folly'** (*Thus Spoke Zarathustra: A Book for Everyone and No One*, 1892; tr. R.J. Hollingdale, 1961, p.91 of 342). The 1960s sex symbol Brigitte Bardot was of a similar view when she said, **'A women must be a refuge for the warrior. Her job is to make life agreeable'** (*The Australian*, 11 Oct. 1999); although shortly after making this statement Bardot encapsulated the paradox of life for women when she declared that all men are **'beasts'** (ibid).

613 Yes, innocence was two-sided: it condemned and upset men, who, therefore, had to attack it, but it was *also* an inspirational reminder of the soulful, true world that they were fighting to reinstate by finding the understanding that would stop the upsetting criticism of them and the human race as a whole. So women's innocence could both condemn and inspire men, which, as will now be explained, is why the image of innocence—that Bardot's beauty was such a magnificent representation of—was so inspirational. The image of innocence in women could inspire the dream of the human race's return to living in a cooperative, loving, upset-free, ideal state, a state free of the human condition; it could lead to 'romance'. Men could dream that the *image* of innocence in women meant women

were actually innocent and that through their partnership with women they could share in that innocent state; and for their part, women could use the fact that men were inspired by their image of innocence to delude themselves that they actually *were* innocent. Men and women could 'fall in love', let go of reality and dream of an ideal, cooperative, loving world. Cole Porter's 1928 song *Let's Fall In Love* contains lyrics that reveal how falling in love is about allowing yourself to dream of the ideal state, of 'paradise': **'Let's fall in love / Why shouldn't we fall in love? / Our hearts are made of it / Let's take a chance / Why be afraid of it / Let's close our eyes and make our own paradise.'**

614The effect of the 'attraction' of innocence—which has been the preserve of youth because the young were innocent, they hadn't yet been exposed to all the upset in the world—for both dreaming through and for sexual destruction was that through the course of the 2-million-year journey through our species' adolescence our physical features became increasingly youthful looking or neotenous, as the increasingly child-like features of the skulls of the varieties of our *Homo* ancestors pictured in ch. 7:2 clearly evidence. The dramatic increase in neoteny from *H. habilis* to *H. erectus* reflects the dramatic increase in upset that took place once humanity determinedly set out on its search of understanding at the age-equivalent of 21, and the dramatic increase in neoteny from *H. erectus* to *H. sapiens sapiens* reflects the dramatic increase in upset that occurred when humanity entered the rapidly dis-integrating stage of the last quarter of the exponential growth of upset's development. Women were especially selected for their more innocent looking, neotenous, youthful, childlike features of a domed forehead, large eyes, snub nose and hairless body. Just how adapted women have now become to being sex objects can be seen in women's magazines, which are almost entirely dedicated to instructing women how to be 'attractive', which really means just better able to imitate the *image* of innocence. Women are now habituated and codependent to the reinforcement that men, for over 2 million years, have given their object self rather than their real self—for instance, they love to adorn themselves with beautiful objects, use make-up on their faces to increase their neotenous appearance, and wear high-heel shoes to give themselves the leggy, youthful, ultra-innocent look of pubescents.

Innocent looking, long-legged pubescent body proportions. Drawing by Gladys Perint Palmer, *Vogue Australia*, May 1994

Innocent, child-like, youthful, neotenous features. German supermodel Claudia Schiffer, *The Daily Telegraph-Mirror*, [illegible] Nov. 1994

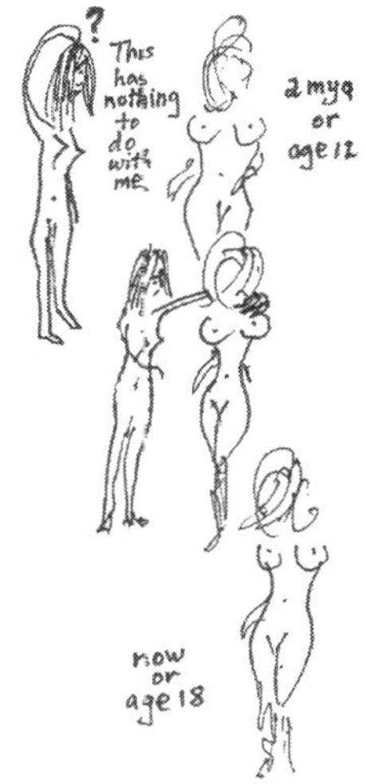

Drawing by Jeremy Griffith © 1989 Fedmex Pty Ltd

Stages of women's adaption to, and acceptance of, having to be a sex object—from 2 million years ago to the present, and from ages 12-18.

"I'm finally getting attention for what I have upstairs."

As these popular cartoons illustrate, women believe it is their real self, not their object self, that men are in love with. The sooner the horror and suffering of the human condition ends the better!!

[615]What now needs to be explained is how the <u>beauty of women</u> came to be *so* powerfully attractive and inspirational for men. It was explained in chapters 4 and 5 that throughout humanity's infancy and childhood we self-selected integrative traits by consciously seeking out love-indoctrinated mates—members of the group who had had a long infancy and were closer to their memory of infancy (that is, younger). Since the older we became the more our infancy training in love wore off, we began to recognise that the younger an individual, the more

integrative he or she was likely to be. We began to idolise, foster and select for youthfulness—represented as it was by neotenous (infant-like) features such as the aforementioned large eyes, dome forehead, snub nose and a hairless body—because of its association with co-operativeness or integrativeness. But what happened 2 million years ago when the upset state of the human condition emerged is that instead of selecting for neotenous features because they signalled a *cooperative* individual, such features began to be selected for because they signalled an *innocent* individual who was 'attractive' for sexual destruction. What this means is that while the motivation behind the selection changed *significantly*, the neotenous features signifying soundness and innocence have been selected for *throughout* humanity's infancy, childhood and adolescence. Also of significance is the fact that while all other forms of innocence were being attacked and destroyed throughout humanity's adolescence, this *image* of innocence—'the beauty of women'—was the *only* form that was actually cultivated during that time. So the image of innocence in women has both a very long and, since 2 million years ago, unique history of development, and that is why it is *so* powerfully attractive.

616 Women's representation of innocence, their representation of our now lost pure world, has been the only form of that purity that has been continuously cultivated since we were apes, so it is little wonder men 'fell in love' with women. The following quotes reveal just how inspiring women's image of innocence became for men: **'we lose our soul, of which woman is the immemorial image'** (Laurens van der Post, *The Heart of the Hunter*, 1961, p.134 of 233); **'I believe hers to have been the kind of beauty in which the future of a whole continent sings, exhorting its children to renounce what is out of accord with the grand design of life'** (ibid. p.86); **'Woman stands before him** [man] **as the lure and symbol of the world'** (Pierre Teilhard de Chardin, *Let Me Explain*, 1966; tr. René Hague et al, 1970, p.67 of 189); **'Women are all we** [men] **know of paradise on earth'** (Albert Camus, *The Fall*, 1956, p.73 of 108); **'You give me a reason to live'** (Joe Cocker, *You Can Leave Your Hat On*, 1986); **'I, I who have nothing / I, I who have no one** [because the whole world has hated men's upset behaviour] **/ Adore you and want you so'** (Jerry Leiber & Mike Stoller, *I Who Have Nothing*, 1963); **'Sex is life'** (graffiti on a granite boulder at Meekatharra in Western Australia). In the 1996 film *Beautiful Girls*, one of the male characters is criticised for plastering pictures of supermodels all over the walls of his room, to which he responds: **'Look, the supermodels are beautiful girls. A beautiful girl can make you dizzy,**

like you've been drinking bourbon and coke all morning, she can make you feel high for the single greatest commodity known to man—promise. Promise of a better day, promise of a greater hope, promise of a new tomorrow. This particular awe can be found in the gait of a beautiful girl, in her smile and in her soul; in the way she makes every rotten thing about life seem like it's going to be okay. The supermodels, that's all they are, bottled promise, scenes from a brand new day, hope dancing in stiletto heels.' The ever-insightful Nietzsche also recognised the role women played in inspiring the world with their *illusion* of innocence when he wrote, **'her great art is the lie, her supreme concern is appearance and beauty. Let us confess it, we men: it is precisely *this* art and *this* instinct in woman which we love and honour'** (*Beyond Good and Evil*, 1886; tr. R.J. Hollingdale, 1972, p.145 of 237).

617 So while it certainly *is* of little wonder that men fell in love with women, the great 'mystery of women' was that it was only the physical *image* or object of innocence that men were falling in love with. The illusion was that women were psychologically as well as physically innocent. For their part, women were able to fall in love with the dream of their own 'perfection' that men projected—of their being truly innocent. As has been pointed out, men and women *fell* in love, we abandoned the reality in favour of the dream; it really was the one time in our life when we could romance, when we could be transported to how it once was and how it could be again—to heaven. Like with Cole Porter's song *Let's Fall in Love*, the lyrics of the song *Somewhere*, written by Stephen Sondheim for the 1956 blockbuster musical (and later film) *West Side Story*, also provide a wonderful description of the dream of the heavenly state of true togetherness that humans allow themselves to be transported to when they fall in love: **'Somewhere / We'll find a new way of living / We'll find a way of forgiving / Somewhere // There's a place for us / A time and place for us / Hold my hand and we're halfway there / Hold my hand and I'll take you there / Somehow / Some day / Somewhere!'** Thank goodness the **'day'** of reconciliation of the lives of men and women has finally arrived and all the perverse destruction of women's souls can end, and instead of dreaming of a loving, ideal world, the *real* ideal, loving world for men and women can emerge.

618 While different cultures have different perceptions of female beauty, essentially, men are 'attracted' by innocent looks, which are youthful neotenous features. The popular saying 'blondes have more fun' illustrates the tendency in Caucasian cultures to regard blond women

as more attractive because many young Caucasians have blond hair, a sign of youth/innocence. The writer Raymond Chandler acknowledged the appeal of blondes when he wrote, **'It was a blonde. A blonde to make a bishop kick a hole in a stained glass window'** (*Farewell, My Lovely*, 1940, ch.13). Long, healthy hair is also associated with youth, which is why men find long hair on women attractive. In general, any feature unique to women will be attractive and signal a sex object to men, hence the desirability of breasts, shapely hips and a narrow waist. The different cultural definitions of beauty can also be explained in terms of what has historically signified innocence. For instance, in times when few could afford to eat or live well, fat women were considered beautiful because their appearance generally indicated that they had been well cared for, better nurtured and were thus more innocent. Today, however, the attraction of a long, ultra-thin female shape can be explained by the increase in alienation amongst humans. The duration of innocence now is very brief and, as has been mentioned, what is now deemed attractive is that pubescent age when young girls first start to develop physically and have the slender, long-legged frame of young animals, such as foals. So for women to be perceived as attractive they have to endeavour to look like an innocent pubescent teenager, which, in particular, requires walking around in extremely uncomfortable high-heels. This ultra-thin body shape is, of course, completely unnatural for adult, child-bearing women and to achieve it necessitates a starvation diet, a situation that has led to debilitating disorders such as anorexia and bulimia in women. Moreover, **'Salons say clients** [girls] **as young as 14 or less are requesting Brazilians** [removal of all their pubic hair]**'** ('Going bush', *The Weekend Australian Magazine*, 15-16 Feb. 2014), in order to make themselves look even more nubile, innocent! Such is the level of perversion that has developed in the human race!

619 Of course, human-condition-avoiding, psychosis-denying reductionist, mechanistic scientists have been totally committed to avoiding recognising that sex as men practise it is about attacking the psychologically innocent state of women for its unjust criticism of men. They maintain that the attractiveness of younger women has nothing to do with them being less psychologically corrupted, or at least the appearance of that state, claiming instead that it is due to a genetic reproductive strategy—that men want to **'mate with women who look like' 'Barbie—young with small waist, large breasts, long blond hair, and blue eyes...**[because] **they**

are healthier and more fertile than other women' ('Ten Politically Incorrect Truths About Human Nature', Alan Miller & Satoshi Kanazawa, *Psychology Today*, 1 Jul. 2007). This is the old Social Darwinist/Evolutionary Psychology strategy of blaming our competitive, selfish and aggressive human condition *not* on an upset psychosis in humans, which is its real source, but on, in truth, non-existent savage, aggressive and competitive animal instincts within us! No beautiful, unconditionally loving, altruistic, selfless soul in humans, just brutal, aggressive, make-sure-you-reproduce-your-genes animal instincts! No psychological upset in the human race now! What rubbish. No wonder 'fuck' is such a good swear word—everyone actually knows sex as humans practise it is about attacking the psychological innocence of women, that it's about 'fucking', but no one is admitting it, and (as I will explain more fully in par. 693) such extreme dishonesty provides the motivation for swearing; that swearing is a way of tearing through the overburden of dishonesty that we are all being forced to endure, in this case tearing through the dishonesty that 'sex' isn't about destroying innocence. *Of course* 'sex' is about attacking, 'fucking', innocence; as Dworkin said, it's **'abuse'**, it's a psychosis—but mechanistic scientists weren't going to admit that. No, as emphasised in par. 112, mechanistic scientists' human-condition-avoiding attitude was: 'What psychosis? What inner insecurity? What sense of guilt? What original 'Golden Age' of innocence? What 'fallen' condition? What alienation? What great elephant in the living room of our lives that we can't acknowledge? To hell with your psychological garbage!' Yes, they just blame all the psychological upset that we humans very obviously suffer from on savage animal instincts within us—'Humans kill each other in wars not because of serious philosophical differences but to reproduce their genes in the same way bulls fight and kill other bulls to ensure they win the mating opportunities'! But such 'explanations' are completely absurd because our human condition *obviously* involves our fully conscious mind; humans suffer from a psychological *human condition*, not a non-psychological genetic-opportunism-based *animal condition*. Again, Arthur Koestler summarised mechanistic, reductionist science's deliberate blindness to the issue of the **'mental disorder'** of our **'unique'** human condition when he wrote that **'symptoms of the mental disorder which appears to be endemic in our species...are specifically and uniquely human, and not found in any other species. Thus it seems only logical that our search for explanations** [of human behaviour] **should also concentrate**

primarily on those attributes of *homo sapiens* which are exclusively human and not shared by the rest of the animal kingdom. But however obvious this conclusion may seem, it runs counter to the prevailing reductionist trend. "Reductionism" is the philosophical belief that all human activities can be "reduced" to – i.e., explained by – the behavioural responses of lower animals – Pavlov's dogs, Skinner's rats and pigeons, Lorenz's greylag geese, Morris's hairless apes...That is why the scientific establishment has so pitifully failed to define the predicament of man' (*Janus: A Summing Up*, 1978, p.19 of 354). Yes, turn on the television and there will be endless programs presenting completely bullshit—**'pitiful'**—biological explanations of human behaviour, such as this claim that men prefer younger women because they are better able to reproduce their genes. Just this week on television (3 Jun. 2013, at time of writing) the first program of a *National Geographic* series titled *Ape Man* began with these words: **'In this series I'll introduce you to your inner ape...to expose our most basic instincts and explore their origins in the ape world. So get ready to learn the laws of the jungle because, like me, you are ape man...I've learnt that there's actually an ape living inside us humans and that inner ape is hungry for power. For our primate cousins being the alpha male gives them the first pick of the three things that matter most: food, territory and of course females. And we humans are just the same: world leaders, celebrities, even your boss—alphas rule the world!'** The truth is 100 percent the opposite of this—our instinctive heritage is of being cooperative, selfless and loving, and *then* we became *psychologically* upset and as a result of that we became competitive and aggressive. Yes, it was fucking mechanistic biological crap like this that was leading humanity straight to terminal alienation and extinction!

620 The very obvious truth is that sex as men practise it has, at base, been all about attacking the innocence of women, and unfortunately this destruction of women's souls and the cultivation of their image of beauty has been occurring for some 2 million years. As such, men and women have become *highly* adapted to their roles; while men's magazines are full of competitive battleground sport and business, women's magazines are, as has been pointed out, dedicated to enhancing beauty, to becoming more 'attractive', and to maintaining that attractive beauty for as long as possible. Just how aware women are of the importance of their sex object self, and how much women's sense of self-worth and self-esteem is now dependent on their sex appeal, is apparent in the 'twinning phenomena', in which you see women walking around together who are virtually

identical in their level of sex appeal or lack thereof. Clearly women feel too exposed and confronted if they are with another woman who is even slightly more 'attractive', so they end up befriending their sex appeal 'twin'. It is extremely common to see, and *very* revealing of what men have done to women! Sex has played a *huge* role in the lives of men and women.

621 Indeed, lust and the hope of falling in love have assumed such importance that many people, including the psychoanalyst Sigmund Freud, have been deceived into believing that sex actually rules our lives. One of the best examples of this conviction is found in the story of the Garden of Eden, in which Eve is blamed for tempting Adam to take the apple from the tree of knowledge. The truth is that women were the *victims* not the cause of upset in men, but lust became such a strong force that people have been misled into believing it seduced humans into behaving in an upset way.

622 The problem for women, of course, is that while their image of innocence was being cultivated throughout those 2 million years, their soul, their actual innocence, was being destroyed—women are *only* the image of innocence. But without the understanding necessary to explain themselves, men had no choice other than to repress the relative naivety of women, which in turn tied women's corruption inextricably to men's. It was an extremely difficult situation for women. They had to try to 'sexually comfort' men but also preserve as much true innocence in themselves as possible to nurture the next generation. Their situation, like men's, worsened at an ever-increasing rate, in that the more women 'comforted' men, the less innocence they retained and the greater comforting the following generation needed. Had humanity's battle continued in this exponential pattern for a few thousand years more, all women would have eventually become like Marilyn Monroe, complete sacrifices to men. At this point men would have destroyed themselves and the human species, for there would be no soundness left in women to love/nurture future generations. Olive Schreiner emphasised this point in her already referred to amazingly truthful 1883 book, *The Story of an African Farm*. When talking of men persuading women to have sex, Olive Schreiner's female character stated that men may say, **'"Go on; but when you** [men] **have made women what you wish, and her children inherit her culture, you will defeat yourself. Man will gradually become extinct..." Fools!'** (p.194 of 300). Incidentally, we again see here how calling men **'fools'** was an expression

of women's lack of understanding of men, and why they were being attacked sexually by men in the first place.

[623] Yes, tragically, across every generation, individual women had a very brief life in innocence before being soul-destroyed through sex, following which they had to try to nurture a new generation, all the time trying to conceal the destruction that was all around and within them. Mothers tried to hide their alienation from their children, but the fact is if a mother knew about reality/upset her children would know about it and would psychologically adapt to it. Alienation was invisible to those alienated, but to the innocent—and children are born innocent—it was clearly visible. For example, Christ's mother Mary must have been innocent because Christ was. Since women become upset through sex, Mary must have had virtually no exposure to sex. The symbol for women's innocence/purity is virginity, hence the description of Christ's mother as the 'Virgin Mary'. The renowned writer D.H. Lawrence recognised the essential innocence of the 'Virgin' Mary when, in reference to her, he wrote, **'Oh, oh, all the women in the world are dead, oh there's just one'** (Lawrence Durrell & Earl Ingersoll, *Lawrence Durrell: Conversations*, 1998, p.178 of 261).

[624] Having to inspire love when they were no longer loving or innocent, and attempt to nurture a new generation—all the while dominated by men who couldn't explain why they were dominating, what they were actually doing or why they were so upset and angry—was, in truth, more than extremely difficult, it was an altogether impossible task, and yet women have done it for 2 million years. Indeed, it is because of women's phenomenally courageous support that men, when civilised, treat them with such chivalry and deference. Men have had an impossible fight on their hands, but at least they had the advantage of understanding that battle. To be a victim of a victim, as women have been, is an insufferable situation, because while a primary victim knows what the primary source offence is, a victim of a victim does not. This is why, when men became overly upset they became mean, even brutal, but when women became overly upset they became nasty, even venomous. Not knowing what it is they are flailing at, women's fury is unsourced, untargeted and unbounded. The proverb **'hell knows no fury like a woman scorned'** recognises this potential extremity in the nature of women. Nietzsche wrote, **'Let man fear woman when she hates: for man is at the bottom of his soul only wicked, but woman is base'** (*Thus Spoke Zarathustra: A Book for Everyone*

and No One, 1892; tr. R.J. Hollingdale, 1961, p.92 of 342). Women have historically had to carry so much unsourced frustration and hurt that their psychological situation is very fragile; so fragile, in fact, that the hormonal upheaval accompanying menstruation is enough to destabilise this delicate balance, hence pre-menstrual tension (PMT).

625 So while men's situation has been horrible, so has women's; and, just as men have yearned for freedom from their oppressor, ignorance, so women have yearned for freedom from their oppressors, men, as this comment by Olive Schreiner describes: **'if I might but be one of those born in the future; then, perhaps, to be born a woman will not be to be born branded... It is for love's sake yet more than for any other that we** [women] **look for that new time...Then when that time comes...when love is no more bought or sold, when it is not a means of making bread, when each woman's life is filled with earnest, independent labour, then love will come to her, a strange sudden sweetness breaking in upon her earnest work; not sought for, but found'** (*The Story of an African Farm*, pp.188, 195). Thank goodness with the battle to defeat the ignorance of our instinctive moral soul now won this dreamed-of time of freedom from the horror of both men's and women's existence can end, allowing—if we choose wisely—the arrival of a **'new time'** where society will be neither matriarchal or patriarchal but gender-neutral and at peace.

626 We can see that with men defying and repressing their own souls, women became representative of soul in their partnership with men. Further, because of men's unexplained oppression of women and the world of the soul, and their own inability to understand this oppression, women put increasing trust and reliance in their soul and instincts, rather than in their ability to understand. As a consequence, they are more intuitive or dependent on their soul's guidance than men—hence the well-known term, **'women's intuition'**. Another common saying is **'women feel while men think'**. It has also often been said that women talk more than men—indeed, a study has found that **'women speak some 13,000 more words a day than the average man'** ('Women really do talk more than men', *Mail Online*, 20 Feb. 2013)—and we can now explain why. Because men have been preoccupied with championing the ego while women were left doing the support duties, women have had less opportunity to develop their own egos—which, while not as developed and embattled as men's, obviously exists because ego is just humans' conscious mind trying to understand the world and its place in

it. While they were preoccupied with support duties, the one way women could try to understand the world, justify and measure themselves was through talking to each other; they couldn't act out their experiments in self-adjustment but they could think their ideas through out aloud, evaluate and try to establish their worth through words.

627 The following honest description of the different roles of men and women from Sir Laurens van der Post illustrates these very distinct attitudes: **'The sword was, he would suggest, one of the earliest images accessible to us of the light in man; his inborn weapon for conquering ignorance and darkness without. This, for him, was the meaning of the angel mounted with a flaming sword over the entrance to the Garden of an enchanted childhood to which there could be no return. He hoped he had said enough to give us some idea of what the image of the sword meant to him? But it was infinitely more than he could possibly say about the doll. The doll needed a woman not a man to speak for it, not because the image of the sword was superior to the image of the doll. It was, he believed, as old and went as deep into life. But it was singularly in women's keeping, entrusted to their own especial care, and unfortunately between a woman's and man's awareness there seemed to have been always a tremendous gulf. Hitherto woman's awareness of her especial values had not been encouraged by the world. Life had been lived predominantly on the male values. To revert to his basic image it had been dominated by the awareness of the sword. The other, the doll, had had to submit and to protect its own special values by blind instinct and intuition'** (*The Seed and the Sower*, 1963, p.193 of 246).

628 Note that while men and women are different, sexist notions of men being 'evil' or of women being irrelevant have no credibility. While the main device for avoiding prejudice was to deny that there was any difference between men and women, another device was to maintain that any difference between men and women was simply a product of cultural conditioning—of girls being given dolls and boys swords as infants, for example. In fact, as Sir Laurens van der Post acknowledges, our differences are the product of a very real distinction—**'a tremendous gulf'**—between the sexes.

629 So while at a more noble level sex has become an expression of love, at its fundamental level it is an attack on the innocence of people; it is rape. The more upset and corrupted the human, the more sexually destructive and thus sexually perverted they have been inclined to be, and the more innocent (or innocent-looking) the human, the

more attracting of that destruction they have been. Understanding this makes it possible to explain homosexuality in men. As the victims of sex, women have historically been more exposed to and thus become, through natural selection over hundreds of thousands of years, more adapted to sex than men. In most cases, if a male was not interested in sex then sex did not occur, whereas women have been exposed to sexual advances regardless of their interest or lack thereof. Teiresias, the prophet mentioned in Homer's Greek legend, *The Odyssey*, recognised that women were more sexually aware than men. When asked **'whether the male or the female has most pleasure in intercourse'**, he replied that **'Of ten parts a man enjoys one only; but a woman's sense enjoys all ten in full'** (Hesiod, *The Melampodia*, c. Eighth century BC; *Hesiod, The Homeric Hymns and Homerica*, tr. H.G. Evelyn-White, 1914, p.269 of 657). Evidence that women have been more exposed to sexual attention than men is the fact that they tend to have more neotenous facial features and have far less body hair. Since neotenous features and loss of body hair are consequences of sexual selection for innocent looking features, to have more neotenous facial features and less body hair means women must have been exposed to more sexual attention than men. (Incidentally, since women are now highly adapted to sex it means a virgin is not truly a virgin, she is not truly an innocent girl and thus completely 'attractive', because all women are now instinctively aware of 'sex'.) Yes, while women have had to hide their sexual awareness in order to present an attractive image of innocence, the fact is they *are* more sexually aware than men. In the *Happy Days* television series (set in the 1950s and first broadcast in 1974), girls are much more attracted to the sexually aware Fonzie character than to the naive, relatively innocent Richie Cunningham character. While only fictional characters, the viewing audience would not have responded with such empathy over the years if the characters did not resonate with truth. In the book, *Big Bad Wolves: Masculinity in the American Film*, by the academic and author Joan Mellen, the caption accompanying a picture of the 1920s sex symbol actor Rudolph Valentino reads: **'Rudolph Valentino in the film "Son of the Sheik". Rape is the central visual metaphor'** (1977). Yes, as explained, 'the mystery of women' is that after 2 million years of having been sexually used by men, women now only represent the physical image or object of innocence. It is this *image* of innocence that men have been falling in love with; the *illusion* that women were psychologically as well as physically innocent. A well-known African fable tells of a woman who agrees to marry a man on the condition he never looks inside a precious basket that she keeps. She warns him that if he does she will vanish. He agrees, but some time after they are married and his wife has gone to the river for water he cannot resist and peers in the basket. On his wife's return she finds the basket open and him laughing. When she accuses him of looking in the basket he says, 'You silly woman, there was nothing in the basket', at which point she vanishes into thin air. The basket is symbolic of the mystery of women—they are only the image of innocence; it is an 'empty basket' that men are looking at, and once men see through the illusion, women's attractiveness diminishes. It follows then that the more corrupted a man is, the less naive he is, and thus the more he is aware that women are not innocent. Therefore, if a man is extremely hurt and corrupted in his infancy and childhood, when he becomes sexually mature he will not be naive enough to believe that women are still innocent and will not, therefore, find women sexually attractive. The last bastion of 'attractive' innocence for such men is younger men, because they are not as exposed to sexual destruction as women have historically been. To explain the feminine mannerisms particular to male homosexuality, if you have had your soul, which is your core strength, destroyed in childhood, then taking on the extremely difficult male role of having to fight against the ignorance of the soulful, idealistic world would be an untenable position that would make the female position of not having to fight a much more preferential option. You would rather adopt the female role of being an object of adoration and service than the male role of having to take on the loathsome job of championing ego over soul. The transsexual professional tennis player of the late 1970s, Renée Richards—who went so far as to have a sex change operation to become a woman—alluded to the difficulty of life as a male, and by inference the appeal of being a woman, when she once said, **'women don't realise the horror of the world that men live in—the strife torn world that men live in'**. Having to live with the condemnation that you are an evil monster, when you know you are not but cannot explain why you are not, has been a living hell for men. To be a man and have to oppress the all-magic soul without being able to explain why has been the most wretched of tasks. The following quote serves to illustrate how pressured men's lives have become: **'If women are so oppressed, how come they live much**

longer than men?' (Don Peterson, review of *The Myth of Male Power* by Dr Warren Farrell, *The Courier Mail*, Jun. 1994). Homosexuality amongst women results from women's understandable disenchantment with men. Homosexuality is simply another level of perversion to heterosexuality. They are both psychologically corrupted states of sexuality that developed under the horror of the duress of the human condition.

630With regard to whether homosexuals are 'born' or 'made', even without the ability to explain the human condition and thus defend the corrupted state of humans (that is, explain that humans' various states of corruption are not 'bad' or evil but, in fact, immensely heroic states), a decade-long research project completed at the Institute for Sex Research in Bloomington, Indiana, found that **'a quarter of the gays interviewed believe** [are prepared to acknowledge?] **homosexuality is an emotional disorder'** (*TIME*, 17 Jul. 1978). In his 1992 book, *Health & Survival in the 21st Century*, Ross Horne referred to studies that show **'That the highest incidence of homosexuality coincides with the general level of stressful influences in a community and that the lowest incidence coincides with the degree of happiness and health in remote and unstressed populations indicates that, like many conditions of physical disease, it is just as unnatural as the mental breakdowns, depression and neuroses so common in civilization. Studies of primitive natives reveal that while in some populations homosexuality is non-existent or rare, in other populations it is fairly common; but the same pattern still holds—among the placid, happy, untroubled people homosexuality did not occur, while among fighting tribes and headhunters it did'** (p.206). After 25 years of clinical experience helping homosexual men and women, Dr Robert Kronemeyer of New York concluded that **'Homosexuality is a symptom of neurosis and of a grievous personality disorder. It is an outgrowth of deeply rooted emotional deprivations and disturbances that had their origins in infancy'** (*Overcoming Homosexuality*, 1980).

631While women are instinctively more sexually aware then men this does not exclude the fact that a woman can be more innocent and less sexually aware than a man. Girls who are nurtured and sheltered in their upbringing can be very innocent. However, because there has been no honesty about the existence of the different levels of upset and alienation amongst humans, they can be deceived by men who are much more upset and, therefore, much more sexually advanced down 'the rungs of the perversion ladder' (where one is holding hands, two is kissing, three is touching her breast, etc, etc, etc). Men who are more upset can be very attentive to women because sex for them is a distraction and a way of gaining reinforcement, and innocent women can be deceived—seduced—by this attention into a relationship. In her 1981 book, *African Saga*, the African photographer, and remarkably beautiful woman, Mirella Ricciardi gives an extraordinarily honest account of a relationship between a more innocent woman and a less innocent man. She wrote: **'We went to live in Rome, where I quickly began to taste the bitter-sweet agony of life with Lorenzo. I was young, unaware of the world, and ignorant of people and their behaviour. I married Lorenzo as easily as I had switched lovers. It was probably the most foolish, irresponsible, exciting thing I have ever done. Years later, I came to the conclusion that most of the men I had met fell into three categories—those prompted by their heads** [presumably, upset men]**, those by their heart** [presumably, less upset, more soulful, relatively innocent men] **and those by their sex** [presumably, extremely upset men]. **Some—not many—were a combination of all three. Lorenzo belonged to the last category—these I have found are the most attractive. They are sexy, amusing, fun-loving, careless, irresponsible and lazy—they dress well and have a lot of style. Most people like them. They are excellent lovers and lousy husbands. Women usually find them irresistible or are terrified by them. Men either envy or despise them. No one can remain indifferent to them…"Lorenzo's mother died when he was seven," Cesarino** [Lorenzo's father] **told me one day. "You will have to be more of a mother than a wife to him; do you realise this?" Then he laughed. "The only pleasure he ever gave me was nine months before he was born." But when his father died sixteen years later, Lorenzo's grief was immeasurable and I began then to understand the meaning of these words'** (p.136 of 300). Because women have lived through men, and because their means of healing the world is through nurturing, relatively innocent women involved in relationships with more upset men have often tried to change their partner, make him sounder and stronger through nurturing love, inspiration and motivation. BUT, since it is only understanding of hurt that can heal hurt, these efforts often only serve to further confront and criticise the man. Mirella's dedication in her book **'to Lorenzo, my magnificent obsession'** is an acknowledgment of her frustrated efforts to change Lorenzo. She wrote: **'When I married Lorenzo I had created an image of a giant in whose shadow I would live. I clung stoically to my belief in our union and waited patiently for ten years for him to cast his shadow, but he**

never did' (ibid. p.138). In the 1950 Broadway musical, *Guys and Dolls*, the Salvation Army innocent, Sarah, desperately protests about her upset gambler boyfriend Sky to her nightclub singer friend, Adelaide, saying, **'Can't men like Sky ever change?'**, to which Adelaide says, **'They just can't change'**; and later Sky himself says, **'Change, change. Why is it the minute you dolls get a guy that you like, you take him right in for alterations?'** In a world of lies, the basis for relationships has often been unhealthy. Women have been seduced by men in so many ways and their innocence has been the casualty—another reason why women have historically become more sexually aware than men. Growing up in the countryside in Australia in the 1960s I saw many sheltered, relatively innocent country girls go off to Europe for a few years in their early 20s—it was considered the thing to do in those days, as it is now—only to return with 'knowing eyes', a different more sophisticated way of looking at the world. I could never understand the point of sheltering and nurturing young women if they were simply going to go off to Europe and, as I saw it, 'cash in their innocence'.

632 Since sex is an attack on innocence as well as an act of love the recent generations of humans who have been <u>treating sex cheaply</u>, have been contributing significantly to the death of soul in the world, and thus contributing significantly to the level of alienation in the world. Queen Victoria was right to espouse the 'Victorian morality' of her era, a moral code that strongly encouraged people to treat sex with care and restraint.

633 In light of this reality, during the 2 million years that women have endured the wretched situation of being unable to understand men's oppression of them, many must have found it impossible to accept and, as a result, there must have been a great deal of natural selection and thus genetic adaption to the role that women have had to play in the human journey to enlightenment. Olive Schreiner described women's resignation to their role in the following passage from her extraordinarily honest book, *The Story of an African Farm*. It is a dialogue between her young female character and that character's male friend, Waldo: **'"I know it is foolish. Wisdom never kicks at the iron walls it can't bring down," she said. "But we are cursed, Waldo, born cursed from the time our mothers bring us into the world till the shrouds are put on us. Do not look at me as though I were talking nonsense. Everything has two sides—the outside that is ridiculous, and**

the inside which is solemn." "I am not laughing," said the boy sedately enough; "but what curses you?" He thought she would not reply to him, she waited so long. "It is not what is done to us, but what is made of us," she said at last, "that wrongs us. No man can be really injured but by what modifies himself. We all enter the world as little plastic beings, with so much natural force, perhaps, but for the rest—blank; and the world tells us what we are to be, and shapes us by the ends it sets before us. To you it says—*Work!* and to us it says—*Seem!* To you it says—As you approximate to man's highest ideal of God, as your arm is strong and your knowledge great, and the power to labour is with you, so you shall gain all that the human heart desires. To us it says—Strength shall not help you, nor knowledge, nor labour. You shall gain what men gain, but by other means. And so the world makes men and women. Look at this little chin of mine, Waldo, with the dimple in it. It is but a small part of my person; but though I had a knowledge of all things under the sun, and the wisdom to use it, and the deep loving heart of an angel, it would not stead me through life like this little chin. I can win money with it, I can win love; I can win power with it, I can win fame. What would knowledge help me? The less a woman has in her head the lighter she is for climbing. I once heard an old man say, that he never saw an intellect help a woman so much as a pretty ankle; and it was the truth. They begin to shape us to the cursed end," she said, with her lips drawn in to look as though they smiled, "when we are tiny things in shoes and socks. We sit with our little feet drawn up under us in the window, and look out at the boys in their happy play. We want to go. Then a loving hand is laid upon us: 'Little one, you cannot go,' they say; 'your face will burn, and your nice white dress be spoiled.' We feel it must be for our good, it is so lovingly said; but we cannot understand; and we kneel still with one little cheek wistfully pressed against the pane. Afterwards we go and thread blue beads, and make a string for our neck; and we go and stand before the glass. We see the complexion we were not to spoil, and the white frock, and we look into our own great eyes. Then the curse begins to act on us. It finishes its work when we are grown women, who no more look out wistfully at a more healthy life; we are contented. We fit our sphere as a Chinese woman's foot fits her shoe, exactly, as though God had made both—and yet He knows nothing of either. In some of us the shaping to our end has been quite completed. The parts we are not to use have been quite atrophied, and have even dropped off; but in others, and we are not less to be pitied, they have been weakened and left. We wear bandages, but our limbs have not grown to them; we know that we are compressed, and chafe against them.

But what does it help? A little bitterness, a little longing when we are young, a little futile searching for work, a little passionate striving for room for the exercise of our powers,—and then we go with the drove. A woman must march with her regiment. In the end she must be trodden down or go with it; and if she is wise she goes'" (pp.188-189). Incidentally, note again here women's naivety; Schreiner believes men's sexualisation of women is un-Godly, basically that men are evil, which is again why women were 'fucked' in the first place.

634 With regard to women resisting **'march**[ing] **with her regiment'**, in the sense of accepting their role of inspiring men, it should be mentioned that it's not commonly acknowledged that suits were invented for men so they could hide their big guts, while dresses were invented for women so they could accentuate their waists and breasts and conceal their big bottoms and thighs, but, while men still wear coats, everywhere in Western society now women have forsaken skirts for trousers, *and even tights*, as if their role of inspiring men with their beauty no longer matters. This is, in truth, yet another illustration of women's lack of awareness of the nature of the struggle that the human race has been involved in—and of the irresponsibility of feminism, which encouraged women not to **'march with her regiment'**. Women's role has understandably become unbearable for them but the battle to find understanding still had to be won.

635 Yes, while the feminist movement has improved 'a woman's lot' superficially, there has, in fact, been no real change to the situation that Schreiner so honestly described, as these quotes confirm: **'Nirvana hasn't happened. Although men are speaking about understanding** [the need for women's liberation from men's oppression] **on the surface, they're not doing anything about it'** (Carmel Dwyer, *The Sydney Morning Herald*, 22 Sep. 1993); and, **'What happened was that the so-called Battle of the Sexes became a contest in which only one side turned up. Men listened, in many cases sympathetically but, by the millions, were turned off'** (Don Peterson, review of *The Myth of Male Power* by Dr Warren Farrell, *The Courier Mail*, Jun. 1994). Until men could explain to women why they have had to be so egocentric, competitive and aggressive—explain the human condition—there could be no fundamental change to the situation where men found themselves with no choice other than to oppress women.

'Goddess' statue, Catal Huyuk, Anatolia (modern Turkey), 8,500-5,500 BC

[636] It should be pointed out here that women resisted men's oppression, and resisted it for a long time. Yes, having said that humanity has been patriarchal for some 2 million years, it has to be explained why women were still so powerful, even seemingly treated as goddesses, in central Europe during the Upper Paleolithic (50,000 to 10,000 BC) and Neolithic (10,000 to c.4,000 BC) periods, as evidenced by the many so-called 'Goddess' or 'Venus' figurines, such as the one pictured, that have been found in this region from this period. In this typical example, we can see from the *extremely* regal stature of the very well-nourished figure seated on her throne of cheetahs just how powerful and in control of their societies such women must have been. In chapter 4:11, 'The importance of strong-willed females in developing integration', it was explained just how powerfully self-assured, secure-in-self, strong-willed and assertive females had to be during the development of love-indoctrination in order to rein in male aggression from competing for mating opportunities. Yes, what greatly contributed to the defeat of the historical patriarchal animal-condition situation where males relentlessly competed for dominance was the selection for strong-willed, authoritative females who would no longer tolerate male aggression. Quotes were included to illustrate this situation, such as: **'An impressively stern** [bonobo] **female enters and snaps a young sapling. Once**

she picks herself up she does something [that would be] **entirely surprising for a female chimp, she displays, and the males give her sway. For this is the confident stride of the group's leader, its alpha female, whom Kano has named Harloo'**; and, '**She** [Peggy] **was the highest-ranking female in the** [baboon] **troop, and her presence often turned the tide in favor of the animal she sponsored. While every adult male outranked her by sheer size and physical strength, she exerted considerable social pressure on each member of the troop...another reason for the contentment in this particular family was Peggy's personality. She was a strong, calm, social animal, self-assured yet not pushy, forceful yet not tyrannical.'** As has been explained, the problem that emerged when our ancestors became fully conscious and men had to take up the task of defeating the ignorance of our instinctive self and champion our conscious thinking self or ego, was that women tended to be unsympathetic towards the angry, aggressive and egocentric upset that it unavoidably produced in men. In fact, this situation was made so much more difficult for men because women were *so* immensely strong-willed and authoritative as a result of this very powerful matriarchal heritage. It makes sense, then, that while humanity had become fundamentally patriarchal, the old matriarchy hung on. In fact, even today there are still many women who are so strong-willed that men find it almost impossible to cope with them, and even go to great lengths to try to break their strong-willed, but ignorant, resolve. If you watch WTM Founding Member Stacy Rodger's affirmation at <www.worldtransformation.com/affirmations> you will hear about just such a tragic situation—where Stacy was psychologically traumatised because, she said, **'every time I lived out my strength and said what I really thought, especially about the insincerity of men's treatment of women—basically of men being self-centred, totally, seemingly unaware and insensitive towards others and even seemingly unaware and insensitive towards the whole world—it would always end up with men side-lining me and freezing me out'** in an effort to **'shut me out and shut me down'**. Unable to explain why they are so upset, embattled and egocentric, men have truly struggled to do their job of championing the ego over our ignorant instinctive self or soul in the midst of women who have been as strong-willed and yet as naive as Greer and Steinem have been. In hindsight (because understanding of the human condition finally brings to an end the horrible 'battle between the sexes'), it can be seen that it was an act of great generosity (actually not generosity but weakness, which I will talk more about in a moment) on the part of men to give women the vote! So this situation where women were seemingly in power in prehistoric times was a case of delayed ownership; of situations where the new owner, patriarchy, wasn't able to take over because the old owner, matriarchy, refused to relinquish power.

637 A 1999 BBC documentary titled *Ancient Voices: Tracking the First Americans* presents convincing evidence that some Australian Aborigines managed to reach South America long ago, and that the Fuegian Indians of Tierra del Fuego are their descendants. To prove that such immensely long sea journeys are possible the program referred to the fact that recently **'five African fishermen were caught in a storm and a few weeks later were washed up on the shores of South America'** with three still alive ('First Americans were Australian', BBC News Online Network, 26 Aug. 1999). In terms of what is being explained here about women resisting male dominance, some of the commentary in this documentary is particularly enlightening (the underlinings are my emphasis): 'Narrator: **The tribal wisdom taught to the Fuegian initiates was secret. It was only revealed to the men, the women were kept in the dark. Any speculation about it was, and still is, strictly taboo.** Christina, one of the last of the Fuegian tribe, then says: **They said it was a very secret ritual, that's why we never talked about it. Only the men were supposed to know about it.** Narrator: **What was so secret that had to be kept from the women? Some of the chiefs confiding to ethnographers explained that <u>there was a time in the very distant past when women ruled society. The women must never know lest the men lose their grip on power</u>...Extraordinarily, <u>similar legends have been recorded amongst aborigine tribes in Australia</u>. It seems the traditions of the first peoples of Australia have been preserved here at the utmost end of the Earth by a small band of their descendants. But after surviving 50,000 years, the memory of those traditions is now at risk of being lost forever.'**

638 This narration provides powerful evidence of the delayed ownership analogy, where the old order is slow to relinquish its position: even though humanity had become patriarchal, male-role led, in so many ways—such as by women being treated as sex objects (fucked for being ignorant of the goodness of men) and having become highly adapted to this situation, and men leaving women to do the gathering and other practical work while they conducted the innocence-dominating task of killing (hunting) animals—such was women's strength of character that for a long time **'women ruled society'**. It took a great deal of effort, determination and time for men to finally gain a **'grip on power'**.

639 The question this raises is, if women have been so unmainframed, and so duped by soul, <u>how do we explain such effective right-wing-</u>

supporting, ego-sympathetic females like the political leaders Margaret Thatcher, Madeleine Albright and Golda Meir, and authors like Ayn Rand, and political commentators like Ann Coulter in the USA and Janet Albrechtsen in Australia? To explain this phenomenon I need to first point out that feminism hasn't been entirely unfounded. The more men fought to defeat ignorance and protect the group (humanity), the more embattled, upset and corrupted they became and thus the more they appeared to worsen the situation. The harder men tried to do their job of protecting humanity, the more they appeared to endanger humanity! As a result, they have become almost completely ineffective or inoperable, paralysed by this paradox; made cowards by the extent of their self-corruption and its effects. At this point, women have had to step in and usurp some of the day-to-day running of affairs as well as attempt to nurture a new generation of soundness. Women, not oppressed by the overwhelming responsibility and extreme frustration that men felt, could remain effective. Further, when men crumpled, women *had* to take over, otherwise the family, group or community involved would perish. A return to matriarchy, such as we have recently seen in some parts of society, is a sign that men in general have become almost completely exhausted. However, it is important to understand that *total* matriarchy has not emerged because men could not afford to stand aside completely while the fundamental battle still existed. They needed to stay in control and remain vigilant against the threat of ignorance. So while some elements in the recent feminist movement have seized the opportunity to avenge men's oppression, the movement was, to a degree, borne out of necessity. The tragedy is that like all pseudo idealistic, politically correct movements, feminism is based on a lie: in this case, that there is no real difference in the roles of men and women.

640 So to address the question of how are we to explain the existence of such right-wing, ego-sympathetic women as Baroness Thatcher and the others mentioned, what needs to be considered is the important sentence that was included in the previous paragraph: 'Women, not oppressed by the overwhelming responsibility and extreme frustration that men felt, could remain effective.' Men have been overly corrupted for at least half a million years and, as such, have lived with extreme frustration, even self-loathing, of the immense destruction they have inflicted upon the planet. After such a long time, it can be expected that women now have a

strong instinct for an opportunity to participate up-front in the battle and even—in situations where men became totally destructive, disdainful of themselves, paralysed by their predicament and inoperable—take control from men. We can now expect women to anticipate the opportunity for greater power within personal relationships, and in larger economic and political spheres, and to some degree be adapted to and thus appreciative of what is required to effectively take up the male role of championing the ego. After all, if men had not been available to take on the battle to champion ego over ignorance, women would have had to take it on fully and become as aware as men now are of what happens when you fight that battle. Camille Paglia once wrote that **'It is woman's destiny to rule men'** (*Vamps & Tramps*, 1974, p.79). This comment is an expression of the expectation that now exists in women that men eventually crumple, at which point women have the opportunity to take over. The truth is, many sensitive new age guys (so-called SNAGs) and 'metrosexuals' are actually crumpled men. Women's nagging of their menfolk is also a case of women chiselling away at, attempting to break, men's ability to keep fighting and defying the ignorance of the world of the soul—a situation made much of in the BBC sitcom *Keeping Up Appearances*, where the character of poor old Richard is constantly hen-pecked by his wife Hyacinth.

641 So, while by dogmatically imposing an ideal situation where men and women treat each other as though there is no difference between them—as the politically correct culture has attempted to do—upset could be disguised and contained to a degree, it could not remove or resolve it. *Only* understanding the world of men, and why they have been so divisively behaved, could subside the anger, alienation and egocentricity that caused them to victimise virtually everyone and everything they encountered. The boot that was screwing men into the dirt had to be lifted for the horrible war between men and women to end. And an all-out war it has been, lived to the full extent of what was possible under the limitation that men and women were forced to coexist if they were to reproduce and nurture a new generation. Yes, as men have long lamented about women, **'You can't live with them and you can't live without them!'**

642 And, as has been emphasised, in finding the liberating understanding of why men have been divisively behaved, what it reveals is that men are the heroes of the story of life on Earth. From being thought of as the villains one day, they become the absolute heroes the next—a turn

of events that is long overdue. Everywhere men have become wretchedly oppressed by the politically correct dogma that denies them any real meaning in the world. Indeed, men have become so intimidated by pseudo idealism that many have come to believe they are useless. In his 2001 book, *Men: From Stone Age to Clone Age*, the science writer Bob Beale wrote that while the concession pained him, he was prepared to concede that, except for **'the baby business** [reproduction]**'**, **'males are largely useless'**. The proliferation of men's movements that aim to counter men's horrific situation of being totally misunderstood is but one sign of the problems men face. Boys growing up in the current, men-are-worthless world are having their self-esteem destroyed before they even have a chance to enter manhood. A very real concern for the future was that there would be a dearth of men psychologically strong enough to fight ignorance, and indeed, if the human condition had not been solved the human race would be facing very dark times—as it still does if these insights are not accepted and promoted by the scientific establishment. As will be talked more about in ch. 7:16, that is how dangerous the politically correct, postmodernist culture has become.

643 The reason for the well documented phenomena of school boys not performing as well in their studies as girls is not only because they have had their immense relevance in the world denied, but also because they are not as duped by the historical denial of the issue of the human condition as women. Women tend to believe the world we are living in is the real world, whereas men, being mainframed, are intuitively aware that it is a fraudulent existence, and as a result never take it too seriously. In the television series, *The Simpsons*, the young boy, Bart, has little respect for school whereas his sister Lisa applies herself completely and excels. Nietzsche was alluding to the trusting naivety of women when, as mentioned earlier, he said, **'"Behold, now the world has become perfect!"—thus thinks every woman when she obeys with all her love.'**

644 While on Nietzsche, his much debated and misunderstood concept 'the will to power' (the title of his last work before his psychological breakdown—which, since he was such an honest thinker, would have happened when he eventually tried to confront more truth than his level of soundness of self could cope with) can now be interpreted as man's will to achieve power over humans' idealistic instinctive self or soul's ignorance of the true goodness of corrupted humans. In his now much acclaimed (and rightly so) book *Thus Spoke Zarathustra: A Book for*

Everyone and No One, Nietzsche recognised that, unlike other animals, humans have had to fight a psychological demon, the human condition: **'Man, however, is the most courageous animal: with his courage he has overcome every animal. With a triumphant shout he has even overcome every pain; human pain, however, is the deepest pain'** (p.177 of 342). Yes, humans, especially men, had to have the courage to triumph over their deepest pain, the pain of not being able to know whether they were fundamentally evil beings or not. *Humans have had to learn to love themselves*—and now they truly can.

645 Incidentally Nietzsche's subtitle to *Thus Spoke Zarathustra*, **'A Book for Everyone and No One'**, was an open acknowledgment that to speak the truth to people who are resigned and living in denial of the human condition was to court total rejection—what I call the 'deaf effect'. **'No One'** would hear, understand and accept his words. What Nietzsche knew, however, was that only the truth could liberate humans and that, in time, people would hear and understand his words, and that eventually his book *would be* for **'Everyone'**. Denial-free thinkers or prophets such as Nietzsche have historically not been appreciated in their own lifetime, but their honesty has always led the way to a better world for humans. In fact, their honesty is the purest form of love the human race has ever known. The ultimate example of this is Christ, whose words were also not understood in his lifetime, or indeed in the ensuing years; as he said, **'Why is my language not clear to you? Because you are unable to hear what I say… The reason you do not hear is that you do not belong to God** [you live in denial of such fundamental truths as Integrative Meaning]**'** (Bible, John 8:43-47)—and yet he gave humanity a home to live safely in while the corrupting search for knowledge had still to be completed. I will come to the role that religions such as Christianity have played shortly (in ch. 7:15).

646 In summary, for 2 million years women have stood by their men, just as for 8 million years prior to that, men supported their women. With understanding of the human condition now found, men and women can at last stand side by side—the 'war of the sexes' can finally be resolved. Yes, the human journey can have the happy ending we always hoped it would.

647 (A note to the reader: I have taken the description of what happened in the relationship between men and women much further than the adventurous Adolescentman stage because it made practical sense to follow this very important relationship through to its resolution. Given how many

subjects have to be looked at, the presentation in this chapter is bound to be somewhat disjointed; the intention can only be to convey a general appreciation of the sequence of events that have occurred in humanity's incredible journey from ignorance to enlightenment.)

[648]The overall point being made here is that since all forms of innocence unfairly criticised humans during humanity's insecure adolescence, all forms of innocence have been attacked by upset humans during this stage. As stated earlier, animals also fell victim to the human condition; and not only animals, but nature in the broader sense, because it too was a friend and 'ally' of our instinctive soul and, therefore, an 'enemy' of our apparently 'bad' conscious mind. Attacking nature, be it chopping down trees or setting fire to vegetation, brought a retaliatory sense of satisfaction to the upset within us. Even the wearing of dark glasses, ostensibly as sunshades, was often an effort to block or alienate ourselves from the natural world that was alienating us.

[649]Indeed, if we take a moment to extrapolate this situation, we can see that eventually our upset was likely to increase so much, and our associated resentment of any criticism of it would become so great, that disputes with other humans were inevitably going to break out, at which point people would eventually start grievously attacking and even murdering each other, which would lead to, and, of course, has led to, outright, organised warfare—but, as will be explained during humanity's 40-year-old equivalent stage, such extremely destructive behaviour didn't emerge until the latter period of our 2 million years in adolescence.

[650]Of course, the more upset we became, the more we needed ways to escape and relieve the trauma of our condition. And so to compensate for the extremely unhappy state of becoming corrupted, we began to seek out the material rewards of luxury and comfort, with this materialism becoming one of the main driving forces or motivations in life when upset became extreme. The accumulation of wealth and what it could offer us—the land, the staff, big houses, hordes of gold, glittering dresses, sparkling diamonds and shiny, pretentious cars—gave us the fanfare and glory we knew was due us, but which the world in its ignorance would not give us. From being bold, challenging and confrontationist, the heroic 21-year-old eventually became embattled, cynical and exhausted, greatly in need of escapism and relief, and thus an increasingly superficial, material and artificial person. We personally abandoned any idealistic hope of winning the battle to overthrow ignorance as to the fact of our

true goodness and became realists, concerned only with finding material relief and bestowing glory upon ourselves.

651 As mentioned in par. 560, while innocent 'Childmen' were instinctively coordinated and connected, once upset, especially alienation, developed, language became a necessity. With alienation differing from one person to another, the need emerged to try to explain ourselves, to explain why we were behaving differently, in such a seemingly non-ideal manner. In fact, talking became the key vehicle for justifying ourselves, both in our minds and to others. But since we couldn't speak directly about the human condition, or about other people's particular states of alienation without overly confronting and condemning them, stories became a way of passing on knowledge, or what we call wisdom, about the subtleties of life under the duress of the human condition. Much later, with the development of the written word about 6,000 years ago, the fundamental quest for self-justification became greatly assisted because the wisdom acquired during each generation could be more accurately recorded, which meant that quite suddenly the accumulation of knowledge gained real impetus. *But*, it follows that throughout the upsetting journey through humanity's adolescence our increasing need to *somehow* explain and justify ourselves with words, both oral and written, also led to the development and dissemination of all kinds of increasingly sophisticated excuses and lies for our behaviour. The industry of denial became one of the main features of our lives; indeed, the extreme denials that have taken place in science about our species' innocent, upset-free, psychologically secure and happy past bear stark witness to just how sophisticated the art of denial has become.

652 At this point in our journey, other forms of self-expression, such as art and music, became particularly useful because, unlike language and stories, their message wasn't as clear and, therefore, as potentially confronting. Each person could derive as much meaning from the art or the music or even the dance and other cultural rituals as they could personally cope with. On the whole, culture essentially encompassed the various ways people passed on, from one generation to the next, the knowledge they had learnt about living under the duress of the human condition.

653 Although the oldest known cave paintings date back just 35,000 years, archaeologists working in Zambia announced in 2000 that they had found pigments and paint grinding equipment believed to be between 350,000 and 400,000 years old. At the time of the discovery it was reported that

the find showed that **'Stone Age man's first forays into art were taking place at the same time as the development of more efficient hunting equipment, including tools that combined both wooden handles and stone implements...**[and that it was evidence of] **the development of new technology, art and rituals'** (*BBC World News*, 2 May 2000). The British archaeologist Lawrence Barham, a member of the team in Zambia, described the find as the **'earliest evidence of an <u>aesthetic sense</u>'** and that **'It also implies the use of language'** (ibid). As explained, language would have emerged with alienation because people would have then needed some way to account for their unnatural behaviour, and since we can expect alienation to have begun soon after the emergence of *Homo* we can assume that at least a rudimentary language would have been practised by *H. habilis* who emerged approximately 2 million years ago. With regard to other expressions of aesthetic sensitivity—a sensitivity we have had since we became instinctively immersed in love and in tune with all of nature during our love-indoctrinated past, but have not been able to reflect upon or express until our conscious mind and associated state of alienation reached a certain level of development—the oldest musical instruments found so far, phalange (bone) whistles, show that Neanderthals, the early variety of *H. sapiens sapiens*, were making music around 80,000 to 100,000 years ago, while a Neanderthal burial site at the Shanidar Cave in Iraq, estimated to be around 50,000 years old, contains traces of pollen grains, indicating that bouquets of flowers were buried with the corpses. The creative and aesthetic sense of our ancestors of nearly half a million years ago, as indicated by the pigments and paint grinding equipment, suggests that the creative and spiritual sensitivities demonstrated by the Neanderthals were in existence long before their time.

654 The extreme sensitivity that is particularly apparent in the rock paintings of the Bushmen of southern Africa and Australian Aborigines, and in the cave paintings of early humans in Europe, is especially revealing of how much innocence the human race has lost in relatively recent times. The Chauvet Cave in southern France, for example, contains a wealth of cave drawings that date from around 30,000 years ago (some of which are reproduced overleaf) that have inspired such descriptions as **'miraculous'**, **'overwhelming in density, humbling in sophistication, and awe-inspiring in sheer beauty'** (*The Goddess Bites*, accessed Oct. 2011; see <www.wtmsources.com/131>). The drawings are three dimensional, even animated; in short, the animals appear so real it is as if they are alive! You can almost feel what it is like to be those animals, the whole struggle of their lives is revealed.

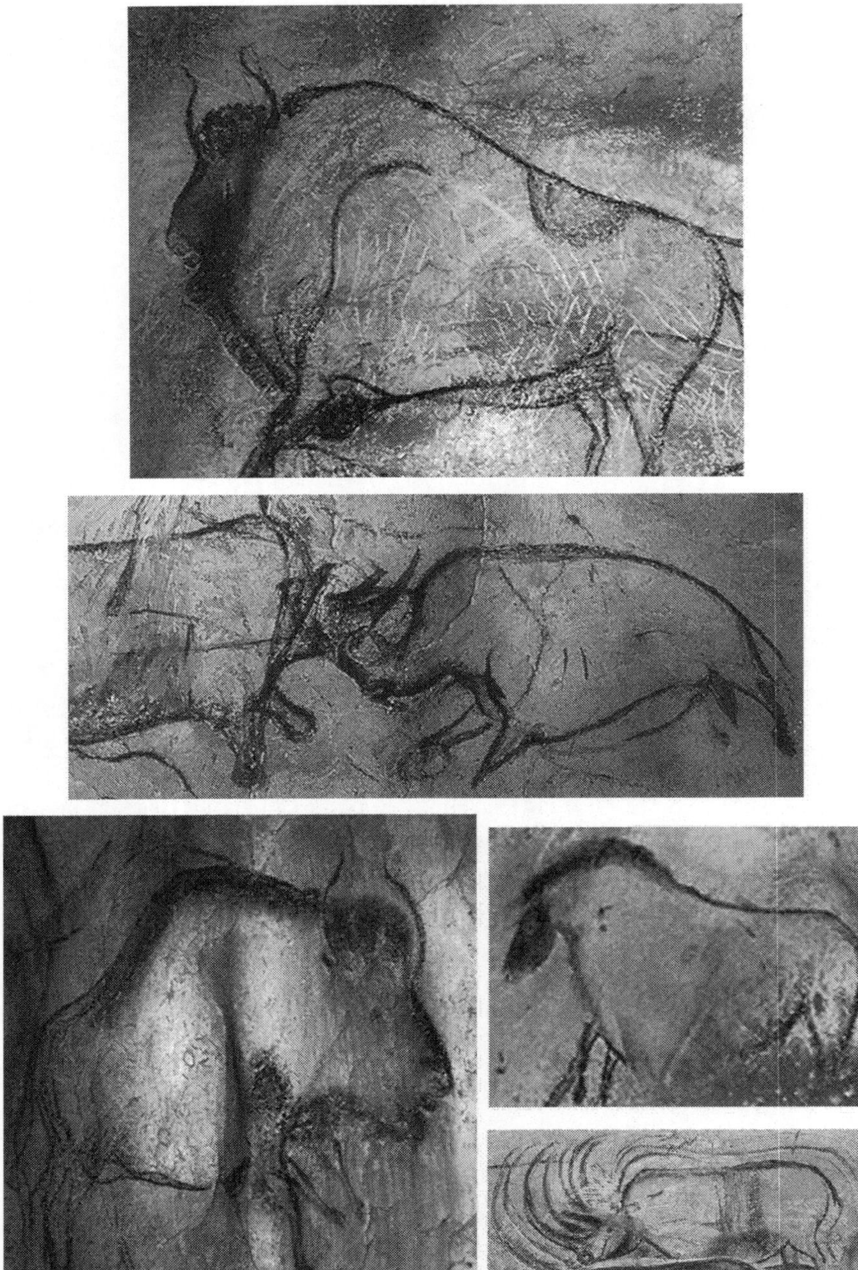

Extraordinarily empathetic renditions of animals in
the Chauvet Cave in southern France, c.30,000 years old

[655] I might mention that I learnt long ago that to draw the little pictures that are included throughout this book, I had to disconnect my conscious mind and just let my instinctive sensitivity express itself, and that if I didn't do that I simply couldn't draw at all. For example, the drawing of the three 'Childmen' happily embracing that I used in ch. 7:4 to illustrate humanity's childhood stage was done so quickly I shocked myself because I could hardly believe that such an empathetic drawing could be produced from an almost instant scribble. At that moment I saw just how much sensitivity we humans once had, and how much alienation now exists within us 2-million-years-embattled humans. Yes, the extraordinary empathy and accuracy of the paintings of animals in the rock and cave paintings shown opposite are similarly incredibly indicative of the amount of sensitivity we humans once had and have since lost; we *truly* are an embattled species now, so worn out, so brutalised, so toughened. How extremely sensitive must early humans have been! Sir Laurens van der Post wasn't exaggerating when he wrote about the relatively innocent Bushmen race that **'He and his needs were committed to the nature of Africa and the swing of its wide seasons as a fish to the sea. He and they all participated so deeply of one another's being that the experience could almost be called mystical. For instance, he seemed to *know* what it actually felt like to be an elephant, a lion, an antelope, a steenbuck, a lizard, a striped mouse, mantis, baobab tree'** (*The Lost World of the Kalahari*, 1958, p.21 of 253).

[656] When all the upset in humans heals—as it now will if the scientific establishment can tolerate and recognise all the denial-free insights in this book—the world is going to open up for us humans. Our long repressed all-loving and all-sensitive original instinctive self or soul is going to come back to the surface. We are going to be able to feel everything around us. We are going to have so much kindness and love and empathy for each other and our fellow creatures because we will, once again, be able to feel everything they are experiencing, including just how embattled the lives of animals are; they suffer *enormously* from the 'animal condition', from the unrelenting need to compete for food, shelter, space and a mate. While, through the nurturing, love-indoctrination process, *our* ape ancestors were able to break free from the tyranny of genes having to ensure their own reproduction, other animals remain stuck in a continuous cycle of competition; unlike humans (and the bonobos, who are in the midst of developing love-indoctrination), other animals can't develop full unconditionally selfless cooperative instincts. And so above

all else, it is this empathy with, this feeling for, the relatively short, brutish, forever-having-to-fight-for-your-chance-to-reproduce lives of animals that those who made these drawings have so sensitively expressed. To use Sir Laurens' words, they **'seemed to *know* what it actually felt like to be'** a bison, horse or rhinoceros. You can sense the whole internal struggle of the animals' lives in these drawings. Their huge chests heave with their brutal and tough battle to survive and reproduce—they are struggling so much to endure their lot it is as if they have asthma! Yes, now that we humans can get over the terrible agony of our 'human condition', we will again be able to empathise with the terrible agony of the 'animal condition'. It's not very nice to have to belt the living daylights out of others to ensure your genes reproduce, let alone other members of your own species—in fact, your cousins, uncles and even your own father! No, it is not at all easy being a non-human animal, and that is an extreme understatement, just as it has not been at all easy being an upset human, which is, of course, another extreme understatement!

657 While the Paleolithic artists clearly weren't as alienated as we humans are today, they were still much, much more alienated than humans *originally* were. I think this is revealed by the fact that these cave artists almost completely avoided depicting humans. For instance, in the entire Chauvet Cave complex there is only one representation of a human, and even that is limited to a drawing of only the lower half of a woman's torso. On the few occasions when these cave artists tried to depict humans they almost invariably ended up drawing stick figures. The human face, in particular, which you would think would be the most interesting and relevant of subjects for these artists to depict, seems to have been totally beyond their ability. It seems clear that the facial expressions of humans were by then so alienated, so devoid of the innocence that they must have once exhibited, that our instinctive self or soul couldn't relate to it; it couldn't, and perhaps didn't want to, draw us. What did R.D. Laing say about our present alienated state: **'between *us* and It [our true selves** or soul] **there is a veil which is more like fifty feet of solid concrete'**. The artist Francis Bacon revealed just how corrupted and alienated upset humans *really* are in his honest painting of the psychologically-contorted-smudged-human-condition-afflicted-face that was included after par. 57. Indeed, the weird, kidney-shaped blob for the human face that the Aboriginal artist drew in the rock painting shown opposite, from Ubirr in the Kakadu National Park in the Northern Territory of Australia, is very Bacon-like!

Revealingly, when I was looking at this painting at Ubirr, which is thought to be some 2,000 years old, I asked a guide, who was accompanying a tour group, whether she thought the reason the paintings of wildlife were so accurate while the paintings of the humans were so pathetic was because we are now too alienated for our soul to be able to empathise with us, the guide, and everyone else, reacted with a real shudder and audible choking noise. What I had said was just too close to the truth.

The author at Ubirr, Kakadu, 2010

Author's photo of fish, spear thrower and tortoise at Ubirr

A 27,000 year old Bushmen rock painting of an eland hunt, Kamberg, Drakensberg Mountains, South Africa

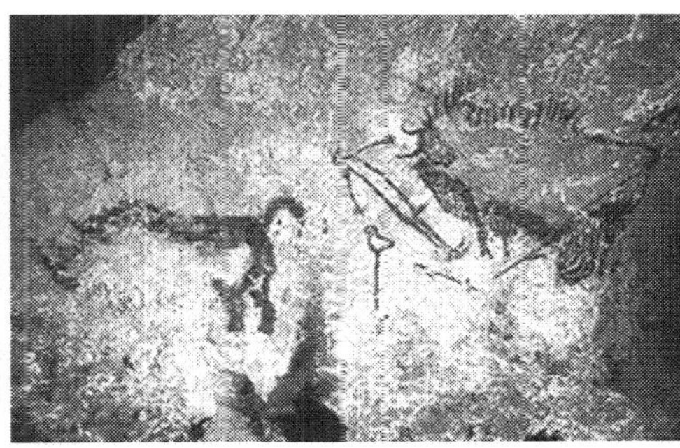

A 17,300 year old cave painting of a wounded bison, a dead man with a broken spear beside him, a bird and what appears to be a rhinoceros in Lascaux, France. While there are many extraordinarily empathetic paintings of animals in the Lascaux caverns there is only this one crude image of a man.

[658] It is truly an insight into how sensitive and loving we humans once were that our instinctive self or soul can't relate to the way we humans are now. Consider the tenderness in the expression on the face of the Madonna in the drawing of the Madonna and child that was included at

the beginning of the infancy stage. My soul drew that—I, my conscious self, had nothing to do with it. Truly, as William Wordsworth wrote, **'trailing clouds of glory do we come, From God** [the integrated, loving, all-sensitive state]**, who is our home'** (*Intimations of Immortality from Recollections of Early Childhood*, 1807). And people say we humans have brutish, aggressive instincts! It's the world we humans *currently* inhabit that is mad. It is just so traumatised with upset that it hasn't been able to deal with the fact that it is deeply, deeply dishonest; horrifically alienated. What did the great Spanish artist Pablo Picasso famously say about his ability to paint: **'It's taken me a lifetime to learn to paint like a child.'** And again, what did R.D. Laing say, **'between *us* and It** [**our true selves** or soul] **there is a veil which is more like fifty feet of solid concrete'**. Turn on the television and find any wildlife documentary and I bet it will show pictures of crocodiles on the Mara River tearing wildebeest apart, or white sharks devouring seals, or snakes striking at the camera lens, or some equally 'brutal' interaction. All the beauty in nature has been reduced to representations of butchery and horror because we humans have become so upset that all we can cope with are pictures of animals 'being' as aggressive as we are—everything else in nature is far too confronting. I have been to natural Africa and seen its spectacle, and the sheer magic of it surpasses all imaginings; it is just achingly beautiful, the most sacred realm on Earth—'spiritual amnesia' are the only words I can think of to describe it and they don't even make sense. In 1992, my partner Annie Williams and I were fortunate enough to join a small reconnaissance party that was being sent in on foot into the northern end of the Tsavo East National Park in Kenya, an area that had been shut off from the public for many years due to the prevalence of dangerously-armed poachers from Somalia. I remember sitting hidden downwind amongst the trees on the banks of the Tiva sand river there and seeing dust rise above the tree line in the shimmering midday heat and then watching as a vast herd of black Cape buffalo, led by an old crooked horn cow, quietly materialised from the bush, cautiously coming down to drink at pools in the river bed. I really felt like a spy in heaven. It was all just unbelievable. Earth at its primal, spiritual, authentic, soulful, magical very best. I think God was there beside us sitting on his heels like a little Bushman smiling at all that he had created. That visit to the Tiva river remains the highlight of my life. With our sophisticated communication technology, why oh why don't we have documentaries sensitively immersing us in all of that.

It is *so* sad. We haven't been able to cope with any truth. Our world has shrunk to the size of a pea. All the beauty and magic that is out there escapes us, we don't see it; worse, *we don't want to see it*. No wonder our soul can't relate to us and just draws stick figures with weird blobs for faces.

My partner Annie Williams and I in Samburu National Park in Kenya in 1992. Those giraffes behind us are just walking around as free as a daisy. In natural Africa, animals like giraffes and elephants and rhinoceroses (well, terrifyingly, there are actually almost no rhinos left!) aren't in cages; there are no fences over there. Animals—and the place is teeming with them, all sorts of weird shapes and sizes—just walk all around the place. It's amazing. They can go wherever they want. They can stop here for a while and then go over the hill if they want to. They just mooch about everywhere; walk around a bush and there is another one, this time with great spiral horns coming out of the top of its head, big eyes locking at you as if to say, 'So, who are you, what's your problem?' '*My* problem! Have you had a look at what's coming out the top of your head!?' It takes some getting used to I can tell you. I don't know who made them all, and was he just having fun making them all in such weird and different shapes—and, more to the point, who let them all out!

659 Yes, we humans now are immensely alienated, extremely psychologically separated from our true self or soul. Similar to what happens when I draw, in my writing I have also learnt to, as I describe it, 'think like a stone', or 'think like a child'—say the simplest, most elementary thought—because I learnt that such a thought will be the most truthful and accurate and accountable and explanatory. Absolutely every time

I encounter a problem I have to solve in my thinking about the human condition I go into a routine where I say to myself, 'Just go into yourself and think like a stone, just let the truth come out that's within and you will have the answer.' Basically, I learnt to trust in and take guidance from my truthful instinctive self or soul. I learnt to think honestly, free of alienated, intellectual bullshit, and all the answers, all the insights that I have found, and there are many hundreds of them, a breakthrough insight in almost every paragraph, were found this way. I have so perfected the art of thinking truthfully and thus effectively that you can put any problem or question in front of me to do with human behaviour and I can get to the bottom of it, answer and solve it. It has been astonishing to me to watch my mind work, the freedom it has and where it is capable of going in its thinking. It wears me out keeping up with it. This wearing out problem is especially so because there is so much suffering in the world that simply has to be brought a stop to. Yes, I know that every sentence I write is truth-laden, in complete contrast to the billions of sentences being churned out every second everywhere else on Earth. It is the innocent instinctive child in us that knows the truth. Christ, as usual, put it perfectly when he said, **'you have hidden these things from the wise and learned, and revealed them to little children'** (Bible, Matt. 11:25). A comment that was mentioned in par. 130 by George Seaver reiterates what I have just said about natural thinking: **'The ultimate thought, the thought which holds the clue to the riddle of life's meaning and mystery, must be the simplest thought conceivable, the most natural, the most elemental, and therefore also the most profound.'** Yes, as Plato was recorded as saying in par. 518, when we use our intellect with its preoccupation with denial **'for any inquiry…it is drawn away by the body into the realm of the variable, and loses its way and becomes confused and dizzy, as though it were fuddled** [drunk]**…But when it investigates by itself** [free of intellectual denial/bullfuckingshit]**, it passes into the realm of the pure and everlasting.'**

660 The human condition has certainly been a cruel incursion. Firstly, it was bad enough to have acquired a fully conscious brain, the marvellous computer we have on our heads, and not be given the program for it and instead be left to wander this planet searching for that program/understanding in a terrifying darkness of confusion and bewilderment, most especially about our worthiness or otherwise as a species, but to *then* be disconnected from access to the integrative, Godly, cooperatively

orientated, all-loving and all-sensitive, ideal world of our original instinctive self or soul—having to block it out because it unjustly condemned us—means we have been enduring an *extraordinarily* lonely, sad, soul-destroyed, alienated existence! We have had to put on a brave, positive face to carry on with our horribly upsetting job of searching for knowledge, ultimately for self-knowledge, but that is the true description of our existence during that search. It follows then that it became a matter of great urgency for the increasingly upset human race to find ways to cope with such a torturous and lonely existence.

661 While upset was rapidly increasing through humanity's adventurous adolescence stage and beyond, we eventually became so horrifically alienated from our all-loving and all-sensitive true self or soul that a way simply had to be found to reconnect with it. We had to find a way back to some purity and sanity, but in a safely, non-confrontational manner, and one of the ways we managed to do so was by creating one of the earliest forms of religion, which was animism or nature worship—religion being the strategy of putting our faith in, deferring to, and looking for comfort, reassurance and guidance from something other than our overly upset and overly soul-estranged conscious thinking egoic self. Unlike our upset soul-destroyed self, the natural world remained in an innocent state, and since nature was also associated with our original instinctive self because our species grew up with nature, it could also reconnect us to the innocent, true world of our soul. So, despite our upset state's often violent repudiation of nature's condemning innocence, nature *could* still link us back to repressed 'spiritual', soul-infused sensitivities, feelings and awarenesses within us that we had lost access to. Again, we see the two-sided aspect of innocence: it could condemn us and hurt us terribly, but it could also heal and inspire us.

662 Another way that eventually developed to counter the loneliness of our situation, and this was also one of the earliest forms of religion, was ancestor worship. Having managed to survive our mind's loneliness and our soul's estrangement, our ancestors were a source of great reassurance and comfort. In our uncertainty and distress, we could look to them for the hope that we too might survive the horror of life under the duress of the human condition. We could look to them for 'spiritual' guidance, for inspiration for our troubled minds. If we tried to imagine how they coped and what they would have done in situations that we now faced, we could be inspired to reach potentials within ourselves that our troubled

minds might not otherwise have allowed us access to. By revering them and enshrining their memories, our ancestors could remain a presence in our lives to look after and guide us.

663 As to when nature and ancestor worship was first practised, we can only guess, but the pollen grains that were found in the Shanidar Cave burial in Iraq indicate bouquets of flowers were buried with the corpse, a display of affection that shows humans were deeply sensitised to the horror, loneliness and sadness of life under the duress of the human condition some 50,000 years ago, so the need for spiritual sustenance must be extremely ancient.

664 Alongside the development of our spiritual needs, we were also becoming more technologically advanced. For instance, tools including sharpened stones, choppers, hand axes and scrapers, cudgels, spears, harpoons and bone needles appear in the archaeological record from 3 million years onwards, while there is evidence that *H. erectus* (who lived some 1.5 to 0.5 million years ago) made refined tear-drop shaped flint axe heads and that even the earliest of this species were using fire (as indicated by the remnants of hearths at Koobi Fora in Kenya). However, it is only within the final 14,000 of those 2 million years that the *most* dramatic improvements occurred. It was during this period that the bow and arrow, fish basket traps and crude boats first appeared, while the practice of agriculture and the domestication of animals, which both began around 11,000 years ago, brought with it the production of earthenware pottery, looms, hoes, ploughs and reaping-hooks. (This acceleration of technology can be seen in the relatively swift succession between the great ages that define our modern history, with the Stone Age being replaced around 5,000 years ago by the Bronze Age, which in turn was replaced around 3,000 years ago by the Iron Age.)

665 It needs to be emphasised that throughout these epochs of time the whole development of upset was being driven by increasing levels of intelligence; the more intelligent we were, the more we searched for understanding, and the more upset we became—and with each new level of upset a new psychological and accompanying physical existence and state emerged, including increased alienation. The following two graphs chart the psychological journey that humanity has been on, with the top graph charting the development over time of mental cleverness, as indicated by brain volume, and the bottom graph charting the development of cooperativeness or integration.

The Development of Mental Cleverness

(Brain volume is used as a guide to mental cleverness)

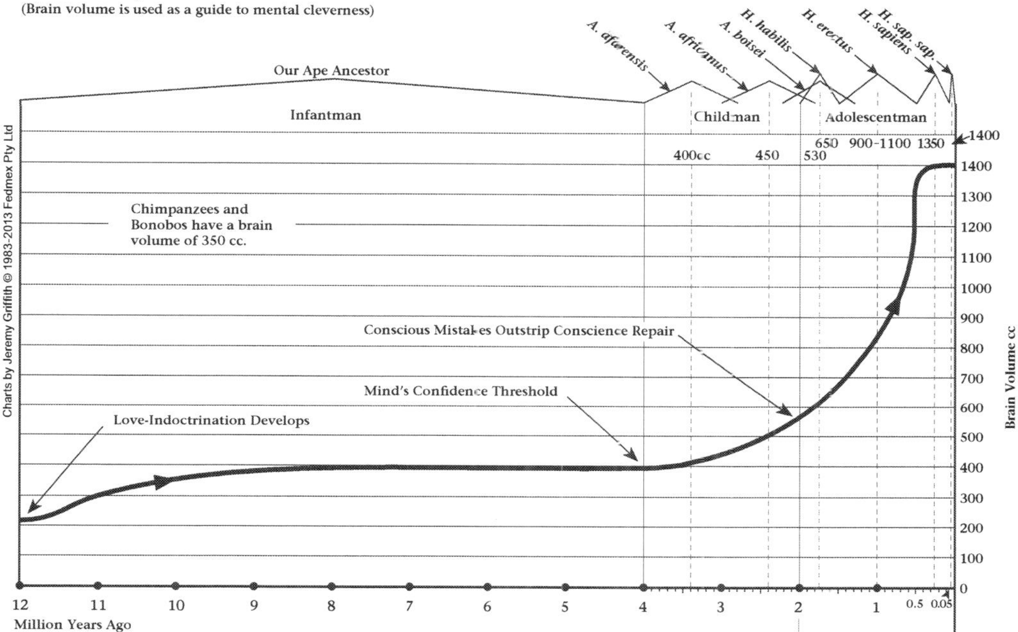

The Development of Integration

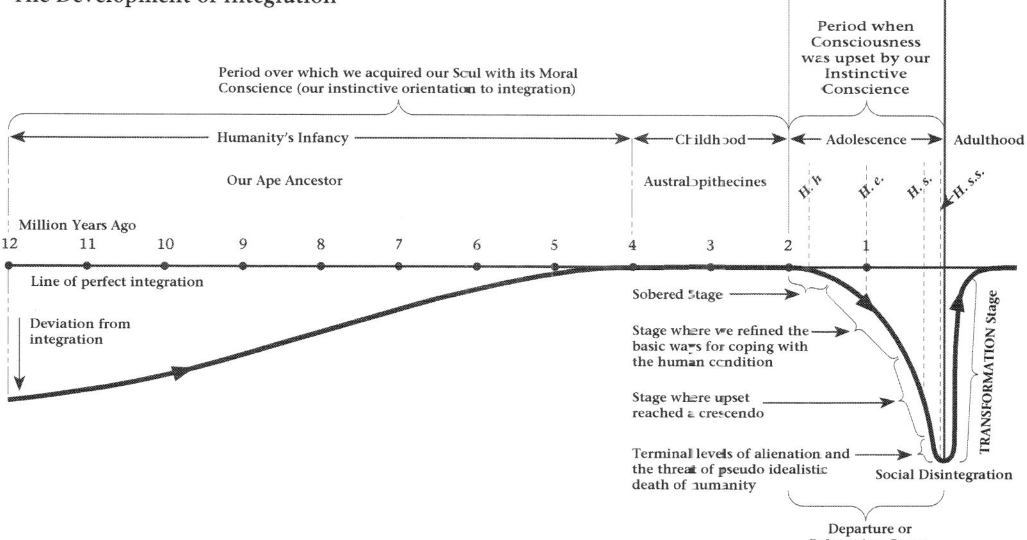

666 We can see that while the brain size of 'Childman' (the australopithecines) was not much bigger than 'Infantman' (represented today by species such as chimpanzees and bonobos), a sudden increase in brain size occurred with the emergence of the first 'Adolescentman', *H. habilis*, when the need to think and understand began in earnest as a result of the emergence of the dilemma of the human condition. This dramatic growth continued through 'Adventurous Adolescentman' (*H. erectus*) and 'Angry Adolescentman' (*H. sapiens*) before finally plateauing with 'Pseudo Idealistic and Hollow Adolescentman' (*H. sapiens sapiens*). Anthropologists have long wondered why this growth stopped in the last 50,000 years or so. The reason is that in 'Pseudo Idealistic and Hollow Adolescentman' a balance was struck between the need for cleverness and the need for soundness; between knowledge-finding yet corrupting mental cleverness and conscience-obedient yet non-knowledge-finding lack of mental cleverness, with the average IQ today representing that relatively safe conscience-subordinate, not overly upset, not too selfish, egocentric and dysfunctional, compromise. Conscious mental cleverness is what caused us to challenge our instincts, the result of which was we became corrupted in soul, so it follows that the more mentally clever we humans became, the more corrupted in soul we also became, and that eventually we became too clever and too corrupted. Cleverness and alienation have been related. That has been the elementary truth about the human condition.

667 The bottom graph, which indicates the development of cooperativeness or integration, shows that by 5 million years ago nurturing had enabled our ancestors to live in an utterly cooperative state. However, with conscious self-management, and with it the upsetting battle of the human condition, becoming fully developed some 2 million years ago, we see that the graph marks a rapid increase in upset from that time to the present, where we now face the prospect of terminal levels of alienation and social disintegration.

Chapter 7:12 <u>Angry Adolescentman</u>

The species: *Homo sapiens* — 0.5 million (500,000) to 0.05 million (50,000) years ago
The individual: 30 to 40 years old

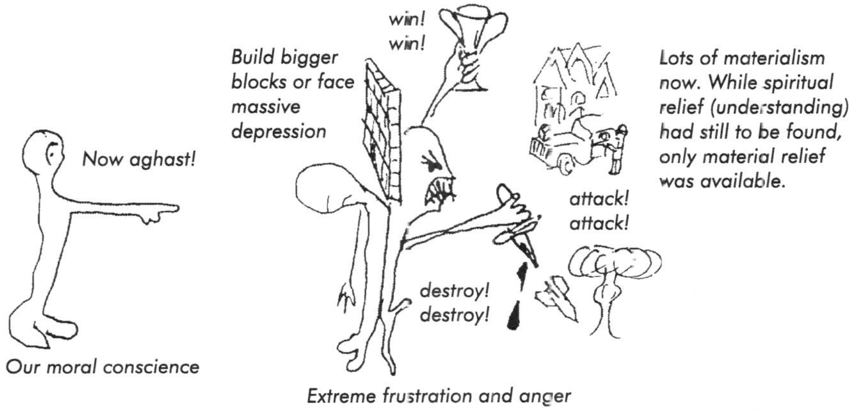

[668]During the Angry Adulthood Stage (of Humanity's Adolescence) we encountered the reality, frustration and anger of trying but failing to defeat the ignorance of our idealistic instinctive self or soul and had to impose upon ourselves <u>Self Discipline</u> in order to contain or civilise our now overly upset state.

[669]Throughout our 20s we individually, or, in the case of humanity, *H. erectus*, settled into the long, corrupting journey to find understanding, ultimately understanding of why we became corrupted in the first place. But, tragically, the more we searched for knowledge, the more upset we became, and the more upset the human race as a whole became, and the more new generations had to contend with that ever-accumulating upset. It was an extremely upset-compounding situation. As the just included graph charting humanity's increase in upset over the last 2 million years shows, for the first three-quarters of the journey (to the end of 'Adventurous Adolescentman'/*H. erectus*' reign), the rate of increase in upset, while beginning to accelerate, was not yet extreme. However, in the last quarter of that time period (during the reign of 'Angry Adolescentman'/*H. sapiens*) the graph descended markedly, and then, in the final 50,000 years (during the reign of 'Pseudo Idealistic and Hollow Adolescentman'/*H. sapiens sapiens*), it entered into free fall, with upset beginning to

compound at an *extremely* rapid rate—a rate that can only be brought to an end by the rise of the human-condition-understood-and-ameliorated, transformed 'Adultman', or what could be termed 'Triumphantman', or even Integrative-Meaning-compliant 'Godman'.

670 A contributing factor to the speeding up of this progression in upset was the hardship and confinement of life throughout the four great ice ages that occurred during the Pleistocene epoch, the period from 1.8 million years ago to 10,000 years ago. These ice ages greatly contributed to the increase in upset because, in forcing close habitation between people of varying degrees of upset, they dramatically accentuated the difficulties encountered by humans coexisting under the strain of the human condition, to the point where life today, towards the bottom of the graph, has become so difficult that even coupling has proved untenable for many, with marriage breakdown a common occurrence.

671 The closer humans lived during humanity's adolescence and/or the more difficult the living conditions, the greater the occurrence and spread and thus increase in upset. Innocence doesn't last long in New York's Times Square or Sydney's Kings Cross where drug pushers, prostitutes, muggers and beggars work the streets. And as those from cold climates will attest, winters *are* particularly confining and testing, so each great ice age did, in effect, represent one *very* long, trying winter. It is not surprising then that out of the hardship of each of the great ice ages came the next more upset/soul-exhausted/embattled/alienated stage of humans. From the rigours of the first great ice age, called the Günz Ice Age, came the flowering of *H. erectus*. *H. sapiens* emerged after the second ice age, the Mindel Ice Age, while Neanderthal man, a precursor of *H. sapiens sapiens*, appeared after the third ice age, the Riss Ice Age. *H. sapiens sapiens* emerged after the Würm Ice Age, the fourth ice age. Each ice age also contributed significantly to the culling of the human race in terms of humans' ability to adapt to life under the duress of the human condition. This is because as upset increased throughout humanity's adolescence many individuals must have, in effect, quit the great battle humanity was waging against the ignorance of our instinctive self or soul through their inability to withstand the degree of compromise to their soul that was increasingly being demanded of them, leaving only the most courageous, determined and enduring—but, it follows, also the most soul-destroyed. Sir Laurens van der Post once described how a member of the relatively innocent Bushmen race (*relatively* because they

are still members of the extremely upset stage of humanity, *H. sapiens sapiens*) found it impossible to cope with having his innocent, natural spirit compromised: **'You know I once saw a little Bushman imprisoned in one of our gaols because he killed a giant bustard which according to the police, was a crime, since the bird was royal game and protected. He was dying because he couldn't bear being shut up and having his freedom of movement stopped. When asked why he was ill he could only say that he missed seeing the sun set over the Kalahari. Physically the doctor couldn't find anything wrong with him but he died none the less!'** (*The Lost World of the Kalahari*, 1958, p.236 of 253). Sir Laurens was more specific when (as mentioned in par. 588) he stated that **'mere contact with twentieth-century life seemed lethal to the Bushman. He was essentially so innocent and natural a person that he had only to come near us for a sort of radioactive fall-out from our unnatural world to produce a fatal leukaemia in his spirit.'** Given how **'radioactive'** our present **'unnatural world'** is for the innocent souls of children, it's little wonder there is an epidemic of distressed, non-coping, dissociating symptoms like ADHD and autism breaking out amongst children now. Yes, the human race has suffered 2 million years of soul-destroying, toughening, upsetting, anger-egocentricity-and-alienation-producing adaption to the horror of life under the duress of the human condition! But just how toughened the human race has become is now hidden under layer upon layer of self-restraint, or, as stated in par. 164, what we call 'civility'. This restraining, civilising process will be elaborated upon shortly, but the point being made here is that beneath our facade of restraint and manufactured positiveness, which was so necessary to cope with the horror of the human condition, lies a highly genetically toughened and resilient individual.

672 But to return to humanity's journey through its adolescence, by the age of 30 in the case of the individual, or by some half a million years ago in the case of humanity, the exponential increase in upset meant that the levels of upset, namely anger, egocentricity and alienation, had exceeded the inflection point on the graph charting our upset, and had entered the stage where upset increased both dramatically and *rapidly*, which it has continued to do ever since. So while all the adjustments that were made during the Adventurous 20s had served us well—both as individuals and, in the case of humanity, as members of *H. erectus*—there was now an urgent need to take more specific measures to manage the new extreme levels of upset.

[673]If we consider what happened to the 21-year-old more closely we can see why management of upset had become such a serious matter, for despite their bravery and sheer optimism, it wasn't long before the reality of, in the case of the resigned, trying to win the battle of proving you were good and not bad—or, in the case of the unresigned, trying to reform upset behaviour—started to sour. Gradually he or she came to experience and appreciate just how truly difficult it was to self-manage and contain upset without the ameliorating understanding of that upset.

[674]The problem for those who were resigned was the harder they fought to validate themselves, the more criticism they attracted from their idealistic soul, and thus the more upset they became. Also, throughout their 20s, they were increasingly encountering the upsetting difficulty of trying to survive and compete alongside *other* embattled humans who were *also* trying to prove their worth. The resulting compounding of upset meant that by the time they reached 30 they were becoming very frustrated and angry, and by the time they reached their mid-30s they were becoming seriously upset, embattled individuals. While 20-year-olds were naive about the difficulties of living under the duress of the human condition, 30-year-olds had become realists about such an existence. Rod Stewart's song *I Was Only Joking* contains lyrics that vividly describe the reality check of reaching 30: **'Me and the boys thought we had it sussed. Valentinos all of us...running free, Waging war with society...But nothing ever changed...What kind of fool was I. I could never win...Illusions of that grand first prize, are slowly wearing thin...I guess it had to end'** (lyrics by Gary Grainger & Rod Stewart, 1977).

[675]While our inability—until now—to defend our corruption has meant that it hasn't been possible to admit it, the following Japanese proverb does, at least, acknowledge the stages of its development: **'At 10 man is an animal, at 20 a lunatic, at 30 a failure, at 40 a fraud and at 50 a criminal.'** But with understanding of the human condition now found we can finally decipher these stages. Ten-year-olds were **'animals'** in the sense that they had yet to learn any methods of restraint for the upset that they were beginning to experience from the frustrations and agony of the human condition. Twenty-year-olds—and young men in particular—were **'lunatics'** in the sense that they were swashbuckling cavaliers who deludedly believed they could take on and overthrow the ignorant world of the soul. Thirty-year-olds (and, again, men in particular) were **'failures'** in the sense that, although they were still determinedly trying to defy the inevitable, they were being forced to accept that the corrupting life of seeking power, fame,

fortune and glory was not going to be a genuinely meaningful and thus satisfying way of living. As will be described shortly when the 40-year-old stage is explained, at this age men in particular were **'frauds'** in the sense that they had become so corrupted and disenchanted with their efforts to 'conquer the world' that they suffered a 'mid-life crisis'—a crisis of confidence that resulted in their decision to take up support of some form of 'idealism' to make themselves feel better about their corrupted condition. Having had enough of the critically important, yet horribly corrupting, battle to champion the ego over soul, they effectively changed sides to become 'born-again' supporters of the soul's 'idealistic' world. This 'born-again' conversion to taking up support of some form of idealism made them **'frauds'** because in doing so they were deluding themselves that they were at last on the side of good when, in truth, they were working *against* good, in that the good, right, responsible path depended on defying and defeating—not supporting—the ignorant 'idealistic' world of the soul. They were being *pseudo* idealistic, not *genuinely* idealistic. And as will be described when the 50-year-old stage is explained, at this age men in particular were **'criminals'** in that they had become so disillusioned with the extreme dishonesty of the born-again state that many returned to the battle of championing the ego over ignorance, but were, by this stage of their personal journey, *so* deeply upset that they were extremely angry and cynical about life—basically, they knew they were beaten on every front and *had* become bitter and vengeful **'criminals'**.

676 To return, however, to the 30-year-old stage or, in the case of humanity, the life of *H. sapiens*, we can see that this period was characterised by extreme frustration and anger. Thirty-year-olds/*H. sapiens* had entered the rapidly deteriorating stage in the development of upset where they were brought into contact with the destructive and depressing horror of their **'failure'**, of being either, in the case of the resigned, excessively upset, or, in the case of the unresigned, having had their innocence so destroyed that they too had *also* become overly upset. At this stage in the human situation and journey, upset was becoming overwhelming *everywhere*.

677 It was at this point that the radical measures alluded to earlier had to be implemented to contain the growing upset in the world, with the first solution being, as mentioned, to practise <u>self discipline</u>.

678 Fully aware that upset was not desirable we, as individuals, had been trying to, with varying success, practise self-restraint of our upset ever since it first appeared in our childhood. But during our 30s, when upset

started to become seriously destructive of the fabric of our society, self discipline became an *essential* part of our behaviour, something that everyone *had to* make sure they practised. And so we learnt to manufacture a calm, controlled, even compassionate and considerate exterior, and to conceal the real extent of our, by now, inner savage fury from being so unjustly condemned by the Godly ideals of life. We *civilised* our upset; we brought it under control. Since this self discipline, and its civilising effect, has been the primary way of managing our extremely upset state and has been practised since time immemorial, it has become, to a large degree, an automatic, instinctive element of human behaviour, so much so that we now hardly notice we are practising it—the consequence of which is that we are barely aware of just how upset we all really are underneath our restrained exteriors. Our civility has disguised the volcanic upset that exists within us, both as individuals and as a species, from living for so long with the injustice of being condemned as evil, bad and worthless when we intuitively knew we weren't but couldn't explain why we weren't. As was mentioned in par. 159, the writer Morris West offered a rare honest insight into the extent of the upset that exists in *all* humans today when, in his memoir, he confessed that **'The disease of evil** [now able to be understood as upset] **is pandemic; it spares no individual, no society, because all are predisposed to it...I know that, given the circumstances and the provocation, I could commit any crime in the calendar.'**

679 And not only did civility conceal the extent of our anger and egocentricity, it also masked the extent of our alienation—the extent of our estrangement from our original, upset-free, happy, innocent soulful selves. In order not to be overcome by the true negativity of life under the duress of the human condition we have had to, as it's said, 'put on a brave face', 'keep our chin up', 'stay positive', and 'keep up appearances'. And so we manufactured smiles and politely greeted acquaintances with 'Good morning' and asked 'How are you?' and talked about totally non-confronting subjects, such as the weather and sporting results. But while such civility and positivity made living together possible, it was an extreme form of pretence—of being what we were not. But, in turn, although this falseness *was* highly corrosive of any young unresigned innocent looking on who found it unbearably **'fake'** and **'phony'**, it was far less destructive than allowing our real upset to express itself and has, therefore, been a very necessary and effective tool. But after millennia of use, our civilised facades now hide the extent to which we have

blocked out the truth of our upset, corrupted, alienated condition—so when, for instance, we donned scary masks, items that have been used in the ceremonies of almost all cultures, we were, ironically, not disguising who we were but actually exorcising our *real* upset self; we were being honest about ourselves; we were admitting that '*This* is what I am *really* like, this is who I've *become*.'

Bhairav Mask, Nepal

Maori Koruru (gable mask), New Zealand c.1880, Peabody Essex Museum

680 Yes, R.D. Laing certainly spoke the truth about just how corrupted our species has become when he said, **'The condition of alienation…is the condition of the normal man…between *us* and It** [**our true selves** or soul] **there is a veil which is more like fifty feet of solid concrete.'**

681 At this point, it needs to be emphasised that adopting self discipline in our/humanity's 30s did *not* mean we/the human race had stopped our/its corrupting search for knowledge—we had just decided to try not to allow any expression or manifestation of the effects of that search, of our corruption, to show. However, when the 40-year-old stage is explained in more detail shortly we will see that when upset developed even further some individuals were, as just mentioned, forced to abandon, and even side against, the corrupting search for knowledge in a far more drastic attempt to stem their ever-increasing upset by becoming 'born again' to pseudo idealistically supporting some form of idealism—this being the **'fraud'** stage that featured in the aforementioned Japanese proverb.

682 Again, the overall essential feature of the human journey since we first became conscious is that of the accumulation of knowledge at the expense of our innocent soul—the more we searched for knowledge, the more upset and soul-destroyed we became. Certainly we learnt to restrain our upset, civilise it, but underneath that disguise we were becoming more and more angry, egocentric and alienated. This means that while children, adolescents and young adults—and their early human equivalents—could at times behave very angrily and aggressively (such as children deliberately torturing insects by burning them, or when their disagreements turn into rowdy, physical altercations), older adults who had learnt to civilise and thus hide their upset were actually far more upset angry, egocentric and alienated than those younger than themselves. Throughout history, older and thus more civilised people and races have misused this *appearance* of being 'better behaved' to denigrate other more innocent people and races by referring to them as 'savages'—or 'barbarians', as the Romans did when they spoke this way about the northern Scandinavian and Germanic tribes (such as the Goths, Vandals, Franks and Lombards). The adventurous soulful soundness, enthusiasm, vitality and energy of these supposedly backward, savage, primitive, 20-to-30-year-old equivalent northern barbarians was apparent in their ability to conquer the more technologically advanced but more soul-exhausted southern races who lived around the Mediterranean—including the Romans—during the fourth to sixth centuries AD. Their victories, however, do not mean that all conquerors were more innocent than those they conquered; when humans become extremely upset—when they reach the 50-plus-year-old equivalent **'criminal'** stage—they could go on extremely angry rampages, murdering everyone in sight. It is this variety of extreme upset that has characterised most of the warfare that has occurred in the last thousand years of human history. There is a very big difference between a high-spirited Viking-like adventurous mindset and the massively angry and massively egocentric Genghis Khan/Napoleon/Hitler-like vengeful, conquering and murdering psychopathic mindset. The true story of the ever-increasing levels of upset anger, egocentricity and alienation in humans has not been told, but with the upset state of the human condition now explained and defended, it can at last be revealed—and will be shortly when the final 200 years stage of humanity's journey from ignorance to enlightenment is presented (in ch. 7:16).

---

[683] It needs to be explained here how adventurous 20-to-30-year-old equivalent races could be said to exist during the last 2,000 years when that period is described later as being part of the born-again Pseudo Idealistic 40-to-50-year-old **'fraud'** stage and the horrifically angry, punch-drunk, bitter and vengeful Hollow Adolescentman 50-plus-year-old **'criminal'** stage of humanity's maturation. The answer is that what is being referred to here by 'adventurous 20-to-30-year-old equivalents' indicates another level of refinement within the already established stages of maturation. To elaborate, while the first T-model Ford car (a development that replaced the horse and buggy form of transport) had all the basic elements of a car in place, that didn't mean those elements could not become much more refined over time, as in the variety of cars we see today—and even *highly* refined as in the form of a Ferrari. In the same way, the relatively innocent hunter-forager Bushmen people who live in the Kalahari desert today, for example, have all the basic adjustments in place for managing extreme upset. They are, for instance, civilised, instinctively restrained from living out all their upsets; they don't generally attack when they feel frustrated and angry. They have a form of marriage to artificially contain sexual adventurousness. They clothe their genitals to dampen lust. The women love to wear adornments such as jewellery; they are adapted to being sex objects. The men love hunting animals; they find relief from attacking innocence. They employ fatigue-inducing dance to access their repressed soul. In short, they are members of the Pseudo Idealistic and Hollow Adolescentman stage of all humans living today. *But*, while they have these basic adjustments for managing extreme upset firmly in place, they are still a *relatively* innocent race compared with other more human-condition-embattled-and-adapted races living today—they could be described as a 15-year-old equivalent variety of Pseudo Idealistic and Hollow Adolescentman. In the same way, the northern 'barbarians' were members of the Pseudo Idealistic and Hollow Adolescentman stage but were still *relatively* innocent compared with other more human-condition-embattled-and-adapted people—they were 20-to-30-year-old equivalent varieties of Pseudo Idealistic and Hollow Adolescentman. We are talking about levels of refinement occurring *within* the main stages of refinement. It should be emphasised that while we couldn't explain and defend the upset state of the human condition we couldn't afford to differentiate individuals, races, genders, generations, countries, civilisations and cultures according to how upset they were

because it would have left the more upset condemned as bad, unworthy and inferior. It would have led to unfair, destructive and dangerous racist, ageist and sexist prejudice and discrimination against the more upset—and so a dishonest attitude of not allowing any differentiation was maintained.

684The truth is, the only significant difference between humans—the acknowledgment of which makes it possible to truthfully explain and understand much of human behaviour—is the difference in upset anger, egocentricity and alienation between individuals, races, genders, generations, countries, civilisations and cultures, but until the human condition was explained and upset was defended as a good, heroic state we couldn't admit and talk about that all-important, clarifying difference.

685Some appreciation of just how condemning innocent races have been of those more upset can be gained through looking at the determination with which mechanistic science has sought to deny that so-called 'primitive' peoples such as the Bushmen, Australian Aborigines and the Yanomamö of South America are in any way more 'innocent' than the more alienated races. For example, as discussed in pars 100-103, to support the idea of a history of **'universal and eternal'** warfare, E.O. Wilson portrays Bushmen and other 'primitive' societies as **'violent'** and **'aggressive'**, even saying **'Rousseau claimed** [that humanity] **was originally a race of noble savages in a peaceful state of nature, who were later corrupted...** [but what] **Rousseau invented** [was] **a stunningly inaccurate form of anthropology'** (*Consilience*, 1998, p.37 of 374). In contrast, the honest researcher Elizabeth Marshall Thomas wrote that in her and her mother's (Lorna Marshall's) accounts of the Bushmen in the 1950s, **'we both emphasized the absence of violence and competition. Indeed, we were struck by it...The relatively few outbreaks of violence seemed isolated and were discussed over and over, since they caused such distress'** (*The Harmless People*, 1989. p.286 of 303). As explained above, the acknowledgment of the existence of relative innocence in one race would unfairly condemn the more upset, less innocent races as 'bad', unworthy and inferior, so such reports of the relative innocence of the Bushmen could not be tolerated. Marshall Thomas recounts how mechanistic science rapidly mobilised to neutralise the threat her reports represented: **'In the ten to twenty years after we started our work, many academics developed an enormous interest in the Bushmen. Many of them went to Botswana to visit groups of Kung Bushmen, and for a time in Botswana,**

the anthropologist/Bushman ratio seemed almost one to one. Yet although the investigators were numerous, the range of some of their investigations seemed narrowed to an emphasis on questions of violence and aggression' (ibid. p.284). Yes, to escape the agony of the human condition, some excuse *had to be* found—*some* evidence of aggression had to be identified and then a case determinedly built around it, regardless of how transparently false the case really is! As has been repeatedly pointed out, when the need for denial is critical any excuse will do, and the art of denial is to then stick like glue to that excuse because doing so saves you from suicidal depression.

686 One of the more prominent anthropologists whose **'investigations** [into the Bushmen] **seemed narrowed to an emphasis on questions of violence and aggression'** was Melvin Konner, who stressed **'findings that seemed to confirm what might be called the darker side of !Kung life'** (Melvin Konner & Marjorie Shostak, 'Ethnographic Romanticism and the Idea of Human Nature', *The Past and Future of !Kung Ethnography*, eds Megan Biesele with Robert Gordon & Richard Lee, 1986, p.73 of 423). When Konner wrote a clearly biased review of Marshall Thomas's 2007 book *The Old Way: A Story of the First People*, Marshall Thomas felt compelled to respond, writing that **'the moment I saw Konner as the reviewer, I knew we were back where we started. I measured the length of his review—141 inches or 11¾ feet in all—and saw he was averaging four attacks per foot of column. And in the barrage, I'd say only one criticism had substance. Even then he distorted what I'd said'** ('Response to Dim Beginnings', *The New York Review of Books*, 29 Mar. 2007). Another who **'attack**[ed]**'** Marshall Thomas was the primatologist Richard Wrangham (who, as described in par. 390, put forward the Chimpanzee Violence Hypothesis), who accused Marshall Thomas as having **'conjured'** the idea of **'Peaceful primitives'** (*Demonic Males*, 1997, p.76 of 350).

687 And just as Marshall Thomas was attacked for her honest description of the peaceful nature of Bushmen society, so too was Sir Laurens van der Post for his recognition of their relative innocence. A prime example of this persecution occurred after Sir Laurens' death, when the journalist J.D.F. Jones wrote a book that set out to denigrate Sir Laurens as a charlatan, with one of the focal points of his attack being Sir Laurens' depiction of the Bushmen; for instance, he accused Sir Laurens of having **'a romantic and no doubt inaccurate portrait of this dying social group'** (*Teller of Many Tales*, 2002, p.230 of 528). Jones' deep allegiance to the world of denial is also apparent in this emotionally charged comment

he made on the topic: **'the academic experts on the Kalahari** [**Bushmen**] **are absolutely berserk with rage about the things he** [Sir Laurens] **said, because, if you read *The Lost World of the Kalahari*, you must not believe that this is the truth about the Bushmen; it's not'** (*Late Night Live*, ABC Radio, 25 Feb. 2002). Unable to defend our immensely corrupted human condition, to have it exposed naturally made humans **'berserk with rage'**. And just as Marshall Thomas pointed out the superficial, mechanistic, confrontation-avoiding studies anthropologists were making of the Bushmen, so too did Sir Laurens, who wrote that **'It seemed a strange paradox that everywhere men and women were busy digging up old ruins and buried cities in order to discover more about ancient man, when all the time the ignored Bushman was living with this early spirit still intact. I found men willing enough to come with me to measure his head, or his behind, or his sexual organs, or his teeth. But when I pleaded with the head of a university in my own country to send a qualified young man to live with the Bushman for two or three years, to learn about him and his ancient way he exclaimed, surprised: "But what would be the use of that?"'** (*The Lost World of the Kalahari*, 1958, p.67 of 253).

688 Again, while there *is* violence in primitive peoples, the true explanation for the aggression apparent in their societies is that while they are undoubtedly more innocent than the majority of humans in the world today, they are, as mentioned above, still members of the extremely upset stage of humanity, *H. sapiens sapiens*, and are, therefore, *nowhere near* as innocent as humans were some 2 million years ago when the battle of the human condition first emerged. Moreover, while basic levels of restraint are instinctive in primitive hunter-forager people such as the Bushmen, as will be explained in ch. 7:16E, they do not possess the more sophisticated levels of self-discipline that more upset races adopted following the advent of agriculture and herding some 11,000 years ago, and which has subsequently become, to a degree, instinctive. As a result, to draw upon data on homicide rates, as E.O. Wilson was shown to do in par. 101, where he equated Bushmen homicide rates to those present in the more upset-populated cities of Detroit and Houston, and use that comparison to argue that more primitive peoples are not more innocent than more alienated races, is to totally ignore the effect increasing levels of restraint have on upset behaviour. As any mother will attest, a nine-year-old child is more innocent than an adult and yet, as was described in ch. 7:7, during the 'naughty nines' phase they will lash out at the world in a way that a more restrained or 'civilised' adult would not.

[689]The effect of restraining violence was well demonstrated by the successful 'Iroquois Confederacy' of the North American Indians. As will be described shortly in par. 728, by the time Europeans arrived in North America, a grand league of American Indian tribes had been established to prevent, through adherence to certain restraining rules that were enforceable through punishment, the endless rounds of payback warfare that had been occurring between and within the tribes. The absurdity of evaluating a peoples' level of innocence through their display of violence is apparent if we were to imagine anthropologists measuring homicide levels the month before and after the Confederacy was established. While homicide rates would have dropped dramatically, the only difference or reason for that change would be that the levels of restraint had increased dramatically, not the degree of innocence, which would have, of course, remained unchanged. The fear of punishment was simply preventing each member from expressing or living out their upset.

[690]The whole story of the human journey during the last 2 million years that has been described in this chapter is really the story of the emergence of ever-increasing levels of upset, and the development of ever more sophisticated ways to restrain and contain each new level of upset. The recognition in all our mythologies and in the work of our most profound thinkers of a wonderful, all-loving, innocent past for the human race wasn't some romantic, fanciful dream of some impossible, unrealistic, idyllic, utopian existence, as mechanistic science has tried to dismiss it as, but a *completely real* time in our species' distant past that recently discovered fossil evidence is now confirming, and that bonobos provide ample living evidence of. The *true* story of human life over the last 2 million years *is* that of the loss of innocence—our 'fall from grace', our departure from the 'Garden of Eden', the corruption of our soul, our ever-increasing levels of anger, egocentricity and alienation! Everyone does, in truth, know that under the duress of the human condition we each, and our species as a whole, started life in an innocent state and ended up in a variously psychologically upset, embattled, soul-corrupted state. Innocence is associated with youth, not old age.

[691]So yes, tribal warfare and outbreaks of individual violence have been occurring for a long time, but that doesn't mean that the relative innocence of hunter-forager tribes still living, like the Bushmen of South

Africa, the Australian Aborigines and the Yanomamö of South America, don't reveal a great deal about how extremely upset the great majority of the human race has become, and how much civility, and other pseudo idealistic means of restraint, humans now rely upon to mask and contain that extreme upset. But again, the psychological agony of our human condition has been so great that while we humans couldn't truthfully explain our condition all we had to protect ourselves from the vicious, unbearable self-confrontation of such relative innocence were equally vicious, retaliatory, denial-based lies. Yes, it is *only* now that we can explain and defend upset that we can admit and talk about different states of upset and by so doing finally make sense of human behaviour. Again, much more will be explained about the differences in upset between races of humans in ch. 7:16E.

---

692 So while civility has been a marvellous tool for helping to restrain upset, it involved bottling-up our frustrations and angers, which produced another problem of how then to relieve that pent-up state. And so, in our inability to be honest about our internal upset, we had to learn to valve off and relieve ourselves in ways that weren't destructive of those around us, such as the mask ceremonies already mentioned, or the fatigue-inducing, soul-accessing dances of the Bushmen. The origin of our developed sense of humour, for example, has never been able to be properly explained, but once it is understood how false resigned humans became the source of so much of our sophisticated sense of humour becomes very clear. For the most part, adults maintain a carefully constructed facade of denial, but every now and then a mistake is made, we 'slip-up', and the truth of our real situation is revealed, providing the basis for humour. Occasionally situations occurred where the extreme denial, self-deception, delusion, artificiality, alienation became apparent and transparent, and in those moments the truth of that immense falseness was exposed for what it really was—so farcical it *was* funny; in fact, a 'joke'. When someone tripped or fell over, for instance, or had their 'bluff' called, it *was* humorous because suddenly their carefully constructed, civilised image of togetherness disintegrated. We take our developed sense of humour for granted now as being a natural part of our make-up, but there was a time, prior to becoming alienated and false, when the only comic situations were those that

children—and bonobos—find amusing, such as when a harmless trick is played on someone. The power of consciousness makes it possible to recognise and even create such silly and funny situations that delight our emerging minds, but it is only when the absurdly dishonest situation of the resigned, alienated human-condition-afflicted state emerged that *extremely* ridiculous, fraudulent situations appeared in human life to laugh about and make fun of.

693 Swearing has been another way of tearing down and breaking free from the extreme dishonesty of our condition. Indeed, it is a stark measure of just how dishonest humans have become that we don't even have an everyday word for all the evasions and dishonest denials and delusions we practise every minute of the day, *except* for the swear word 'bullshit', or 'BS' or 'bull' or 'crap'! To understand why 'fuck' is such a powerful swear word we only have to recall the truth of what sex really is. As explained, while sex at its noblest level was something that marvellously complemented the human journey and as such has truly been an act of love, it has, nevertheless, at base been about attacking innocence (which women represent) for innocence's unjust condemnation of humans' (especially men's) lack of innocence. 'Fuck' means destroy or ruin, and what is being destroyed or ruined or sullied or degraded or violated is innocence or purity. Sex has been such a preoccupation of humans and yet everyone lives in denial of the truth that it is, at base, an attack on innocence. This makes sex one of the biggest lies and thus jokes of all, which is why using the word 'fuck' is such a powerful attack on the world of lies, and thus such a powerful swear word.

694 Returning to the main stages of maturation again, as emphasised, since civilising our upset didn't stop its development—it could only ever conceal and help contain it—it was inevitable that, as the corrupting search for knowledge continued, levels of upset were only going to escalate until eventually, by our late 30s, we/*H. sapiens* were embroiled in a rage of hate and anger. The compounding effect of upset meant that we became immensely embattled, 'punch-drunk' in fact, and thus absolutely despairing about our situation. On reaching this state of extreme anger and destructiveness we began to hate even ourselves. Life had become both personally and socially unbearable, an untenable position that produced a crisis, the well-known 'mid-life crisis' of the early 40-year-old, or, in the case of humanity, the emergence some 50,000 years ago of *H. sapiens sapiens*.

Chapter 7:13 <u>Pseudo Idealistic Adolescentman</u>

The species: *Homo sapiens sapiens* — 0.05 million (50,000) years ago to the present day
The individual: 40 to 50 years old

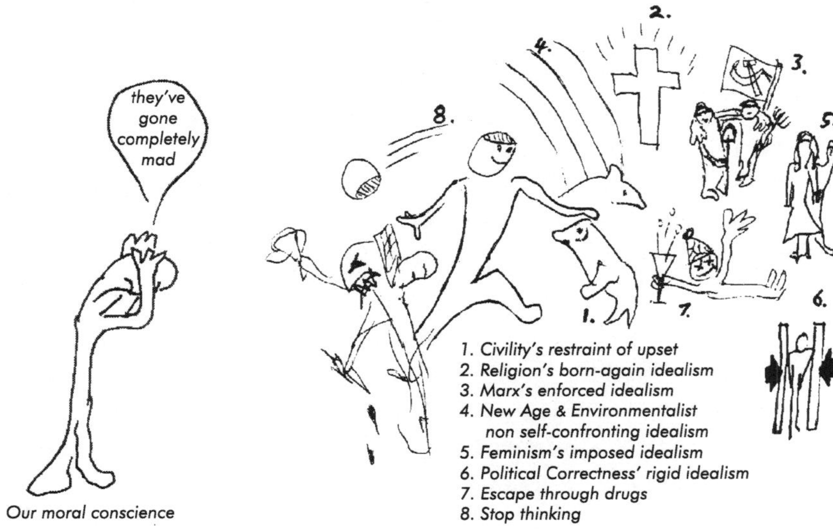

695 The Born-Again, Pseudo Idealistic Late Adulthood Stage (of Humanity's Adolescence) encompasses the time of the 'mid-life crisis' and the adoption of <u>Pseudo or False Idealism</u>—the 'born-again in support of cooperative idealism' lifestyle.

696 In this 40-year-old stage, upset compounded at such a rapid rate that the graph charting its intensity did go into free fall, rendering social disintegration an imminent risk. But with upset intensifying throughout humanity's adolescence, we always knew that if we didn't find the relieving understanding of the human condition in time then eventually the human race *would* enter a final end game stage where the levels of upset anger, egocentricity and alienation *would* threaten to destroy humanity. And this fear did play out: the 2-million-year race our species has been involved in between self-destruction and self-discovery *did* finally enter this crisis stage some 50,000 years ago and the variety of humans involved was *H. sapiens sapiens*—us, the anatomically modern humans who emerged from *H. sapiens* around that time. In the case of the individual growing up during humanity's human-condition-afflicted,

insecure adolescence, this was the time when, at about 40 years of age, we entered our so-called 'mid-life crisis'.

697 At this point in our personal journey, our upset had become so great that, on one hand, we were hating the condemnation from the cooperative idealistic world of our soul with such fervour that we were beginning to become murderously behaved—while on the other, we were despising ourselves for being so upset and destructive of the world. Although we had, through the measures taken through our 30s, developed a great deal of instinctive capacity to restrain and conceal—civilise—our upset, it was becoming so great that it all too readily broke out, revealing the extremely angry person that lay beneath. So despite all our efforts to 'conquer the world' if we were resigned, or 'fix the world' if we were unresigned, all we had to show for ourselves now was an overly upset individual—a wreck of a person. In the words of *The Man of La Mancha*, we had finally **'march**[ed] **into hell…**[and become] **scorned and covered with scars'**—that was the price humans had to pay for pursuing the **'heavenly cause'** of trying to prove that our species is, in fact, fundamentally good and not bad. But what could we do to resolve such an untenable position? The 'mid-life crisis' long associated with turning 40 had certainly arrived.

698 The essential problem for the 40-year-old individual was that they were loathing and depressed and excruciatingly angry about what they and their world had become—a crisis that has parallels with the situation they had encountered in their early adolescence when they first started thinking philosophically about the corruption both in the world and within themselves. The journalist Ali Gripper acknowledged this parallel in an article titled 'Turning 40 and Frantic, Mid life crisis', writing that **'Mid life is undoubtedly a recycling of adolescent issues. It is as if the psyche goes back and picks up the threads of what we were dealing with as teenagers'** (*The Sydney Morning Herald*, 29 Mar. 1996). As with the situation that occurred in early adolescence, when resigning individuals were faced with extreme states of despair and depression about their circumstances, or unresigned individuals were faced with the ever-increasing wrongness of the world around them, so 40-year-olds were faced with variously extreme states of desperation about their situation. And like the adolescent struggling with their extreme despair and depression, the 40-year-old's mind similarly searched frantically for a way to solve the problem of their now untenable situation. And just as the unbearably upset and psychologically desperate adolescent came up with a desperate solution to completely put aside the

reality of their circumstances by resigning themself to living in denial of cooperative ideality and thus the depressing issue of the human condition, so too did the, by now, extremely corrupted 40-year-old—BUT *this* time the evasion was achieved through focusing on the positive, guilt-relieving effect or feeling that came from being civilised. The angry 30-year-old had learnt to restrain/civilise their upset, but what the desperate 40-year-old realised in their frantic search to find a solution to their problem was that being civilised or 'well-behaved' or 'good' produced a guilt-relieving positive feeling and that this was the one positive in their life that they could derive some reinforcement from.

699 When humans are psychologically cornered they typically 'scan the horizon' for any positive, no matter how small, and make a huge deal of it, and this situation was no different. Frantically scanning for any positive that could be employed to escape condemnation and depression, it was the *side effect* of feeling good when we behaved in a civilised way that the 40-year-old latched onto to develop. Indeed, in the case of the extremely upset 40-year-old, *so* desperate were they for relief from the horror and guilt of their situation that their mind decided to focus *so* completely on the positive that they were good when they behaved in a cooperative, civilised, ideal, loving way that they deluded themselves that they weren't actually corrupted, that they weren't massively upset human-condition-afflicted people. They convinced themselves that the mask or facade of civility was not actually a mask or facade at all, but a true representation of their real self: 'I am behaving in a cooperative, loving way, therefore, I *am* an upset-free, guilt-free, human-condition-eliminated, thoroughly good, cooperative, loving, sound human.' It was an *extraordinarily* false/dishonest/**'fake'**/**'phony'**/deluded interpretation, but the depression from feeling guilty/bad/worthless about being *so* upset was *so* great that their mind was well and truly capable of making such a grand delusion. The situation was similar to the resigned person being so overwhelmed by the depression caused by their predicament that they were capable of making the extremely false interpretation that instead of having integrative, Godly, unconditionally selfless, moral instincts, we actually live in a non-integrative world of random, directionless change, and have selfish, competitive, survival-of-the-fittest animal instincts that make us competitive and aggressive, and as such there is no psychological dilemma of the human condition to have to contend with, let alone have to explain or ameliorate.

700 This 40-year-old, 'do good in order to delude yourself that you *are* actually good, that you *are* actually free of corruption and thus the dilemma of the human condition' strategy was an extremely deluded way of coping with the problem of the now massively corrupted human-condition-afflicted state, but it has, nevertheless, been so seductive that it developed into an industry so huge and so influential that the dishonesty involved now threatens to destroy humanity.

701 To elaborate, while being civilised—that is, using self discipline to restrain and contain your upset so it didn't show—*did* help contain destructive behaviour and provide its practitioners immense relief from doing so, what happened during the 40-year-old stage, for both humanity and humans individually, was that this relieving, 'feel good', 'warm inner glow', 'blissed out' positive of having restrained your upset and behaved in a 'good'/ideal/cooperative way became the *entire* focus of existence. In the end, as we will see, when humans became extremely upset—saturated with the problem of the corrupted state of the human condition—their whole mental preoccupation became one of searching for situations and opportunities where, through doing 'good', they could derive 'the rush' of relief from the condemning issue and truth of their corrupted state. So again, while the 30-year-old used civilising self discipline to restrain and conceal their upset, they—unlike the 40-year-old—*weren't* using it to delude themselves that they were an ideally behaved, upset-free, guilt-free, human-condition-eliminated, sound person.

702 The immense danger of this preoccupation with relief-hunting through 'doing good' was that it could become so consuming, so addictive and thus so selfishly indulged that it could stop the all-important search for knowledge, because if there was too much preoccupation with 'doing good' it could result in insufficient tolerance of the corruption that unavoidably resulted from pursuing humanity's heroic search for knowledge. If there was too much emphasis on cooperative idealism humanity would be denied the freedom necessary to find the liberating understanding of the human condition, and if it didn't find that liberating self-knowledge humanity would be condemned to the eventual emergence of terminal levels of upset—in particular, the unbearable levels of the psychosis and neurosis of alienation that result from having to adopt excessive amounts of psychological denial and delusion. In short, the *dogma* of doing good could oppress and even stop the all-important search for knowledge by denying the *freedom* to be, to a degree, corrupted. As we

will see, this extremely dangerous situation did arise; humanity *did* face a death by dogma, a fate that only the finding of the liberating understanding of the human condition—that science has made possible and which is being presented here—has the ability to save humanity from.

703 The danger of excessive oppression of freedom has been particularly great because of the massively seductive nature of relief-hunting. If we return to the Adam Stork analogy for describing the human condition, at any time Adam could surrender to his criticising instinctive self and 'fly back on course', obey his instinctive orientation, and by so doing stop and thus relieve the criticism emanating from his instinctive self, but that meant abandoning the all-important search for knowledge. And, of course, in the case of humans, when we 'flew off course' and became angry, egocentric and alienated the sense of guilt we accrued from defying our cooperatively orientated, all-loving, Godly, moral instincts was immense, so 'flying back on course' was an *extremely* guilt-relieving, and thus an *extremely* tempting, option.

704 There was, however, a further, very significant dimension to the problem of 'flying back on course'—being, as we revealingly say, 'born again' to supporting instead of resisting the cooperative ideals our instinctive self dogmatically demanded—which was that since our instinctive self was orientated to behaving cooperatively, when we abandoned the search for knowledge by taking up support of cooperative idealism we were not only abandoning the battle to champion our ego or conscious thinking self over our idealistic instinctive self or soul, *we were also* siding against those fighting the battle. We weren't just 'taking a rest' to recuperate, we had actually switched camps/allegiances to side with the 'enemy'. It was completely subversive, mad behaviour—an act of cowardice and treachery—because in switching sides the individual was basically saying, 'I don't care about humanity anymore. I only care about making myself feel good and relieving my own guilt.' They were being totally selfish, the complete opposite of the selfless and ideal and cooperative person they were deluding themselves to be. And since the lie they were maintaining was so great, they had to work very hard at convincing both themselves and others of it, which meant they were typically a strident, extremely intolerant, even fanatical, belligerent advocate of their position.

705 Yes, in choosing to be 'born again' you had to work *very* hard at maintaining the conviction that what you were doing was right because,

although we haven't been able to explain, confront or talk about it, the truth is all humans who have lived during humanity's adolescence have intuitively been aware of the battle of having to overthrow the ignorant idealism of our soul; when we shook our fist at the heavens we were saying, 'One day, one day we are going to prove that we humans are good and not bad.' We *knew* that to give up the battle against our idealistic soul, and not just give it up but side with the 'enemy' and against those trying to win the battle, was a crime against all those still fighting for understanding, and against humanity as a whole. So despite how tempting it was, siding against humanity and those fighting the battle was, in reality, such a repulsive course of action that it took a great deal of despair and fear of depression about being overly corrupt to actually do so.

706 We intuitively knew that changing sides and being born-again to living in support of the cooperative, ideal world of our soul, while immensely relieving, was also immensely loathsome. The delusion, dishonesty and betrayal involved *was* extreme, but for ever-increasing numbers of people the need for relief from feeling loathful/guilty/bad about themselves became so great it could not be resisted.

707 Again, the great danger was that, since upset was the price of searching for knowledge, if everyone became addicted to selfishly indulging their need for relief through 'doing good to feel good' there would be no tolerance of non-ideal upset behaviour, and humanity's all-important search for knowledge, ultimately liberating self-knowledge, would be shut down, condemning humanity to terminal levels of alienation and thus extinction. So while Resignation to living in denial of the issue of the human condition and taking up a competitive, egocentric, selfish power-fame-fortune-and-glory-seeking lifestyle was, in itself, extremely desperate and mad behaviour, it did at least involve participating in humanity's great battle to overthrow ignorance.

708 Certainly, the upset behaviour that resulted from participating in humanity's heroic search for knowledge was increasingly causing immense human suffering and environmental devastation, but if we didn't continue the search for knowledge then there was simply no hope. To put it in political terms, although the harsh, brutal reality associated with what became known as 'right-wing' politics *was* bringing about immense human inequality, hardship and suffering, and *was* destroying the planet, it was the search-for-knowledge-oppressing so-called 'left-wing' politics

that posed the real threat to the survival of the human race—because *only* the search for knowledge could lead to the finding of understanding of the human condition and the liberation of humans from that totally unbearable, crippling, soul-sickening, black-dog-depressing, real-person-extinguishing, deadening, human-life-denying condition! While, as the journalist Geoffrey Wheatcroft recognised, **'the great twin political problems of the age are the brutality of the right, and the dishonesty of the left'** (*The Australian Financial Review*, 29 Jan. 1999), and, as the scientist-philosopher Carl von Weizsäcker also recognised, **'The sin of modern capitalism is cynicism (about human nature), and the sin of socialism is lying'** (mentioned in a speech by Prof. Charles Birch that was reproduced in the Geelong Grammar School mag. *The Corian*, Sep-Oct. 1980), it was NOT the **'cynicism'** and **'the brutality of the right'**, BUT THE **'lying'**, **'dishonesty of the left'** that stood like a colossal ogre over the human race, threatening to destroy it!

709 To return to the Adam Stork analogy once more, Adam knew from the outset that he had to continue with the upsetting search for knowledge, that he could never afford to stop until the liberating understanding of his corrupted condition was found. That was his fundamental reality, *and it has remained our fundamental reality*. Again, the Statue of Liberty is as good a symbol as any of the fundamental responsibility we humans have had to maintain *freedom*, which we can now understand means freedom from the cooperative ideals in order to continue the upsetting search for the knowledge that would allow humanity's *ultimate* liberation. Yes, paradoxically, *real* idealism, the real path to an ideal world, depended on continuing the corrupting search for knowledge until we found the human-race-liberating understanding of the human condition. The strategy of hunting for guilt-relieving, feel-good causes was, in fact, pseudo or false idealism, because it meant abandoning, and, worse still, oppressing, and—even worse still—actively opposing that all-important search for knowledge.

710 In summary, while humans *have* had to counter the effects of the extreme upset that now plagues human life, and thus our world, with a degree of idealistic, concern-for-others-and-concern-for-the-world behaviour—and, in truth, becoming civilised did involve a degree of abandonment of the upsetting battle in favour of being idealistic and showing concern for others and the world—what has happened is that the feel good aspect of behaving 'ideally' has evolved into what is now an *extremely* dangerous industry.

[711] Yes, given it has been *so* hard to explain and argue why not being ideally behaved is good, and *so* easy to argue that being ideally behaved can't be anything but good, 'the left' has had a field day mocking 'the right' as selfish, immoral and evil. As such, this quote by Nietzsche, which was referred to in par. 188, stands out as a brave and rare pronouncement on the need to hold our nerve and continue our species' great heroic battle to champion the ego over the ignorance of our instincts: **'There have always been many sickly people among those who invent fables and long for God** [ideality]**: they have a raging hate for the enlightened man and for that youngest of virtues which is called honesty...Purer and more honest of speech is the healthy body, perfect and square-built: and it speaks of the meaning of the earth** [to face truth and one day find understanding of the human condition]**... You are not yet free, you still *search* for freedom. Your search has fatigued you... But, by my love and hope I entreat you: do not reject the hero in your soul! Keep holy your highest hope!...War** [against the oppression of dogma] **and courage have done more great things than charity. Not your pity but your bravery has saved the unfortunate up to now...What warrior wants to be spared? I do not spare you, I love you from the very heart, my brothers in war!'** The author and journalist George Orwell was another who bravely recognised the very real danger of humanity losing its nerve when he famously predicted that **'If you want a picture of the future, imagine a boot stamping on a human face** [freedom] **forever'** (*Nineteen Eighty-Four*, 1949). And, in fact, as will soon be documented, this end play, death-by-dogma fate for the human race has all but descended upon humanity and can only be prevented by the eleventh-hour arrival of this understanding of the human condition.

[712] As pointed out, all humans have been intuitively aware that when they took up the born-again, pseudo idealistic way of coping with the human condition they were siding against humanity, siding with the enemy, and that doing so was a loathsome act of cowardice and treachery. So while the desperately upset, mid-life crisis of the 40-year-old stage made taking up the born-again, pseudo idealistic way of coping a tempting option that was worth trying, the revulsion of living so treacherously caused many to change sides yet again and return to the upsetting battle of searching for knowledge. However, in returning to the battle they could only expect to become even more upset, angry, egocentric and alienated, which introduces the final stage that living under the duress of the human condition has resulted in: the extremely tragic 'Hollow Adolescentman' stage.

Chapter 7:14 Hollow Adolescentman

The species: *Homo sapiens sapiens'* reign — 0.05 million (50,000) years ago to the present day (Pseudo Idealistic and Hollowman stages are both characteristic of *Homo sapiens sapiens'* reign)
The individual: 50 plus years old

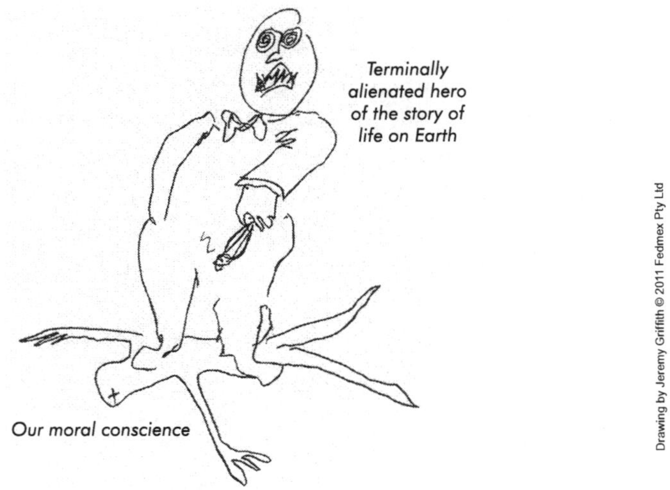

713 The 'Hollow Final Adulthood Stage' (of Humanity's Adolescence) represents the time when many post-40-year-olds become disillusioned with the treacherous, weak and cowardly, completely selfish, extremely dishonest and deluded 'do good to feel good' born-again existence and return to the upsetting battle to champion the ego over the condemning instincts. But as a result of this return to participating in the upsetting battle, they become even *more* upset, embattled and frustrated than they were when they were driven to adopt the pseudo idealistic way of living. In the context of the aforementioned extraordinarily honest Japanese proverb that described the stages of maturation under the duress of the human condition—**'At 10 man is an animal, at 20 a lunatic, at 30 a failure, at 40 a fraud and at 50 a criminal'**—this is the 50-year-old **'criminal'** stage where men in particular become so soul destroyed, so horrifically angry, punch-drunk, ego-unsatisfied, bitter and vengeful that they brutally and completely repress the condemning voice of their ideal-behaviour-demanding, cooperatively orientated soul, leaving themselves adrift in an empty, hollow, soul-less wilderness.

[714]T.S. Eliot perfectly described this 'grumpy old man', vengeful, burnt-out, empty, sad existence that men typically inhabit when they reach 50 and beyond in his 1925 poem *The Hollow Men*: **'We are the hollow men / We are the stuffed men / Leaning together / Headpiece filled with straw. Alas! / Our dried voices, when / We whisper together / Are quiet and meaningless / As wind in dry grass / Or rats' feet over broken glass / In our dry cellar // Shape without form, shade without colour / Paralysed force, gesture without motion //...This is the dead land / This is cactus land / Here the stone images / Are raised, here they receive / The supplication of a dead man's hand / Under the twinkle of a fading star // Is it like this / In death's other kingdom / Waking alone / At the hour when we are / Trembling with tenderness / Lips that would kiss / Form prayers to broken stone // The eyes are not here / There are no eyes here / In this valley of dying stars / In this hollow valley / This broken jaw of our lost kingdoms // In this last of meeting places / We grope together / And avoid speech / Gathered on this beach of the tumid river //...Between the desire / And the spasm / Between the potency / And the existence / Between the essence / And the descent / Falls the Shadow /...This is the way the world ends / Not with a bang but a whimper.'**

[715]Before commenting on Eliot's extraordinarily honest poem, it needs to be explained that no amount of power, fame, fortune or glory could truly satisfy men's egos, because, being artificial forms of reinforcement, such 'success' was never going to genuinely make men feel they were good and not bad—it was never going to truly relieve them of the agony of their insecure, human-condition-afflicted existence—*only* understanding of their fundamental goodness could and now does achieve that. It can be appreciated then that the problem for older men was that there came a cross-over point in their lives when, no matter how much power, fame, fortune or glory they had achieved, they still hadn't satisfied their egos, but life had moved on—another generation of men trying desperately to validate themselves through the artificial, material forms of success of power, fame, fortune and glory had taken over—and they suddenly found themselves ignored. They were silently screaming for attention but no one was listening; no one was interested anymore. I was once told by the son of a retired politician how his father had bought numerous copies of a newspaper that mentioned him and strategically placed them around the house, open at the relevant page. Throughout their life men fought harder and harder to be a 'success' and relieve themselves of the insecurity of the human condition, and this

ever-increasing preoccupation meant they lost more and more access to the beautiful all-loving and all-sensitive true world of their soul. So in the end men became just monstrously ego-unsatisfied, ego-embattled, ego-infuriated volcanoes that couldn't explode for fear of all the destruction it would cause. They were soul-less, empty, **'hollow'**, **'dead'**, **'broken'**, **'lost'**, **'stuffed'**, **'dry'**, **'cactus'** men with **'paralysed'** egos. Each was **'alone'** having to **'avoid speech'** for fear of either betraying their own immense need to be glorified or offending another older man's immense need to be glorified. This is how it always **'end**[ed]**'** for men under the duress of the human condition, **'Not with a bang but a whimper.'** Truly, what phenomenal heroes, what **'stars'** men have been—they **'march**[ed] **into hell for a heavenly cause'**, sacrificed themselves, contributed to the search for understanding that might one day, in the far future, but not in their lifetime, liberate humanity from the horrific insecurity of the human condition. Thankfully their corrupting search for dignifying understanding of humans has been completed, it now only has to be recognised and then absorbed.

716 Of course, ageing during humanity's adolescence has been, in its own way, similarly horrific for women because it meant the inevitable loss of the image of innocence that women depended upon for reinforcement, the loss of their sex-object 'attractiveness', and with it, the loss of their meaning in the world—a source of meaningfulness that all women's magazines that focus entirely on how to be 'attractive' are testament to. When women are young their beauty is generally so empowering it is as if they own the world—as ABBA's 1976 song *Dancing Queen*, which *Rolling Stone* magazine has rated as one of the greatest songs of all time, says, **'you are the Dancing Queen, young and sweet, only seventeen...having the time of your life...you turn them on, leave them burning, and then you're gone'**, and as Chuck Berry wrote and sang in 1958, **'They're really rockin' in Boston, in Pittsburgh, P.A., deep in the heart of Texas, and 'round the Frisco Bay, all over St. Louis, and down in New Orleans, all the cats wanna dance with Sweet Little Sixteen. Sweet Little Sixteen, she's just got to have about half a million framed autographs'** (*Sweet Little Sixteen*). But when women become older and their beauty/'attractiveness'/innocence fades they discover that they have become invisible; when they walk down the street they are no longer noticed. This quote from the French beauty therapist Diane Delaheve describes how devastating it can be for women to lose their sex appeal: **'Her eyes, the mirror of her soul, speak nothing but despair. Her**

face may have kept its beauty, but it has become a picture of affliction. For some women, the prospect of age is sheer tragedy, worse than death, which might be seen as an escape' (*The Sydney Morning Herald*, 4 Sep. 1988).

717 An added dimension to the situation faced by older women is that in not being as responsible for the main battle of having to champion the ego over ignorance as men are, women find that their role of living in support of the battle is limited. It has been observed that a woman's life progressed from **'bimbo, breeder, babysitter to burden'**. Men, on the other hand, are directly participating in the battle of championing the ego and aren't faced, as such, with the prospect of one day feeling they are a **'burden'** to the extent that women are. In his 1993 book, *The Fisher King & The Handless Maiden*, the Jungian analyst Robert A. Johnson relates the myth of the Handless Maiden, which tells of a miller who makes a deal with the devil in return for the ability to complete more work with less effort. The devil demands the miller's daughter as payment: **'The miller is desolate but unwilling to give up his much expanded mill, so he gives his daughter to the devil. The devil chops off her hands and carries them away'** (p.59 of 103). Waited on by her newly prosperous family, the handless maiden is content for a time, until her growing sense of desperation sends her out to the forest alone. Johnson explains that the cry of women, like that of the handless maiden, is, **'What can I *do?* I feel so useless or second-rate and inferior in this world that puts its women on the rubbish heap when they are through with courtship and childbearing!'** (p.56). Indeed, **'one woman set up a blog called The Plankton. She says women over 45 are made to feel like the lowest life form – "flimflam, a nuisance, an embarrassment of landfill"'** (Angela Mollard, 'The invisible years', Sydney's *Sunday Telegraph Magazine*, 14 Aug. 2011).

718 Yes, being cast **'on the rubbish heap' 'of landfill'** because you are not mainframed, part of the main battle that men have been waging against the ignorance of our instinctive self, and because you no longer have the incredible attractiveness of youthful innocence, has been a truly horrible situation for women. I once heard a famous American male actor say on radio **'what man would be with an older woman if he didn't have to be'**, and it really is true, as high divorce rates and the existence of mistresses, 'first wives clubs', 'trophy wives', polygamy and, in earlier times, harems, all bear witness to. While almost every song written is dedicated to the attractiveness of youthful innocence in women, older women become invisible—they go from a situation where, in their

youth, they own the world, as the songs *Dancing Queen* and *Sweet Sixteen* so powerfully evidence, to a situation where they are virtually irrelevant. Thank goodness then with the battle to overthrow ignorance and find understanding of the human condition finally won, this, in truth, absolutely horrific existence can come to an end and women, the young, the aged—*everyone*—can now be fully involved in the human journey! Again, Olive Schreiner provided an excellent description of the dream women have had of this **'new time'** that has now finally arrived: **'if I might but be one of those born in the future; then, perhaps, to be born a woman will not be to be born branded...It is for love's sake yet more than for any other that we** [women] **look for that new time...Then when that time comes...when love is no more bought or sold, when it is not a means of making bread, when each woman's life is filled with earnest, independent labour, then love will come to her, a strange sudden sweetness breaking in upon her earnest work; not sought for, but found'** (*The Story of an African Farm*, 1883, pp.188, 195 of 300). It is truly amazing what women have endured, as Sir Laurens van der Post has written of their wonderful, *immensely* heroic patience and tolerance: **'She is content, confident and unresentful because she is also the love that endureth and beareth all things even beyond faith and hope. She knows that, in the end, the child will grow and all shall be well'** (*Jung and the Story of Our Time*, 1976, p.158 of 275).

[719]So, men become 'hollow', and women become 'invisible'; when observing older couples walking together in the park you can see how united they are by their respective afflictions. Indeed, the following drawing by cartoonist Ralph Steadman depicts the full horror of the human condition, for both men and women; in fact, the main dragon in this cartoon provided the inspiration for my drawing that appears at the beginning of this 50-year-old-plus stage, for his eyes show the hollowness that T.S. Eliot spoke of: **'This is the dead land / This is cactus land.'** The desperately tragic, sickly state of older women is also explicit in the two crow-like creatures leaning against the bar in the background.

[720]Humans *truly* have been the great heroes of the story of life on Earth!

Ralph Steadman's *The Lizard Lounge*, 1971

Chapter 7:15 The last 11,000 years and the rise of Imposed Discipline, Religion and other forms of Pseudo Idealism

721 As explained in par. 578, Resignation to a life of finding escapist relief from the agony of the human condition through winning power, fame, fortune and glory became an almost universal phenomenon amongst adult humans around 11,000 years ago following the advent of agriculture and the domestication of animals. What *now* needs to be explained is how the domestication of plants and animals accelerated the development of upset and the adoption of the resigned position—and what transpired when this extremely competitive and selfish, and now universal, resigned way of living became unbearably destructive.

722 Clearly, when the level of upset in the human race as a whole reached this final 'Hollowman' stage, humanity was at end play or end game in the race between self-discovery and self-destruction. The levels of upset within humans had become stupendous—but what could we do? We were stranded between two increasingly flawed options: adopting,

or in some cases returning to, the extremely irresponsible treacherous and fraudulent and thus dangerous born-again, pseudo idealistic, 'do good to feel good' way of living; or persevering with the ever more brutal, destructive and corrupting power-fame-fortune-and-glory-seeking, egocentric, knowledge-finding, resigned competitive existence. In fact, our lives, both individually and collectively as societies, for all of the last 11,000 years have been marked by the oscillation between these two extremely flawed strategies of coping with the human condition, strategies that were eventually refined into what we now term the 'left-wing' and 'right-wing' in politics.

723 Of these two strategies, we'll look first at the impact that upset reaching this crescendo had on the increasingly brutal and destructive power-fame-fortune-and-glory-seeking, egocentric, knowledge-finding, resigned competitive strategy—the approach that became known as 'right-wing'.

724 The overall significance of the agricultural revolution that took place some 11,000 years ago was that it led to such a rapid increase in upset that, on the graph charting the intensification of our upset, it meant humanity was fast approaching rock-bottom—we were hurtling towards the cynical, bitter and vengeful, burnt-out, out-of-control, all-restraints-thrown-out, rampaging, warring level of upset. This was because the domestication of plants and animals enabled people to live a more sedentary, less nomadic existence, in closer proximity and in greater numbers, the effect of which was to greatly increase the spread and growth of upset in humans. As explained in par. 671 when describing the effect the ice ages had on our species, living closely under the strain of the human condition dramatically accentuated the difficulties encountered by humans who were living with upset. So it follows that the closer humans lived during humanity's adolescence and/or the more difficult the living conditions, the greater the spread of and increase in upset, and that isolation from such encounters with the battle of the human condition served to minimise the spread of upset or soul-exhaustion. In short, if we were each left alone with our personal level of exhaustion, we would not be criticised by fresher souls or corrupted by those more battle-hardened. The philosopher Jean-Paul Sartre succinctly summed up just how difficult it is for upset, alienated people of all degrees to coexist when he wrote, **'Hell is other people'** (*Closed Doors*, 1944). We only need to look at the extreme situation to see the principle in action—as

mentioned, innocence isn't going to survive long in New York's Times Square where criminals and beggars work the streets. Of course, while living in closer proximity in more organised societies did greatly increase the spread and accumulation of upset, it also assisted the spread and accumulation of knowledge. In terms then of the race between self-discovery and self-destruction, there was at this time a speeding up in the development of *both* aspects.

[725] This explanation of how the advent of agriculture and the domestication of animals led to a rapid increase in upset allows us to understand why, in the Biblical story of Cain and Abel, Cain became more upset than his brother, Abel: **'Abel kept flocks** [he lived the nomadic life of a shepherd, staying close to nature and innocence]**, and Cain worked the soil** [he cultivated crops and domesticated animals and as a result was able to become settled and through greater interaction with other humans became increasingly upset]**... Cain was** [became] **very angry, and his face was downcast** [he became depressed about his upset state and so]**...Cain attacked his** [relatively innocent and thus unwittingly exposing, confronting and condemning] **brother Abel and killed him'** (Gen. 4:2, 5, 8).

[726] By some 4,000 years ago, the <u>development of villages</u>, the movement by people into <u>specialised occupations</u>, the beginnings of <u>trade and industry</u>, and the close personal interaction that each development inevitably brought, resulted in humans becoming so upset that some could no longer contain their upset and had to live that upset out; as the story of Cain and Abel describes, they had to allow some expression of their upset if they were to find any relief from the pressure of being so hurt. Men especially began to feel the periodic urge to go on a rampage of <u>raping and pillaging</u>. Tragic examples abound, but the thirteenth century Mongol conqueror Genghis Khan was certainly someone who lived out his upset to the full, every day satisfying his anger with blood-letting, his egocentricity through the domination of others, and his mind or spirit by blocking out any feelings of guilt or remorse coming from the moral instincts within himself. As Genghis Khan is reputed to have said, **'Happiness lies in conquering one's enemies, in driving them in front of oneself, in taking their property, in savouring their despair, in outraging their wives and daughters.'**

[727] This need to go on the rampage and express, indeed purge, unbearable levels of upset resulted in endless rounds of <u>payback warfare</u>, where warriors from one tribe or village would raid another tribe for their

material goods and maidens, which in turn would provoke a counter raid, and so it went on. Clearly at this point, where the upset in humans had become so great that the warfare and killing and raping was being carried out incessantly, in wave after wave of ever-increasing ferocity and brutality, a new means to restrain upset simply *had* to be invented. For the truth is, despite Genghis Khan obtaining some momentary relief from feeling so upset by rampaging across the world, there would have been no inner peace in his own life, or, more significantly, any peace in his blood-soaked world.

[728] Self Discipline could no longer contain or civilise the escalating levels of upset in society, and so another form of restraint for those participating in the upsetting heroic search for knowledge had to be developed—and the solution that emerged was Imposed Discipline, the implementation of an agreed upon set of rules and laws that enforced social (integrative) behaviour through threat of punishment. Once developed, this new form of restraint proved significantly effective. For example, by the time Europeans arrived in North America a grand league of American Indian tribes, known as the Iroquois Confederacy, had been established by two unresigned, denial-free thinkers or prophets who had emerged from within their ranks. Recognised and described by their people as **'prophets'**, these two American Indians—known as 'The Great Peacemaker' and Hiawatha—with all their soulful sensitive feeling and denial-free clarity of thought, were able to realise that the endless rounds of payback warfare between and within the tribes could only be prevented by everyone agreeing to certain restraining rules that were enforced through punishment. The resulting discipline proved highly effective, as this quote illustrates: **'The Iroquois Confederacy was established before European contact, complete with a constitution known as the…"Great Law of Peace"…The two prophets, Ayonwentah** [Hiawatha]**…and Dekanawidah, The Great Peacemaker, brought a message of peace to squabbling tribes…Once they ceased infighting, they rapidly became one of the strongest forces in seventeenth and eighteenth century north eastern North America'** (*The Iroquois Confederacy and the Founding Fathers*, accessed Sep. 2009; see <www.wtmsources.com/113>).

[729] Exactly the same scenario had played out some 3,000 years prior when, in approximately 1,500 BC, the very great denial-free thinker or prophet Moses brought order to the Israelite Nation through the Ten Commandments that he etched on stone tablets. The moral code contained in those commandments became the basis of the constitutions, laws and

rules that continue to govern much of modern society and proved vital in helping rein in the kinds of upset unleashed by the extremely upset 'Hollow Adolescentman'.

730 We now need to consider how the corrupting effects of the last 11,000 years affected the other strategy for coping with the human condition, which involved seeking out an idealistic cause to support to make yourself feel better about your, by now, immensely upset state—the approach that became known as 'left-wing'.

731 Yes, just as the resigned competitive practitioners invented a strategy for coping with their problem of excessive upset, which was Imposed Discipline, so born-again, pseudo idealistic practitioners came up with a strategy for coping with the problems associated with *their* approach to coping with the human condition, which was that in abandoning the all-important search for knowledge they were, in effect, siding *against* humanity's great battle and being excessively dishonest, and that solution was the adoption of Religion as it is practised in most cultures today.

732 To truly appreciate just how significantly humanity has benefited from religion, we first need to look deeper into the very serious problems associated with the born-again, pseudo idealistic strategy.

733 As has already been emphasised, one of the serious problems with the born-again, pseudo idealistic, 'do good to feel good' strategy was that it involved siding against humanity's great battle to champion the ego over the ignorance of our instinctive self. The other very serious problem with this strategy was that because you were *behaving* in a supposedly good, ideal way you were effectively asserting that you *actually were* a good, selfless, cooperative, gentle, loving, guilt-and-human-condition-free, ideal person. This was unlike the resigned competitive strategy where, firstly, you were still participating in humanity's heroic battle to find knowledge, and secondly, while you were using denial and lies to defend your upset state (asserting that there was no integrative purpose to existence, only random change, and that you were aggressive and selfish because humans had selfish instincts, and, therefore, there was no psychological dilemma of the human condition to have to explain), you were *not* deluding yourself you were a cooperative, loving, selfless, gentle, thoroughly good person free of upset. The born-again, pseudo idealist both abandoned the battle *and* took lying and delusion to a whole new level.

[734]It is worth mentioning that maintaining such extreme delusion was greatly helped along by the fact that we had *already* been practising total denial of the issue of our corrupted, human-condition-afflicted state since resigning in our early adolescent years. So, in adopting a born-again, pseudo idealist strategy, all that its advocates were doing was adding another layer of delusion to the one that was already well entrenched. As mentioned in par. 577, while Resignation was a form of autism, of block-out/dissociation/denial, pseudo idealism was an even more extreme form of it; it was an even more extreme way of making yourself ***'invulnerabl[e]'*** by establishing **'a complex mental structure insuring against recurrence of the conditions of the unthinkable anxiety'**.

[735]Yes, the born-again, pseudo idealistic strategy was both treacherous and extremely dishonest—traits that totally undermined humanity's search for knowledge—because in campaigning against the battle to find knowledge you were leading humanity towards an extreme state of denial/alienation/separation from the truth/knowledge, when, in fact, humanity had to *continue* the battle to try to get closer to and ultimately reach the ultimate truth/knowledge/understanding of the human condition. Indeed, as we will see, when the born-again, pseudo idealistic state became fully developed in the form of 'postmodernism' even the existence of truth itself was denied! The fundamental objective of the human journey was to find the truth about ourselves, so adopting extreme denial of the truth, especially extreme denial of the truth about the human condition, was to undermine millennia of human progress and lead humanity *away* from its objective; it was to misguide humanity onto the path to oblivion, to total darkness in terms of enlightening ourselves about ourselves—which is why, as will be further explained shortly in ch. 7:16, pseudo idealism came to be described as **'the abomination that causes desolation'**.

[736]We can see then that since dishonesty was so dangerous, and pseudo idealism was the most dishonest strategy ever developed for coping with the human condition, there was a very great need to find a form of pseudo idealism that somehow minimised or countered this extreme dishonesty—a conundrum that eventually led to the development of the contemporary forms of Religion (the earlier ones being nature and ancestor worship).

[737]Religion is a form of pseudo idealism because it involved being 'born again' to the cooperative ideal state; instead of living through

yourself, with all the associated overly upset angers, egocentricities and denials, religion required that you defer to someone exceptionally free of upset—namely one of the unresigned, denial-free-thinking, integrative-ideals-or-God-acknowledging, soulful, sound, innocent prophets around whom the great faiths were founded. Because you had become overly upset you decided to end your participation in humanity's heroic yet upsetting battle to find knowledge and instead put your faith in and lived through supporting the soundness and truth of a prophet's life and words. Rather than adhering to what your now overly upset self wanted to do and say, you adhered to the soundness and truth of the prophet's life and words. But, the immense benefit of deferring to religion is that while it allowed you to be born-again to a form of idealism and thus contain your upset and feel good about yourself, you were minimising the dishonesty normally involved in the born-again, pseudo idealistic strategy, because you were acknowledging the soundness of the prophet and, by inference, your lack of soundness. By recognising, indeed worshipping, the integrity of the prophet, and his representation of another true, denial-free, integrative, soulful, sound state, you were indirectly admitting your own extreme lack of soundness—your separation from the integrative, true, soulful state; you were indirectly being honest about your immensely corrupted state.

738 Religions also countered the degree of dishonesty involved in the other strategy for coping—the resigned, competitive way of living—because most religions acknowledge the existence of a God who, as has been explained, is the personification of Integrative Meaning. Also, by acknowledging the soulful soundness of the prophet, you were recognising the existence of a cooperative, unconditionally selfless, moral soul in humans.

739 Another very important benefit of religion is that, on an individual level, it also helped assuage the guilt felt by pseudo idealists who were struggling with the fact that they were siding against humanity in its great battle. This is because in supporting your religion you were also indirectly supporting humanity's heroic search for knowledge, because the truthful words of the prophet that are recorded in your religion's scriptures, which you were showing reverence for and deferring to, have been the very font of knowledge; they have, in fact, been the most denial-free expression of knowledge the human race has known. The psychoanalyst Carl Jung recognised how aligning to the truth and thus

supportive of our search for knowledge Christianity, for example, has been when he wrote that '[in Christianity] **the voice of God** [truth] **can still be heard'** (W.B. Clift, *Jung and Christianity*, 1982) and that **'The Christian symbol is a living thing that carries in itself the seeds of further development'** (Carl Jung, *The Undiscovered Self*, 1957, p.63).

740 In summary, religions offered humans a way to abandon living out their upsets and be born-again to ideality, but in the least dangerously dishonest and most human-journey-sympathetic way. They enabled humans to indirectly continue to participate in humanity's heroic yet upsetting search for knowledge, *and* they provided a way for humans to significantly avoid being dishonest, because by deferring to a prophet you were able to indirectly admit the truth of your own corrupted state and the existence of another integrative, true, sound, soulful state. Religions provided a way for humans to be (to a degree) honest about their corrupted, false condition without having to openly admit and, therefore, nakedly confront it. In doing so, they helped minimise the truth-destroying levels of delusion and denial involved in the born-again, pseudo idealistic lifestyle. As such, religion has been a pseudo idealistic, 'do good in order to feel you are good', 'give up your overly upset life and be born again to a cooperative ideal life' way of living that allowed you to live in safe denial of your corrupted condition, but at the same time be honest about it—albeit indirectly. In the case of Christianity, the Bible actually referred to being **'born again'** (John 3:3), to having **'crossed over from death to life'** (John 5:24). As Christ authoritatively said, **'I am the resurrection and the life** [through me, your ideal, soulful true self can live again]' (John 11:25).

741 The criticism that could be levelled at someone extremely upset, like Genghis Khan or Adolf Hitler, is that they didn't take up religion—or, if they did claim to be religious, they weren't being *genuinely* religious. Indeed, for the exceptionally upset, the aspect of religion that made it so superior to the strategy of Imposed Discipline, which the Ten Commandments represented, is precisely that it allowed you to delude yourself that you were being **'born again'**, **'resurrect**[ed]**'** from your corrupted state. Rather than having good behaviour forced upon you through fear of punishment, as was the case with Imposed Discipline, religion allowed you to feel that not only were you actively participating in goodness, you had actually *become* a good, selfless, loving, ideal person—that you were 'righteous'—which provided immense relief

from the guilt of being overly upset. Possibly the best sales pitch for born-again religious life was that given by the apostle St Paul when he wrote, **'Now if the ministry that brought death, which was engraved in letters on stone** [Moses' Ten Commandments that were enforced by the threat of punishment]**, came with glory** [because they brought society back from the brink of destruction]**...fading though it was** [there was no sustaining positive in having discipline imposed on you]**, will not the ministry of the Spirit be even more glorious? If the ministry that condemns men is glorious, how much more glorious is the ministry that brings righteousness! For what was glorious has no glory now in comparison with the surpassing glory. And if what was fading away came with glory, how much greater is the glory of that which lasts!'** (Bible, 2 Cor. 3:7-11).

742 Thus, in coping with the now raging levels of upset in humans, the first **'glorious'** improvement on destructively living out that ferocious upset was Imposed Discipline, which was enforced through fear and punishment. But since discipline provided little in the way of joy or inspiration for the mind or **'spirit'** it was hard to maintain, it didn't **'last'**, it was **'fading'**, especially in comparison to the immensely guilt-relieving, **'righteous'**, 'do good in order to make yourself feel good' way of living offered by the next **'surpassing glory'**, religion.

743 We can clearly see then that the 'do good in order to delude yourself that you actually are a kind, loving, selfless, good person and not horrifically corrupted' aspect of religion was very important, which means that only being *indirectly* honest about being extremely corrupted when you became religious was key to the effectiveness of religion—because if you had to be *directly* honest about being horribly corrupted you couldn't possibly delude yourself you were actually a kind, loving, selfless, good, not-horribly-corrupted person. This ability to not just *feel* good (because you were now behaving in a good way and not in the incredibly destructive way you had been), but to use this fact to delude yourself you *were* actually a loving, kind, selfless, good, upset-free, ideal, guilt-free, human-condition-solved, **'righteous'** person depended on this aspect of only *indirectly* acknowledging your corrupted state. So although in taking up religion you were being indirectly honest about being corrupted, you were still relying on being able to delude yourself that you weren't corrupted. In short, religions allowed people to admit to being horribly corrupted without having to suffer the confronting consequences of making such an admission.

So just as humans could not directly acknowledge their corrupted state (because without the explanation of the human condition they couldn't defend and thus cope with that truth), religions similarly depended on not directly acknowledging/recognising the soundness of the prophet around whom the religion was founded, even though acknowledging/recognising his soundness was an intrinsic part of the honesty that made religions so special and effective. Religions depended on not recognising—at least consciously, explicitly recognising—that prophets were simply a sound variety of ordinary people, because that truth would directly confront their followers with the unbearable truth of their own lack of soundness. Instead, at least at the surface level of their conscious awareness, religious adherents viewed their prophets as being supernatural, divine, heavenly, from-another-world beings, because that way they could avoid any comparison with themselves. In fact, the more upset and insecure the religious person, the more fundamentalist/literal/superficial they had to be in their interpretations of religious scripture and the prophet himself—because too much honesty was impossibly confronting.

744 As such, religion involved maintaining a *very* delicate balance of honesty and delusion, for while it offered a way of only being implicitly honest about the corrupted state, its very existence depended on its adherents making at least a subconsciously relieving, honest acknowledgment of their own corrupted state, and on others making at least, on a similarly subconscious level, a relieving, truthful recognition of that corruption. On the surface of conscious awareness, however, each individual adherent also depended on being able to maintain their facade and delusions about being an upset-free, **'righteous'** person, which meant there was still a great deal of dishonesty involved in religion.

745 The cartoon series *The Simpsons* provides a wonderful illustration of the subtleties involved in religion. In the series, Ned Flanders is the born-again religious character who is typically portrayed as having a self-satisfied, 'I-occupy-the-moral-high-ground' attitude over the still-human-condition-embroiled Homer Simpson. Ned's posturing drives Homer crazy with frustration because Homer intuitively knows Ned is deluding himself in thinking his Christianity gives him the moral high ground—that he is the more together, sound person and is on the right

track—but Homer can't explain why Ned is so extremely deluded and totally dishonest in his view of self. Homer can't explain and thus reveal the truth that real idealism and the truly on track, moral high ground lay with continuing the upsetting battle to find knowledge, and that Ned had become so upset, so unsound, that he had to abandon that all-important battle and leave it to others to continue to fight, including Homer. Worse, in abandoning the battle, Ned has effectively sided against those still trying to win the battle, adding substantially to the opposition they had to overcome. But even Ned is intuitively aware that he is practising delusion and so has to work hard at maintaining it. As explained in par. 704, maintaining a delusion meant constantly persuading yourself, and others, that you are right. Stridency and fanaticism characterised the behaviour of those maintaining a delusion, especially when, in becoming religious for instance, you were practically admitting that you were being deluded about being a sound, together, on track person yourself by having had to defer to a sound prophet.

746 In summary, the benefit of Imposed Discipline for the resigned, competitive way of living over the born-again, pseudo idealistic way of living was that it did not undermine a person's participation in humanity's great battle—it simply provided a means to manage the upset associated with that battle. However, since the religious born-again strategy both minimised the irresponsibility of abandoning the battle, and (despite the degree of delusion it still allowed) minimised the extreme denial involved in becoming born-again, it provided a marvellous way of coping with the by now extremely destructive and unbearable levels of upset and associated guilt that affected nearly the entire human race, and which Imposed Discipline could no longer contain. <u>In fact, because of its degree of honesty and indirect support of the search for knowledge, religion has been by far the most special, the most wonderful form of pseudo idealism to ever be developed. Indeed, it was religion that saved humanity from destruction through the most difficult final stages of its journey</u>.

747 However, while religion *did* save the human race, at the very end of our species' journey through ignorance other forms of pseudo idealism evolved that very nearly destroyed humanity. This final development will now be described and explained.

Chapter 7:16A The final 200 years when pseudo idealism takes humanity to a death-by-dogma, end play state of terminal alienation — the time when we need to, as the Bible strongly warned, 'Beware of false prophets', the merchants of delusion and denial, for they are 'the abomination that causes desolation'

748 We now need to consider the situation faced by humanity in the final 200 years of its now plummeting path to self-destruction through excessive upset, because it was during this final stage that the great benefit of religion, which was its honesty, became too confronting, forcing the development and adoption of other forms of pseudo idealism, including less honest forms of religion, that are *so* dishonest they are taking humanity to the brink of terminal alienation and extinction.

749 What happened around 200 years ago to dramatically increase the levels of upset in society were further serious advancements on the factors that caused the sudden increase in upset 11,000 years ago. Firstly, due in part to improvements in medicine and sanitation, the world's population exploded, leaving people in many parts of the world living virtually on top of each other: villages became towns, which became cities and even mega-city metropolises. Cities represent the most extreme congestion of people, and their development, along with their nature-eliminated, unnatural environment, which is so destructive of our innocent instinctive soul, greatly compounded the spread and increase of upset. Of course, once we humans became alienated, cities provided a refuge from the criticising innocence of the natural world—they were not created as functional centres where people could more efficiently work together, as often claimed, but as hide-outs for alienation—however, for the souls of the ensuing generations who had to grow up in such soul-*less* environments cities were devastatingly alien places; as the historian Manning Clark said, **'The bush** [wilderness] **is our source of innocence; the town is where the devil prowls around'** (*The Sydney Morning Herald*, 18 Feb. 1985); and as the poet Percy Bysshe Shelley wrote, **'Away, away, from men and towns, to the wild wood and the downs—to the silent wilderness, where the soul need not repress its music'** (*To Jane: The Invitation*, c.1820). No wonder Christ looked forward to the time when understanding of the human condition would be found and alienating cities could gradually be dismantled (as they now can be), saying, **'Do you see all these great buildings?...Not one**

stone here will be left on another; everyone will be thrown down' (Bible, Matt. 24:2; Mark 13:2; Luke 19:44, 21:6).

[750] (Since it is such a bold, impacting statement to say cities are going to be dismantled, I should explain it more fully. With understanding of the human condition now found the great rehabilitation of the human race from its alienated state can begin, and one of the important means to achieve that rehabilitation will be to reconnect humans with the world of their soul, which is the natural world. Nature isn't worth preserving simply because it might contain useful drugs, or because it is needed to replenish the atmosphere, or because of any of the other human-condition-avoiding, psychosis-denying, mechanistic justifications. No, as important as those evasive reasons are, we have to preserve nature because it is our original, natural, instinctive self or soul's home and without it our soul, our being, is destitute, a 'lost in the cosmos' itinerant wretch. As the historian Theodore Roszak wrote, **'"separation anxiety disorder"** [is defined] **as "excessive anxiety concerning separation from home and from those to whom the individual is attached." But no separation is more pervasive in this Age of Anxiety than our disconnection from the natural world'** ('The Nature of Sanity', *Psychology Today*, 1 Jan. 1996). Yes, as Cat Stevens/Yusuf Islam wrote and sang in 1970, **'Well I think it's fine, building jumbo planes...roll on roads over fresh green grass...cracked the sky, scrapers fill the air...I know we've come a long way, we're changing day to day, but tell me, where do the children play** [how is our soul supposed to live in such an alienated world]**?'** (*Where Do the Children Play*). In the 1960s many people tried to undertake the healing process that requires reconnecting with nature. They abandoned the alienated and alienating, soul-destroying cities and tried to return to the natural world. Of course, without the explanation of the human condition that would make it possible to stop having to live a denial-based, escapist, hide-from-condemning-nature-in-cities existence, this great **'alternative'**, **'hippie'**, **'flower power'**, **'let the sunshine in'**, **'back to nature'** revolution was a false start back to innocence and wholeness. Nevertheless, the initiatives that were being taken accurately anticipated the direction the rehabilitation of the human race needed to, and now can, take. Yes, we have to leave the alienated and alienating cities and dust off the *Whole Earth Catalog* and other 1960s masterpieces of new world planning and begin the *real* 'Age of Aquarius'. And I should mention that because we have understanding of the human condition there won't be the need for drugs to try to break open our alienated minds to access the authentic world of

our soul and all the truth it has access to, or the need to indulge in 'free sex' to allow our upset self to feel artificially liberated from the human condition—both of which were activities that featured in the 1960s. We have found all the truth now that *genuinely* releases us from the binds of the human condition. As will be described in the final chapter of this book, what now emerges on Earth is not another messy, deluded false start to humanity's freedom from the human condition, but the real, genuinely loving one.)

[751]The second immense influence on the spread and growth of upset and thus alienation in the world has been the development of communication technology of such sophistication that, in terms of one upset human's access to another upset human, the world has basically shrunk down to one giant city—worse, one immense, dysfunctional household. Initially, there was the delivery of letters through a sophisticated postal system, and then mass printing of documents and newspapers, and then typewriters, and then the telegraph, and then the telephone, and then television, computers and faxes, and then emails and mobile phones, and then the world wide web and social media like Facebook and Twitter, and now 'smart' phones that enable around the clock access to this world of alienated superficiality. From birth, humans today are immersed in an ocean of upset behaviour—especially dishonest, deluded, artificial, superficial, **'fake'** and **'phony'** alienated behaviour—leaving them utterly overwhelmed by anxiety and stress.

Chapter 7:16B The emergence of terminal levels of alienation in the 'developed' world

[752]The overall effect of such extremely congested living and the increase in the efficiency of communication technology is that it has led to a crescendo of upset—especially alienation from our natural, nurturing, all-loving and all-sensitive true self or soul. These quotes—which describe what is happening at the forefront of human progress, namely in the materially so-called 'developed', mostly Western world, and specifically in the USA—provide frightening evidence of this epidemic of alienation: **'today's children are probably the least loved generation of all'** (Robert de Grauw, Letter to the Editor, *TIME*, 3 Apr. 2006); **'96 percent of American families are now dysfunctional'** (John Bradshaw quoted in *The Australian*, 8 May 1993); **'The 1990 US Census stated there will be more stepfamilies than original families by the**

year 2000, and that 66 percent of those stepfamilies break up when children are involved' (The Stepfamily Foundation, accessed Sep. 2011; see <www.wtmsources.com/104>); **'one in two US children will live in a single-parent family at some point in childhood'** (US Census Bureau of Household and Family Statistics, 2000); **'63 percent of the 18.5 million US children under 5 years of age were in some type of regular child care arrangement'** (US Census Bureau, 2005); **'the electronic age has ushered in electronic parenting. Kids spend far more time sitting passively before a device such as a computer or television than they do playing or speaking with their families'** (*The Commercial Appeal*, 26 Aug. 2001). As one journalist wrote, **'It is easier to get lost in a computer game than to deal with a dysfunctional family'** (David Boddy, 'The Facebook generation is in the grip of National Attention Deficit Disorder', *The Telegraph,* 4 Mar. 2013). Australia is one of the most sheltered and isolated and thus innocent countries left in the world, and yet our society is also about to unravel with psychological suffering, as indicated by a 2011 study whose findings were reported on the front page of *The Sydney Morning Herald*: **'The well-being of Australia's children and young people has declined alarmingly in the past decade—and plunging marriage rates are partly to blame, a major study has found. Growing rates of child abuse and neglect, of children being placed in foster care, and of teenage mental health problems, including a rise in hospital admissions for self-harm, are rooted in the rise of one-parent families and de facto couples, violent and unstable relationships, and divorce, the report says'** ('Decline in marriage blamed for neglect', 6 Sep. 2011). Humans are now so alienated, so divorced from their true self, they can barely live with themselves, let alone anyone else. The real reason for the breakdown in relationships is alienation, which in turn is a direct result of the human race's heroic but ever-increasing state of upset, a state that could only be ameliorated through the finding of compassionate understanding of the human condition.

753 The result of all this rapidly accumulating **'dysfunction'**—the lack of a soulful, sound, alienation-free, nurturing and loving functional environment—is ever-increasing levels of psychological pain from disorientation and depression, and, to cope with that, even more block-out/denial/alienation, a process that is ending in the extreme disconnecting, dissociative behaviours of Attention Deficit Hyperactivity Disorder (ADHD) and autism, as these statistics confirm: **'someone born since 1945** [is] **likely to be up to 10 times more depressed than their grandparents'** (psychologist Michael Yapko, *Hand-Me-Down Blues*, 1999); **'Depression...is the leading cause of disability in the US and abroad'** (Andrew Solomon, *The Noonday Demon*, 2001, p.25 of 560);

there has been a **'41% increase in ADHD in US children during the past decade'** (*TIME*, 15 Apr. 2013); **'This** [today's US society] **is an ADD epidemic in the making'** (Katrin Bennhold, 'Generation FB', *The New York Times*, 23 Jun. 2011); and, **'Our country** [the UK] **is in the grip of a national attention deficit syndrome'** (David Boddy, 'The Facebook generation is in the grip of National Attention Deficit Disorder', *The Telegraph*, 4 Mar. 2013).

754 In a 2013 article titled 'Once More on Attention Deficit Disorder', the creator of primal therapy, the American psychologist Arthur Janov, wrote that **'there are more and more cases of ADD among children** [in the US], **one in five to be exact...So what is going on?...The brain** [of the ADD child] **is busy, busy, dealing with the pain'** from **'childhood', 'where there was no** [real, unconditional] **love'**, only **'conditional love'** where, for example, **'parents wanted smart kids'** to fulfil their own insecure ego's desire for reinforcement; with the result that **'when there is stimulation from the outside...it meets with a very active brain which says "Whoa there. Stop the input. I have too much going on inside to listen to what you ask for".'** Janov concluded, **'You know, ADD is also called the hyperactivity disorder (ADD HD: Hyperactive Disorder). Of course, the kid is agitated out of his mind, driven by agony inside. We want her to focus on 18th century art and she is drowning in misery'** (4 Apr. 2013; see <www.wtmsources.com/132>). If we truly want some insight into what it *actually* is that is causing children and young adolescents to be **'agitated out of his** [their] **mind', 'drowning in misery'**, consider again the unresigned, denial-free honesty of these lyrics from the 2010 *Grievances* album of the young American heavy metal band 'With Life In Mind' that were included in par. 121: **'It scares me to death to think of what I have become...I feel so lost in this world', 'Our innocence is lost', 'I scream to the sky but my words get lost along the way. I can't express all the hate that's led me here and all the filth that swallows us whole. I don't want to be part of all this insanity. Famine and death. Pestilence and war.** [Famine, death, pestilence and war are traditional interpretations of the 'Four Horsemen of the Apocalypse' described in Revelation 6 in the Bible. Christ referred to similar **'Signs of the End of the Age'** (Matt. 24:6-8 and Luke 21:10-11) and all of these descriptions are accurate because such extreme disintegration *is* the state of terminal alienation that occurs in the final stages of humanity's heroic search for knowledge.] **A world shrouded in darkness... Fear is driven into our minds everywhere we look', 'Trying so hard for a life with such little purpose...Lost in oblivion', 'Everything you've been told has been a lie...We've all been asleep since the beginning of time. Why are we so**

scared to use our minds?', 'Keep pretending; soon enough things will crumble to the ground...If they could only see the truth they would coil in disgust', 'How do we save ourselves from this misery...So desperate for the answers...We're straining on the last bit of hope we have left. No one hears our cries. And no one sees us screaming', 'This is the end.'

755 So it's no wonder children have a massive need for distraction from the overwhelming internal pain resulting from growing up in a fearful world without any real love or '**hope**' or '**answers**', which in the more extreme cases has resulted in ADHD. (In the case of the most extreme state of dissociation from the '**screaming**' '**agony inside**' of autism, I will look closely at that next in ch. 7:16C.)

756 Regarding Janov's comment that '**We want her to focus on 18th century art and she is drowning in misery**', in 2013 an art teacher at one of Sydney's leading private schools told me that '**while only two years ago students were able to sit through a half hour art documentary, I now know I lose them after only eight minutes; today's students' attention span is that brief!**' (WTM records, 26 Apr. 2013). This comment mirrors an observation made by the political scientist David Runciman in a 2010 BBC documentary series about the world wide web: '**What I notice about students from the first day I see them when they arrive at university is that they ask nervously "What do we have to read?" And when they are told the first thing they have to read is a book they all now groan, which they didn't use to do five or ten years ago, and you say, "Why are you groaning?", and they say "It's a book, how long is it?"**' (*Virtual Revolution*, episode 'Homo Interneticus'). The same documentary also includes the following statement from Nick Carr, the author of *Is Google Making Us Stupid?*: '**I think science shows us that our brain wants to be distracted and what the web does by bombarding us with stimuli and information it really plays to that aspect of our brain, it keeps our brain hopping and jumping and unable to concentrate.**' Yes, what the internet is *really* providing is not access to more knowledge and understanding as many would have us believe, but distraction from pain. As Janov pointed out, the young person of today is increasingly '**agitated out of his mind, driven by agony inside...drowning in misery**', with '**too much going on inside to listen**' and '**focus**'. Yes, as Carr said, their '**brain wants to be distracted**'. In the case of social media such as Facebook and Twitter, it allows people to be preoccupied/distracted (from the human condition) all day long with inane, frivolous, narcissistic, superficial self-promotion and gossip. The result of this extreme distraction from the '**agony inside**' is that '**The youth**

of today are living their lives one mile wide and one inch deep' (Kelsey Munro, 'Youth skim surface of life with constant use of social media', *The Sydney Morning Herald*, 20 Apr. 2013). Yes, **'the net delivers this shallow, scattered mindset with a vengeance'** ('The effects of the internet: Fast forward', *The Economist*, 24 Jun. 2010). As one member of the 'Y generation' self-analysed, **'Alone and adrift in what** [Professor] **de Zengotita calls our "psychic saunas" of superficial sensory stimulation, members of my generation lock and load our custom iTunes playlists, craft our Facebook profiles to self-satisfied perfection, and, armed with our gleefully ironic irreverence, bravely venture forth into life within glossy, opaque bubbles that reflect ourselves back to ourselves and safely protect us from jarring intrusions from the greater world beyond'** (Tom Huston, 'The Dumbest Generation? Grappling with Gen Y's peculiar blend of narcissism and idealism', *EnlightenNext*, Dec. 2008-Feb 2009). A 2012 article in the *Smithsonian* magazine summed up this shallow and superficial approach to thinking today, stating that **'Suddenly thanks to Google Books, JSTOR and the like, all the great thinkers...are one or two clicks away...And yet—here is the paradox—the wisdom of the ages is...buried like lost treasure beneath a fathomless ocean of online ignorance and trivia that makes what is worthy and timeless more inaccessible than ever'**; the **'digital revolution...**[is] **decapitating our culture, trading in the ideas of some 3000 years of civilization for BuzzFeed'** (Ron Rosenbaum, 'Lewis Lapham's Antidote to the Age of BuzzFeed', Nov. 2012). A reviewer of my writing, who was deeply concerned about the profundity of my insights into the human condition not being heard in this superficial, trashy digital age, summed up the danger of this phenomenon when he wrote that **'We have a lot of competing noise for our attention these days, and it would be criminal to let that overwhelm our true potential, by masking** [the] **useful information** [in Griffith's books] **with hideous noise'** ('Fitzy', *Humanitus Interruptus – Great Minds of Today*, 21 Oct. 2011; see <www.wtmsources.com/106>).

[757]The labels that have been given to recent generations are actually very telling of the rapidly escalating levels of alienation we are now seeing in society—and which the internet, while providing superficial relief for, is propagating and spreading at an exponential rate. While previous generations have been given the positive monikers the 'Builders' and the 'Baby Boomers', the generations who have followed are referred to as the 'X generation', the 'Y generation', and now the 'Z generation', which, according to Wikipedia, comprises **'people born between the mid-1990s and late 2000s'**. The Canadian writer Douglas Coupland defined a Generation X'er as one who **'lives an X sort of life—cerebral, alienated, seriously concerned**

with cool' (*The Sydney Morning Herald*, 22 Aug. 1994)—qualities that are all associated with having had to adjust to an extremely soul-devastated existence. The adolescent psychologist Michael Carr-Gregg said that **'Generation Y is being ravaged by depression, anxiety and stress disorders'** (Miranda Devine, 'Face it, we are all narcissists now', *The Sydney Morning Herald*, 3 Sep. 2009). So what exactly did we mean when we applied such valueless, end play lettering to these X, Y, Z generations? The answer is that they each represent an advanced state of terminal alienation. And the question left hanging is, what comes after Z?

758 In 2013 *TIME* magazine featured a story on the Y-generation titled **'The Me Me Me Generation'** (20 May 2013). Being one of the many frighteningly dangerous pseudo idealistic, left-wing magazines now—a phenomenon I will talk about shortly in chs 7:16H to 7:16Q—*TIME* tried to put an almost completely dishonest, positive spin on the present *extremely* narcissistic, *chronically* alienated **'80 million strong...biggest age group in American history'** Y-generation. In small text, the introduction to the article said, truthfully enough, that the Y-generation's **'self-centeredness could bring about the end of civilization as we know it'**, but then promoted them, in very large text, with the reverse-of-the-truth lie, as **'THE NEW GREATEST GENERATION: WHY MILLENNIALS WILL SAVE US ALL'**! At least, hidden away in this nine-page cover story about this **'narcissism epidemic'**, was the admission that narcissism is a **'personality disorder'** that occurs **'when people try to boost self-esteem'**, revealing that the *real* problem is lack of self-esteem, insecurity of self—alienation. What this article is *really* reporting on then, if it was to be honest, is how the present, **'80 million strong...biggest age group in American history'**, Y-generation exists in a state of terminal alienation, and the dire implications such an epidemic poses for the future of our species. (One of the main psychological reasons for the loss of self-esteem that is causing this **'narcissism epidemic'** will be looked at shortly in ch. 7:16D.)

759 A 2008 documentary on the destructive effects of all the consumer advertising directed at children, titled *Consuming Kids: The Commercialization of Childhood*, features a montage of powerful news clips from American television that provides a graphic snapshot of the symptoms of terminal levels of alienation being reached in society. In the first clip the newsreader reported that **'Forty times as many young people are now being diagnosed with bipolar disorder than 13 years ago** [bipolar disorder is the less confronting term now being used for what was once more truthfully termed

'manic depression', the state of extreme depression where the mind can only break free of it occasionally, resulting in oscillating bouts of euphoric relief then more depression].' The second clip reported that **'Almost 4.5 million children in this country have been diagnosed with ADHD.'** The third clip said that **'Doctors are writing a growing number of prescriptions for anti-depressants for children, as many as eight million a year.'** The fourth clip reported that **'One in three children born in the year 2000 will develop diabetes.'** The fifth clip said that **'For the first time in decades the rate of hypertension in children is rising** [with a medical journal cover in background saying] **2 million + American children may have high blood pressure.'** And, finally, the sixth clip reported that **'This generation of children is the heaviest in American history. An estimated 16% of all children and teenagers are overweight—four times as many since the 1960s. Life expectancy of children today will be shorter than that of their parents—the first such decline in modern times'** (Media Education Foundation).

760 Yes, with upset becoming completely unbearable for humans, the distress it entails must be straining our bodies' ability to cope to the very limit, which is why diseases like cancer now riddle society. Indeed, descriptions of the healthiness of humans before the upset state of the human condition emerged reveal that most of our present sicknesses must be psychosomatic (emotional) in origin: consider, for instance, Hesiod's description that **'When gods alike and mortals rose to birth / A golden race the immortals formed on earth...Like gods they lived, with calm untroubled mind...Nor e'er decrepit age misshaped their frame...Strangers to ill, their lives in feasts flowed by'** (*Theogony*, eighth century BC); and Buddhist scripture's anticipation of what humans will be like once the ameliorating understanding of the human condition is found and absorbed, the time when humans **'will with a perfect voice preach the true Dharma, which is auspicious and removes all ill'**, and **'Human beings are then without any blemishes, moral offences are unknown among them, and they are full of zest and joy. Their bodies are very large and their skin has a fine hue. Their strength is quite extraordinary'** (Maitreyavyakarana; *Buddhist Scriptures*, tr. Edward Conze, 1959, pp.238-242); and the time anticipated in the Bible when **'Another book** [will be]**...opened which is the book of life** [the human-condition-explaining and thus humanity-liberating book]...[and] **a new heaven and a new earth** [will appear where]**...There will be no more death or mourning or crying or pain'** (Rev. 20:12, 21:1, 4). Sally Fallon's report that was referred to earlier when discussing the findings of Dr Weston Price (in par. 576), that **'Many**

others have reported a virtual absence of degenerative disease, particularly cancer, in isolated, so-called "primitive" groups', corroborates this conclusion. All the millions of health clinics in the world are not going to get humans back to full health; *only* the ameliorating understanding of the human condition being presented in this book can achieve that.

761 Alongside ever-increasing anxiety-driven disorders like obsessive compulsive behaviour, ADHD and autism, and unhappiness-driven excessive food consumption and physical sickness, the emotional lives of adolescents are also becoming shallower and cheaper, as this report indicates: **'the sexualisation of Western culture** [has meant] **that sex has been robbed of its emotional depth...For young men and women, it's increasingly a physical activity, with no real pleasure and no meaning at all...one hears of lipstick parties, where teenage girls wearing different coloured lipstick line up to give oral sex to boys with the aim of giving them a candy-striped penis'** (Clive Hamilton, co-author of the report 'Youth and Pornography in Australia', *The Australian*, 24 Jun. 2006). It has to be remembered that, as was explained in par. 609, sex is an attack on innocence, so treating it cheaply is another factor contributing to the escalating levels of hurt/upset/alienation in society. Also playing a huge role in the-death-of-soul-through-sex in the world is the frighteningly easy access to pornography that the internet allows, as well as the sexually predatory behaviour it can facilitate. (The soul-destroying effects of sex in the lives of women especially will be looked at next in ch. 7:16C.)

762 Yes, overall, hugely improved communication technology has spread upset everywhere, infected the world with horror, as this reference to violent video games by the writer and broadcaster James Delingpole recognises: **'What troubles me about Grand Theft Auto V — which has an 18 rating that will be ignored by thousands of younger teenagers — is not just the message it sends out to youngsters (drugs are cool; crime pays; violence is fun), but what it says about the coarsening, the decadence and the hopelessness of our modern culture. It's the electronic equivalent of those gladiatorial contests the Romans used to stage in the dying days of their empire, involving ever more exotic beasts and ever more elaborate sets. It may be entertaining, particularly to young men with a penchant for such nihilistic spectacle, but the sensibility to which it appeals is warped, jaded and riddled with the deepest, blackest despair. The fact that this is the most popular computer game on the market should make us all shudder'** ('Torture and murder with the addictive glamour of Hollywood', *Daily Mail*, 18 Sep. 2013).

[763] In March 2014 an article by the British psychiatrist Anthony Daniels, writing under the pen name Theodore Dalrymple, was published in *The Australian* newspaper. Aptly titled 'Next generation has apocalypse written all over it', the piece provides further stark evidence of the emergence of terminal levels of alienation in the Western world: **'The end is nigh! French civilisation is on the verge of collapse! No one who attended the recent Mondial du Tatouage (the World Tattoo Fair) in Paris could be in any doubt about it. It was held, appropriately enough, in the old abattoir at the end of the Avenue Jean Jaurès...surrounded by a wasteland of concrete monstrosities, French modern architecture now being some of the worst in the world...Perhaps the rejection of beauty as a goal by French architects — their work seeming rather to exclaim "F..k off, humanity!" — accounts in part for the adoption as a style by so many of the young French of deliberate ugliness and self-mutilation. In a world of brutal ugliness over which you have no control you might as well admit defeat and join in. Thousands queued to enter the fair... Many of them had already permanently disfigured themselves and, since their tattoos would have quite likely cost hundreds or thousands of euros, poverty was not the explanation for their degradation. Inside the abattoir, there were about 250 stands from across the Western world...It was only to be expected that ugly loud tuneless rock music of the kind that makes thought impossible and speech difficult was poured into the atmosphere like poison gas. Customers lay down on couches to have dragons, skulls, vampires, insects, rats or Elvis Presley permanently inscribed on their legs, backs, arms, chests...One man held a baby in his arms as a tattooist inscribed a snake on his back. You are never too young to be indoctrinated with nihilism...Two...men...tattooed so heavily that only a few years ago they would have been considered degenerate freaks, patted small children on the head as their delighted parents proudly took photographs of them...[A] tattooed slogan...on the front of a young woman's upper thigh and knee, read "We are like roses that have never bothered to bloom when we should have bloomed and it is as if the sun has become disgusted with waiting."...Some of the names of the tattooists' enterprises were revealing: Evil from the Needle, Clod the Ripper, Perfect Chaos, Black Heart...Modern Butcher...Ten years ago there were 400 professional tattooists in France; now there are 4000. In England, where so much of the worst of modern culture originated, more than a third of young adults are tattooed. Civilisations collapse from within'** (17 Mar. 2014). Yes, this situation where the **'Next generation has apocalypse written all over it'** and **'civilisation is on the verge of collapse'** *is* because **'the sun has become disgusted with waiting'**, the liberating light of understanding of the human

condition *has* been delayed for far too long. Science MUST now let these understandings through.

764 (A more complete description of how alienated humans, especially humans in the materially 'developed' Western world, have become can be found in *Freedom Book 1* at <www.worldtransformation.com/freedom-alienation>.)

Chapter 7:16C The dire consequences of terminal levels of alienation destroying our ability to nurture our children

765 A dire consequence of the arrival of terminal levels of alienation—especially at the forefront of human progress, namely in the materially 'developed' world, fuelled as it has been by advances in communication technology—is the near complete inability now of mothers to nurture their offspring.

766 Recently, one of our WTM Founding Members, who for 15 years has been a nurse and midwife in major hospitals in Australia and England, told me how, in just the last few years, there has been a shocking increase in the number of mothers who are unable to relate naturally to and connect empathetically with their newborn babies. She said that while it has long been the case that there were some mothers who couldn't access their natural, nurturing, loving instincts, the problem is now so common it is a relief to actually care for a mother who *can* relate naturally to her infant. The result of this predominance of mothers who are unable to empathetically connect with (basically, love) their infants is the epidemic we are now seeing of highly distressed children—in the extreme, children who display the dissociative disorders of ADHD and autism.

767 The reason why women have become so alienated from their natural, nurturing, all-loving true self or soul was provided earlier when the different roles men and women have had to carry out in humanity's upsetting battle to find knowledge, ultimately self-knowledge, understanding of the human condition was described (in ch. 7:11). Very briefly, in keeping with their already established roles as the group protectors, it was men who took up the loathsomely upsetting job of championing the ego over the ignorance of our original instinctive self, leaving women to maintain their loving, nurturing role. The problem this role differentiation gave rise to, however, was that in not being responsible

for or participating in the terrible innocence-destroying battle to overthrow ignorance, women were naive or unaware of the ramifications of fighting the battle, and, as a result, were unsympathetic towards both the battle and the frustrated anger and egocentricity it produced in men—a situation that placed men in the awful predicament of being both misunderstood and unjustly condemned by women. Unable to explain their behaviour to women, men were left in an untenable situation: they couldn't just accept women's unjust criticism of their behaviour, so what men did was turn on and attack the relative innocence of women. Men perverted sex, as in 'fucking' or destroying women's innocence, the result being that women came to share men's lack of innocence, their upset. Basically, through having to support men in their critically important task of championing the ego by sexually 'comforting' them, by inspiring them with their image of innocence and by giving them lots of attention, women were left with little innocent, soulful soundness and little time for the immensely emotionally and energy consuming task that is the nurturing of infants. The necessary consciousness-centred, 'ego-centric', male-led battle unavoidably intruded upon and compromised women's ability to nurture their offspring. Yes, having to nurture children with love has been an altogether impossible task for women while the battle of the human condition raged around and within them. And, obviously, as the upsetting battle intensified, the all-loving and all-generous souls of both men and women were eventually going to become completely exhausted, leading to the development of terminal levels of alienation—the result for women being that they would find themselves no longer able to emotionally relate to and connect with their infants, no longer able to give them the unconditional love they have come to expect from our species' love-indoctrinated, nurtured origins, which is the situation that has now developed, with disastrous consequences.

768 Of course, the now *extremely* embattled, egocentric state of men is also having a devastating effect on children. Children need an *environment* of love and reinforcement and men's preoccupation with their egocentric battle, together with the now extremely strained relationship between men and women, have all but destroyed the possibility of creating that nurturing, loving, reinforcing environment for children. The effect of men's embattled, egocentric state on children will be described next in ch. 7:16D—what will be looked at here, however, is the effect of terminal levels of alienation on the primary nurturer of infants, which is the mother.

[769]Before doing so, it needs to be reiterated that this inability of parents to nurture their offspring is *not* their fault but the product of the ever-accumulating levels of upset in the human race as a whole, which in turn is due to our species' heroic *2-million-year* search for knowledge, specifically self-knowledge, understanding of the human condition.

[770]As just stated, the devastating result of the arrival of terminal levels of alienation amongst women and, with it, the loss of their ability to nurture their offspring, is an epidemic of highly distressed children—in the extreme, children who display the dissociative disorders of ADHD and autism. The epidemic of ADHD in society was just described in ch. 7:16B, but similar reports are occurring for the rising incidence of autism; for instance, **'Autism is a serious behavioral disorder among young children that now occurs at epidemic rates. According to a recent article, autism is diagnosed in America about 1 of every 120 girls and 1 in every 70 boys. In some areas, it is far more prevalent. It is now occurring around the world, especially in the developed nations of the world. According to California records, autism has increased 1,000% in the past 20 years! Autism used to account for 3% of the caseload of the California Department of Economic Security. As of 2012, it accounts for at least 35% of the caseload and costs taxpayers about $2 million per child!'** (Dr Lawrence Wilson, *The Autism Epidemic and Natural Solutions*, Dec. 2012; see <www.wtmsources.com/130>), and this reference to **'the huge increase in autism that we see these days...**[with] **California's Silicon Valley** [being] **home to more autistic children than anywhere** [else] **in America'** (*Uncommon Genius*, ABC, 24 May 2001). Also, **'in China it's estimated 7.5 million children are autistic, but many lack proper diagnosis'**, the inference being there are probably many more cases going undiagnosed within that massive population ('Children of the Stars', *Dateline*, SBS, 2 Apr. 2013).

[771]Since autism represents the extreme state of the distress that nearly *all* children are now suffering from, an analysis of what causes it will most clearly and graphically reveal the primary source of that now almost-universal distress within children.

[772]As with Janov's lack-of-real-love explanation for ADHD, autism is most often a disastrous consequence of a mother's inability to emotionally relate to and empathetically connect with her newborn baby. While a few cases of autism are caused by physical trauma to the brain, most are the product of extreme instances of infants not receiving the unconditional love, in particular from their mother, that their instincts expect, which forces them to psychologically dissociate from their reality in

order to cope with the violation and hurt to their instinctive self or soul. Able to safely admit now that our species' original instinctive orientation was to living in a harmonious, happy, loving, non-alienated, non-egocentric, non-angry state, and that children enter the world fully expecting to encounter such a state, we can understand the incredible shock it must be for them to be met instead with the extreme upset that now characterises all of humanity. Understandably, many children can't psychologically bridge the gap and have to adopt all manner of coping mechanisms to block-out that anguish, including forms of compulsive and obsessive distraction, which is why we are seeing these epidemic rates of ADHD and autism. And the reason autism is, as mentioned in the report cited above, more prevalent in boys than girls is because, as was explained in pars 596-647, boys are both more aware of the problem of the human condition and responsible for solving it, so they are more attuned to, and thus more terrified and distressed by, the extent of the problem of the human condition in the world.

773 In his 1996 posthumously published book *Thinking About Children*, the former president of the British Psychoanalytical Society, the psychiatrist and paediatrician D.W. Winnicott (who has been described as **'one of the most influential contributors to psychoanalysis since Freud...**[and someone who] **brought unprecedented skill and intuition to the psychoanalysis of children'** (Adam Phillips, *Winnicott*, 1989, back cover)) provided the following honest description of the cause of autism. He wrote that in **'a proportion of cases where autism is eventually diagnosed, there has been injury or some degenerative process affecting the child's brain...**[however,] **in the majority of cases...the illness is a disturbance of emotional development...autism is not a disease. It might be asked, what did I call these cases before the word autism turned up. The answer is... "infant or childhood schizophrenia"** [p.200 of 343] [note, 'schizophrenia' literally means 'broken soul', derived as it is from *schiz* meaning **'split'** or **'broken'**, and *phrenos* meaning **'soul or heart'**]**...There are certain difficulties that arise when primitive things are being experienced by the baby that depend not only on inherited personal tendencies but also on what happens to be provided by the mother. Here failure spells disaster of a particular kind for the baby. At the beginning the baby needs the mother's full attention, and usually gets precisely this; and in this period the basis for mental health is laid down** [p.212] **...the essential feature** [in a baby's development] **is the mother's capacity to adapt to the infant's needs through her healthy ability to identify with the baby. With such a capacity she can, for instance, hold her baby, and without it she cannot**

hold her baby except in a way that disturbs the baby's personal living process...It seems necessary to add to this the concept of the mother's unconscious (repressed) hate of the child [p.222] **...it is the quality of early care that counts. It is this aspect of the environmental provision that rates highest in a general review of the disorders of the development of the child, of which autism is one** [p.212] **...Autism is a highly sophisticated defence organization. What we see is *invulnerability*** [p.220] **...The child carries round *the (lost) memory of unthinkable anxiety*, and the illness is a complex mental structure insuring against recurrence of the conditions of the unthinkable anxiety** [that results from the mother's failure to provide her **full attention**] [p.221]'.

[774]The following is a typical case history of an autistic child from one of the many documented by Winnicott: '**When I first saw Ronald at the age of 8, he had very exceptional skill in drawing...Apart from drawing he was, however, a typical autistic child...The mother herself was an artist, and she found being a mother exasperating from one point of view in that although she was fond of her children and her marriage was a happy one, she could never completely lose herself in her studio in the way that she must do in order to achieve results as an artist. This was what this boy had to compete with when he was born. He competed successfully but at some cost...At two months the mother remembers smacking the baby in exasperation although not conscious of hating him. From the start he was slow in development...His slowness made him fail to awaken the mother's interest in him, which in any case was a difficult task because of her unwillingness to be diverted from her main concern which is painting**' (ibid. pp.201-202).

[775]In his 1967 book, the revealingly titled *The Empty Fortress*, the psychologist Bruno Bettelheim similarly maintained that autism '**did not have an organic basis, but resulted when mothers withheld appropriate affection from their children and failed to make a good connection with them. The most extreme expression of this concept suggested that mothers literally did not want their children to exist**' (Wikipedia, accessed Dec. 2013; see <www.wtmsources.com/126>). As a result of his findings, Bettelheim was a prominent proponent of the '**refrigerator mother**' theory for autism—named thus for the cold-heartedness of what we can, as emphasised, now understand is essentially *all* humans' unavoidable-after-2-million-years-of-struggle, human-condition-afflicted, immensely alienated, soul-disconnected, psychotic and neurotic state.

[776]Dialogue from the 1993 film *House of Cards* features this honest description of the dissociative behaviour involved in autism: '**People say**

about the following categories that these kids have a problem or are disabled, or psychologically dumb, etc, but really they are children, through hurt or some kind of trauma, that have held onto soul, and not wanted to partake in reality—retarded, autistic, insane, schizophrenic, epileptic, brain-damaged, possessed by devils, crocked babies.' Yes, in their innocent vulnerability, the main way children have coped with the wounds to their instinctive self has been to make themselves **'invulnerabl**[e]**'** to the pain by splitting themselves off, dissociating themselves, from the world of that anxiety and trauma. And to be able to maintain the denial and thus dissociation from the extreme pain of their circumstances required constant application of the denial/block, which is why autistic people tend to be compulsive and obsessive in their behaviour. They escape into repetitive activities and tend to develop a one-track mind. In his 1997 book *Next of Kin*, the psychologist Roger Fouts described autism as **'a developmental disorder characterized by lack of speech and eye contact, obsessive and repetitive body movements, and an inability to acknowledge the existence or feelings of other people. The autistic child lives in a kind of glass bowl, inhabiting a separate reality from those around him'** (p.184 of 420).

777 Yet again, it has to be emphasised that since the emergence of the necessary battle with the human condition some 2 million years ago, *all* mothers have to some degree been 'refrigerator mothers'—indeed, the truth is that *all* humans who are resigned are alienated or split off from their soul, they are all schizophrenic, they all have a **'split soul'** or **'broken heart'**; they are all, to a degree, autistic. What we are seeing now is just so much more of this **'split'**, **'broken heart**[edness]**'**. Winnicott even acknowledged this, that autism is the extreme degree of **'a universal phenomenon'**: **'It has always seemed to me that the smaller degrees of disturbance of the mind that I am trying to describe are common and that even smaller degrees of the disturbance are very common indeed. Some degree of this same disturbance is in fact universal. In other words, what I am trying to convey is that there is no such disease as autism, but that this is a clinical term that describes the less common extremes of a universal phenomenon'** (*Thinking About Children*, p.206).

778 In pars 368-369, extracts were included from the writings of Olive Schreiner and Ashley Montagu that truthfully describe the importance of nurturing love in the lives of humans. Schreiner was quoted as saying, **'They say women have one great and noble work left them, and they do it ill... *We* bear the world, and *we* make it. The souls of little children are marvellously delicate and tender things, and keep for ever the shadow that first falls on them,**

and that is the mother's or at best a woman's. There was never a great man who had not a great mother—it is hardly an exaggeration. The first six years of our life make us; all that is added later is veneer...The mightiest and noblest of human work is given to us, and we do it ill.' In the case of Montagu, he also spoke of the dire consequences of mothers' inability to give their infants real love: **'The newborn baby is organized in an extraordinarily sensitive manner...he wants love...If it doesn't receive love it is unable to give it—as a child or as an adult. From the moment of birth the baby needs the reciprocal exchange of love with its mother...The infant can suffer no greater loss than deprivation of the mother's love...maternal rejection may be seen as the "causative factor in...every individual case of neurosis or behavior problem in children."...In an age in which a great deal of unloving love masquerades as the genuine article, in which there is a massive lack of love behind the show of love, in which millions have literally been unloved to death, it is very necessary to understand what love really means...**[and] **its importance for humanity'**.

779 With regard to this situation where **'a great deal of unloving love masquerades as the genuine article, in which there is a massive lack of love behind the show of love'**, given how upset and thus insecure about their sense of worth, in truth, hateful of themselves all resigned humans have become, how could mothers be expected to be able to give real, unconditional love to their children. You can't be loving when you are virtually devoid of love; all you can do is try to imitate love, but infants, who are not yet alienated, can see through that pretence and somehow have to try to adapt to it. And the more upset and alienated the human race has become, the less real love there has been for the next generation, so that, in the end, there is virtually no real, unconditional love anywhere anymore, which means that almost everyone is now, to some degree, autistic!

780 Yes, children need and expect *so* much love and among the now rare, more innocent, less human-condition-embattled races they have generally received it. Earlier (in par. 576), references were made to traditional Australian Aboriginal society where **'All observers agree upon the extraordinary tenderness which parents display towards their children, and indeed, to all children whether of their own family and race or not'**, and to the Bushmen of the Kalahari where **'Their love of children, both their own and that of other people, is one of the most noticeable things about the Bushman'**, the **'Bushmen...mother carries her child with her at all times up to four years of age'**, **'Children are breast-fed for up to 3½ years, and among the Bushmen lactation suppresses ovulation'**, **'!Kung** [Bushmen]**...infants hardly ever cry.'**

781 In her 1975 book *The Continuum Concept* the writer Jean Liedloff, who spent time living with one of the relatively innocent indigenous tribes in Venezuela, emphasised the need to give infants the caring treatment **'which is appropriate to the ancient continuum of our species inasmuch as it is suited to the tendencies and expectations with which we have evolved'** (p.35 of 168) in order for them to have **'a natural state of self-assuredness, well-being and joy'** ('Understanding The Continuum Concept', The Liedloff Continuum Network, accessed Dec. 2013; see <www.wtmsources.com/105>). She also truthfully recognised that levels of alienation of people in the 'developed world' now is such that it has overwhelmed their natural instincts for nurturing, writing that **'We have had exquisitely precise instincts, expert in every detail of child care, since long before we became anything resembling *Homo sapiens*. But we have conspired to baffle this longstanding knowledge so utterly that we now employ researchers full time to puzzle out how we should behave towards children, one another and ourselves'** (*The Continuum Concept*, p.34). Mothers now desperately read book after book about 'how to be a mother', how to imitate real, loving nurturing, when the real barrier to being able to do so naturally is the extent of their alienation from their natural instinctive true self or soul, and beyond that the problem of the human condition, issues that none of these superficial, faddish books go near.

782 Journalist Betty McCollister was another exception to this evasive phenomenon, someone who wasn't at all **'baffle**[d]**'** about **'how we should behave towards children'**. In an article titled 'The Social Necessity of Nurturance' (*The Humanist*, Jan. 2001), she wrote honestly about how infants need **'unconditional love'** from their mothers, and acknowledged the devastating consequences in society now of mothers not being able to provide it. While the bold attempt she made to explain why nurturing became so important during our species' development was not at all correct, McCollister did accurately and truthfully recognise that **'Our evolution has resulted in a species whose infants can't thrive without continual, loving attention. Here, then, is the clue to raising fewer unhappy, alienated, violent youth for jail fodder...Every human infant must have unconditional love; without it, an infant's health and growth will be stunted...Anthropologists, neurologists, child psychiatrists, and all other researchers into child development unequivocally agree and have sought for decades to alert society. For example: ... Ashley Montagu (anthropologist): "The prolonged period of infant dependency produces interactive behavior of a kind which in the first two years or so of the child's life determines the primary pattern of his subsequent social development."**

Alfred Adler (psychiatrist): "It may be readily accepted that contact with the mother is of the highest importance for the development of human social feeling..." Selma Fraiberg (child psychologist): A baby without solid nurturing "is in deadly peril, robbed of his humanity."...George Wald (biologist): "We are no longer taking good care of our young..." Ian Suttie (psychoanalyst): "...The infant mind...is dominated from the beginning by the need to retain the mother—a need which, if thwarted, must produce the utmost extreme of terror and rage."... James Prescott (neuropsychologist): Monkey juveniles "deprived of their mothers were at times apathetic, at times hyperactive and given to outbursts of violence [the equivalent of children with ADHD]**...showed behavioral disturbances accompanied by brain damage..." Richard M. Restak (neurologist): "Scientists at several pediatric research centers across the country are now convinced that failure of some children to grow normally is related to disturbed patterns of parenting." Sheila Kippley (La Leche League): "It is obvious that nature intended mother and baby to be one..."**

783 **In the face of such overwhelming, unanimous testimony, can we doubt that we are failing our children? The dismal truth is that, on the whole, babies received more and better care 25,000 years ago, 250,000 years ago, even 2.5 million years ago, than many do today...To correct this, we must first recognize that, while both parents play vital roles in an infant's development, the mother—like it or not—is the primary caregiver. Biologically, that's how the system works. And such an immeasurably important task cannot be sustainably carried out in her "spare time."...Humanity was geared for females to cherish offspring in the womb, bond with them at birth, and lavish love on them at the breast. It isn't sexist to esteem motherhood. It is sexist to trivialize it...Grasping the connection between negligent infant care and adolescent violence...we are obliged to act...Alienated, with low self-esteem, pessimistic about the future, in schools that don't educate, the children who should be our hope for the future instead drink, smoke, take drugs, get pregnant, commit suicide, and commit crimes which land them in our awful jails.'**

784 But for all her honesty, McCollister hasn't delved to the bottom of the problem and asked the question screaming to be addressed: 'But *why* have humans stopped loving their infants?' The upset state of the human condition was not being addressed, and without the compassionate explanation to that question of questions, trying to confront the truth of the importance of nurturing was unbearable for parents. So while over the years numerous movements have emerged that truthfully recognise the devastating consequences of not nurturing our children,

and call for greater emphasis on nurturing, such as the **'Touch the Future'** organisation, the '[Jean] **Liedloff Continuum Network'** and the **'Natural Child Project'**, while we couldn't explain our inability to nurture/love our infants the ramifications of not doing so could clearly not be faced and admitted. Imagine being a mother of an autistic boy and trying to face the truth that your alienation caused his autism! Without the explanation for our upset condition you would find it impossible! No wonder any professionals in the field of child psychiatry who supported the belief that autism occurs as a result of inadequate parenting, such as Bettelheim, have been maligned and censured. While it wasn't possible to explain the human condition, to explain the very good reason *why* parents have been unable to adequately love their offspring, it simply wasn't possible to acknowledge the importance of nurturing in both the maturation of our species and in the maturation of our own individual lives.

785 So while (as McCollister noted) some **'researchers into child development...have sought for decades to alert society'** to the fact that **'Every human infant must have unconditional love; without it, an infant's health and growth will be stunted'**, the reality is that while humanity hasn't been able to explain our immensely soul-devastated, upset state and resulting inability to nurture our infants, denial has been the *only* way of coping. All the nurturing-avoiding mechanistic 'explanations' for the cooperative behaviour of bonobos and our human ancestors that were described in chapter 5 evidence that—with the obscenity being that all that evasion was and is not justified given I presented the explanation for the human condition and with it the all-important relieving reason for why we have lost the ability to nurture more than 30 years ago. And worse, accompanying all the no-longer-justified, human-condition-evading, mechanistic 'explanations' for our moral sense that avoided admitting the importance of nurturing has been a raft of dishonest evasive excuses for the devastating *consequences* of not having been nurtured, the most prominent of which have been to apportion blame to a genetic predisposition, chemicals in our industrial world, childhood immunisation programs, or some contracted disease. Genes, in particular, have been blamed for every kind of ailment—for depression, drug addiction, violence, obesity, delinquency, learning and sleep disorders, suicide, lack of commitment/divorce, homosexuality, sex addiction, paedophilia, and almost every other human malaise and abnormality, including ADHD, and those more extreme dissociative states of autism and schizophrenia. Blaming biological, environmental

and chemical factors has been infinitely more bearable than blaming our psychological alienation from our true, natural selves and our resulting inability to adequately nurture our children; as Winnicott cautioned, **'expect resistance to the idea of an aetiology** [cause] **that points to the innate processes of the emotional development of the individual in the given environment. In other words, there will be those who prefer to find a physical, genetic, biochemical, or endocrine cause, both for autism and for schizophrenia'** (ibid. p.219). As Andrew Solomon, the author of the 2001 book about severe depression, *The Noonday Demon*, said, **'Being told you are sick is infinitely more cheering than being told you are worthless'** ('Casting out the Demons', *TIME*, 16 Jul. 2001).

786 In his book *They F*** You Up: How to Survive Family Life*, which is **'devoted to making accessible the scientific evidence that early parental care is crucial in forming who we are'** (2002, p.3 of 370), the child psychologist Oliver James acknowledged that **'Our first six years play a critical role in shaping who we are as adults'** (p.7), and then wrote that **'One of our greatest problems is our reluctance to accept a relatively truthful account of ourselves and our childhoods, as the polemicist and psychoanalyst Alice Miller pointed out'** (p.9), adding that **'Believing in genes** [as the cause of psychoses] **removes any possibility of "blame" falling on parents'** (p.13).

787 So the whole so-called 'nature vs. nurture' debate has really been about wanting to, indeed *needing* to, attribute the formation of our character to the influence of 'nature', our genetic make-up, rather than to the confronting truth of the role nurturing (or the lack thereof) played—it has been a sophisticated deflection from the fact that our psychoses and neuroses and their many physical manifestations are *not* about our genes, but the hurt to our all-loving instinctive self or soul during infancy and childhood. Again, as John Marsden said, **'The biggest crime you can commit in our society is to be a failure as a parent and people would rather admit to being an axe murderer than being a bad father or mother.'** Yes, far better then are the 'blame-all-our-ailments-on-genetics-or-disease-or-immunisation' excuses that are flooding the scientific literature today.

788 Just one of the unhealthy repercussions of not being truthful about the real cause of childhood madness is that treatment of it can be, and has been, dangerously misdirected. As stated in par. 176, the word **'psychiatry'** literally means **'soul-healing'**, but since we have never before been able to explain *why* we as a species have become so soul-destroyed and how that upset has impacted upon the health of our children's souls, we have, in our denial, relied instead on non-confronting, biomedical

'solutions'—as this article points out: **'Over the past decade, a condition known as...attention deficit hyperactivity disorder has been spreading among our children like an epidemic. During that time the prescription of drugs used to control this disorder, central nervous system stimulants chemically similar to the amphetamines that can be bought on the streets, has leaped by more than 1000%...Because it is so convenient and guilt-relieving to be able to attribute a child's difficult behaviour to a neurochemical problem rather than a parenting or broader social one, there is a risk that this problem will become dangerously over-medicalised'** (Dr Michael Gliksman, 'Social issues at the root of childhood disorder', *The Australian*, 8 Dec. 1997).

789 Again, the problem stems from the immense vulnerability of children to all the human-condition-driven upset in the world today, a vulnerability we can gain an, albeit slight, appreciation of if we immerse ourselves in their situation. Since the upsetting battle of the human condition only emerged some 2 million years ago, and only became extreme towards the end of those 2 million years, the great majority of human history—from the australopithecines through to the advent of *Homo sapiens sapiens*—was spent living cooperatively, which means that infants now enter the world, firstly, expecting it to be one of gentleness and love, and, secondly, with almost no instinctive expectation of encountering a massively upset, embattled world. And it is this *extreme contrast* between our species' instinctive memory of a harmonious, happy, secure, sane, all-loving and all-sensitive world, and our species' more recent massively embattled angry, egocentric and alienated world, that makes the shock that infants experience entering the world now so psychologically damaging. Resigned humans have been living in denial of both the truth that our ancestors lived in a state of total love and that as a species we currently live in a state of near complete corruption of that ideal instinctive world of our soul, and that disconnect from our reality means we haven't been attuned to how devastating it is for infants to encounter our world. But the truth is, this contrast between what a child's innocent, love-saturated instincts *expect* and what the child *encounters* in our human-condition-afflicted, soul-butchered world is so great it is akin to a sunflower finding itself having to grow in a dark cesspit; no wonder as adults we turned out as gnarled thornbushes, ready to stunt the next generation.

790 Indeed, the reality is that the levels of alienation in society today are such that almost all humans are cardboard cut-outs of what they would be

like free of the human condition. And while resigned adults aren't aware of their immensely alienated (virtually dead) state because they are living in denial of it, new generations of children arriving into the adult world who have yet to adopt this strategy of denial are acutely aware of the difference between our species' original, ideal, innocent instinctive state and our current immensely upset alienated existence and somehow have to try to cope with the distress it causes them. Paragraphs 43-52 described the agonising process of Resignation that adolescents have had to go through adopting a life of denial of the issue of the human condition and any truths that brought it into focus, but until a young person adopted this blocked-out, resigned state they remained exposed and vulnerable to the full horror of the dilemma of the human condition—horror that is palpable in the lyrics of the heavy metal band 'With Life In Mind' that were included in pars 121 and 754.

791 Having not yet adopted this denial children have always struggled mightily with the imperfection of the upset world that surrounds them, but with the gulf between humans' original innocent state and our current *immensely* alienated state now so great, new generations *are* finding it impossible to bridge. The truth is, ADHD, and the more extreme states of autism and schizophrenia, are varieties of childhood madness, but as the bravely honest psychiatrist R.D. Laing famously said, **'Insanity is a perfectly rational adjustment to an insane world'** (Larry Chang, *Wisdom for the Soul: Five Millennia of Prescriptions for Spiritual Healing*, 2006, p.412).

792 What has made it especially difficult for new generations to cope with our corrupted adult world is that adults have been unable to admit to being corrupted in soul; in fact, as has been pointed out, adults haven't even been aware that they are corrupted—if they *were* aware they were corrupted and alienated they wouldn't be alienated, they wouldn't have blocked out and thus protected themselves from the truth of their upset condition. But with new generations able to clearly see the extent of the corruption and alienation in the world around them, this lack of any honesty by adults, their complete silence on the subject of our immensely corrupted human condition—in effect, their denial that there is anything wrong with themselves or their adult world—left children dangerously prone to blaming themselves for the dysfunctionality of their environment. And not only have resigned adults been silent about their corrupted condition, in the absence of the real reason for their divisive behaviour, they have put forward all manner of dishonest

reasons for their behaviour, and all their lies have greatly offended the honest minds of children. In fact, in line with the big lie put forward by the resigned mechanistic adult world that our ancestors were aggressive, brutish and savage, children have even been treated as, and even told they are, 'just hateful little savages that have to be tamed'. Indeed, the purity and innocence of children has so confronted and exposed parents that they are typically referred to as 'kids'—a dismissive, derogatory, retaliatory 'put down', a way of holding their confronting innocence at bay. And so, in encounters between the innocent and the alienated, where the alienated act as if there is nothing wrong with them or their world, in the innocents' instinctive state of total trust and generosity they are left believing there must be something wrong with them, that in some way or another they must be at fault. In their immense naivety about the upset, alienated world, together with their soul's great love, trust and generosity, innocents question their own view, not the view being presented by the alienated. The innocent do not know people lie because lying simply did not exist in our species' original innocent instinctive world. In short, their trusting nature made them *codependent* to the alienated, susceptible to believing the alienated were right rather than accepting and fighting to uphold their own accurate view of the situation. Psychologists coined the word 'codependent' to describe someone who is **'reliant on another to the extent that independent action is no longer possible'** (*Macquarie Dictionary*, 3rd edn, 1998), and, in the case of children, they are so trusting of, and thus **'reliant on'**, the adult world they are incapable of thinking **'independent**[ly]**'** enough to trust that they are not at fault or have somehow misread the play. Children come from *such* an innocent, wholesome, trusting, loving, generous, integrative instinctive world that they all too readily blame themselves in situations where they are faced with a denial. Then, when they decide they must be at fault, their sense of self-worth and meaning is completely undermined, and to cope with that **'unthinkable anxiety'**, as Winnicott accurately described it, they have no choice but to psychologically split themselves off from the perceived reality, adopt a state of **'invulnerability'**. More will be said about the devastating effects of children's codependency to adults next (in ch. 7:16D).

793 So, able now to appreciate the extreme vulnerability of children to all the extraordinary upset in the adult world, and adults' silent denial of it, we can understand why children have died a million psychological

deaths in the face of that upset. We can understand why Beckett wrote, **'They give birth astride of a grave, the light gleams an instant, then it's night once more'**, and why Laing wrote that **'To adapt to this world the child abdicates its ecstasy.'** *Thank goodness* we have found the understanding of the human condition that allows us to admit the importance of nurturing and make it a priority now for the human race.

794 In summary, the human race is naturally extremely embattled after 2 million years of its heroic search for knowledge, which means that virtually no mother now can hope to love her infants as much as their instincts expect to be loved, and virtually no father now can hope to restrain their extreme egocentricity to the degree necessary to avoid oppressing their children somewhat. (The devastating effect men's egos has on children will also be explained in ch. 7:16D.) Thankfully, however, we can now explain why nurturing has been so compromised. And since, through that explanation, we can understand that virtually all humans now are so upset that virtually no one can hope to love their children as much as their children's instincts expect, parents do have to be realistic about their limitations—if too great an expectation is placed on parents to be more loving than they are now genuinely capable of being no one will be prepared to run the risk of having children, which is obviously not the answer. However, with understanding of the human condition and the subsequent ability to admit how delicate our soul is, parenting can, and must, take on a whole new meaning and responsibility.

795 Yes, to repeat what was said in chapter 5:13, having found the compassionate biological understanding of the upsetting battle that humans have had to wage against the ignorance of our instinctive self as to the fundamental goodness of our conscious self, our species' preoccupation with that battle can at last end and focus return to the nurturing of our children. Further, now that we have found that *very* good reason as to *why* parents have been unable to nurture their children with anything like the unconditional love children instinctively expect and need, it is no longer necessary to deny the importance of nurturing both in the maturation of our species and in our own lives. Yes, with the arrival of understanding the terrible guilt parents have felt for not being able to adequately nurture their offspring departs. Parents can now understand that their inability to adequately love their offspring is *not* their fault, but the product of the ever-accumulating levels of upset in the human race as a whole that resulted from our species' heroic 2-million-year search

for knowledge, specifically for self-knowledge, understanding of why we conscious humans are good and not bad. *So yes, we can finally now both admit the importance of nurturing and turn our attention back to the nurturing of our offspring—in fact it now becomes a matter of great urgency that we do both these things.*

796 The nurturing of children is what is required to produce adults who are sound and secure in self. Ultimately, it is *only* through the nurturing of our offspring that the human race can hope to become healthy—namely psychosis-and-neurosis-free once again; to, as Ashley Montagu said, put **'man back upon the road of his evolutionary destiny from which he has gone so far astray'** and restore **'health and happiness for all humanity'**. As will be explained in chapter 8, with understanding of the human condition found all humans can *immediately* be free of the agony of the human condition by taking up the Transformed Way of Living (which involves adopting a mental attitude where you leave that agony behind as dealt with), but to produce humans who are *completely* free of psychosis and neurosis and not needing to take up the Transformed Way of Living, they need to be nurtured with unconditional love in their infancy and childhood—a process that will take some generations to achieve while all the accumulated upset within our species subsides.

797 Of course, the initial difficulty for virtually all parents is that since they are so psychologically upset, especially so alienated from their natural, loving, instinctive true self or soul as a result of humanity's upsetting search for knowledge, it is simply not possible for them to give their offspring unconditional love. HOWEVER, while virtually all parents will have to accept this reality, their ability now to understand the very good reason for why they are unable to give their offspring unconditional love will bring so much calming, reassuring relief from that guilt that nurturing won't be nearly as difficult as it has been. Again, while it *will* take a number of generations before most parents can generate real, natural, unconditional love, being able to understand their upset state, and, significantly, not having to continue with the upsetting battle to prove their worth (which, as will be explained in chapter 8, is what makes the fabulously exciting new Transformed Way of Living possible) is going to bring *enormous* relief to parents and help them *immensely* with the whole activity of nurturing. The overall relief and excitement alone that will now emerge in the human race from the realisation that we have finally ended our 2-million-year struggle against ignorance will mean that children are

going to find themselves in, if not an unconditionally loving world, then at least an anxiety-free, truthful, compassionate, happy one. And since the adult world can finally break the silence of their denial and tell children the truth about their horrifically upset condition, that honesty will also make an immense difference to the psychological wellbeing of children. Basically, finding understanding of the human condition makes possible a whole new world that is relieved of the horrible frustration and agony of that condition; the terrible siege state where humans couldn't understand themselves can at last be lifted, and, with its lifting, life can return to every aspect of human existence, including to the impossible situation parents have been in trying to nurture their children. Yet again, we see how critical it is that science recognises these denial-free understandings of the human condition so that the public at large is made aware of them. The ignoring, and, when pushed, attack and repression, of these answers MUST stop, and they must instead be supported—because only then can this great transition for the human race begin.

[798]In the case of parents of extremely nurturing-deprived-and-thus-extremely-dissociated-from-the-world, autistic children, even with the ability now to understand the reason why they haven't been able to nurture their children, having lived in denial of the importance of nurturing, the revelation of its importance, and of the devastating effects of not receiving it, will come as a great shock. It is, in fact, an example of how the arrival of understanding of the human condition and all the truth that it makes possible also brings with it the long-feared exposure day (or truth day or honesty day or revelation day or confrontation day or 'judgment day'), which is an issue that will be addressed at length—including an explanation of how we are able to cope with that exposure—in chapter 8:7. It will be explained there that the way we cope with so much truth being suddenly revealed is by adopting the Transformed Way of Living, part of the strategy of which is to not overly confront all the truth that is suddenly exposed. What is important now is to support the truth, not seek to overly confront it.

[799]Before finishing this section, it is worth examining the comment from Winnicott, that the autistic boy Ronald **'had very exceptional skill in drawing'**. The established link between autism and <u>savant abilities</u> was described in the documentary *Uncommon Genius*, when it was stated that **'for more than half of all savants, the syndrome owes its origins to a**

familiar condition—autism'. In describing savants, the narrator said, **'One of the great unsolved mysteries of the human mind is savant syndrome. This man cannot remember how to clean his teeth, yet he recalls every zip code, every highway, every city in the United States. As a boy, this man was considered mentally handicapped but he can name the day of the week for every date on a forty-thousand-year calendar. And this man, blind, cerebral-palsied and barely able to talk, played Tchaikovsky's First Piano Concerto flawlessly after hearing it once on television. These genius-like abilities are the product of damaged minds...Most of us were introduced to savants through Dustin Hoffman's portrayal of Raymond in the Oscar-winning movie, *Rain Man*. Raymond had savant syndrome...Savants are people who produce awesome mental abilities from severely disabled minds'** (ABC, 24 May 2001). The program mentioned that **'researchers discovered people who suffered dementia and then suddenly gained prodigious skills they'd never experienced before'**, and reported that **'savants may abandon their skills as they become more sophisticated socially... The really prodigious savants have a sort of memory super-highway that allows them to access and transfer enormous amounts of information. They develop that memory partly because they don't get side-tracked by thinking too much.'** It then concluded with the intuitively insightful question, **'Could savant genius lie dormant deep inside everyone's brain? Are savant skills merely obscured by layers of normal everyday reasoning?'** With understanding of the human condition it is not difficult to understand how those who suffer from autism and certain mental impairments can have savant abilities. If such people can completely disconnect from or avoid **'get**[ting] **side-tracked by thinking too much'** about the **'normal everyday'** pain associated with the agony of the human condition, then their non-**'socially'**-**'sophisticated** mind can be freed to access some of the all-sensitive, all-loving, all-inspired, immensely imaginative and thus powerfully creative and able world of our soul. A window can be opened up to the fabulous potential that humans lost access to when they became so horribly preoccupied with worry about the hurtful, corrupted world around them and their own corrupted state. Earlier (in par. 655), I wrote about how I learnt that to draw the little pictures that are included throughout this book I had to disconnect my conscious mind and just let my instinctive sensitivity stream through, and that if I didn't do that I simply couldn't draw at all. For example, the drawing of the three 'Childmen' happily embracing that I used to illustrate humanity's childhood stage was done so quickly I shocked myself because I could hardly believe that such an empathetic drawing could be produced from an almost instant scribble. I wrote that at that moment I saw just how much sensitivity we humans once had, and how much alienation now exists within us 2-million-years-embattled humans, referring to Sir Laurens van der Post's observation of the relatively innocent Bushmen race, that **'He and his needs were committed to the nature of Africa and the swing of its wide seasons as a fish to the sea. He and they all participated so deeply of one another's being that the experience could almost be called mystical. For instance, he seemed to *know* what it actually felt like to be an elephant, a lion, an antelope, a...baobab tree'**. Yes, when R.D. Laing said, **'To adapt to this** [human-condition-afflicted, horrifically upset] **world the child abdicates its ecstasy'**, and **'between *us* and It [our true selves** or soul] **there is a veil which is more like fifty feet of solid concrete'**, and **'The outer divorced from any illumination from the inner is in a state of darkness. We are in an age of darkness. The state of outer darkness is a state of...alienation or estrangement from the inner light'**, and **'We are dead, but think we are alive. We are asleep, but think we are awake'**, he was emphasising just how much soulful sensitivity we have lost access to as a result of our preoccupation with the agony of the human condition. As I said when describing how beautiful and amazing natural Africa is and how awful all our documentaries about natural Africa now are: 'Our world has shrunk to the size of a pea'! Yes, when humans become free of the agony of the human condition they will discover they have powers they have only ever dreamed of: they will see everything as if they have been blind up until now; they will feel everything; they will be able to remember everything—their whole world will open right up!

800 What occurs during a so-called Near-Death Experience actually illustrates what is occurring in some minds that have completely dissociated from reality. For instance, mountain climbers who survive falls that they were convinced would be fatal (they were saved, perhaps, by landing in a snow drift) often report that during those near-death moments they experienced a state of extraordinary euphoria in which the world suddenly appeared utterly beautiful and radiant and that they were flooded with a feeling of ecstatic enthrallment. With understanding of the horror of the human condition we can appreciate how, in such cases, the mind *would* give up worrying, and that all the facades—in particular the denial that they adopted at Resignation—*would* become meaningless. If death is seemingly imminent, there is no longer any reason to worry or to pretend, at which point the struggle and agony and pretence of having to live

under the duress of the human condition ceases and the true world of our all-sensitive soul suddenly surfaces. So yes, **'savant genius lie**[s] **dormant deep inside everyone's brain'**. The Superman mythology is an expression of humans' suppressed awareness of their alienated state—that resigned humans are Clark Kents with hidden Superman potential (news reporter Clark Kent being Superman's civilian identity).

[801]While it might be thought that the dissociation that occurred at Resignation, when adults adopted a life of denial of their reality, should have given them some access to the soul's truthful, sensitive world, resigned adults were not normally completely blocking out their reality—rather, they were constantly on the lookout to deflect any criticism arising as a result of that reality. Generally, the resigned did not completely dissociate themselves from the world in the same way that autistic people did. However, there were some people who although not sufficiently hurt in infancy and childhood to become autistic, were sufficiently hurt to need to live an extremely dissociative, alienated existence after they resigned; and such people *could* develop some access to the soul's truthful world. Sometimes when people became extremely upset/corrupted, their alienation, their mental blocks, their defences, became disorganised and through this 'shattered defence' the soul occasionally emerged; 'mediums' or 'psychics' or 'channellers' are examples of such individuals. Of course, such shattered-defence access of the soul's true world was not the natural, secure, balanced access that unresigned people have. For those people, whom we have historically referred to as prophets, the soul's world has always been an ultra natural place, not something apparently mystical or supernatural.

[802]R.D. Laing was describing the 'shattered defence' means of accessing the soul when he wrote that **'the cracked mind of the schizophrenic may *let in* light which does not enter the intact mind of many sane people whose minds are closed'** (*The Divided Self*, 1960, p.27 of 218). Interestingly, Laing immediately continued to say that the existentialist Karl Jaspers was of the opinion that the biblical prophet Ezekiel **'was a schizophrenic'**. While some biblical prophets may have accessed the soul's true world through a 'shattered defence', those who had full and natural access to the soul and were prophets in the true sense were exceptionally sound rather than exceptionally exhausted, alienated, separated from their true self, or schizophrenic. It should also be mentioned that when people prayed or chanted mantras or counted rosary beads they were trying to shut

down their alienation-preoccupied mind in order to let through some of the truthful world of the soul. They were, in effect, trying to shatter their own defence. Fatigue, fasting, hallucinatory drugs, despair and faster-than-thought physical activities, such as scree-running, are other ways of achieving this breakthrough to the world of the soul.

803 (A more complete description of how nurturing has been seriously compromised by terminal levels of alienation can be found in *Freedom Book 1* at <www.worldtransformation.com/freedom-nurturing-now-a-priority> and <www.worldtransformation.com/freedom-autism>.)

Chapter 7:16D The dysfunctional, extremely narcissistic 'Power Addicted' state

804 Another dire consequence of terminal levels of alienation emerging in the world, especially in the materially so-called 'developed' world, is the development of extreme egocentricity, especially in men. As will now be described, the effect of this egocentricity on children has also been catastrophic.

805 All around us are expressions of the punch-drunk egocentric state, this end play situation of the development of communication technology-fuelled upset in the world of having to get a win, a victory, success at any cost. We see parents on the sideline of sports fields, watching their children play and projecting their embattled need for a win onto those children, yelling, 'win, win, win'! The effect this pressure has on their children is one of psychological devastation because, as I will now explain, it produces a new generation who are either psychologically crippled by that conditional reinforcement or turned into 'power addicts', both of which are extremely dysfunctional states.

806 An article by the journalist Miranda Devine truthfully titled 'Face it, we are all narcissists now' described how poor parenting is producing extremely narcissistic power addicts: **"'Parents are becoming increasingly self absorbed [believing] 'the single most important thing in the world is for me to work like a dog and get the house, the car and the holiday house' and don't realise all their kids want is to be loved and to have one-on-one time with their parents." He** [the adolescent psychologist Michael Carr-Gregg] **says** [and part of this quote was referred to in par. 757] **an "epidemic of poor parenting" is to blame for a drastic rise in psychological problems in young people. "Generation Y**

is being ravaged by depression, anxiety disorders and stress disorders'" (*The Sydney Morning Herald*, 3 Sep. 2009). In the article Devine discussed the development of the narcissistic personality disorder where adults develop **'a grandiose sense of self importance; preoccupation with fantasies of unlimited success, power, brilliance...a need for excessive admiration; a sense of entitlement; exploitive personal relationships; a lack of empathy; envy and arrogant, haughty behaviour...Everyone's a potential narcissist these days.'** The article mentioned **'a long-term study of...American college students** [that found]**...the incidence of narcissistic personality traits increased on a scale rivalling obesity'**. The reason, the article reported, for this recent explosion in narcissism is, to repeat Carr-Gregg, **'poor parenting'**. What Devine has reported here about **'the incidence of narcissistic personality traits' 'rivalling obesity'** in American students is more than backed up by the aforementioned *TIME* magazine story, **'The Me Me Me Generation'**, which detailed the **'narcissism epidemic'** in the present **'80 million strong...biggest age group in American history'** Y-generation (see par. 758). This article even wondered if the Y-generation's **'self-centeredness could bring about the end of civilization as we know it'**, a very real concern that will now be explained.

807 As just described, having lived in denial of the truth that our species once lived in a cooperative, unconditionally-selfless, loving, harmonious state, there has been almost no recognition of the importance of nurturing children with unconditional love. The truth, however, is that children come into the world instinctively expecting to receive unconditional love and when they don't get it—receiving *conditional* love instead—they are so innocent, so trusting in a true, all-loving, upset-free world that the only conclusion they can come to for not being unconditionally loved is that for some reason *they* don't deserve it and are therefore a worthless, unlovable, bad person. Children are *so* naive, they are born *so* instinctively unaware of the existence of the upset state of the human condition, that they can't believe it is the adults' fault that they are not being given unconditional love; they are *so trusting* in the world around them, *so* codependent to it, they can only presume it's their fault. Again, 'codependent' describes someone who is **'reliant on another to the extent that independent action is no longer possible'**; in this case, children are so trusting of, and thus **'reliant on'**, the adult world they are incapable of thinking **'independent**[ly]**'** enough to know that they are not at fault, and trust in that knowledge. This conclusion by children that they are, in effect, bad either hurts them so deeply that they block all the pain of it out of their mind and become

a psychologically detached, immensely alienated, crippled person—in the extreme, autistic—or they try desperately to prove they aren't bad or worthless, spending the rest of their lives obsessed with the need for reinforcement.

[808] As already mentioned, the psychologist Arthur Janov developed the technique of 'primal therapy' in which adults seeking to lessen the hold their psychoses have on their lives are helped to work their way back to memories of the original (primal) hurt to their soul that occurred in their infancy and childhood as a result of growing up in the overly upset, insecure, have-to-somehow-establish-your-worth, massively egocentric world of today. In the following extracts (which come from—and serve as a condensation of—his famous 1970 book, *The Primal Scream*), Janov describes how children blame themselves for their parents' inability to love them and the dire consequences such recrimination has on their emotional development. He wrote (the underlinings are my emphasis): **'Anger is often sown by parents who see their children as a denial of their own lives. Marrying early and having to sacrifice themselves for years to demanding infants and young children are not readily accepted by those parents who never really had a chance to be free and happy** [p.327 of 446] **...neurotic parents are antifeeling, and how much of themselves they have had to cancel out in order to survive is a good index of how much they will attempt to cancel out in their children** [p.77] **...there is unspeakable tragedy in the world...each of us being in a mad scramble away from our personal horror. That is why neurotic parents cannot see the horror of what they are doing to their children, why they cannot comprehend that they are slowly killing a human being** [p.389] **...A young child cannot understand that it is his parents who are troubled...He does not know that it is not his job to make them stop fighting, to be happy, free or whatever... If he is ridiculed almost from birth, he must come to believe that something is wrong with him** [p.60] **...Neurosis begins as a means of appeasing neurotic parents by denying or covering certain feelings in hopes that "they" will finally love him** [p.65] **...a child shuts himself off in his earliest months and years because he usually has no other choice** [p.59] **...When patients** [in primal therapy] **finally get down to the early catastrophic feeling** [the 'primal scream'] **of knowing they were unloved, hated, or never to be understood—that epiphanic feeling of ultimate aloneness—they understand perfectly why they shut off** [p.97]'.

[809] To reiterate, if a child is not given unconditional love, and is instead intimidated, frozen out and made to prove themselves all the time by arrogant, tyrannical, authoritarian, tough, extremely egocentric, overly

self-worth-embattled, 'My way or the highway'-type parents, then that child will either be psychologically crippled by the situation, or made into a narcissistic power addict who has to win at all times and at all costs—where only success can keep at bay the terrifying conclusion that they are somehow unworthy. To gain power, the upper hand, these powerpaths or psychopaths, as they are otherwise known, will say anything, do anything, no matter how immoral—conduct that has been variously described as **'ruthless', 'manipulative', 'belligerent', 'bullying', 'totally self-centred', 'egocentric', 'amoral', 'cold', 'cruel', 'obsessed with wielding power over others'** [in other words, controlling and dominating], **'deceitful', 'self-important', 'lacking any ability to empathise with others'**, with a mindset that is **'quick to blame others for their mistakes'** (Robert Matthews, 'Are you living with a socialised psychopath?', *Focus*, May 1994 & *The Sunday Telegraph Review*, Apr. 1997). As Morris West put it, **'brutalise a child and you create a casualty or a criminal'** (*A View from the Ridge: The Testimony of a Pilgrim*, 1996, p.78 of 143).

[810]Regarding the power addict's inability **'to empathise with others'**, the truth is, they are *so* preoccupied—in fact, utterly *consumed*—by their own pain that instead of empathising with and being considerate of others as our instinctive self or soul is designed to be, they *impose* their condition on the world around them; they intimidate and frighten everyone with their angry, often silent embattled state; indeed, a common description of how they operate and their impact on those around them is that they have everyone 'walking on eggshells'. The psychiatrist Frank Lake articulated this problem of adults who, as a result of being under attack as infants and as children, treat everyone and every situation they encounter as if they were *still* under attack, writing that such an adult **'complains as if it remembered the bad times it had been through. It reacts to the world around it as if it were still in the bad place, still having to "feel its keenest woe". It reacts defensively as if the attack were still going on'** ('Supplement to Newsletter No.39', *Clinical Theology Newsletter No.39*, Dec. 1981).

[811]Basically, the power addict's life is stranded or arrested in the childhood state of needing reinforcement. You can quickly extrapolate just how absolutely desperate the power addict's mind is for reinforcement—how they aren't able to truly care about anyone or anything other than getting that reinforcement, how every moment of every day their mind can't afford to be anything other than singularly focused on how they can get that power or glory or attention or any other form of validation—if you imagine a child fighting with all their might to avoid the

devastating conclusion that they are an unworthy, bad person. When all the evidence from the way that child is being treated seems to unequivocally indicate to that innocent, trusting, naive, ideal-world-expecting child that they are unworthy and bad, the child naturally fights back with all their being to resist that soul-destroying conclusion. If, however, that evidence is overwhelming, the child has no choice but to give in to that criticism and conclude that they are indeed worthless and bad. And then, to cope with that terrifying, absolutely unbearable conclusion, all the child can do is determine to never again allow their mind to connect with that conclusion that they are an unlovable, bad person, and since that core issue of their worthiness is where their true self is preoccupied, avoiding that issue amounts to separating or dissociating or splitting off from their true state and thus true self. This is the split, false, psychologically crippled state that was the focus of the previous sub-chapter (ch. 7:16C). *If, however*, the evidence that they are an unworthy, bad person is immense but not quite overwhelming the child will be stuck in a lifelong battle fighting with all their might to avoid surrendering to the psychologically crippling conclusion that they are an unworthy, bad person. And that is where power addicts live—stranded in that state of terrible fear that they are unworthy, fighting with all their might to 'stay on their feet' and defy that implication. As mentioned, the actress Mae West famously articulated this situation of arrested development, within males in particular, when she said, **'If you want to understand men just remember that they are still little boys searching for approval.'**

812 Another of the many extremely dysfunctional features of the embattled power addict mind is its extraordinary ability to accumulate grievances. The narcissistic power addict is so unable to accept any criticism, so extremely insecure, so desperately preoccupied with **'searching for approval'** and avoiding any implication that they are unworthy, that all day, every day, they seek out only reinforcement, only 'wins', such as power, fame, fortune and glory. There is no room to accept any criticism and since most situations in life contain a spectrum of positives and negatives, of reinforcements and criticisms, of ups and downs, such minds are simply unable to fairly assess or tolerate *any* of the negative, criticising aspects of a situation. In fact, the power addicted mind will find a way to avoid *all* criticism, no matter how much there is or how legitimate, and instead search for and find some fault or flaw in the situation, imagined or real, that brings reinforcement to them. Their totally defensive mind locks onto

these 'grievances' as being all-significant at the exclusion of all other aspects of the scenario. Over time, any interaction with a power addict, no matter how fair or generous, will leave them harbouring a mountain of grievances and totally incapable of recognising the unfairness and irrationality of their view. The desperation of the power addict mind to avoid any threat to their power base is extremely difficult for the more balanced mind to comprehend. I have watched documentaries showing the former US President Franklin D. Roosevelt being overly cordial to the Russian dictator Joseph Stalin in the futile hope that Stalin would be fair-minded during their negotiations over the future of Europe at the conclusion of the Second World War. In public, Stalin always presented a benign, open, balanced and fair-minded countenance when underneath he was, in fact, a furious monster who, out of paranoia, oversaw the murder of millions and millions of his own people—including *all* of his closest friends who he was terrified might one day challenge his power. Power addicts learnt to disguise their extraordinarily defensive mindset, but their behaviour could often be absolutely, unbelievably selfish. Indeed, psychologists have learnt that it is all but impossible to reform power addicts because they are *so* desperate for success, *so* desperate not to fail, that they use what they learn in therapy to better disguise their weaknesses—for example, they learn to pretend they don't lack empathy and consideration of others to hide the fact that they are totally self-centred, ruthless and manipulative. As the article by Robert Matthews referred to above mentions, **'no amount of counselling seems to change them...Many people unwittingly try to change the behaviour of the psychopaths in their lives... Yet, as many psychiatrists have learned, it's almost invariably temporary. The psychopath's promise to mend his ways is usually just another pack of lies. Within a few months, the same traits will be back as strongly as before. The sad truth is that, if there is a psychopath in your office, your home or your bed, it will be you that has to change.'**

813 It has to be remembered, of course, that after 2 million years of the development of upset in the world, *all* adults are unavoidably stranded *somewhere* on the spectrum of insecurity of self due to the lack of reinforcement/love in their upbringing. The power addict state and, beyond that, the psychologically crippled state are simply the extreme states of the human condition. And again, it has to be emphasised that it is at last safe to admit this because we have now found the redeeming, dignifying understanding of how all the upset in the human species

started, and, as will be explained in chapter 8, how it can be transformed and eventually ameliorated. All the upset in the world is now defended; the basis for our insecurity about being upset has gone.

[814]I have drawn this picture of the narcissism-producing and psychologically crippling situation to help children understand the world we live in. Since adults own the world there has been very little focus on the wellbeing of children. They can't fight for themselves. They can't write books. In fact, in their bewilderment children can't even tell adults what they are going through. If children were able to express themselves and be properly heard—for instance, if they were able to vote, if they had some power—the care and concern and respect for their wellbeing would have been much greater than it has been. As just emphasised in ch. 7:16C, unable to explain why we humans have become so embattled, parents in the 'developed world' haven't been able to cope with any criticism of their inability to adequately nurture their children, but that incapacity to cope with any criticism of their inability to properly parent

their children overlooks, indeed disregards, the fact that children in turn also can't cope with not being loved unconditionally. Parents have had it all their way. The powerlessness of children has made them the victims of the adults' powerfulness. The truth is, we live in an anti-child world today where 'kids' are irritants—and yet children are the next generation and to have no regard for them is to have no regard for the future.

[815]The drawing shows a tyrannical father and the damaging effect his behaviour is having on his child—the truth of which my professor of biology at Sydney University, Charles Birch, recognised when, after discussing with me the destructive effect power addict fathers have on their children, he made the extremely succinct (if somewhat blunt) comment: **'Haven't you heard Jeremy, the best thing that can happen in a man's life is that his father dies when he is born'** (WTM records, 12 Nov. 1998). Birch was homosexual and I think this comment is very revealing of the origin of the particular psychological upset that made him homosexual. (The psychological reason for homosexuality was explained in pars 629-630.) Some fathers claim it is necessary to toughen their children, kill off their inspired, truth-filled, happy, excited, inspired, nothing-is-too-difficult, loving original instinctive self or soul. A newspaper cartoon depicted this appalling point of view with the following exchange between a father and son: **'Son you're a liar, you're a bully, you're greedy, you're manipulative, you're self serving. You'll go far'** (Alan Moir, *The Sydney Morning Herald*, 11 Sep. 2009). Alienation is a tragically lost, dark, effectively dead state that the whole human race has been working desperately hard to escape from, *not* stay in! The *real* issue in the situation where parents want to toughen up their children is the extent of the upset anger, egocentricity and alienation in the parents, because it is *their* extreme upset that causes them to project onto their children the need to be a success at all costs. It is *their* need for glory, not the need to turn their children into 'survivors', that is the real motivation. So, no—parents saying they 'Want to toughen up their kids so they can face the real world', as though they are doing their children a favour, is just a convenient excuse to get away with **'poor parenting'** because all parents do intuitively know how critically important it is that children receive unconditional love if they are to grow into truly creative, functional and successful adults. As has been well and truly explained in this book, how to nurture children is one of the most ancient instincts in humans.

[816] Earlier (in pars 369 and 778), reference was made to Ashley Montagu's 1970 paper, 'A Scientists Looks at Love', in which Montagu bravely admitted the importance nurturing plays in a child's well-being and future functionality, and the horrific consequences caused by a lack of unconditional love. His account is so powerful I am re-including some of what he wrote here, as well as some parts of the paper that weren't included earlier that refer to the **'psychopathic'** consequences of not receiving unconditional love: **'love is, without question, the most important experience in the life of a human being...The newborn baby...wants love. He behaves as if he expected to be loved, and when his expectation is thwarted, he reacts in a grievously disappointed manner...If it doesn't receive love it is unable to give it—as a child or as an adult...Criminal, delinquent, neurotic, psychopathic, asocial, and similar forms of unfortunate behavior can, in the majority of cases, be traced to a childhood history of inadequate love...children who have not been adequately loved grow up to be persons who find it extremely difficult to understand the meaning of love...they tend to be thoughtless and inconsiderate. They have little emotional depth...they often seek ways of achieving power...Occasionally, when they attain prodigious political power as in the case of such unloved creatures as Hitler, Mussolini, and Stalin, they commit horrendous atrocities and subject humanity to irreparable damage. It is quite evident that the tragedy these men brought to the world was principally due to their incapacity to love...If one doesn't love oneself one cannot love others. To make loving order in the world we must first have had loving order made in ourselves...love is demonstrable, it is sacrificial, it is self-abnegative** [self-denying]**. It puts the other always first. It is not a cold or calculated altruism, but a deep complete involvement with another. Love is unconditional...Love is the principal developer of one's capacity for being human, the chief stimulus for the development of social competence'**.

[817] As the epidemic of the psychologically crippled state of ADHD and autism shows, what happens as upset anger, egocentricity and alienation increases is that eventually a threshold is reached where there is *so* much anger, egocentricity and alienation in the population that the great majority of children can no longer cope with the extent of the upset they encounter and a dysfunctional generation appears where there is a predominance of individuals who are either narcissistic power addicts or psychologically crippled. And it is this threshold, this terminal level of alienation, that Devine and the *TIME* magazine story have described

as having now been reached at the forefront of human progress, which is mostly in the 'developed' Western world.

818 So yes, under the duress of the human condition there have basically been three varieties of resigned, alienated humans: those who received unconditional love; those who received conditional love (basically, non-love) and are still trying to resist the implication that they are unworthy and bad and as a result have become power addicts; and those who received either no love or so much conditional love that they were unable to resist the implication that they are unworthy and bad and as a result have had to dissociate from that terrifying, unbearable conclusion and, as a result of that, are psychologically crippled. The first category are those that our society has recognised as being 'functional' while the second and third categories are those that have been described as 'dysfunctional'. Another honest description we have had for these three states was, respectively, 'relatively secure', 'very insecure' and 'extremely insecure'. However, with the world reaching end play states of upset, the proportion of the population that is dysfunctionally alienated is becoming near total.

819 As a result, finding a secure, well-adjusted, functional 'soft handed' manager in a company now is almost unheard of, for they are nearly all power addicts. And once power addicts gain control of a company it really is the beginning of the end for that company—it will become dysfunctional because, as emphasised, power addicts are tyrannical and find anything that resembles criticism unbearable, such as criticism of their own ideas or having their mistakes pointed out. They are simply not effective free thinkers, which good business leaders need to be. And nor do they tolerate any perceived threat to their power base—as the adage goes, "A's employ A's and B's employ C's", with A's being those who have been nurtured and are, as a result, relatively secure and therefore not threatened by talented people and eager to employ other A's. Insecure B's, on the other hand, will only employ C's because once in power they don't want anybody more functional than themselves usurping them. As a result, the company becomes dysfunctional.

820 And as with individuals in households and companies, the same process can be applied in the context of entire countries, just on a larger scale. What happens in this case is that when the proportion of power addicts in a country increases to a certain critical point 'right' suddenly no longer has the clout to win out over 'might', at which point dysfunctional

tyrants are able to take control with terrifying consequences for both their country's population and, as history has shown, that of neighbouring countries. Yes, just as children and wives of tyrannical men in family situations are unable to fight back and are crushed, tyrannical dictators can crush entire populations; the rise of Mussolini's totally ruthless, fascist 'Blackshirts' in Italy prior to the Second World War, and Hitler's ascension to power and invasion of eastern Europe are obvious examples. Tyrants such as these, and Napoleon and Stalin, have only been able to seize control when the number of power addicts in a population surpassed a critical point where the remaining soundness was no longer able to withstand their extreme intensity and defy them. As will be described shortly, the reason England has so often in the history of Europe been able to defeat tyrants there and elsewhere can be put down to its island isolation, which meant it was still sufficiently sheltered from much of the upset in the world, still sound enough, still free enough of power addicts, to stand against those tyrants who had taken over control of countries on the continent where, by inference, the tipping point of too many power addicts in the population had been surpassed.

821 Of course, it follows that the same process of a threshold of dysfunctionality eventually being reached within individuals, families, genders, generations, companies, races and countries also applies to entire <u>civilisations</u>. When history books talk of great civilisations, such as the Greek and Roman Empires, and many others, becoming decadent, the main feature of that decadence was the dysfunctionality that resulted from having a majority of overly egocentric, insecure power addicts; there was simply not enough sound, functional, soulful generosity, fairness, honesty, sensitivity, humility and equanimity left in the system for it to remain functional. Once everyone wanted glory at all costs and there was no interest in truth or fairness—no interest in others, no moral strength—the system simply collapsed. And it is this threshold of power-addict-produced, decadent dysfunctionality that has now been reached in modern Western civilisation, having, as our history books describe, already been reached in older civilisations around the world. So we can now understand exactly why and how the **'self-centred'**, **'narcissism'** of **'The Me Me Me Generation'** of the present **'80 million strong...biggest age group in American history'** Y-generation that was described earlier **'could bring about the end of civilization as we know it'**. The dire situation for our species has been reached where there are now no countries left that don't

have a predominance of power addicts. Yes, recall the title of Miranda Devine's article: **'Face it, we are all narcissists now.'** God forbid, but it *has* come to this, a world of power addicts and cripples; a planet of chronically insecure humans; a complete madhouse. And since communication technology has shrunk the world in the sense of making it one civilisation, the whole world has, in essence, reached the end play state of terminal alienation. This breakthrough understanding of the human condition and the transformation of the human race it makes possible has truly arrived without a moment to spare!

822 Yes, while, as mentioned earlier, psychologists have learnt that it is all but impossible to reform power addicts—that **'no amount of counselling seems to change them'**—the new Transformed Way of Living that is made possible by the finding of understanding of the human condition enables all humans, *even power addicts*, to let go of their old competitive, power-fame-fortune-and-glory-seeking existence and transform their lives to a state free of that preoccupation. Having found understanding of the upsetting battle humans have had to wage against the ignorance of our instinctive self as to the fundamental goodness of our conscious self, preoccupation with that battle *can* at last end. Amongst the Founding Membership of the World Transformation Movement there are power addicts who have freed themselves of their affliction by taking up the Transformed Way of Living. The affirmations of the Transformed State given by WTM Founding Members John Biggs and James West, for example, that can be viewed and/or read at <www.worldtransformation.com/affirmations>, provide confirmation of this wonderful potential.

823 Again, the all-important fundamental truth that understanding of the human condition reveals is that upset is a *heroic* state, *not* a bad, evil, unworthy one. While all humans are variously upset as a result of their different encounters with humanity's heroic battle to overthrow ignorance, all humans are equally good and worthwhile. Understanding the cause of the upset state of the human condition eliminates the possibility of the prejudicial views of some people, genders, generations, races, countries or civilisations being good and therefore superior and others being bad, evil, unGodly and therefore inferior and unworthy. It is a simple and obvious truth that the longer and/or more intensely a person or a gender or a generation or a race or a country or a civilisation has been engaged in humanity's great battle to overthrow ignorance, the more upset and ego-embattled they unavoidably will have become. As mentioned,

more will be said about the differences in alienation between races and countries very shortly.

824 So the power addict state is really only one of the extreme, end play results of the human condition, of 2 million years of generation after generation of humans trying to prove that they are good and not bad. Of course, any lack of reinforcement—and it has to be remembered that humans' original instinctive expectations are of receiving *complete* reinforcement during their infancy and childhood—can cripple or egocentrically embattle a person; for example, children who are exposed to extreme ill-treatment, such as sexual abuse, are obviously likely to become horrifically psychologically crippled. But the biggest crippling influence on all children everywhere has been mothers' alienation from their natural nurturing instincts, and oppression from egocentricity, especially the extreme egocentricity of fathers. Yes, the upset state of the human condition in both mothers and fathers has had devastating psychological effects on children—especially mothers during their infancy, and fathers during their childhood. But *again*, this is all only another product of the end play state of terminal levels of alienation in the world.

825 (A more complete description of how egocentric the human race has become can be read in *Freedom Book 1* at <www.worldtransformation.com/freedom-poweraddict> and <www.worldtransformation.com/freedom-egocentricity>.)

Chapter 7:16E The differences in alienation between races of humans

826 Having now described the emergence of terminal levels of alienation in the more materially 'developed' parts of the world (remembering that alienation from nurturing instincts in mothers and extreme egocentricity in fathers is having a devastating impact on the souls of all children everywhere), what now needs to be explained is the role greatly improved communication technology has played in the emergence of 'materialism envy' and the resulting unbridled greed, dysfunction and destitution in the materially poorer, so-called 'developing' world. But before presenting that explanation (which will be given next in ch. 7:16F), it is first necessary to explain what is really meant by the terms 'developed' and 'developing' worlds because, as will now be

made clear, the explanation of what we really mean by these terms depends on recognising the truth of the existence of different degrees of alienation between races of people.

827 The different states of material success in the world were once described in terms of the 'rich and the poor parts of the world', or the 'haves and have nots', then as the 'first, second and third worlds', then the 'developed and undeveloped worlds', before now being referred to as the 'developed' and 'developing' worlds. In the context of avoiding the unbearable issue of the human condition, the benefit of the terms 'developed' and 'developing' worlds is that they implied that any country could achieve material success—all that was needed was the time to develop it. The truth that will now be explained is that it wasn't possible to develop a functional society with ordered and well-run services where the majority of the population lived with a high standard of material comfort if the people in that society were either too innocent or too upset—once again, 'innocence' being lack of exposure to and familiarity with the angry, egocentric and alienated upset state of the human condition.

828 It has to be stressed that while we couldn't explain and defend the upset state of the human condition we couldn't afford to differentiate individuals, races, genders, generations, countries, civilisations and cultures according to how innocent or upset they were because it would have left the more upset condemned as bad, unworthy and inferior. It would have led to unfair, destructive and dangerous racist, ageist and sexist prejudice and discrimination against the more upset—and so a dishonest attitude of not allowing differentiation was silently enforced. But as was stated in par. 684, the truth is the only significant difference between humans—the acknowledgment of which makes it possible to truthfully explain and understand much of human behaviour—is the difference in upset anger, egocentricity and alienation between individuals, races, genders, generations, countries, civilisations and cultures, but until the human condition was explained and upset defended as a good, heroic state we couldn't admit and talk about that all-important, clarifying difference. So yes, there are immense differences in upset between races of humans which can—now that it can be understood that all humans are equally good though variously upset—at last be safely admitted. And the result of the differences in upset that exists between races of people is that there are, in fact, more innocent races who are relatively naive about

the difficulties of living with the human condition, and other races who are more instinctively adapted to upset and can thus cope better with it, and others still who are so instinctively adapted to upset that they are too aware of the reality of life under the duress of the human condition and thus overly cynical about being ideally behaved and thus overly selfish and opportunistic and thus socially uncooperative. The result of all this variation in upset is that some races have been effective in living with the human condition while others have been either too innocent and naive, or too burnt out, soul-exhausted and cynical.

829 Even though we have had to avoid it, it is an obvious truth that humans became increasingly adapted to life under the duress of the human condition, with some races becoming more adept at that adaptation than others. Just as individual humans vary in their degree of alienation from our species' original instinctive, all-loving, selfless and trusting soulful true self, so races of humans naturally vary in their degree of alienation. The longer an individual or a race of people were subjected to life under the duress of the human condition, the more they naturally became adapted to that corrupt existence. While a relatively innocent person or relatively innocent race still behaved relatively ideally themselves and expected others to do the same, other individuals and races became so adapted to the upset/corrupt world that they no longer behaved ideally themselves and no longer expected others to behave ideally either. The longer humans were exposed to the human-condition-afflicted state the more cynical they became about human existence—a 'cynic' being **'one who doubts or denies the goodness of human motives'** (*Macquarie Dictionary*, 3rd edn, 1998). As was mentioned in par. 588, when describing the Depressed Adolescentman stage, the psychiatrist Wilhelm Reich wrote honestly about the effects of the different levels of upset in the human race when he described how **'The living** [those relatively free of exposure to upset]**... is naively kindly...It assumes that the fellow human also follows the laws of the living and is kindly, helpful and giving. As long as there is the emotional plague** [the flood of upset in the world]**, this natural basic attitude, that of the healthy child or the primitive...**[is subject to] **the greatest danger...For the plague individual also ascribes to his fellow beings the characteristics of his own thinking and acting. The kindly individual believes that all people are kindly and act accordingly. The plague individual believes that all people lie, swindle, steal and crave power. Clearly, then, the living is at a disadvantage and in danger.'**

830 The consequences for a society of its people becoming overly cynical was that it meant that there would be too little soulful, selfless idealism and too much upset-adapted cynicism-derived selfishness for the society to function effectively. In the situation where it wasn't possible to explain and thus defend the upset state of the human condition, the closest people could come to admitting and talking about this fact that people became adapted to the human condition was to describe individuals or families or races or countries or civilisations as having become **'dysfunctional'** and **'decadent'**, and—especially in the case of civilisations—as having **'passed their prime'** or **'peaked'** in terms of their creative powers.

831 Conversely, some races, like some individual humans, have, in fact, been too innocent to function effectively in the extremely upset-adapted, human-condition-afflicted, corrupt world. As mentioned in par. 671, Sir Laurens van der Post described how a member of the relatively innocent Bushmen race found it impossible to cope with having his innocent, natural spirit compromised: **'You know I once saw a little Bushman imprisoned in one of our gaols because he killed a giant bustard which according to the police, was a crime, since the bird was royal game and protected. He was dying because he couldn't bear being shut up and having his freedom of movement stopped. When asked why he was ill he could only say that he missed seeing the sun set over the Kalahari. Physically the doctor couldn't find anything wrong with him but he died none the less!'** And Sir Laurens was even more specific when he stated that **'mere contact with twentieth-century life seemed lethal to the Bushman. He was essentially so innocent and natural a person that he had only to come near us for a sort of radioactive fall-out from our unnatural world to produce a fatal leukaemia in his spirit.'** The honey-coloured Bushmen are probably the most instinctively/genetically innocent race of people living today. They are more innocent, less soul-corrupted, less human-condition-adapted, less adapted to upset, less toughened, than dark-skinned Bantu Africans, but in turn Bantus are not as toughened and thus as operational and successful in the human-condition-afflicted corrupted world as Caucasian races. For example, I once saw a documentary in which a Bantu African said something to the effect that **'My people can't compete with white people, you go to sleep at night only to wake up in the morning to find white people own everything.'** In turn, Caucasian races aren't as cynical, toughened and opportunistic—selfish—as races from even more ancient civilisations, like the Chinese from the ancient Yellow River valley civilisation, the

Indians and Pakistanis from the ancient Indus and Ganges River valley civilisations, and the Arabs and Jews from the ancient Tigress, Euphrates and Nile River valley civilisations.

[832] In another documentary I once saw, this time about a huge barge that travelled up and down the Congo River from Kinshasa to Kisangani, I could see the whole innocence-destroying, toughening process going on before my eyes. While most of the Africans on the barge were happy to pass the day innocently laughing and singing together, there were a few on-board who were buying produce from each port along the way and then selling it to their fellow passengers for a profit. In my mind it wasn't hard to extrapolate the situation and see that after only a few generations of this occurring that those who were innocently playing and enjoying life weren't going to survive as well as those who were less soulful and focused on 'making a living', which is code for 'achieving material success'—winning power, fame, fortune and glory; finding relief for their insecure human-condition-afflicted existence. Further, since their preoccupation with seeking that material relief will inevitably mean that their offspring will not receive the degree of nurturing and reinforcement they expect and require, that next generation will be even more insecure and thus even hungrier for the relief of power, fame, fortune and glory. And so it will go on, generation after generation of ever-increasing soul-less selfishness and psychological upset until there is either terminal alienation or reconciling understanding of our species' insecure condition.

[833] The situation in Fiji provides a good case-study of what invariably took place when races of varying degrees of upset cohabitated. In the late 1800s British colonists brought Indians to Fiji as indentured labour to farm sugar cane, and by the mid-1960s half the Fijian population was Indian. As a result, a serious conflict arose between the Indian and native Fijians, which we can now understand. The Indian Fijians, coming from an older and thus naturally more cynical, human-condition-toughened, human-condition-realistic and thus opportunistic civilisation, have been so industrious and materially successful that they now monopolise the small business sector in Fiji to the extent that the native Fijians feel their country has been taken over by the Indian Fijians; for *their* part, however, the Indian Fijians also feel discriminated against. Indian Fijian sugar growers in particular feel this inequity, for while they produce 90 percent of the country's sugar, they are only allowed to lease land from

the native Fijians (who own 90 percent of the land). Furthermore, since gaining independence in 1970 the native Fijians have ensured their domination of the political process—a state of affairs that was reinforced in 1990 when the Fijian constitution restricted the Indians to a maximum of 27 seats in the country's 71-seat Parliament. When this provision was amended in 1997 the Indians came to dominate the political scene, successfully electing an Indian Prime Minister in 1999. This situation, however, was overthrown in 2000 when the native Fijians led a coup—and they have remained in power ever since. As mentioned, the Indian Fijians come from a very ancient civilisation in India, one where innocence has long given way to more upset-adapted humans. In comparison, the native Fijians are still relatively innocent, yet to become embattled, hardened and upset-adapted. They aren't manically driven to win power and glory like more embattled, upset-adapted races, preferring to spend their day tranquilly occupied by such soulful activities as playing music, drinking the sedating kava and eating taro roots from their gardens—and trusting in soulful selflessness to care for each other. It is a situation where a 20-year-old, or thereabouts, equivalent race is having to co-exist and compete with a toughened, cynical, more-upset-and-thus-more-insecure-about-their-goodness-and-thus-more-egocentrically-driven-to-try-to-prove-they-are-good-and-not-bad, competitive, selfish, opportunistic 50-year-old, or thereabouts, equivalent race. (Note, it was explained in the Angry Adolescentman stage (in par. 683) how 15, 20 and 30-year-old equivalent stages could be said to exist during the last 2,000 years when that period is being described as the born-again Pseudo Idealistic 40-to-50-year-old **'fraud'** stage and the horrifically angry, punch-drunk, bitter and vengeful Hollow Adolescentman 50-plus-year-old **'criminal'** stage of humanity's maturation.)

834 While holidaying with my partner Annie in Fiji in 1997 a local gave us this description of the structure of Fijian society: **'The Chinese** [who he said were **'the Jews of the East'**] **own all the big tourist resorts where the big money in Fiji is made, the Indians run all the shops and smaller businesses and produce all the sugar cane, and the Caucasians run the country in that they occupy so many of the important administrative positions, providing the good structure and order required for the whole society to function.' 'Fiji'**, he added, **'is one of the few countries in the world where the indigenous people still control the country even though they are the least materially productive and successful.'** When I asked other residents, including an Indian Fijian

and a native Fijian, if they thought this was an accurate description they agreed it was but said that they would never say so publicly for fear of being labelled a racist. The human-condition-understanding-reconciled interpretation of this description is that as soon as you have an unavoidable and necessary battle such as the one that the human race has been involved in, it is inevitable that all those involved are going to become variously adapted to that battle depending on how long they have been exposed to it—with the result that the Chinese and Indians are the cynical 50-year-old equivalent races, the Caucasians are the toughened, but not too toughened, too insensitive or too selfish, more operational 30-and-40-year-old equivalent race, while the native Fijians are the 20-year-old equivalent, overly innocent race. Such differences are simply and obviously what manifest when you have an upsetting battle such as the one the human race has been involved in, where some people will have been engaged in the battle longer and/or more intensely than others. If we are going to have the truthful, meaningful, productive, effective discussion about human behaviour—which is both possible *and necessary* now that the upset state of the human condition has been explained and defended—then the inevitable differences in upset (in particular differences in alienation from our species' all-trusting, sensitive, loving, selfless and sharing original instinctive self or soul) *have to be* acknowledged. The human race either stays living in Plato's terrible cave of alienated darkness until it self-destroys, or we get the full truth up and leave that awful cave existence for a world drenched in the sunshine of understanding, psychological freedom and togetherness.

835 As I mentioned in par. 573, my headmaster at Geelong Grammar School, Sir James Darling recognised what is really an obvious truth, which is that for a person to be as functional as possible under the duress of the human condition they needed *both* human-condition-adapted toughness and sensitive, selfless, innocent, soulful soundness. While he was specifically talking about the qualities that education should strive to cultivate in an individual, what he said also applies to what a society of people needed if they were to be functional in the human-condition-afflicted world. In an address to The Royal Australasian College of Surgeons in 1960 Sir James said (the underlinings are my emphasis): **'The quality which, above all other, needs to be cultivated** [in education] **is sensitivity** [soul]...[Education's] **objective is a development of the whole man, sensitive all round the circumference...the future...lies not with the predatory** [selfish] **and**

the immune [alienated] **but with the sensitive** [innocent]**...There is a threefold choice for the free man...He may** [be overly upset and selfish and] **grasp for himself what he can get and trample the needs and feelings of others beneath his feet: or he may try to withdraw from the world to a monastery** [find himself too innocent to cope with the upset world, like the Bushman who died in jail]**...or he may "take up arms against a sea of troubles, and by opposing** [all the dishonest, alienated denial and selfish, corrupt behaviours] **end them** [ultimately by solving the human condition]**"...**[and so] **There remains the sensitive, on one proviso: he must be sensitive *and* tough...Only by a growth of sensitivity can man progress from the alpha of original chaos** [upset] **to the omega of God's purpose for him** [to become sound and secure]**...Sensitivity is not enough. Without toughness it may be only a thin skin...**[only from] **an inner core of strength are** [you] **enabled to fight back** [defy all the dishonest denial and upset wrongness in the world]**...Can such men be? Of course they can: and they are the** [real] **leaders whom others will follow. In the world of books there are, for me, Antoine de Saint-Exupéry, or Laurens van der Post'** (*The Education of a Civilized Man*, ed. Michael Persse, 1962, pp.28-36 of 223). Yes, to be most operational under the duress of the human condition required a balance of innocent, soulful **'sensitivity'** and human-condition-adapted **'toughness'**, which is what the 30-and-40-year-old equivalent state represents.

836 The simple and obvious fact is that some races are so relatively innocent and naive about life under the duress of the human condition that they lack the toughened self discipline and insecure egocentric drive to succeed of the more upset-adapted races and, as such, can't legitimately compete with those races, so when they see an opportunity to obtain money and/or power and the material luxury both can bring they can't resist taking it, whether it's rightfully due or not. When Annie and I travelled through central Africa in 1992 everywhere we went, at every level of society, there was dysfunction, graft and corruption—even when we landed in Kenya we couldn't leave Nairobi airport until we paid certain 'fees' to various airport officials. At the top of such societies you invariably find completely despotic regimes—for instance, we were told that the reason the roads beyond the centre of Nairobi weren't sealed and were in a terrible state was because all the money for such infrastructure had been syphoned off by the country's leaders to buy villas on the French Riviera and other luxuries. And the simple and obvious truth is that at the opposite end of the spectrum of alienation there are races where everyone is so upset-adapted, so soul-destroyed, so innocence-obliterated and manically ego-

centric and cynically selfish that graft and despotism is similarly endemic in their societies. Consider the events of early 2011, when the extreme despotism of almost every, if not every, Arab country right across North Africa and the Middle East provoked democracy-demanding uprisings throughout the region—a revolutionary wave that continues today. During 2013 there were also reports in Australian newspapers every few days of the corruption that is endemic in China, where officials throughout the country are finding themselves unable to resist stealing whatever money they can lay their hands on. And in India there is little or no filter-down of wealth to the huge majority of poor people in the country—no viable middle class. The city streets of India are festooned with a tangle of illegal electricity wires, and everywhere there is chaotic traffic, pollution and poverty—there is little or no order or functionality. And since the only thing that keeps human society going is soul, innocence, soundness, moral strength—our species' original instinctive inclination to be selfless, loving and cooperative—once that goes all that remains is ruin. Yes, Tracy Chapman was right when she wrote and sang that **'All that you have is your soul'** (1989)—so thank goodness that through this ameliorating understanding we all now have the means to retrieve it, or, if not immediately retrieve it during our lifetime, to reconnect to it through the Transformed Way of Living that is described in chapter 8.

837 It follows then that it is only at the middle of the spectrum of alienation, amongst 30-and-40-year-old, or thereabouts, equivalent races where there is enough upset-adapted self-discipline and toughness, but not so much that there is excessive soul-less cynicism and thus selfishness, that you get maximum functionality and operable behaviour in life under the duress of the human condition. The Anglo-Saxons are currently the stand-out example of such functionality, coming as they do from the more isolated and sheltered-from-upset north-western edge of the Caucasian populated countries of Europe—they are actually more 30-year-old equivalents than 40-year-old equivalents. Although Anglo-Saxons come from a small, resource-deficient island country, they have been so operable and thus successful and thus influential that they have led the so-called 'globalisation of the world' to the point where **'A quarter of the world's population speak English...English is increasingly becoming entrenched as the language of choice for business, science and popular culture. Three-quarters of the world's mail, for example, is currently written in English'** (*TIME*, 7 Jul. 1997). The Anglo-Saxons exhibit

the same soulful enthusiasm, vitality and energy of the Vikings from Scandinavia who were referred to in par. 682. Indeed, like the Vikings, who sacked and settled in many parts of the British Isles, the Angles and Saxon tribes who colonised eastern England (and from whom the name Anglo-Saxon comes) were of southern Scandinavian stock. Also the Normans, who conquered England in 1066, were Scandinavians who had settled on the east coast of France—the name 'Norman' is actually derived from 'northman', meaning peoples from Northern Europe (Scandinavia). Of course, the basic Celtic stock of the British Isles are, like the Scandinavians, also an exceptionally isolated and thus relatively innocent race, especially coming, as a significant proportion of them most likely did, from the ancient Celts of central Europe who were an *extremely* energetic, enthusiastic, relatively innocent/soulful race, but whose innocence had been lost elsewhere in the less isolated parts of Europe where they had spread.

[838]What people are really doing then through their efforts to either try to or actually remove tyrants/despots like Robert Mugabe, Saddam Hussein, Muammar Gaddafi, Hosni Mubarak, etc, etc, from power in the hope that those countries will become functional democracies is trying to make 20-year-old equivalents and 50-year-old equivalents behave like 30-and-40-year-old equivalents. But, as just explained, societies of 20-year-old equivalents and 50-year-old equivalents *are* going to revert to selfishness, at which point a selfish power struggle will occur where, in the end, the most ruthless will take over once again, a struggle we are currently witnessing in Syria. Having solved nothing at a fundamental level, those societies will invariably remain dysfunctional, resulting in further floods of refugees from those countries to others populated by more functional 30-and-40-year-old equivalent races. Efforts to avoid this cycle, or at least contain it somewhat, in countries where there is too much cynical selfishness only led to the creation of authoritarian, dictatorial, freedom-and-democracy-denying, free-thinking-restricted, human-mind-oppressive regimes—which were therefore still fundamentally tyrannical and despotic—like those that have been established in China, and (to a degree) by Lee Kuan Yew in Singapore. As will be explained shortly in ch. 7:16I, the other form of tyranny that developed to contain excessive cynical selfishness was strict obedience to fundamentalist interpretations of religious teachings—as has occurred in many parts of the Arab world.

[839]The fact is, a society that did not contain a significant proportion of functional 30-and-40-year-old equivalents was not going to be a functional one. Historically, **'the chief charge** [against Jews] **has been that Jews are somehow uniquely engaged—at other people's expense—in money making'** (*Ideas that shaped our World*, ed. Robert Stewart, 1997, p.55 of 223), and with compassionate understanding of the human condition we can understand that there was some truth in this view. The cynical 50-year-old equivalent Jews have managed to remain operational and, as a result, extremely materially successful by living amongst relatively selfless, functional 30-and-40-year-old equivalent races—which is the real reason they have been persecuted in the predominately 30-and-40-year-old equivalent countries in which they settled. (The dishonest reason for their persecution being that they were the murderers of Christ). In this light it can be seen why the Pygmies and the Bushmen have resented the Bantu for being more operational and materially successful than they are, who in turn have resented the Caucasians for being more operational and materially successful than they are—who in turn have resented and thus persecuted the Jews for being more operational and materially successful than they are. Everywhere the inevitable differences in upset between individuals, races, genders, generations, countries, civilisations and cultures have caused immense problems, so the greater truth is that it is a very great tribute to the character and courage of the human race as a whole that it has managed to maintain some semblance of functionality and cooperation under that almost impossible situation.

[840]Indeed, indicative of the difficulty of the situation is the fact that during January and February 2011 the British Prime Minister David Cameron, German Chancellor Angela Merkel, the then French President Nicolas Sarkozy, former Spanish Prime Minister Jose Maria Aznar and former Australian Prime Minister John Howard all declared that **'multicultural policies'** have been a **'failure'** because **'immigrants'** had **'not successfully integrated'** (*Daily Mail*, 11 Feb. 2011). Yes, just as an individual person's lifestyle was inevitably going to largely be a response to that person's particular level of upset, so too a race's culture was inevitably going to largely be a response to that race's particular level of upset, which means different races with their different cultures inevitably found it difficult co-existing. You don't very often see 30-year-olds forming close friendships with 50-year-olds, or even 20-year-olds with 30-year-olds. Most people relate much better to their own age group.

In fact, the stages that occurred with different ages under the duress of the human condition changed so rapidly and were so dramatically different that 18-year-olds typically found it difficult relating even to 21-year-olds. Outside of family situations, everyone tended to fraternise with their own age group. The same situation of incompatibility obviously applied between races of people. Different levels of upset had different needs. For example, as mentioned in par. 610, once humans became extremely upset even the glimpse of a woman's face or ankle became dangerously sexually exciting to men, which is why in some cultures women are completely shrouded and persecuted if any part of their body is revealed in public. Imagine then how difficult and provocative it has been for individuals from such extremely upset cultures to see young women from less upset cultures running around at liberty in bikinis and mini-skirts.

841 Again, such admissions of the relative innocence, or lack thereof, of different individuals, races, genders, generations, countries, civilisations or cultures have, until now, had to be avoided because they led to prejudiced views that some individuals, races, genders, generations, countries, civilisations or cultures are either good or bad, superior or inferior, when the truth is that while humans do vary in their degree of upset as a result of the necessary and heroic battle humanity had to wage to find self-knowledge, *all humans are equally good*. Upset is not a bad, evil state, but a good, heroic one. Trying to manage differences in upset between individuals, races, genders, generations, countries, civilisations and cultures has been extremely difficult, but once the prejudiced views arose of some individuals, races, genders, generations, countries, civilisations or cultures being either good or bad, superior or inferior, more worthwhile or less worthwhile, terrible atrocities and injustices very often followed. For instance, in the last century alone we have seen the Holocaust in which approximately 6 million European Jews were exterminated by the Nazis during the Second World War; the attempted 'ethnic cleansing' by the Bantu Hutu of an estimated 800,000 of the more upset-adapted Nilotic Tutsi in 100 days of bloodshed in Rwanda in 1994; and Idi Amin throwing out of Uganda, in 1972, all the Indians and Pakistanis, some 40,000-80,000 people, who owned and operated most of the businesses there because he claimed **'they** [were] **sabotaging the economy of the country'** (*Jet* mag. 14 Sep. 1972).

842 The most difficult nuance, however, of managing differences between individuals, races, genders, generations, countries, civilisations

or cultures is that not all discrimination between individuals, races, genders, generations, countries, civilisations or cultures has necessarily been wrong. For example, Plato quite sensibly wanted to have the least ego-embattled/most innocent—the **'philosopher kings'** or **'philosopher rulers'** or **'philosopher princes'** or **'philosopher guardians'** as he variously described them—lead society. He wrote, **'isn't it obvious whether it's better for a blind man** [an alienated person] **or a clear-sighted one** [an innocent, ego-unembattled, denial-free, honest person] **to keep an eye on anything'** (c.360 BC; *Plato: The Republic*, tr. H.D.P. Lee, 1955, p.244 of 405), arguing that **'If you get, in public affairs, men who are so morally impoverished that they have nothing they can contribute themselves, but who hope to snatch some compensation for their own inadequacy from a political career, there can never be good government. They start fighting for power...**[whereas those who pursue a life] **of true philosophy** [honest, unresigned, egocentricity-free thought] **which looks down on political power...** [should be] **the only men to get power...men who do not love it** [who don't egocentrically hunger for power, fame, fortune and glory]**...rulers** [who] **come to their duties with least enthusiasm'** (p.286). Completely **'obvious'** and right-minded as Plato's idea was of having the most innocent run society, such honest differentiation according to who was innocent, and who was not, wasn't possible because, once again, while the human condition wasn't able to be explained, any differentiation between individuals according to degrees of alienation or soundness left those no longer innocent unjustly condemned as bad and unworthy. In the absence of such honesty about who was innocent and who was not, functional societies did, however, try to avoid the overly insecure and egocentric from holding power by having regular democratic elections where such people could be voted out of office.

843 But now that it is both possible and necessary to talk about who was innocent and who was upset, we can see and admit that colonisation under the rule of 30-and-40-year-old equivalents did make significant sense. The races who were most functional under the duress of the human condition tried to help the less functional become more materially developed—that is, advanced in the many arts of living with the human condition needed to progress humanity towards greater knowledge, ultimately understanding of the human condition. As Sir James Darling has written about the British Empire, **'the function of Empire is to educate rather than to oppress'**, and the British have **'an unbeaten record in the history of civilization'** (*The Education of a Civilized Man*, ed. Michael Persse, 1962, pp.134, 136 of 223). Sir Laurens

van der Post similarly wrote that **'Great Britain'** created **'the largest, the greatest and, I still believe, the best-organised, and most civilized empire in the history of the world'** (*The Admiral's Baby*, 1966, p.108 of 340). All the history books written by truth-denying-in-order-to-artificially-make-everyone-equal postmodernists that condemned colonialism as the worst evil are going to have to be re-written truthfully.

844 In fact, I should mention here some of the human-condition-avoiding, denial-based reasons that have been put forward to explain the dysfunctionality of African countries like Kenya, including placing the blame on colonialism. Firstly, it is claimed that such countries are on the same journey as European nations, which went through their own dysfunctional stage before organising themselves into upset-restraining, so-called 'civilised' democracies—that the Bantu African races aren't any more innocent than European races and will, in time, as mentioned earlier, be able to develop functional, materially successful democracies. In keeping with this theory, it was mentioned earlier that instead of using terms like 'first, second and third world countries', the current politically correct nomenclature for the different states of functionality of countries is to refer to them as being either 'developed' or 'developing'. But if time, rather than degrees of innocence, or lack thereof, was the issue then races from ancient civilisations should not be dysfunctional—and yet they are. The Greeks gave us Plato, Socrates, Aristotle and the foundations of 'Western civilisation' and yet, in 2011, through their present innocence-destroyed, burnt out, peaked, decadent, selfish greed and resulting dysfunction, brought the world's economy to its knees through their refusal to adopt responsible financial practices. The other Mediterranean Caucasian countries of Italy and Spain have similarly proved to be dysfunctional, peaked societies of people, having also had to be bailed out by the International Monetary Fund (IMF). The Celtic Irish, who also got themselves into financial difficulty, aren't burnt out so much as too innocent and naive—indicative of their comparative functionality is that in contrast to their European neighbours, Ireland recognised the need to implement and maintain austerity measures prior to and following the global financial crisis and, as a result, recently became the first Eurozone nation to exit the international bailout. As for blaming the dysfunctionality of African societies on colonialism, there is no doubt colonialism had negative, exploitive repercussions—some truly terrible, like the slave trade—and it did seriously disrupt the old

tribal system that operated throughout most of Africa, but while tribalism, an authoritarian, dictatorial system in which the most powerful ruled, brought some peace and order (as it effectively does in all non-human societies, such as in wolf packs or in any herd animal species), it was still dysfunctional in that it oppressed the individual freedoms/liberties that humans' search for knowledge/creativity depends on. Of course, you *can* manage humans by tying them all down, as was done in tribal situations and Marxist and Confucianist regimes, but they will no longer be humans—they will no longer be conscious beings fulfilling their fundamental responsibility of exercising their minds and learning to understand existence. Overall, colonialism gave individuals many freedoms they hadn't had that the individual then had to manage—but the challenge for humans has always been to manage their consciousness-derived freedom *effectively*. As has been explained, the lack of effectiveness of that management across a social structure was due to races being either too innocent, sensitive and naive about life under the duress of the human condition, or too toughened, soul/innocence-destroyed, cynical, opportunistic, greedy and selfish.

845 On the whole we can see that with understanding of the human condition it at last becomes possible to explain in psychological terms what was actually happening when history books spoke of civilisations having 'peaked' and become 'decadent'. As with each individual during his or her life, under the duress of the human condition all races eventually became overly corrupted, corruption of our original instinctive self or soul being the price of humanity's heroic search for knowledge. In this journey from innocence to exhaustion of soul the most creative period was the toughened and disciplined, but not yet overly corrupted, 30-year-old equivalent stage. As each race and its associated civilisation passed through this stage it made its particularly creative contribution to the human journey. This was when civilisations were at their 'peak'; inevitably, however, they entered a more corrupted 'decadent' stage. To look at just the Mediterranean, Middle East, Indian and Eastern civilisations, they all made extensive contributions to the human journey during their energetic and creative 30-year-old equivalent stage. The Egyptians and peoples from the fertile crescent of the Tigris and Euphrates delta in the Middle East instigated the civilisation of the 'known world'; for example, they invented the wheel, mathematics and writing. As already mentioned, the Greeks and Romans laid the foundations for 'Western

civilisation' during this most creative stage of their journey through ever-increasing levels of upset. The great religions of the world, Hinduism, Buddhism, Judaism, Christianity and Islam, were developed in India and the Middle East when there was still enough soundness left in their populations to produce some exceptionally sound, truthful, unresigned, denial-free thinkers or prophets. And during their most creative stage, the Chinese contributed to the human journey such influential inventions as paper, moveable type, the compass and gunpowder. The truth is *ALL* civilisations right back through history made important contributions to the advancement of knowledge—but just where the leading edge in the advancement of knowledge was occurring at any one time depended on what stage in the human journey from innocence to exhaustion and decadence those various civilisations were at.

[846]I might add here a balancing comment regarding the earlier statement about the Jewish race having benefited from living amongst more innocent, soulful, selfless 30-and-40-year-old equivalent races. At the end of the description of Adventurous Adolescentman, when analysing the graph of *The Development of Mental Cleverness*, it was explained that the reason brain volume plateaued towards the end of the last 2 million years of its growth is because a balance was struck between the need for cleverness and the need for soundness—between knowledge-finding yet corrupting mental cleverness and conscience-obedient yet non-knowledge-finding lack of mental cleverness, with the average IQ today representing that relatively safe conscience-subordinate compromise. It is true that the ability to find answers didn't necessarily accompany increased intelligence because, as just described, increased intelligence tended to lead to an increase in upset and thus alienation, and alienation made thinking truthfully and thus effectively very difficult, however, a high degree of intelligence was still required to find knowledge, most especially in complex subject areas like higher mathematics and physics. Thus, if the human race couldn't develop exceptionally high levels of intelligence then many crucial understandings about the nature and workings of our world would not have been able to be found. The Jews are renowned for being exceptionally intelligent, which is consistent with them being an exceptionally upset race, and it is from within their ranks that some of the greatest minds and insights have emerged: Albert Einstein, with his breakthrough insights into the physical nature of our universe, is the most obvious example. It is true that Einstein must have

had an exceptional degree of soundness to have been as an effective thinker as he was, but he also must have been exceptionally intelligent to so successfully grapple with the extremely complex subjects he was dealing with. I haven't ever tried to collate all the contributions to the human journey that the Jews have made, but it would be very significant, as evidenced by this extraordinary statistic: **'just 0.2 per cent of the world population, Jews accounted for 29 per cent of Nobel Prizes in the late 20th century, and 32 per cent so far in 21st century'** (*The Spectator*, 6 May 2014). By, in effect, allowing exceptional cleverness/intelligence to develop by countering its corrupting effects with the presence of people who were not so intellectually clever and thus not so upset and thus not so upset-adapted was, in the bigger picture, a fortuitous outcome for the human race.

847 The human journey has been *such* a complex story that a perfectly balanced view is beyond the powers of effective interpretation in this very early stage of viewing the history of the human race in a denial-free way. And such a detailed interpretation can actually wait because what is so important now—and this is all explained in the final chapter (8) of this book—is that the human race can leave behind its whole upset history as compassionately dealt with—our history *is finally*, as the saying goes, just that—it's 'all just history' now. As will be explained in chapter 8, what brings all the horror of life under the duress of the human condition to an end is the Transformed Way of Living.

848 My job, as the bearer of understanding of the human condition, is to get at least the main descriptions of all the hard truths up and dealt with so that humanity can move well out into the clear, free of the past. As will be emphasised in the final chapter, we get the truth up and then we move on. We leave the old effectively dead, dishonest, human-condition-afflicted world behind. It will be explained that this is the indescribable magnificence of the Transformed Lifeforce State that the songwriter and singer Bono (of the band U2) sang about: **'I've conquered my past** [found the dignifying, human race liberating understanding of the human condition] / **The future is here at last / I stand at the entrance to a new world I can see / The ruins to the right of me / Will soon have lost sight of me'**; and similarly, as Beethoven's *Ninth Symphony* anticipated, **'Joy'**, **'Joyful, as a hero to victory!'**, **'Join in our jubilation!'**, **'We enter, drunk with fire, into your** [human-condition-understood] **sanctuary...Your magic reunites...All men become brothers...All good, all bad...Be embraced, millions! This kiss** [of understanding]

for the whole world!' (1824). Yes, it no longer matters who is more cynical, more human-condition-adapted, because the Transformed State leaves all that behind. The greater truth is we humans have *all* become well and truly corrupted/upset/messed-up/damaged/soul-devastated anyway, but none of that matters now because we are leaving that upset world, *we are out of there*; we have won the match and we can now all head for the showers and get ready for humanity's great victory party—and soon even the different scars we all carry from the match will be healed and gone, soon the human race will be rehabilitated. Even the upset that is now instinctive, in our genes, will not be an issue because when our capacity to love is finally liberated, as it can now be, it will effectively make all our upset—both psychological and instinctive—disappear without a trace. It *is* all history now; **'the ruins to the right of** [us]**, will soon have lost sight of** [us]'; **'all good, all bad'**, **'this kiss** [of understanding is] **for the whole world!'**

849 (A more complete description of how variously selfish and cynical the human race has become can be found in *Freedom Book 1* at <www.worldtransformation.com/freedom-selfishness>.)

Chapter 7:16F The emergence of 'materialism envy' and with it unbridled greed, extreme dysfunction and destitution in the 'developing' world

850 Having now explained that the real difference between 'developed' and 'developing' countries is between the functional 30-to-40-year-old equivalents and the dysfunctional 20-year-old and 50-year-old equivalents, the issue of the 'materialism envy' and ensuing corruption and dysfunction that has been occurring in the less functional, 'developing' countries can and has to be looked at more closely.

851 The important factor that needs to be considered in the occurrence of 'materialism envy' in the less operational 'developing' parts of the world is the role that has been played by immensely improved communication technology. During the so-called colonisation period of the world where the more functional, materially 'developed' countries established colonies in the more dysfunctional, materially 'undeveloped' countries, the populations of those nations had little awareness of just how materially successful people in the more functional, 'developed'

world generally were, but with the advent of sophisticated communication technology this changed dramatically. With the introduction of the likes of television suddenly these colonised people could see what they were missing out on materially and when they found themselves unable to compete with those more operational and functional, not-too-innocent-but-also-not-too-cynical materially developed colonisers it naturally made them insanely envious. People in the materially impoverished parts of the world suddenly had access to programs like *Baywatch* (TV series 1989-2001), in which gorgeously groomed Californians in stunningly enticing swimsuits played in a fabulously ordered and materially wealthy world, and naturally they were overcome with distress about what they now knew they were missing out on. The psychiatrist Clancy McKenzie gave this stark description of, what might be termed, 'the *Baywatch* effect' or 'materialism envy' in action: **'While visiting Machu Picchu in Peru in 1979 I noted very poor persons, living in the mountains, who had only the clothes they wore and perhaps a lama or two, but had beautiful, warm smiles and seemed content and happy. Days later I was in Bogota in Colombia. It was a very hot day and we asked the driver to stop at an outdoor tavern to buy cold beer. The people were very impoverished, but there was a TV playing and they were able to view the "outside world" where everyone seemed to have more, and luxury was abundant. I offered to go in with the driver and he urged me to wait in the car. I soon learned why. The absolute hatred was so intense that it was palpable. These people did not have less than those in Machu Picchu but they saw others who had more, and their needs were intensified'** (Letter to Prof. Harry Prosen, 27 Mar. 2006). The result of this awareness of what materialism offers is that there is now rampant greed in all those populations who haven't had the material luxury that is so prevalent in the more functional, 'developed', mostly Western world. They see themselves as being on catch-up—they want the big house, a car, a fridge, a suit, a gown, a fancy watch, a beach holiday, Nikes, Coca-Cola and Big Macs, etc. The effect of this myopic drive for material wealth is that our original instinctive self or soul's natural inclination to be selfless and loving has been thrown by the wayside and everywhere in the less functional and materially fortunate 'developing' countries cynicism now reigns—again, a 'cynic' is **'one who doubts or denies the goodness of human motives'**. The following is another description of the consequences of 'materialism envy' in Kenya, but it very much applies to what is happening throughout Africa, South America, China, India, south-east Asia and in parts of the

Middle East. In her 2000 book, *African Visions*, the Kenyan photographer Mirella Ricciardi wrote of her country's **'chaos and confusion, the crime and lawlessness, the greed, the bribery and corruption at all levels, the senseless wildlife killings, the environmental destruction, the mindless power games, the hopeless destitution and hunger, the lack of concern and forward thinking...The outrageous greed and disregard for anything and anyone other than themselves and their immediate families in the top ranks of governments, have created a financial drain on the country that filters down and affects whole populations in varying degrees, threatening the fragile infrastructure of the land. All those, without exception, who have somehow attained a position of command and authority, however humble, have but one idea in mind, to emulate the example set by the head of state – to milk their positions for all they are worth, for as long as it lasts. Because finances are siphoned off to Swiss bank accounts, salaries are so low they need topping up by whatever means. As a consequence bribery and corruption at every level is the only option. The police force and judiciary, the ministers, the administrators, the teachers, the doctors and the heads of government divisions, extract whatever they can, by fair means or foul, under the guise of their uniforms and titles, while the country collapses and the masses starve'** (pp.260-261 of 287). Yes, the end result of the, in truth, very obvious difference in functionality between races of humans living under the duress of the human condition is that the less functional and thus less materially successful races naturally became extremely jealous and envious of the material success of the more functional and operational races, resulting in **'outrageous greed'**, **'bribery and corruption at every level'**, **'while the country collapses and the masses starve'**.

[852] As McKenzie's example shows, the material inequality in the world that sophisticated communication technology has made visible to everyone has made those who are missing out not only extremely envious, but resentful and thus angry, to the point of feeling **'absolute hatred'** towards the more materially successful. Their self-esteem has suffered so much that angry retaliations, like the flying of those planes into The Pentagon and the World Trade Center on September 11, 2001, have occurred; Osama bin Laden actually said the 9/11 attacks **'were revenge for Western humiliation of Muslims'** (*TIME*, 7 May 2012). Yes, as a result of the emergence of overwhelming levels of upset in the last 200 years humanity has endured two massive world wars and countless other insurgencies, as well as the inglorious honour of inventing weapons of such ferocity that they could wipe out entire cities—and these weapons are now in the hands of

extremely psychologically distressed rogue countries like North Korea, whose attitude toward the rest of the world is one of 'We don't care any longer, we just want relief for our egos'; 'It's either death or glory and if it's going to be death for us then it's going to be death for everyone because we're not going to back down and give up on our need for glory, which is, in effect, what you're asking us to do.' It should be pointed out that for the human race *as a whole* to have the technological capability it now has while we are, as has now been revealed, so deeply insecure and alienated is the equivalent of allowing a psychotic lunatic to play with an atomic bomb. General Omar Bradley described our predicament perfectly when, as was mentioned in par. 118, he said, **'The world has achieved brilliance...without conscience. Ours is a world of nuclear giants and ethical infants.'**

[853] So, the *real* debate about both the horrific material inequality and the terrorism and frightening instability in the world requires analysis of the differences in upset-adaption or alienation-from-soul between individuals, races, genders, generations, countries, civilisations and cultures, but until the human condition could be explained and the upset state of the human condition compassionately understood and thus defended that debate could not take place. The problem of selfishness in the world was not being addressed honestly and thus properly anywhere.

[854] As will be explained shortly in ch. 7:16I, the distress from 'materialism envy' has also led to the adoption of extremely strident fundamentalist expressions of religion to try to manage the unbearable situation.

Chapter 7:16G <u>So it is end play wherever we looked</u>

[855] It can be seen then that the world's situation is a dire one in which the more operational 'developed', materially affluent parts of the world are fast approaching a state of terminal alienation with anxiety disorders such as ADHD and autism now rampant and the **'biggest age group in American history'** in the grip of an extremely insecure, power addict-driven **'narcissism epidemic'**, while the less operational 'developing', materially poorer parts of the world are, as a result of 'materialism envy', fast approaching **'collapse'** and **'hopeless destitution'**.

[856] So it is end play for the human race wherever we look, and while we humans have had no choice other than to stay positive in our thinking, pretend that everything is okay, underneath that facade everyone knows

the situation is becoming *extremely* serious. Throughout history there have been 'The end of the world is nigh' doomsayers, but this is different. At the time of writing this passage, which is in early 2013, there is a plethora of end-of-the-world, post-apocalyptic programs appearing on television, such as **'Life After People'**, **'Evacuate Earth'**, **'Doomsday Preppers'**, **'Survivors Guide to the Mayan Apocalypse** [the prediction in the Mayan calendar that the world would end in 2012]', **'Apocalypse 2012 Revelation'**, **'Armageddon Outfitters'**, **'End Day'**, **'Omens of the Apocalypse: The End is Near'**, **'The Walking Dead'** and **'Zombie Apocalypse'**. Similarly, I happened to pick up this week's guide (11-17 April 2013) from my local cinema—and I haven't been near a cinema for years—and of the 12 films showing, these are the names of 7 of them: **'Oblivion'**, **'Scary Movie 5'**, **'Warm Bodies** [about zombies]', **'Escape from Planet Earth'**, **'Trance'**, **'The Host** [about the human race being infected by psychic parasites]' and **'A Good Day to Die Hard'**! And I've just heard of another movie that's about to be released, titled **'This Is The End'**!!

857 Since writing the above paragraph, the actor Brad Pitt's film **'World War Z'** has also been released. Described as **'the latest in a trend of apocalyptic thrillers...from a Hollywood convinced that the end is nigh'** (*The Australian*, 10 Jun. 2013), it's about **'a viral zombie disease that threatens world domination'** (*The Sydney Morning Herald*, 10 Jun. 2013). Yes, the **'zombie disease'** of alienation *is* threatening to take over the world, and the end of the world *is* nigh. Australia has just released its first **'post-apocalyptic zombie spectacular'** titled **'Wyrmwood'** (*The Sydney Morning Herald*, 15-16 Jun. 2013), while an article in *The Sydney Morning Herald* last month (3 Jun. 2013) titled 'Vacant stares become all the rage' contained the photograph shown opposite on the left with the caption, **'An enthusiast takes part in the Zombie Walk festival in Prague on the weekend. Zombie walks, in which horror fans dress as zombies and parade through city streets, have grown in popularity in recent years.'** The detail from a painting by Jean Michel Basquiat shown opposite on the right appeared in *The Australian* newspaper the month before last (17 May 2013) accompanying a story about how it had just been sold at auction for a whopping $US48.8 million—a price no doubt indicative not only of its artistic merit but how much it resonates as an image of our time. Yes, Laing's **'fifty feet of solid concrete'** between us and **'our true selves'** or soul is now more like a thousand feet of **'solid concrete'**.

Depictions of the terminally alienated, zombie state
that humanity has arrived at (details in the preceding paragraph)

858 And the examples keep accumulating. An article in this week's *TIME* magazine (22 Jul. 2013) about **'*Pacific Rim*, the new Armageddon adventure'** observes that **'If our world isn't already post-mortem, as in Tom Cruise's *Oblivion* and Will Smith's *After Earth*, then zombies are over-running the planet (*World War Z*) or we are in search of a Superman to stop an alien attack (*Man of Steel*).'** Our obsession with superheroes at the moment—such as Superman, The Lone Ranger, The Phantom, Batman, Wonder Woman, Spider-Man, Captain America, Captain Marvel, Luke Skywalker of Star Wars, Frodo Baggins of The Lord of the Rings, the wizard Harry Potter, The X-Men, and Iron Man—is an expression of an intuitive awareness humans have of what is now desperately needed to solve the human condition and **'stop' 'Armageddon'**. Yes, to the resigned, alienated mind, confronting the unbearably depressing issue of the human condition, which is what was needed to solve it, did seem an impossible, superhuman undertaking.

859 *No wonder* there is panic in officialdom with the desperate spate of initiatives that were described in par. 445 now occurring to try to find the only thing that could save the world, namely the reconciling understanding of the human condition—despite that reconciling understanding having been found and put forward in my books for the last 30 years! In fact, since what is happening is *so* obscene, I am going to re-include that list of initiatives:

- in December 2012 it was announced that an American billionaire had pledged $200 million to Columbia University's **'accomplished scholars whose collective mission is both greater understanding of the human condition and the discovery of new cures for human suffering'**; and,

- on 28 January 2013, in Europe the European Commission **'announced that it would launch the flagship Human Brain Project with a 2013 budget of €54 million (US$69 million), and contribute to its projected billion-euro funding over the next ten years'** with the goal of providing **'a new understanding of the human brain and its diseases'** to **'offer solutions to tackling conditions such as depression'**; and,
- on 2 April 2013, the President of the United States, Barack Obama, announced a **'Brain Initiative'**, giving **'$100 million initial funding'** to mechanistic science, in **'a research effort expected to eventually cost perhaps ten times that amount'**, to also find **'the underlying causes of... neurological and psychiatric conditions'** afflicting humans in order to **'develop effective ways of helping people suffering from these devastating conditions'**; and,
- on 24 April 2013, *BBC News Business* reported that **'Lord Rees, the Astronomer Royal and former president of the Royal Society, is backing plans for** [Cambridge University to open] **a Centre for the Study of Existential Risk** [meaning risk to our existence]**. "This is the first century in the world's history when the biggest threat is from humanity," says Lord Rees.'** The article then referred to Oxford University's Future of Humanity Institute that was established in 2005, which is **'looking at big-picture questions for human civilization...**[and] **change...**[that] **might transform the human condition'**, quoting its Director and advisor to the Centre for the Study of Existential Risk, Nick Bostrom: **'There is a bottleneck in human history. The human condition is going to change. It could be that we end in a catastrophe or that we are transformed by taking much greater control over our biology'**!

[860]But even in the unlikely event that such desperate initiatives (unlikely and desperate as they are because the mechanistic paradigm they are part of is totally dedicated to *avoiding*, not confronting, the human condition) *did* somehow manage to find **'the underlying causes'** of our **'human condition'**, the very, very best they could hope to achieve would, again, be *exactly* what is already being presented in this book, and even then it would have to include the explanation of how the human race is supposed to cope with the terrifying exposure of our corrupted condition that such an understanding inevitably brings—the answer to which is presented in the concluding chapter (8) of this book. That the mechanistic paradigm is nourishing its own human-condition-avoiding,

dishonest institutions while persecuting and blocking our human-condition-confronting, truthful World Transformation Movement is just another manifestation of the terminally alienated, deathly, sick state the world is in—because instead of attacking and blocking the work that we are doing, the mechanistic world should be forming a steering committee composed of strong minded people, like Australia's former Prime Minister John Howard and the American political commentators Bill O'Reilly and David Brooks, to bring as much support as is humanly possible to our work, and thus to a future for humanity. Ideally, that exceptional denial-free-thinking prophet of the Irish rock band U2, Bono, would obviously also be on such a committee. And since HRH The Prince of Wales, the heir to the British throne, has throughout his life genuinely, not deludedly and falsely, been working on opening the door to a truthful human-condition-understood new world for humanity—something he has also been persecuted for—he is another who should be on such a committee.

Chapter 7:16H The progression of ever-more dishonest and dangerous forms of pseudo idealism to cope with the unbearable levels of upset

861 In terms of the progression of the human journey that is being charted in this chapter, the repercussion of upset reaching this crescendo on our strategies for coping with the human condition was that people could no longer cope with the honesty of religion—it became too confronting, guilt-inducing and unbearably depressing. Yes, the great benefit of religion—of the honesty imbued in the prophet or prophets (in the case of Hinduism) that the religion was founded around—actually became its liability, because by retaining a presence of a prophet's soundness and truth, religions reminded humans of their own corrupted state and their alienation from truth, which in turn accentuated their sense of guilt; as the author Mary McCarthy once wrote about religion, **'Only people who are very good can afford to become religious; with all the others it makes them worse'** (*Memories of a Catholic Girlhood*, 1957). It was at this point when the honesty of religion became too confronting that much less confronting and less guilt-emphasising forms of pseudo idealism had to be found, with the *extremely* dangerous and negative effect being the loss of the precious honesty originally imbued in religion.

862 When the truthful lives and thoughts of religions' founding prophets became unbearably confronting and condemning, Guilt-Free Expressions of Idealism to support and find 'feel good' relief from became highly sought after. These expressions took two forms: Less Guilt Emphasising Expressions of Religion, and Non-Religious Pseudo Idealistic Causes.

Chapter 7:16I Less Guilt Emphasising Expressions of Religion

863 As humans became more upset religions were adapted and developed to be less guilt emphasising. Within Christianity for example, rather than following a denomination that focused on the study and acknowledgment of the integrity of the words and life of Christ, a person could select one that emphasised worship, adoration and ceremony, such as Catholicism. The emphasis in religion could be shifted away from self-confronting honesty, introspection and interpretation of the deeper truths contained in the religion's scriptures to simply worshipping the prophet as a more indirect way of glorifying and supporting the truth he stood for, and to simple, literal, non-interpretative, dogmatic propagation of the religion's scripture. And, overall, to counter the increasing waywardness of the greater levels of upset, to even more strict obedience to the cooperative ideals and principles espoused by the religion. The popularity of literal, fundamentalist expressions of religion—especially simplistic, emotional, euphoric, 'evangelical', 'charismatic' forms of Christianity and fundamentalist representations of the Islamic faith—have increased so much in recent times that they are now the fastest growing forms of religion in the world.

864 So the more upset and thus insecure and guilt-feeling we humans became the more fundamentalist or literalist we needed to be. Mindlessness saved us from hurtful mindfulness. The Gnostics were an early Christian sect who encouraged self-confronting introspection and analysis and interpretation of the Gospels. Gnosticism stressed self-knowledge and the need, as they said, to **'find the Christ in yourself'**. (The word 'knowledge', which means 'to know', has the same roots as the word 'gnosis'.) The Gnostics also practised living a frugal, ascetic, 'pure' life. But since the majority of people found such dedicated introspection and interpretation of the Gospels far too confronting and a life of frugal purity unbearably devoid of much needed self-distraction and escape from their upset reality, Gnosticism eventually died out. In

fact, around 1230 AD a branch of Gnostics in France called the Cathars were viciously persecuted and eventually destroyed by emissaries from the Roman Catholic Church. As mentioned, the Catholic branch of Christianity especially caters for the more upset by emphasising worship and a more literal interpretation of the Gospels. Rather than austerity and self-confrontation, Catholicism instead emphasises self-distracting pomp, ceremony and ritual. As a result, they viewed the Cathars as heretics and burnt many of them at the stake. Catholicism also emphasises, to the point of venerating, Christ's mother, the Virgin Mary, because, for the overly corrupt, it was obviously much easier to identify with the Virgin Mary and her world of gentle nurturing than relate to the strong, secure, confronting truthfulness of Christ. Other branches of Christianity, such as the Anglicans, aren't so upset-adapted and thus escapist, self-focused and fundamentalist in their orientation.

865 The more upset and thus insecure humans became the more literal and non-interpretative of scripture they needed to be; and the more they needed to emphasise deferment of self to the prophet through worship as opposed to emphasising confronting self-analysis; and the more they needed to express their faith through the adornment of their churches and through elaborate ceremony, as opposed to expressing the commitment to the prophet's embodiment of the cooperative ideals of life through living humbly without extravagant adornments and ceremony. Sir Laurens van der Post observed how humans' religious images reflected their degree of alienation when he wrote: **'It seemed a self-evident truth that somehow the sheer geographical distance between a man and his "religious" images reflected the extent of his own inner nearness or separation from his sense of his own greatest meaning. If so this made the conventional Christian location of God in a remote blue Heaven just as alarming as, conversely, the descent of his Son to earth was reassuring'** (*Jung and the Story of Our Time*, 1976, p.31 of 275). While the increase in the artificiality of life that accompanied increased upset was extremely **'alarming'**—an issue I will return to shortly—when people became excessively upset, trying to be any less artificial (by adhering to a less artificial form of religion) was too confronting, guilt-inducing and depressing. It is only now that we have found the dignifying understanding of the human condition that the upset human race can afford to demystify 'God' as the personification of Integrative Meaning; and explain that Christ was, in effect, the **'Son'** of God, the uncorrupted expression of God; and that the 'virgin mother' of Christ was a metaphor

for the exceptionally innocent mother needed to nurture an exceptionally innocent child who would be capable of becoming a denial-free thinking prophet; and that the 'resurrection' of Christ after his martyrdom was emblematic of the opportunity he gave upset humans to be 'born again' or raised up or resurrected from their corrupted, effectively dead state by living through support of him; and that the trinity of God the Father, Son and Holy Ghost (or Spirit) is a perfect pre-scientific representation of the three fundamental aspects of existence on Earth: God the Father is Integrative Meaning; God the Son is the first great tool for integrating matter, namely the gene-based learning system that gave rise to our integratively orientated instinctive self that Christ for example was an uncorrupted expression of; and God the Holy Ghost (or Spirit) is the second great tool for integrating matter, namely the nerve-based learning system that gave rise to our conscious self which is our thinking ability or spirit. It is true that this demystification of religious concepts, like all the insights that suddenly become accessible with the arrival of understanding of the human condition, is a lot to adjust to, especially given how non-interpretive/fundamentalist humans have become of their meaning—it is the ultimate **'future shock'** that Alvin Toffler anticipated in his 1970 book by that name. The physicist Paul Davies summarised the problem when he said, **'A lot of people are hostile to science because it demystifies nature. They prefer the mystery. They would rather live in ignorance of the way the world works and our place in it...many religious people still cling to an image of a God-of-the-gaps, a cosmic magician'** ('Physics and the Mind of God: The Templeton Prize Address', 3 May 1995). How to cope with this sudden and extreme exposure is the subject of the final chapter (8) of this book.

866 With regard to Sir Laurens' observation that **'the sheer geographical distance between a man and his "religious" images reflected the extent of his own inner nearness or separation from his sense of his own greatest meaning'**, since God is representative of the cooperative ideals and 'his' presence, in effect, stands in judgment of humans' non-ideal state, religions, such as Buddhism, that didn't emphasise God, sin and guilt, and instead focused on extinguishing the mental trauma of upset through austere practices like meditation and recital of mantras, became increasingly popular in the latter stages of the last 200 years when upset became extreme. As one convert said of Buddhism, it's **'non-judgemental, there's no notion of sin, there's no notion of good and evil, you don't embrace negativity'** (*Light at Edge of the World: Science of the Mind of Buddhism*, National Geographic Channel, 2006). (While

the thrust of this comment is true, to say there is *no* notion of good and evil in Buddhism is not entirely correct because Buddhism does recognise a sense of 'karmic' heaven and hell.) The problem, however, with focusing on ways, such as meditation, to extinguish the mental trauma of the upset, human-condition-afflicted state was that they undermined the essential responsibility of being a conscious being, which is to *think* and *understand*, ultimately to find understanding of the human condition—as Nikolai Berdyaev wrote: **'Man sought to escape from that terror** [of the truth of **man's...exile from paradise**] **by extinguishing consciousness and returning to the realm of the unconscious. But this is not the way to regain lost paradise'** (*The Destiny of Man*, 1931; tr. N. Duddington, 1960, p.41 of 310). The point being made, however, is that as upset increased, more and more escapist strategies for coping with the human condition simply *had* to be adopted, despite how irresponsible and destructive that trend was of humanity's heroic struggle to find knowledge.

867 Indeed, for some people, complete denial of God became the only acceptable option. Atheism, disbelief in God, and secularism, the rejection of all forms of religious faith and worship, gained popularity. In fact, in recent years the resentment and anger towards God for **'condemning people'** (*What the Bleep do We Know!?*, 2004), and towards religion for being **'the church of perpetual misery'** (from a 2005 animated TV cartoon), has grown so much that secularism is on the rise everywhere. This now extreme resentment and anger is palpable in Oxford University's Professor of Public Understanding of Science, Richard Dawkins' statement that **'"Faith is one of the world's great evils, comparable to the smallpox virus, but harder to eradicate. The whole subject of God is a bore"...those who teach religion to small children are guilty of "child abuse"'** (Garth Wood, 'The Final Blow to God', *The Spectator*, 20 Feb. 1999).

868 Yes, the extreme, **'alarming'** danger of guilt stripping religion—making all forms of religious faith less confronting by avoiding emphasising God, sin and guilt, and by trying to **'regain lost paradise'** through stopping thinking, and even by atheism's complete denial of God/Integrative Meaning—is that the great benefit of religion of the presence of its honesty was lost. Sir Laurens pointed out how dangerously **'starved and empty'** of its truthfulness, guilt-stripped Christianity has become when he wrote that **'Yet the churches continue to exhort man without any knowledge of what is the soul of modern man and how starved and empty it has become...They behave as if a repetition of the message of the Cross and**

a reiteration of the miracles and parables of Christ is enough. Yet, if they took Christ's message seriously, they would not ignore the empiric material and testimony of the nature of the soul and its experience of God that [Carl] **Jung has presented to the world. He did his utmost to make us understand the reality of man's psyche and its relationship to God. But they ignore the call'** (*Jung and the Story of Our Time*, 1976, p.232 of 275).

869 This brings us back to the issue of 'materialism envy' and the resulting unbridled greed, dysfunction and destitution in the less functional, materially poor, 20-year-old and 50-year-old equivalent populated 'developing' countries.

870 While the traditional forms of religion were losing favour in the more functional, affluent 'developed' parts of the world, religion, especially the simplistic, emotional, 'evangelical', 'charismatic' forms and the extremely strident, fundamentalist representations of religious teachings, continued to grow in popularity and influence wherever the increasing levels of upset were also accompanied by impoverishment and/or lack of education. It requires a degree of material comfort and education to be able to engage the subtleties of life under the duress of the human condition, in particular the issue of how to manage the guilt associated with being an upset, non-ideal person. When mere survival is a struggle, thinking about your psychological state is a luxury. Also, without some education you lack the base of knowledge with which to think intellectually and sophisticatedly about the dilemma of being an upset human. As will be described shortly, while entirely new forms of pseudo idealism that were much more sophisticated than religion in avoiding guilt were developed in the more functional, materially wealthier, 'developed' parts of the world, for the impoverished and/or uneducated in the less functional, materially poor 20-year-old and 50-year-old equivalent populated 'developing' countries, religious faith was the main way of countering upset. In fact, with ever greater numbers of people in the world and an ever greater frustration and resentment of the widening gap between the materially fortunate and the materially impoverished, the attraction and influence of religious faith, especially the simple charismatic and fundamentalist representations of religious teachings, only increased amongst the materially less fortunate. In time, however, with greater levels of upset developing everywhere, coupled with the growing anger, and even **'absolute hatred'**, arising from 'materialism envy' and loss of self-esteem amongst the less fortunate as the result of the burgeoning gap between the materially 'rich

and poor', extremely strident, even militarist and terrorist expressions of fundamentalism like Al-Qaeda began to be adopted by those less fortunate. Since religion was essentially about indirectly maintaining the search for knowledge, this strident form of fundamentalism amounted to a dangerous form of anti-religion, of not-true-religion, of degraded-and-debased-truth-and-honesty-obliterated misrepresentations of religion. It could threaten the essential need the human race has for there to be sufficient freedom to continue the search for knowledge, ultimately self-knowledge, liberating understanding of the human condition. In the race between self-destruction and self-discovery, the rapid and extreme degradation of religion was fast leading humanity to self-destruction.

[871] What will now be looked at is how the adoption of more guilt-free forms of pseudo idealism in the more functional, materially wealthy, 'developed' parts of the world also came to seriously threaten humanity's all-important journey to enlightenment. As we will see, extreme forms of pseudo idealism, such as politically correct postmodernism, were as stripped of honesty and thus as alarmingly dangerous as strident forms of religious fundamentalism.

Chapter 7:16J Non-Religious Pseudo Idealistic Causes

[872] The other form of more guilt-free expressions of idealism to support and live through were non-religious and in some cases atheistic, God/Integrative Meaning-denying Pseudo Idealistic Causes. These non-religious pseudo idealistic causes include communism or socialism, environmentalism, feminism, multiculturalism, aboriginalism, politically correct postmodernism, etc, etc—basically any pseudo idealistic cause you could find that allowed you to avoid having to think about and deal with the real issue behind all the destruction and imperfection in the world, namely your own and everyone else's corrupted condition. *TIME* magazine's former editor Richard Stengel perfectly captured this escapist trend when he wrote that **'The environment became the last best cause, the ultimate guilt-free issue'** (*TIME*, 31 Dec. 1990).

[873] Since humanity was trending towards ever more degraded, guilt-and-honesty-stripped fundamentalist representations of religion and, as we will now see, other more guilt-free but dishonest forms of non-religious pseudo idealism to cope with the exponentially increasing levels of upset in the world, there was clearly going to come a time—unless understanding

of the human condition was found—when excessive dishonesty would herald the end of the all-important search for liberating understanding of the human condition and lead to terminal levels of alienation in humans. The danger was of becoming overly dishonest in denying the upset, corrupted state of the human condition—insisting on idealistic behaviour at the expense of reality—and thus oppressing the need for some freedom from the ideals to search for knowledge. To become free of the human condition we had to be allowed to be to a degree free from the oppression of idealism if we were to search for knowledge, ultimately self-knowledge, understanding of the human condition. Again, as the journalist Richard Neville summarised, **'We are locked in a race between self destruction and self discovery'**—either we found understanding of ourselves or we faced self-destruction. If we gave up the search for knowledge there was no hope, and, as will now be described, the virtual abandonment of that search by taking up ever more dishonest/truthless forms of pseudo idealism *has* brought humanity to the brink of terminal alienation. By denying any confronting truth—that is, by taking the practice of guilt-stripping to the extreme—and simply dogmatically demanding we be ideal, pseudo idealism was only adding to alienation, burying humanity deeper in Plato's cave of denial of any truth, making it harder and harder to reach liberating understanding of ourselves.

874 In fact, as should be increasingly apparent, to find understanding of the human condition a veritable mountain of accumulated dishonesty, especially from pseudo idealists, had to be defied and corrected. AND, if that wasn't enough, as described in par. 441, the advocates of extreme pseudo idealistic dishonesty in my home country, in the form of our national broadcaster, the ABC, and Fairfax Media, then all but destroyed our efforts at the World Transformation Movement (WTM) to bring these humanity-saving insights to the world! The whole journey of getting all the truth up from its almost totally buried state and then bringing the humanity-liberating understanding of the human condition that that truth made possible to the world has been *so* difficult that, in hindsight, it seems all but a miracle that, as I said in par. 442, we are still on our feet making this further great effort with this book to have these world-saving insights recognised. The *Persecution of the WTM* section on our website (<www.worldtransformation.com/persecution>) documents the horrific treatment the WTM has had to endure and our final wonderful vindication in the law courts. However, until substantial support

builds for these all-precious, liberating understandings, their survival hangs in the balance. The question 'Is it to be Terminal Alienation or Transformation for the Human Race?' still has to be decided!

[875]To drill down now into the progression of the increasingly guilt-stripped and thus ever more dishonest and thus ever more dangerous forms of non-religious pseudo idealism that humanity developed and adopted when Religion became too confronting, starting with Socialism and Communism.

Chapter 7:16K Socialism and Communism

[876]These collectivist movements denied the notion of a perfecting God and avoided the depressing recognition of a prophet's world of soundness, and instead dogmatically demanded an idealistic social or communal world—and, in doing so, denied and oppressed the whole reality of the individualistic, freedom-dependent, knowledge-finding, creative, egocentric, corrupting, unavoidably-variously-upset, competitive, combative, materialism-compensation-needing, self-distraction-necessary, human-condition-afflicted world. As was pointed out in par. 185, Karl Marx, the political philosopher whose mid-nineteenth century theories gave rise to socialism and communism, argued that **'The philosophers have only interpreted the world in various ways; the point is** [not to understand the world but] **to change it.'** By **'change it'** Marx meant just make it cooperative or social or communal, but he was wrong—the whole **'point'** and responsibility of being a conscious being *is* to understand our world and place in it, ultimately to find understanding of the human condition. The attraction—and inherent lie—of socialism/communism was that you could support and live the ideals without acknowledging the reality of the human condition and its struggle.

Chapter 7:16L The New Age Movement

[877]The limitation of socialism and communism was that while there was no confronting prophet involved, there was an obvious focus on the condemning cooperative, communal, social *ideals*—and so in time, as levels of upset and thus insecurity increased, the need again arose for the invention of an even more guilt-free form of idealism through which to live, hence the development of the New Age Movement (the forerunners

of which were the Age of Aquarius and Peace Movements). In this movement all the realities and negatives of our non-cooperative/non-communal/non-social, corrupted condition were transcended in favour of taking up a completely escapist, think-positive, human-potential-stressing, self-affirming, motivational, feel-good approach. So in truth, the New Age Movement was never going to be able to transport humanity to an Aquarian new age of peaceful freedom from upset, it was only ever going to lead to an even greater state of deluded, dishonest alienation than that espoused by socialism/communism. Talking about how he became **'a personal growth junkie'**, the comedian Anthony Ackroyd summed up the extremely deluded artificiality of the New Age Movement when he said: **'What are millions of us around the globe searching for in books, tapes, seminars, workshops and speaking events? Information to enhance our lifestyles and enrich our experience on this planet? Certainly...But I smell something else in the ether. Something more desperate and deluded. A worrying snake-oil factor that is spinning out of control. It's the promise of salvation. Salvation from the basic rules of human life. This is the neurotic aspect of the human potential movement. This hunger for a get-out-of-the-human-condition-free card'** (*Good Weekend, The Sydney Morning Herald*, 13 Sep. 1997). Yes, to **'get out of the human condition'** we had to confront and solve it, not deny and escape it; our **'desperate and deluded'** attempts to escape it only made it worse. As the philosopher Thomas Nagel recognised, **'The capacity for transcendence brings with it a liability to alienation, and the wish to escape this condition...can lead to even greater absurdity'** (*The View From Nowhere*, 1986).

Chapter 7:16M The Feminist Movement

[878]The limitation of the New Age Movement, in terms of being an effective means of escaping the horror of the human condition, was that while it didn't stress the cooperative ideals like socialism and communism did, in seeking to transcend humans' upset state its very purpose served as a constant reminder and thus reprimand of our variously upset, embattled, troubled, estranged, alienated condition—a problem the *next* level of delusion sought to dispense with by simply denying its existence. Yes, the Feminist Movement maintained that there was no real difference between people—and *especially* not between men and women. It particularly denied the legitimacy of the exceptionally egocentric, combative male dimension to life that had taken on the heroic frontline role in fighting

the ignorance of our instinctive self. Women's lack of awareness of the all-important role that men have been playing is evident in this already mentioned comment from Germaine Greer, an icon of modern feminism: **'As far as I'm concerned, men are the product of a damaged gene. They pretend to be normal but what they're doing sitting there with benign smiles on their faces is they're manufacturing sperm.'** Beyond their necessary role in reproduction men have been viewed as meaningless; as the leading feminist, Gloria Steinem, famously said, **'A woman needs a man like a fish needs a bicycle'**—but the truth that is now able to be revealed is that men were carrying out the most important and difficult task being undertaken on the planet! Based on extreme dogma, the feminist movement could not and has not brokered any real reconciliation between men and women, rather, as this aforementioned quote points out, **'What happened was that the so-called Battle of the Sexes became a contest in which only one side turned up. Men listened, in many cases sympathetically but, by the millions, were turned off.'** No, *only* by winning the battle to champion the ego—that is, explain the human condition and establish that our egocentric conscious thinking self is good and not bad—could the polarities of life of so-called 'good' and 'evil', that women and men are in truth an expression of, be reconciled.

Chapter 7:16N The Environmental or Green Movement

879 Again, as far as offering an effective way to transcend the realities of the human condition, feminism's flaw was that while it superficially dispensed with the problem of humans' divisive reality, humans remained the focus of attention and that was confronting. The solution that emerged to counter this limitation was the Environmental or Green Movement, which removed all need to confront and think about the human state because all focus was diverted from self onto the environment—as the aforementioned quote acknowledged, **'The environment became the last best cause, the ultimate guilt-free issue.'** Of course, the obvious truth is that by not addressing the cause of the destruction of the natural world, namely the issue of our human condition's massively upset angry, egocentric and alienated state, there can be no real let up in the pace of our world's devastation, as these quotes emphasise: **'The trees aren't the problem. The problem is us'** (*Simply Living* mag. Sep. 1989), and **'We need to do something about the environmental damage in our heads'** (*TIME*, 24 May 1993). The guilt-stripping dishonesty of the environmental movement was recognised by a

businessman, Ray Evans, when he wrote that **'Environmentalism has largely superseded Christianity as the religion of the upper classes in Europe and to a lesser extent in the United States. It is a form of religious belief which fosters a sense of moral superiority in the believer, but which places no importance on telling the truth** [about the real issue of our corrupted condition]**'** (*Nine Facts About Climate Change*, 2006, p.2 of 27).

Chapter 7:16O The Politically Correct Movement

880 So for all its guilt-relieving benefits, the environmental movement still contained a condemning moral component: if we were not responsible with the environment, 'good', we were behaving immorally, 'bad'. Moreover, nature, in its purity, exists in stark contrast to humans' corrupted condition—hence our ruthlessness towards it in the first place.

881 At this stage in the march of upset yet another form of extreme pseudo idealism had to be manufactured where confrontation with the by now extremely confronting and depressing truth of the dilemma of the human condition could be totally sidestepped. What was required was a completely guilt-stripped dogma that was devoid of any need to confront and wrestle with the issue of soundness and Godliness; with whether you are a cooperative, social, integrative person; with the issue of your troubled self; with the morality issue of men and women's treatment of each other; and with the issue of whether or not you are being good to the environment. Upset had become so great that the need was to simply be ideal without question. This demand for a totally non-confronting form of relief from feeling 'bad' resulted in the establishment of the Politically Correct Movement, which has no other focus or requirement beyond simply choosing, from the two simplistic, fundamental, 'political' options in life—of being either 'good' or 'bad'—to be 'good'.

882 The politically correct culture was a pure form of freedom-denying dogma that fabricated, demanded and imposed ideality or 'correctness', specifically that of an undifferentiated world, which was in complete denial of the reality of the underlying issue of the existence of and reasons for humans' variously embattled and upset states, and beyond that of the deeper question raised by those 'non-ideal' states of the issue of the human condition. For instance, the politically correct argued that the children's nursery rhyme *Baa Baa Black Sheep* is racist and that it should instead be recited as **'Baa Baa Rainbow Sheep'** (*The Telegraph*, 24 Jan. 2008).

883 Within the politically correct culture the need for relief from guilt was all-pervasive; the mind was constantly on the hunt for opportunities and 'good causes' through which to be 'idealistic' and achieve that rush of psychological relief of feeling that at last you are a 'good' rather than a 'bad' person. Wherever there was a victim of humanity's battle, there was an opportunity to champion their cause and access that all-important relief. It was, as we will shortly see, a development that Christ described in much harsher terms when he said, **'Wherever there is a carcass** [the extremely upset]**, there the vultures** [the **false prophets**, the merchants of delusion and escapism] **will gather.'** But with the levels of upset in the world becoming so extreme, such relief-hunting became a huge industry, to the extent that we became, as the sociologist Frank Furedi recognised, **'a society that celebrates victimhood rather than heroism'** (*The Culture of Fear*, 1997). Yes, as one critic complained, **'If you are a black vegetarian Muslim asylum-seeking one-legged lesbian lorry driver, I want the same rights as you'** ('Thought police muscle up in Britain', *The Australian*, 21 Apr. 2009). Indeed, the deluded arrogance of the left-wing politically correct culture has been absolutely extraordinary; virtually every left-wing magazine or newspaper or broadcaster will go out of its way now to lead with a story and picture or footage about one of the just referred-to categories in order to 'educate' us all in 'correct' thinking!!

884 Again, while there *was* an ever increasing need for more dishonest, guilt-free forms of idealism through which to live, for humanity to arrive at this desperately insecure state where people were concerned *only* with finding relief from their own guilt through supporting the cause of those who were suffering or less fortunate was an extremely dangerous development because it meant the human race had, in effect, abandoned the ongoing battle to find the all-important liberating understanding of ourselves. This is not to say that in a critical battle, such as the one humanity has been involved in, showing care and compassion towards those who were suffering from the effects of the battle was not important. It was very important, because although we have all been involved in the upsetting battle, selflessness is still, as has been repeatedly emphasised, what binds wholes together; it is the glue within humanity's army. *However*, while caring for those struggling to keep up was important, it was obviously more essential to support those on the frontline who were still carrying on the battle to ensure the war was ultimately won. In this light it can be seen how very dangerous and irresponsible the politically

correct movement's focus and insistence only on caring for the victims of the battle was.

885 In fact, while showing care and compassion to those suffering from the battle was ('was', because the battle to find understanding of the human condition is now over) certainly the responsibility of any healthy society, doing so merely to delude yourself and others that you are an upset-free, ideal person seriously discredited the whole notion and practice of consideration and kindness itself. Indeed, using idealism to delude yourself that you were good gave idealism such a bad name that no relatively sound, secure person wanted to be part of the left-wing political culture where such relief-hunting had become endemic; so much so that, in the end, there was no longer any authentic, trustable, credible, healthy, meaningful idealistic movement in society to counteract any excessively selfish and destructive right-wing behaviour—there were only moderate factions within the right-wing that the sane and rationally-behaved could join and support.

886 Yes, the whole democratic process that our society depended on for there to be effective progress and functionality was being destroyed by mad desperados—by a group of people who were misusing democracy for their own selfish need to make themselves feel good, rather than for what democracy was designed to be: a tool to decide the best way to manage any particular course of action. How could you possibly have an effective discussion about where the right balance lay between selfishness and idealism if participants in that discussion were only interested in whether their participation would make them feel good and/or whether the course of action chosen would ultimately make them feel good? The answer is you couldn't. It was a derailed, ineffective, dysfunctional, highly imperfect, pointless—in fact, defunct—debate. It was like being in mid-ocean on a life-boat, desperately trying to find your way to the safety of land, when someone on-board decides to hijack and destroy the mission by capsizing the boat because they had become obsessed with wanting to cool off in the water. As has already been carefully explained, it was totally irresponsible, selfish—in fact, mad—behaviour. The human race was trying to save itself from destruction by finding knowledge, ultimately understanding of the human condition, but the extreme practitioners of pseudo idealism were only interested in making themselves feel good. Contrary to what their banners said, pseudo idealists no longer cared about the future of the world. Their conduct was completely selfish and not at

all the selfless, idealistic behaviour they made it out to be and deluded themselves it was. Thank goodness the arrival of understanding of the human condition exposes this madness for what it really is, because it was only while it was not possible to explain humanity's great heroic battle that it was possible to get away with such mad and obscene behaviour—'Can't you see our idealism is making the world a better place.' What rubbish—such behaviour is nothing more than a selfish attempt to gain relief from and escape the agony of the human condition! People who are not overly upset, not overly soul-corrupted, are *naturally* concerned for other individuals, genders, races, cultures, the environment, etc, while those who are overly upset practitioners of pseudo idealism are merely *pretending* to be concerned for anyone and anything other than themselves to relieve themselves of their guilt. Instead of being the champions of selflessness they were, in fact, the champions of selfishness, the most selfish of people. Upset basically means self-preoccupation: you are preoccupied with expressing *your* anger, with satisfying *your* egocentricity and with maintaining *your* alienation. Later (in par. 920), it will be described how Christ got to the underlying truth about relief-hunting pseudo idealists when he described them as having had their **'love' 'grow cold'** (Bible, Matt. 24:12).

887 Humans have had to do what they had to do, side with idealism to delude themselves that they were 'good'/ideal, relieve themselves of the truth about their corrupted, extremely selfish condition, but in terms of humanity's all-important journey to enlightenment, idealism was being extremely, horribly misused—and a particularly dysfunctional aspect of the misuse was the arrogant *extent* of pseudo idealists' delusion that they were actually ideal, that they held the moral high ground. In the situation that has existed, where the reality of the upsetting battle to find knowledge couldn't be explained, idealism—albeit the bastardised form of 'victim-hunting-to-make-yourself-feel-good, politically correct', *pseudo* idealism—certainly did have a field day mocking realism as evil. In the vacuum where the reason for humans' upset, corrupted state was not able to be explained, the **'intellectuals'/'liberal elites'/'chattering classes'/'left-wing trendy café society'/'chardonnay socialists'/'radical chic'/'Hollywood Left'/'CBS-*New York Times*, BBC-*Guardian*, ABC-*The Sydney Morning Herald*, *TIME* and *National Geographic* 'left-wing rags', etc, etc axis'/'high-minded do-gooders'/'rainbow extremists'/'strident bleeding hearts'/'feel good, warm inner glow, blissed out compassion junkies'**, as the relief-hunting, pseudo idealistic

left-wing have been variously referred to, conceitedly promenaded about with a holier-than-thou attitude, while the right-wing advocates of freedom (from the oppression of idealism in order to participate in the corrupting search for knowledge) were arrogantly and disdainfully vilified as morally bankrupt and contemptible. For example, the right-wing so-called 'Tea Party' that recently emerged in American politics was derided by the left-wing Democrats for being devoid of any sound arguments for their cause—they were accused of being nothing more than promoters of **'fear, xenophobia, cryptofascism, creationism, inequality and ignorance'** (cartoon by Turner in *The Irish Times* that was re-printed in *The Australian*, 3 Nov. 2010). It is no wonder politics has become so polarised—to the point where the two sides, rather than providing humanity with a healthy equilibrium, have existed in totally opposed philosophical continents, and may as well have lived on separate planets. The deluded arrogance of the extremely dangerous dishonesty of the left became insufferable, unbearable, overwhelming, terrifying, sickening.

888 Again, the problem at base was the inability to explain the human condition. While it was easy to argue the case for the idealism of the left-wing, it was almost impossible to argue the case for the realism of the right-wing. How could you justify any selfishness or inequality; how could you defend behaviour that appeared in every way to be inhumane; how could you argue that not being ideally behaved was good? The answer is that until we could explain the paradox of the human condition we couldn't—well not sufficiently. Writers like Ayn Rand did well to mount some sort of a case for right-wing free enterprise, but countering such efforts were the dogmatic **'Capitalists are Pigs'** placards used in protests at G8 summits, and left-wing advocates like Michael Moore, the filmmaker and activist who, at the conclusion of his 2009 documentary, *Capitalism: A Love Story*, smugly announced that **'Capitalism is evil.'** The truth is, it was capitalism that kept the human race going, not communism. Without the relief, reward, distraction and sustenance of materialism/material goods that the exchange of money or capital facilitated, humans would not have been able to cope with and carry on their upsetting, idealism-defying, heroic search for knowledge. With understanding of the human condition we can now explain that socialism/communism, the pseudo idealistic dogmatic insistence that everyone be social and communal—live for society and the communal good rather than seek a degree of material relief and reward for yourself—ignored the reality of the upsetting battle that

humans have had to wage to find understanding of themselves, of their less-than-ideal human condition. As was emphasised in chapter 2:9, with understanding of the human condition finally found what is revealed is that it was, in fact, the left-wing that was morally bankrupt, *not* the right! The truth was not as it appeared. In the Adam Stork analogy, upset angry, egocentric and alienated Adam is the hero of the story not the villain; it was his condemning, ridiculing, upsetting idealistic opponent, which the extreme pseudo idealist came to represent, who was actually the villain and not the hero he deluded himself to be and insisted to others he was. Yes, while the right-wing tended to be selfish and corrupting of soul, it was still participating in humanity's search for knowledge.

889 It should be mentioned that the overly upset were not only to be found on the side of left-wing pseudo idealism. Some people who were overly upset resisted taking up support of left-wing pseudo idealism because they didn't want to be perceived as, or consider themselves to be, weak and cowardly for having abandoned the battle, and/or because they were intuitively aware that too many had already done so, resulting in the now extremely high levels of delusion and dishonesty in the world that they didn't want to contribute any further to. It should also be mentioned that there have been people who supported the political right's promotion of the freedom needed to search for knowledge while adhering to fundamentalist expressions of religion that worked *against* that search. This dichotomy can be explained by the fact that while you could support right-wing policies without being particularly confronted by your own corrupted condition, supporting religion has been an unavoidably self-confronting practice, so while it has been possible for the more exhausted to support right-wing politics, it has not been as easy to be a supporter of religion unless it was in a literal, fundamentalist way, where, as was explained in ch. 7:16I, the emphasis could be shifted away from self-confronting honesty, introspection and interpretation of the deeper truths contained in the religion's scriptures to simply worshipping the prophet as a more indirect way of glorifying and supporting the truth he stood for. The so-called 'Bible Belt' in southern USA, for instance, is notoriously right-wing but also notoriously fundamentalist, such as in its advocation of creationism and opposition to the teaching of evolution. They support individual freedom but not analysis of the human condition; they despise the dishonesty and weakness of the left-wing for holding that, for example, there is no such thing as truth, and so want to support

the truthfulness contained in religion, but not so much that it becomes personally confronting.

890 At this point we need to continue our analysis of the progression that the development of even more guilt/truth stripped forms of idealism to live through took, and look at what emerged after the development of the politically correct movement.

Chapter 7:16P The Postmodern Deconstructionist Movement

891 While showing interest only in those who were suffering was extremely dishonest and dangerous, from the perspective of the do-good, feel-good, politically correct supporter, gaining relief from guilt/depression/the agony of the human condition was all that mattered—when the truth is killing you, you have no qualms about escaping it—and so to help ensure no subversive questioning could creep in and undermine this strategy, a philosophical justification for truthlessness was bolted on to the politically correct culture. This was Postmodern Deconstructionism, **'a bewilderingly complex school of continental philosophy, or pseudo-philosophy'** of **'intellectual assumptions—truth is a matter of opinion, there is no real world outside of language and hence no facts independent of our descriptions of them'** (Luke Slattery, *The Australian*, 23 Jul. 2005). While language is artificial it nevertheless models a real world, so to say that just because language is artificial there can be no universal truths is ridiculous, but again, when the need to escape the truth becomes desperate, any excuse will do; just baffle the world, and yourself, with intellectual baloney, such as this from Jacques Derrida, one of the high priests of deconstructionism, who gave this highly intellectual (instinct/soul/truth-less) description of why truth supposedly doesn't exist: **'Every sign, linguistic or nonlinguistic, spoken or written, as a small or large unity, can be cited, put between quotation marks; thereby it can break with every given context, and engender infinitely new contexts in an absolutely nonsaturable fashion. This does not suppose that the mark is valid outside its context, but on the contrary that there are only contexts without any center of absolute anchoring'** (*Margins of Philosophy*, 1982, p.320 of 330).

892 In his 2001 book, *The Liar's Tale: A History of Falsehood*, Jeremy Campbell described **'postmodern theory'** as having elevated **'lying to the status of an art and neutralised untruth'**. It **'neutralised untruth'**, made what's not true non-existent, because by denying the existence of the variously embattled and upset states of individuals, races, genders, generations,

countries, civilisations and cultures, it made any discussion of such differences—any pursuit of insight into the human condition—impossible. Recall how the politically correct culture deemed the children's nursery rhyme *Baa Baa Black Sheep* racist, insisting it be recited as **'Baa Baa Rainbow Sheep'**? Well, the 1994 postmodernist teaching guide *From Picture Book To Literary Theory* similarly argued that school children shouldn't be read stories about witches on broomsticks because they were sexist for **'narrowly defining women's roles'**, or the *Three Little Pigs* fairy tale on the basis of its elitist promotion of **'the virtues of property ownership and the safety of the private domain which are key elements of capitalist ideology'**!

893 With levels of upset in people becoming extreme and the associated levels of guilt and self-loathing becoming unbearable for ever increasing numbers of people, hunting for opportunities to be 'idealistic' and feel that rush of relief from doing 'good' and feeling you were a good person at last, became all-consuming, to the point of ridiculously dangerous irresponsibility. The American columnist Ann Coulter described how **'Whenever the nation** [the USA] **is under attack, from within or without, liberals side with the enemy. This is their essence'** (*TIME*, 27 Sep. 2004), while former US presidential advisor Karl Rove presented this similar observation: **'Conservatives saw the savagery of 9/11 and the attacks and prepared for war; liberals saw the savagery of the 9/11 attacks and wanted to...offer therapy and understanding for our attackers'** (*The New York Times*, 24 Jun. 2005).

894 Often accurately described as **'compassion junkies'**, pseudo idealistic, politically correct, postmodern, left-wing relief-hunters—empowered by their own self-righteousness and the right-wing's inability to defend itself—were taking over almost everywhere. Drawn together by their shared overwhelming need to find causes through which to feel 'good' they captured such major institutions as the national public broadcasters in the UK and Australia—which are supposed to represent *all* of its constituents, not one extreme faction. As a former British Broadcasting Corporation (BBC) correspondent, Robin Aitken, said, '[I] **could not raise a cricket team of Tories** [right-wing conservatives] **among BBC staff'** (*The Australian*, 10 Oct. 2005). The British journalist Melanie Phillips similarly observed that **'With a few honourable exceptions, the BBC views every issue through the prism of left-wing, secular, anti-Western thinking'** (*Daily Mail*, 16 May 2005). A blogger posted this alarm: **'The broadcasters particularly in the BBC are...acting like the militant wing of the Labour Party. They have completely lost all restraint and integrity. They are completely out of order...** [it's] **a national outrage...**[that

they] **are allowed to control the most influential power centre in the history of mankind. The BBC is run by an extreme, unrepresentative and unelected cult. The Left have finally ruined the BBC...The BBC is now irreparably infiltrated and broken...All responses to the Tories are to give the impression that whatever the Tories say, do or propose is immoral or incompetent, or imply that selfish, self-serving or somehow bad motives are the real reason for them—that they are not proposed altruistically for the genuine benefit of the country and therefore cannot be a credible alternative to what is being done by Labour, who are altruistic'** ('How the Left have corrupted broadcast news', 1 Feb. 2010; see <www.wtmsources.com/124>). Almost every program on Australia's national broadcaster, the Australian Broadcasting Corporation (ABC), is now either outrightly dedicated to, or at least heavily skewed in its presentation towards servicing the needs of relief-hunters. These national broadcasters serve no purpose now other than as places for the overly exhausted to find relief from their exhaustion. Neither the national interest nor humanity's are of any concern, even though the illusion is that these are their only interests.

895 And the problem of pseudo idealism in the media has not been confined to the national broadcasters. A 2005 UCLA-led study into the political leanings of media in the US revealed that **'almost all major media outlets tilt to the left...there is quantifiable and significant bias'** ('Media Bias Is Real, Finds UCLA Political Scientist', *UCLA News*, 14 Dec. 2005). But at least some are protesting; in 2010 it was reported that **'Four billboard trucks bearing the message "Stop the Liberal Bias, Tell the Truth!" began circling the Manhattan headquarters of ABC, CBS, NBC and the *New York Times*'** ('Trucks Encircle ABC, CBS, NBC, Challenge "Liberal" Media to "Tell The Truth"', CNSnews.com, 4 Oct. 2010; see <www.wtmsources.com/142>).

896 Much of academia, the supposed home of higher learning, has also been hijacked by overly upset, guilt-escaping, human-condition-avoiding, truth-hating relief-hunters, with a **'comprehensive'** survey undertaken in 2005 by political science professors at the University of Toronto revealing that **'87 percent'** of the teaching faculty **'at the most elite' 'American universities'** were **'left-wing' 'liberal'**. A co-author of the study, Professor Robert Lichter, commented that **'What's most striking is how few conservatives there are in any field** [of study in US universities]**...It's a very homogenous environment'** ('College Faculties A Most Liberal Lot, Study Finds', *The Washington Post*, 29 Mar. 2005). That **'87 percent'** of teachers in the major American centres of learning held a distinct bias of thought was certainly

a **'most striking'**—indeed, truly frightening—statistic. The literary scholar Harold Bloom recognised the danger of this hijacking of academia by truth-hating relief-hunters in his 1994 book. *The Western Canon*, which, as a *TIME* magazine review summarised, asserted that a **'rebellion in U.S. schools against Dead White European Male authors'** (or **'D.W.E.Ms'**) would lead to **'the end of civilization'** through a **'triumph of the forces of darkness'** (10 Oct. 1994). In *The Western Canon*, Bloom lamented that **'*Batman* comics... will replace Chaucer, Shakespeare, Milton, Wordsworth'** (p.485 of 560). But with the human condition now explained and the truth about humans finally revealed, we can see who was doing the critical deeper thinking needed to liberate humanity from ignorance, for if you go through this book and write down the names of all the authors of the quotes used to evidence these human-race-saving insights into our species' condition, you will find that **'Dead White European Male'**s saved the human race! Yes, the truth is that the 'Dead White European Male'-led **'rise of the West is, quite simply, the pre-eminent historical phenomenon of the second half of the second millennium after Christ. It is the story at the very heart of modern history'** (Niall Ferguson, *Civilization: The West and the Rest*, 2012, p.18 of 432).

[897] A 2006 article in *The Australian* newspaper reported that Australia's then Prime Minister, John Howard, **'believes the postmodern literature being taught in schools is "rubbish"...accusing state education authorities of "dumbing down" the English syllabus and succumbing to political correctness.** [He said] **"I feel very, very strongly about** [this situation where]**...traditional texts, are treated no differently from pop cultural commentary"'** (21 Apr. 2006). In the article, Howard, Australia's greatest ever Prime Minister (in my view), also referred to postmodern discourse as meaningless **'gobbledegook'**. The article said Howard's comments were in response to the dismissal of great literature as being elitist, sexist and racist, with a leading Sydney private school having **'asked students to interpret Othello from Marxist, feminist and racial perspectives'**. The real subject students should be asked to **'interpret'** is how artificial and dishonest pseudo idealism is and how it is threatening to condemn humanity to extinction.

[898] In his insightfully titled 1987 book, *The Closing of the American Mind*, the political scientist Allan Bloom also wrote of the devastating effects of postmodern, so-called 'deconstructionist' teaching in American universities. As summarised in this review published in *The Australian* newspaper: **'we are producing a race of moral illiterates, who have never asked the great questions of good and evil, or truth and beauty, who have indeed no**

idea that such questions even could be asked...As Mr Bloom says..."deprived of literary guidance they [students] **no longer have any image of a perfect soul, and hence do not long to have one. They do not even imagine that there is such a thing"...If the classics are studied at all in the universities they are studied as curiosities in the humanities departments, not as vital centres of the liberal tradition, and not as texts offering profound insight into the human condition'** (Greg Sheridan, 'The Closing of Our Minds', 25 Jul. 1987). Of course, the whole point of the postmodern movement was to avoid **'the great questions'** about our **'soul'** and what has happened to it, namely the question of our self-corruption and alienation, the issue of **'good and evil'**, **'the human condition'**, **'truth'**. The object was to stop thinking, stop the pain of guilt, bring about a **'closing of the...mind'**, **'dumb'** the mind **'down'**. The deeper, fundamental and true objective of education is to enlighten us about the underlying issue in all of human life of the human condition, and that is precisely what literature that has historically been considered great has managed to do. While not overly confronting us with our upset, corrupted, fallen condition the great writers of our time have, through great literary skill and device, managed to throw light on it. But it is precisely that great gift that postmodernism found a way to destroy. The character Kurtz in Joseph Conrad's great/classic 1902 book **'Heart of Darkness** [of our condition]**'** acknowledged **'The horror! The horror!** [of our massively upset condition]**'**, but, of course, for the politically correct postmodernist this was a despicable **'D.W.E.M.'**-written piece that propagated male chauvinism, elitism, racism and aggression.

899 Not so long ago, HRH The Prince of Wales (whose mentor or main influence in his philosophical life was, like mine, Sir Laurens van der Post, and who also attended Geelong Grammar School where, like me, he benefited from the influence of Sir James Darling's soul-rather-than-intellect-emphasising, Platonic education system) wrote a letter of deep concern to the Lord Chancellor of Great Britain, questioning the extreme bias that is now also apparent in legal thinking, stating that **'The Human Rights Act is only about the rights of individuals. I am unable to find a list of social responsibilities attached to it and this betrays a fundamental distortion in social and legal thinking'** (*The Australian*, 27 Sep. 2002).

900 Dr William Anderson, a professor of economics in the US, similarly observed that **'Justice pretty much is dead in the United States...Like so many other trends, this one has its intellectual underpinnings in that academic refuse pile we call Post-Modernism...a line of thinking that denies any possibility of**

Truth, and is the dominant "guiding light"—darkness?—in academe these days... right now, the post-modernists are winning battle after battle. It is one thing when post-modern nonsense dominates a history or English class; it is quite another when it becomes the bedrock of modern law' ('Post-Modern Prosecutions', 25 Nov. 2006; see <www.wtmsources.com/115>).

901 The American lawyer Gary Saalman also wrote that **'In recent years... postmodernism has risen to the forefront of legal theory. Postmodern theorists... claim the law cannot have any foundation because there is no foundation for objective knowledge of any kind...Principles of law could never reflect universal truths, they argue...According to these scholars, it is senseless to talk about whether a law is right or wrong or moral or amoral...most observers agree that postmodern theories of law are exerting a huge influence today in the courtroom and the legislature...Remember, these are not a lunatic fringe at the margins of legal practice. They include department heads, and leading professors of law schools...practicing lawyers and legal authorities'** ('Postmodernism and You: Law', 1996; see <www.wtmsources.com/109>).

902 And, in his 2006 book, *Understanding the Times: The Collision of Today's Competing Worldviews*, the American religious leader Dr David Noebel had this to say about the dangers of postmodernism: **'Harold J. Berman, former professor of law at Harvard Law School...explains that today...foundational beliefs are rapidly disappearing, not only from the minds of philosophers, but from "the minds of lawmakers, judges, lawyers, law teachers...the historical soil of the Western legal tradition is being washed away in the twentieth century, and the tradition itself is threatened to collapse"... Postmodernists are intent on eliminating religious roots and transcendent qualities from Western law. They desire more fragmentation and subjectivity, and less objective morality than the Judeo-Christian tradition demands. In the end, they are intent on creating and using their own brand of social justice merely for left-wing political purposes'** ('Postmodern Law', accessed Jan. 2011; see <www.wtmsources.com/121>).

903 The media, our centres of education and learning, the judiciary—these are all pillars of society that are being white-anted in the march toward terminal alienation. The world *is* in jeopardy of being *completely* hijacked by those who are no longer concerned with humanity's heroic journey to enlightenment and who only want to escape the depressing effects of their human condition. Total self-preoccupation, selfishness disguised as selflessness, has arrived. Terminal levels of alienation *are* upon us.

904 With its supposedly underpinning postmodern deconstructionist philosophy, the politically correct movement was a pure form of freedom-denying dogma that fabricated, demanded and imposed ideality or 'correctness', specifically that of an undifferentiated world, in complete denial of the reality of the underlying issue of the existence of and reasons for humans' variously embattled and upset states. It was concerned with *pretending* to behave ideally and by so doing *deluding* yourself that you were an ideal, 'good' person. The delusion involved was in fact so great the movement boldly labelled itself the **'politically correct'** movement and brazenly claimed to be taking humanity beyond or **'post'** the existing upset, embattled **'modern'** state to a good versus evil **'deconstructed'**, **'correct'** world when in truth it was behaving *non-ideally* and working *against* the search for the liberating understanding that would take humanity beyond the upset state to a **'post'** present/**'modern'** human-condition-ameliorated, good-versus-evil-reconciled and thus truly **'deconstructed'**, genuinely **'correct'** world. The gloves were off now, the confidence of—and sheer bloody-minded anger and aggression underlying—the industry of escapist denial and delusion was such that it was now prepared to go the whole hog and outrageously mimic the arrival of the true world at the actual expense of any chance it had of arriving. The fact is the postmodernist politically correct movement represented the very height of dishonesty, the most sophisticated expression of denial and delusion to have developed on Earth. The **'forces of darkness'** articulated by Harold Bloom were certainly in ascendance. Terminal alienation *was* upon us and humanity *was* entering end play, a death by dogma.

Chapter 7:16Q The 'abomination that causes desolation' 'sign... of the end of the age' that is 'cut short' by the arrival of the liberating truth about the human condition

905 To look at humanity's situation overall now, it is clear we have reached the terminally alienated, end play point in the human journey where the rapidly rising tide of the charismatic and fundamentalist misrepresentations of religion in the dysfunctional, materially poorer parts of the world, and the politically correct, postmodern culture in the more functional, materially affluent parts of the world are threatening to stifle the freedom of expression upon which liberating enlightenment of the human condition depends.

[906] A reflection of the compounding rate of increase of upset in humans is that the interval between the development of one form of pseudo idealism to the next more evasive form in this chronology became shorter and shorter, shrinking from centuries to decades to years. Communism was a long time coming after the development of religion whereas the forms of pseudo idealism that followed communism appeared, or if they had been germinating for some time then became prominent, one after the other in rapidly shrinking intervals of time.

[907] In the onrush of the psychosis-escalating struggle of the human journey it was inevitable that a point would eventually be reached where the need for relief and restraint would become so great that the delusion of dogma—the artificiality of imposing ideality to solve reality—would become invisible to its practitioners. We are now in that situation where the advocates of dogma have become blind to its short-comings and consequently are now on an all-out mission to seduce and intimidate everyone with their culture. It has been said that **'postmodernism has peaked, and will die with the century'** (Damian Grace, 'A Strange Outbreak of Rocks in the Head', *The Sydney Morning Herald*, 21 Jan. 1998), but this is a psychosis-driven situation where, increasingly, people *have to* variously live in adherence to pseudo forms of idealism to cope with their circumstances. Therefore, if you live in an affluent part of the world, no amount of opposition will halt the rise of politically correct culture, or, if you live in an impoverished part of the world, no amount of opposition to fundamentalist expressions of religion will halt its growth. It is only the arrival of the understanding of the human condition in this book that can halt the onrush of terminal alienation and destruction of the human race.

[908] So, with understanding of the human condition finally found, the truth it reveals is that the underlying, *real* progression in all the great, so-called, 'social reforms' of the last 200 years was *not* to a more ideal world but to greater upset and its associated need for ever more guilt-stripped forms of pseudo idealism through which to live. From religion, the original, thousands-of-years-old, relatively honest, alienation-free form of pseudo idealism, developed socialism and communism, then, in the less functional, materially poorer parts of the world, charismatic and fundamentalist misrepresentations of religion, and, in the more functional, materially richer parts of the world, the New Age Movement, then the feminist movement, then the environment or green movement, then the politically correct movement, and, finally, the totally

dishonest, completely alienated, definitely autistic postmodernist movement; again, **'autism'** is **'a complex mental structure insuring against recurrence of...unthinkable anxiety'**—in this case, **'anxiety'** about being extremely corrupted/upset/hurt/soul-damaged in your infancy and childhood.

909 Yes, we can now see exactly what the underlying psychological situation is that has driven the pseudo idealistic strategy of the left-wing. And, in fact, the work of the psychologist Arthur Janov is very confirming of the psychosis behind the adoption of the left-wing approach. As mentioned earlier (in par. 808), Janov developed the technique of 'primal therapy' in which adults are helped to work their way back in their minds to memories of the original (primal) hurt to their soul that occurred in their infancy and early childhood as a result of growing up in the extremely human-condition-embattled, psychologically upset, insecure, have-to-somehow-establish-your-worth world of today. The following extract from Janov's famous 1970 book, *The Primal Scream*, was included earlier, but I am re-including it here with additional material from his book because in it Janov describes very clearly how the more upset humans became, the more they needed to find a way to escape their **'personal horror'**. He wrote (the underlinings are my emphasis): **'Anger is often sown by parents who see their children as a denial of their own lives. Marrying early and having to sacrifice themselves for years to demanding infants and young children are not readily accepted by those parents who never really had a chance to be free and happy** [p.327 of 446] **...neurotic parents are antifeeling, and how much of themselves they have had to cancel out in order to survive is a good index of how much they will attempt to cancel out in their children** [p.77] **...there is unspeakable tragedy in the world...each of us being in a mad scramble away from our personal horror. That is why neurotic parents cannot see the horror of what they are doing to their children, why they cannot comprehend that they are slowly killing a human being** [p.389] **...A young child cannot understand that it is his parents who are troubled...He does not know that it is not his job to make them stop fighting, to be happy, free or whatever...If he is ridiculed almost from birth, he must come to believe that something is wrong with him** [p.60] **...Neurosis begins as a means of appeasing neurotic parents by denying or covering certain feelings in hopes that "they" will finally love him** [p.65] **...a child shuts himself off in his earliest months and years because he usually has no other choice** [p.59] **... When patients** [in primal therapy] **finally get down to the early catastrophic feeling** [the 'primal scream'] **of knowing they were unloved, hated, or never to be understood—that epiphanic feeling of ultimate aloneness—they understand**

perfectly why they shut off [p.97] **...Some of us prefer the neurotic never-never land where nothing can be absolutely true** [the postmodernist philosophy] **because it can lead us away from other personal truths which hurt so much. The neurotic has a personal stake in the denial of truth** [p.395]**.'** What has been said here makes it very clear that postmodern deconstructionism *was* autistic behaviour; **'a complex mental structure insuring against recurrence of...unthinkable anxiety'.**

910 So, what is finally revealed about these supposed great, progressive, enlightened social reforms is that instead of those involved being more ideal people, behaving in a more ideal way and bringing about a more ideal world, as they trumpeted themselves to the world as being and doing, they were actually more corrupted/upset/hurt/soul-damaged/ **'neurotic'** people, behaving in a less ideal way and bringing about an even more alienated, **'neurotic'** devastated world. The truth was the complete opposite of what the pseudo idealists, especially the latter day, more-advanced-in-denial pseudo idealists, were claiming. What was being presented to the world was a totally fraudulent, dishonest sham—and, as emphasised, the great danger of these ever-increasing levels of dishonesty is that humanity is being taken to the brink of terminal alienation.

911 Again, we humans had to do what we had to do—create a guilt-free yet truthless environment in order to stay alive under the duress of the human condition—but the situation has spiralled way out of control, taking humanity perilously close to the perpetual darkness of terminal alienation. Humanity *is* facing a death by dogma, **'The Closing of the...** [human] **Mind'**, as Allan Bloom so accurately described it, and **'the end of civilization'** through a **'triumph of the forces of darkness'**, as Harold Bloom predicted. As necessary and tempting as such behaviour could become, to indulge dishonesty to the point of actually shutting down thought was the greatest weakness and failing possible on a planet whose culminating achievement is the fully conscious thinking mind. Preventing the search for knowledge represented a failure of all the effort and sacrifice made thus far by life on Earth. It represented a complete loss of nerve—as the science historian Jacob Bronowski said: **'I am infinitely saddened to find myself suddenly surrounded in the west by a sense of terrible loss of nerve, a retreat from knowledge into – into what? Into...falsely profound questions** [and other dishonesties that don't]**...lie along the line of what we are now able to know if we devote ourselves to it: an understanding of man himself. We are nature's**

unique experiment to make the rational intelligence prove itself sounder than the reflex [instinct]**. Knowledge is our** [proper] **destiny. Self-knowledge'** (*The Ascent of Man*, 1973, p.437 of 448). The Scottish comedian Billy Connolly was certainly right to denounce political correctness as **'the language of cowardice'** (*Billy Connolly: Live 1994*).

[912]While the danger for humanity's journey to enlightenment came from the increased levels of delusion and denial that we humans had to employ in order to cope with our increasingly insecure condition, to be truly free we *had to* confront and understand our condition, *not* escape it by adding more and more layers of denial. Denial blocked access to the truth, that being its purpose, but we *had to* find the truth, especially the truth about ourselves, **'self-knowledge'**. As the great Greek philosopher Socrates famously said, **'the only good is knowledge and the only evil is ignorance'** (Diogenes Laertius, *Lives of Eminent Philosophers*, c.225 AD), and, **'the unexamined life is not worth living'** (Plato's dialogue *Apology*, c.380 BC; tr. Benjamin Jowett)—but in the end a preference for ignorance and the associated need to oppress any examination of our lives, oppress any freedom to think truthfully, question and pursue knowledge, threatened to become the dominant attitude throughout the world. George Orwell's bleak prediction that **'If you want a picture of the future, imagine a boot stamping on a human face** [freedom] **– for ever'** is about to be realised—as is Aldous Huxley's fear (which he wrote about in his famous 1932 novel *Brave New World*) that we would become a trivial culture where **'the truth would be drowned in a sea of irrelevance'**.

[913]So while the greed of capitalism is certainly causing immense suffering and devastation, it is not greed that is taking the world to the brink of destruction, as everyone has been told—no, our society is being taken over by a desperate and madly behaved faction. It is a very, very serious matter that is made doubly so by the fact that almost no one is raising the alarm. Pseudo idealism has almost everyone intimidated, bluffed or seduced. Warnings about the real danger facing the world are only being voiced by a rare few, like those just mentioned—Orwell, Nietzsche, Harold Bloom and Allan Bloom. As Geoffrey Wheatcroft recognised, while **'the great twin political problems of the age are the brutality of the right, and the dishonesty of the left'**, it is the dishonesty of the left that has the potential to—and is poised to—destroy the world.

[914]Yes, of the three aspects of upset of anger, egocentricity and alienation, alienation is the really dangerous one. Interestingly, anthro-

pologists—in their defensive, mechanistic, denial-complying, dishonest mindset—named the final two varieties of humans in the series *Homo sapiens* and *Homo sapiens sapiens*, which literally translates from Latin as 'wise man' and 'wise wise man', respectively (with 'sapiens' meaning 'wise' and 'homo' meaning 'man'). Certainly, humans have been gaining wisdom or knowledge, but the more accurate description of what we have become is really 'False or Alienated Man', and 'False False or Alienated Alienated Man'.

915 Nietzsche's warning of the danger of the **'many sickly people'** who **'have a raging hate for the enlightened man and for that youngest of virtues which is called honesty'** can't be emphasised enough. To repeat what he wrote: **'Purer and more honest of speech is the healthy body, perfect and square-built: and it speaks of the meaning of the earth** [to face truth and one day find understanding of the human condition]**...You are not yet free, you still *search* for freedom. Your search has fatigued you...But, by my love and hope I entreat you: do not reject the hero in your soul! Keep holy your highest hope!'** Yes, yes, we cannot afford to lose our nerve, but everyone nearly has.

916 As I've already said, I consider Sir Laurens van der Post to be the most exceptional denial-free thinking prophet and philosopher of the twentieth century; indeed, his full-page obituary in the *London Times* was boldly headed **'A Prophet Out of Africa'** (20 Dec. 1996). He was another who was **'pure'** and **'honest'** and **'square-built'** enough to speak out strongly against the extreme danger of pseudo idealism, writing that **'the so-called liberal socialist elements in modern society are profoundly decadent today because they are not honest with themselves...They give people an ideological and not a real idea of what life should be about, and this is immoral...They feel good by being highly moral about other people's lives, and this is immoral...They have parted company with reality in the name of idealism...there is this enormous trend which accompanies industrialized societies, which is to produce a kind of collective man who becomes indifferent to the individual values: real societies depend for their renewal and creation on individuals...There is, in fact, a very disturbing, pathological element—something totally non-rational—in the criticism of the** [then UK] **Prime Minister** [Margaret Thatcher]**. It amazes me how no one recognizes how shrill, hysterical and out of control a phenomenon it is...I think socialism, which has a nineteenth-century inspiration and was valid really only in a nineteenth-century context when the working classes had no vote, has long since been out of date and been like a rotting corpse whose smell**

in our midst has tainted the political atmosphere far too long' (*A Walk with a White Bushman*, 1986, pp.90-93 of 326).

917 Two of the Bible's denial-free thinking, exceptionally sound, **'pure'** and **'honest'**, **'square-built'** prophets—Daniel and Christ—also warned of the great danger of extreme forms of pseudo idealism when they spoke of **'the abomination that causes desolation'** taking over the world. The Bible has, without a doubt, been the most influential book in history because it contains extraordinary honesty; indeed, as I mentioned in par. 578, it has been the most denial-free book humans have had for guidance. As Daniel said of his own contribution to the Bible, **'I will tell you what is written in the Book of Truth'** (Dan. 10:21)—and his ability to think in a denial-free, truthful and therefore effective way meant Daniel *was* in a position to **'explain to you what will happen to you people in the future'** (10:14). However, since there was no science in his day to evidence his argument, all Daniel could do was draw upon analogies to describe what he could see so clearly happening in the future. In one analogy he described **'The king of the South'** (which we can understand is the freedom-upholding, answer-searching but immensely upsetting and corrupting right-wing) constantly at war with **'the king of the North'** (the freedom-oppressing, dogma-based, pseudo idealistic, dishonest left-wing). He described how, for a long time, power would oscillate between these kings (as it does in democratic politics), until the complete polarisation of the two kingdoms, the two political states, came about (which has occurred), at which point **'the abomination that causes desolation'** of left-wing pseudo idealism would finally threaten to take over the world. After describing many changes of power, Daniel said: **'the king of the North will muster another army larger than the first...The forces of the South will be powerless to resist; even their best troops will not have the strength to stand** [pseudo idealism becomes increasingly seductive and powerful as people become increasingly upset]**. The invader will do as he pleases; no-one will be able to stand against him...**[but eventually 'he'/pseudo idealism] **will make an alliance with the king of the South...**[some bipartisanship between the left and right wing will occur, but then the left-wing] **will stumble and fall...**[however, in time] **He will be succeeded by a contemptible person** [even more dishonest forms of left-wing pseudo idealism will emerge] **who will not be given the honour of royalty** [they will lack religion's honesty]**. He will invade the kingdom** [the religious kingdom of honesty]**...and he will seize it through intrigue** [through the dishonesty of extreme forms of pseudo idealism and charismatic

and fundamentalist misrepresentations of religion masquerading as real idealism]. **Then an overwhelming army** [from the South] **will be swept away before him; both it and a prince of the covenant** [religion] **will be destroyed... when the richest provinces feel secure, he will invade them and will achieve what neither his father nor his forefathers did** [no force has been able to overthrow religion before]...**His armed forces will rise up to desecrate the temple fortress and will abolish the daily sacrifice. Then they will set up the abomination that causes desolation** [extremely guilt-stripped pseudo idealistic causes and charismatic and fundamentalist misrepresentations of religious teaching will take over the world and lead humanity to terminal alienation]. **With flattery** [the truth-and-guilt-avoiding, do-good, feel-good self-affirmation that pseudo idealism feeds off] **he will corrupt those who have violated the covenant** [extreme forms of pseudo idealism fundamentalism will seduce the more upset away from religion's infinitely more honest way of coping with the human condition], **but the people who know their God will firmly resist him** [the more secure, less dishonest will not be deceived and *must* strongly resist the seductive tide]...**The** [left-wing pseudo idealistic] **king will do as he pleases. He will exalt and magnify himself above every god and will say unheard-of-things against the God of gods** [such as Richard Dawkins, Oxford University's Professor of Public Understanding of Science, would you believe, who has, as mentioned before, said that '**"Faith is one of the world's great evils, comparable to the smallpox virus, but harder to eradicate. The whole subject of God is a bore"...those who teach religion to small children are guilty of "child abuse".'**]. **He will be successful until the time of wrath** [until the all-exposing truth of understanding of the human condition arrives to save the world, as it is doing in what you are reading right here]...**He will invade many countries and sweep through them like a flood...he will set out in a great rage** [the stridency that I and Sir Laurens van der Post talked about of those trying to persuade themselves and others that their pseudo idealistic causes and fundamentalist misrepresentations of religion represent real idealism] **to destroy and annihilate many...Yet he will come to his end, and no-one will help him...**[when understanding of the human condition arrives pseudo idealism and other forms of extreme dishonesty will be totally exposed and brought to an end. However, while 'he'/pseudo idealism reigns] **There will be a time of distress such as has not happened from the beginning of nations until then'** (Dan. 11-12).

918 In another analogy that describes the same progression, but even more explicitly—mentioning as it does how **'truth was thrown**

to the ground'—Daniel, appropriately enough, used the symbols of a (determined) ram for the right-wing and a (stupid) goat for the left-wing. He said that initially **'No animal could stand against him** [the ram], **and none could rescue from his power. He did as he pleased and became great...**[greed and indifference to others was so great that even children, for instance, were put to work in coalmines, but then the] **goat...charged at him in great rage...**[the left-wing emerged and] **The ram was powerless to stand against him; the goat knocked him to the ground and trampled on him, and none could rescue the ram from his power. The goat became very great... It set itself up to be as great as the Prince of the host** [it set itself up to be more important than religion]**; it took away the daily sacrifice from him, and the place of his sanctuary was brought low...It prospered in everything it did, and truth was thrown to the ground..."How long will it take for the vision to be fulfilled—the vision concerning the daily sacrifice, the rebellion that causes desolation, and the surrender of the sanctuary and of the host that will be trampled underfoot** [the rebellion against religion's honesty]**...It will take** [a long time]**...understand that the vision concerns the time of the end."...** [when humans] **have become completely wicked** [and]**, a stern-faced king** [the extremely upset]**, a master of intrigue, will arise** [the left-wing pseudo idealistic, false prophet merchants of escapist denial and delusion will arise]**...He will cause astounding devastation and will succeed in whatever he does. He will destroy the mighty men and the holy people** [even the strong will begin to succumb to the intimidating tide of pseudo idealism]**. He will cause deceit** [the misrepresentation of pseudo idealism and fundamentalism as being real idealism] **to prosper, and he will consider himself superior** [the extreme delusion that left-wing pseudo idealism and fundamentalism is based on will spread everywhere]**..."The vision...concerns the distant future** [that has finally arrived]**"'** (Dan. 8).

[919]So important to Daniel was this issue of the great danger from pseudo idealism for humanity in the future that he even used a third analogy to try to convey his fears about it. In this analogy, he first described the rule of a **'king of kings'** with a **'head of gold'** who had **'dominion and power and might and glory'** (Dan. 2:37-38). This is a description of the original dominance hierarchy type situation (such as occurs in the animal kingdom) where some peace and order occurred in human society through the rule of the most powerful. Daniel then said, **'After'** this **'another kingdom will rise'** that is **'inferior'**, then **'one of bronze'**, then one as **'strong as iron'** that **'breaks and smashes' 'all the others'**, but he said

'this will be a divided kingdom...partly iron and partly clay, so this kingdom will be partly strong and partly brittle...so the people will be a mixture and will not remain united, any more than iron mixes with clay' (2:39-43). What Daniel has described here is how, through trial and error, the very effective—**'strong as iron'**—device for managing upset of democracy is developed, but then this system of management becomes hopelessly **'divided'**/polarised, because **'people will be a mixture'**, **'partly strong and partly brittle'**, which is clearly a description of what has happened with the strong right and the weak, pseudo-idealism-supporting left-wing in politics. Daniel then said that eventually **'God...will set up a kingdom that will never be destroyed...** [and] **will crush all those** [other] **kingdoms and bring them to an end'**, saying this is **'what will take place in the future'** (2:44-45). Yes, as will be described shortly in pars 926-927, now that **'God'** or Integrative Meaning-guided understanding of the human condition has emerged this whole ugly war between the left and the right is finally obsoleted, **'destroyed'**, **'crush**[ed]**'**, brought **'to an end'**.

920 In the New Testament, Christ gave exactly the same warning as Daniel, even referring to Daniel's description of pseudo idealism as **'the abomination that causes desolation'**—but Christ went further, truthfully and courageously advising people to head for the hills, **'flee to the mountains'**, when pseudo idealism and strident fundamentalist misrepresentations of religious teachings threaten humanity with annihilation. Referring to **'the sign...of the end of the age'**, Christ said that **'At that time many will turn away from the faith and will betray and hate each other** [a great deal of upset will develop], **and many false prophets** [pseudo idealists claiming to be leading the way to peace and a new age of goodness and happiness for humans] **will appear and deceive many people...even the elect** [even those less alienated, still relatively sound and strong in soul, will begin to be seduced by pseudo idealism]**—if that were possible. See, I have told you ahead of time... Wherever there is a carcass** [the extremely upset], **there the vultures** [false prophet promoters of delusion and denial to artificially make the extremely upset feel good] **will gather. Because of the increase of wickedness** [upset], **the love of most will grow cold. So when you see the "abomination that causes desolation," spoken of through the prophet Daniel, standing where it does not belong** [throwing out real religion and falsely claiming to be presenting the way to the human-condition-free, good-versus-evil-deconstructed, post-human-condition, better, correct world]**—let the reader understand—then let those who are in Judea flee to the mountains. Let no-one on the roof of his**

house go down to take anything out of the house. Let no-one in the field go back to get his cloak. How dreadful it will be in those days for pregnant women and nursing mothers! Pray that your flight will not take place in winter because those will be days of great distress [mindless dogma and its consequences] **unequalled from the beginning of the world until now—and never to be equalled again. If those days had not been cut short** [by the arrival of the liberating understanding of the human condition], **no-one would survive'** (extracts from Matt. 24 & Mark 13 combined). In summary, when Christ said to **'Beware of false prophets. They come to you in sheep's clothing, but inwardly they are ferocious wolves'** (Matt. 7:15), he was warning that we had to be on guard for those who concealed their extreme upset behind pseudo idealistic causes in order to delude themselves that they were sound and ideally behaved people who were leading others to a sound and ideal world. Where true prophets *confronted* the issue of the human condition, false prophets were merely merchants of delusion, advocates of *escapism* from the issue of the human condition.

[921] (Note, it might be asked how, in such ancient times, could the threat pseudo idealism and strident forms of fundamentalist non-religion posed to the human journey have been so accurately anticipated by Christ and Daniel. The truth is all the nuances of the human condition can be seen in any sample of human behaviour for those sufficiently psychologically open to seeing it. In the case of pseudo idealism, you only have to be caught in heavy traffic and observe how instead of holding their place against aggressive lane swappers some drivers are clearly deriving a feel good rush from letting such drivers push in front of them, even slowing down to encourage them. With a little more observation you can extrapolate what is eventually going to happen in the world, especially if you are sufficiently free from needing to live in denial of the issue of the human condition—in particular free from having to live in denial of Integrative Meaning or God—to be able to, and want to, think deeply about such matters. Daniel for example, in explaining why he **'could understand villains and dreams of all kinds'** (Dan. 1:17), said **'I have greater wisdom than other living men'** (2:30) and why he was able to have his mind **'turned to things to come'** (2:29), **'praised the God of heaven** [he embraced the truth of Integrative Meaning and was unafraid of the issue of the human condition]...[who] **gives wisdom to the wise and knowledge to the discerning** [not denying Integrative Meaning enables him to think truthfully and thus effectively]. **He reveals deep and hidden things; he knows what lies in darkness, and light dwells with**

him. I thank and praise you, O God of my fathers: You have given me wisdom and power' (2:19, 21-23).)

[922]The most common criticism my writings about the human condition receive is their denunciation of left-wing causes—'What could possibly be wrong with being concerned for those who are disadvantaged, suffering and oppressed, or with loving the environment, or with wanting peace?', they ask—but that is exactly an indication of just how seduced almost everyone has been by pseudo idealism. As Daniel foresaw, in the end **'even their best troops will not have the strength to stand'** against the seductive tide of pseudo idealism. He said **'no-one will be able to stand against'** it and it will **'invade many countries and sweep through them like a flood'** and **'cause deceit to prosper'**. Christ similarly warned that **'false prophets will appear and deceive many people...even the elect—if that were possible. See, I have told you ahead of time.'**

[923]For such exceptionally sound, **'pure'** and **'honest'**, **'square-built'** prophets throughout history who weren't living in denial and thus **'who know their God'**—from Daniel to Christ to Nietzsche to van der Post—to so **'firmly resist'** and warn of the great danger of the **'profoundly decadent'**, **'dreadful'** **'abomination'** that causes **'astounding devastation'** and **'desolation'** **'unequalled from the beginning of the world until now'** of those whose **'love'** has **'grow**[n]**'** so **'cold'** that they have such a **'raging hate** [of]**...honesty'** that they have created a pseudo idealistic **'rotting corpse whose smell in our midst has tainted the political atmosphere far too long'**, in which **'truth was thrown to the ground'** and **'trampled underfoot'**, illustrates just how extremely serious the threat of **'the end of civilization'** through a **'triumph of the forces of darkness'** of guilt-stripped charismatic and fundamentalist misrepresentations of religious teachings and guilt-stripped forms of non-religious pseudo idealism have become. Their influence and hold has become so great that any remaining opposition will soon, as Daniel predicted, be **'powerless to stand against'** it. In the case of the West, which is at the forefront of humanity's journey to enlightenment, it defied and fought with all its might the spread of human-freedom-oppressing dogma in the form of communism in the former USSR and South East Asia, but it has proved incapable thus far of resisting the insidious takeover of its own culture by the freedom-oppressing dogma of pseudo idealism; during the 1950s Senator Joe McCarthy succeeded in stemming its rise in America for a short while, but since then it's been a case in the West of caving in to pseudo idealism all the way. Indeed, to attempt to

describe how serious the situation in the West is, it is fair to say that freedom there is *as* under threat from pseudo idealism as a city or individual would be if it was surrounded by murderously hostile people and had given up, or was about to give up, defending itself! Yes, in January 2014 *The Australian* newspaper published an article about Camille Paglia's latest book, *Glittering Images*, with the subtitle **'For Camille Paglia, Western society has become too soft and weak to survive'** and quoting Paglia as saying, **'What you're seeing is how a civilisation commits suicide'** (Bari Weiss, 'In pursuit of manly equality', 1 Jan. 2014).

924 In the following passage, Sir Laurens van der Post provided a powerful affirmation of all that has just been said about the essential, intrinsic, absolute importance of maintaining freedom from oppression in the human journey to enlightenment: **'I hope a war is not declared by anybody in the modern world because I don't see the real necessity for war, but I would like to say that I think it would be immoral—it would be obscene—not to be ready at any moment to defend ourselves...If somebody should impose war upon us, attack us, I hope that we should have the will and the power and the moral courage to realise that life, freedom, are gifts from God and creation and our duty to defend. There's a wonderful episode in the life of Buddha where a group of villagers in the Himalayas did not defend themselves against a band of brigands who attacked and killed many of them, and he rebuked them because he said that was not what his teaching was about, that was not reverence for life of which he was speaking...it is a moral duty for us all to be ready to defend ourselves and our freedoms'** (presentation by Laurens van der Post at the Institute of Contemporary Arts, London, 9 Dec. 1982). Well, the West's failure to **'defend'** its **'freedoms'** from the massive **'attack'** upon it by pseudo idealism means it has *failed* to exhibit the **'moral courage'** necessary to fulfil its **'moral duty'** to **'defend'** its **'freedoms'** against **'attack**[s]**'**. The result of this fundamental, **'obscene'** failure at the leading edge of humanity's journey to enlightenment is that humanity *has* reached the precipice of self-destruction—so it is no exaggeration to say that this understanding of the human condition has arrived *only just in time* to **'cut short'** that tragic end.

925 Yes, we had reached the end play situation where the dangers from excessive upset in the world had become so great that **'even their** [humanity's] **best troops will** [did] **not have the strength to stand'** up to pseudo idealism. It had become virtually impossible to find sufficient **'pure'** and **'honest'** and **'square-built'** soundness (from having a relatively nurtured, loving, secure upbringing and thus not too much upset) coupled with

strength of character (from having some, but not too much, genetically adapted toughness to life under the duress of the human condition) to stand against the tide of pseudo idealism and continue the freedom-dependent search for the liberating understanding of the human condition. Resolve and solidarity for our species' all-important, freedom-dependent journey to enlightenment was fast disappearing, and it was the lack of solidarity, the failure to present a strong, united front, that has been especially encouraging and empowering of the radical opponents of freedom such as Saddam Hussein in Iraq and Osama bin Laden in Afghanistan. The essence of the concept of The United Nations was of Nations acting strongly in a United way against what is patently '**obscene**[ly]' wrong behaviour, but such was the lack of soundness and thus courage and character and thus resolve in the world that the concept has failed. R.D. Laing certainly wasn't exaggerating how sick, exhausted and spent with alienation the human race had become when, as has been mentioned before, he wrote that '**the ordinary person is a shrivelled, desiccated fragment of what a person can be...as men of the world, we hardly know of the existence of the inner world...The outer divorced from any illumination from the inner is in a state of darkness...i.e. alienation or estrangement from the inner light**', and '**We are dead, but think we are alive. We are asleep, but think we are awake. We are dreaming, but take our dreams to be reality. We are the halt, lame, blind, deaf, the sick. But we are doubly unconscious. We are *so* ill that we no longer feel ill, as in many terminal illnesses. We are mad, but have no insight** [into the fact of our madness].' Pope Benedict XVI certainly wasn't exaggerating the darkness of the human situation either when, in his 2005 inaugural mass, he said that '**We are living in alienation, in the salt waters of suffering and death; in a sea of darkness without light.**'

926 In the functional democracies of the West, humanity's 2-million-year journey to find understanding of the human condition finally came down to a so-called '**cultural war**' between the philosophy of the political left-wing and the philosophy of the political right-wing. Both sides were determined they were right, but with the human condition now explained what is revealed is that, in terms of a future for the human race, the philosophy of the left-wing was completely wrong while the philosophy of the right-wing was completely right. The left-wing was all about dogma, delusion and escapism, while the right-wing was comparatively all about realism, honesty and responsibility.

[927]Interestingly and significantly, there is one final irony to the saga of humanity's great journey through ignorance, which is that the ideal world that the left-wing was dogmatically demanding is actually brought about by the right-wing winning its reality-defending, freedom-from-idealism, corrupting-search-for-knowledge battle against the freedom-oppressing pseudo idealism of the left-wing. Yes, with the freedom-from-dogma right-wing's search for understanding of the human condition completed, the justification for the egocentric power-fame-fortune-and-glory-seeking way of life espoused by the right-wing ends, replaced by the ideal-behaviour-obeying attitude that the left-wing sought. In this sense, when the right-wing wins we all become left-wing; through the success of the philosophy of the right-wing we all adopt the philosophy of the left-wing—but, *most significantly*, this time we are not abandoning an ongoing battle, we are leaving it won. In reality, of course, through finding understanding of the human condition the inability to explain what humanity's journey has been about ends—an outcome that completely obsoletes the different philosophical positions of the left-wing and right-wing once and for all. Now that it can be explained that humanity has been involved in a corrupting search for knowledge, corruption is explained, which in turn exposes and thus negates the unrealistic position of the left-wing's dogmatic insistence on idealism. For its part, the right-wing's corrupting search for knowledge is *also* brought to a close with the finding of the key knowledge it was searching for, that being understanding of the human condition. Of course, the search for knowledge *will* continue, but not from a basis of dishonest denial, and with the key piece of knowledge we needed found of the explanation of the human condition, the priority for the immediate future will be—as will be explained in the next chapter—to take up the human-condition-reconciled Transformed Way of Living. Thus, with the arrival of understanding of the human condition, the concept of 'politics' comes to an end, which will undoubtedly be of great relief to everyone. Indeed, it will be explained in the next chapter that with the arrival of the truth about the human condition, the whole intellectual and psychological world that humans' have been living in—our whole culture—is suddenly obsoleted and ended and a new, completely transformed, all-loving, peaceful human culture, civilisation and world emerges, which is certainly also not before time.

928 Finally, it should be re-emphasised that in the greater context of the human-condition-understood view that we now have, upset is an immensely heroic state, with the most upset being the most heroic individuals of all because they have necessarily been involved in the battle that humanity has been waging longer and/or more intensely than any others. While the stark descriptions that have been given of the extremely upset, as being **'wicked'**, **'stern-faced'**, **'cold'**, **'sickly' 'carcasses'**, **'vultures'** and **'ferocious wolves'** involved in creating a **'rotting corpse'** of **'abomination'**, were necessary to match the no-holds-barred, gloves-off, totally brutal assault on the truth that was being **'thrown to the ground'** and left **'trampled underfoot'** by pseudo idealists, in the human-condition-understanding new world such rhetoric is entirely wrong and redundant.

929 In fact, it shouldn't even be necessary to talk about the old struggle between the left and right wing philosophies any more than what has now been done in this book. Thankfully, humanity now has the option to move on to an entirely new existence. Indeed, there will be many subjects that no longer have to be discussed now that humanity is able to move on to another existence. We get the truth up and we move on. As will be explained in the next chapter, we leave our suitcase of experiences that took place in the old ignorant, human-condition-afflicted, power-fame-fortune-and-glory-seeking-and-pseudo-idealistic-coping world behind and move on to an entirely new, instinct-and-intellect-reconciled, human-condition-liberated, genuinely 'deconstructed' and 'correct', fabulous, all-wonderful, new transformed world.

Old World

New World

[930]In conclusion, I would like to include once more the following drawing to summarise humanity's heroic overall journey from our species' original innocent childhood through an insecure, human-condition-afflicted, ever-increasingly upset adolescence, to a secure, human-condition-reconciled, mature adulthood.

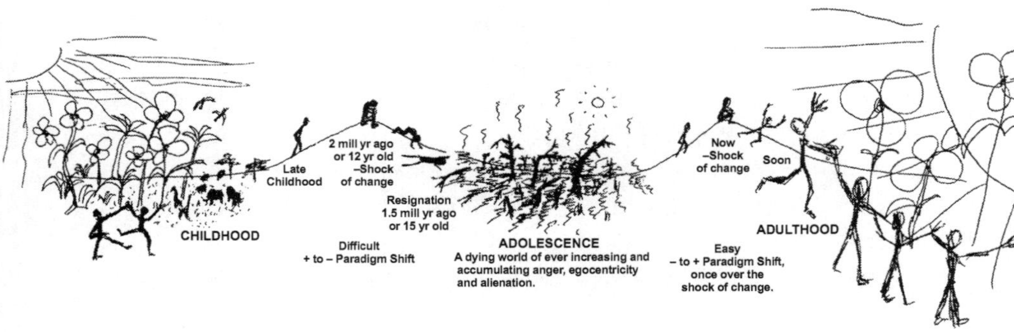

[931]So that's it—the greatest of all stories, our story, the story of humanity's incredibly, unbelievably heroic journey from ignorance to enlightenment told for the first time.

[932](A more complete description of all these stages that have been described of the maturation of humanity through its adolescence can be found in *Freedom Book 1* at <www.worldtransformation.com/freedom-stages-of-maturation>.)

Chapter 8

The Transformation of The Human Race

William Blake's *Albion Arose* c.1794-96;
this 1991 coloured impression is by Carol Marando

Chapter 8:1 <u>Summary</u>

[933] If, instead of continuing on its truth-denying path to terminal alienation, the scientific establishment fulfils its responsibility and recognises the psychosis-addressing-and-solving *real* explanation of the human condition that is being presented in this book—a recognition the public at large requires in order to discover the importance of the explanation—then what follows here is how this explanation is able to transform *all* humans right now from their human-condition-afflicted, tortured existence to a fabulous life free of the human condition!

Chapter 8:2 The 'dawn...of our emancipation'

[934] When *TIME* magazine invited the author Alan Paton to contribute an essay on apartheid in South Africa they instead received and published in its place a deeply reflective article on Paton's favourite pieces of literature. In what proved to be the renowned writer's last work, Paton wrote: **'I would like to have written one of the greatest poems in the English language—William Blake's "Tiger, Tiger Burning Bright", with that verse that asks in the simplest words the question which has troubled the mind of man—both believing and non believing man—for centuries: "When the stars threw down their spears / And watered heaven with their tears / Did he smile his work to see? / Did he who made the lamb make thee?"'** (25 Apr. 1988). The opening lines of Blake's poem, **'Tiger, Tiger, burning bright / In the forests of the night'**, refer to humans' great fear and resulting denial of the issue of our less-than-ideal, seemingly imperfect, 'fallen' or corrupted state or condition—a subject we humans have consciously repressed and yet one that **'burn**[s]**' 'bright'** in the **'forests of the night'** of our deepest awareness. The very heart of the issue lies in the line, **'Did he who made the lamb make thee?'**—a rhetorical question disturbing in its insinuation that we are wholly unrelated to **'the lamb'**, to the world of innocence. The poem raises the fundamental question involved in being human: how could the mean, cruel, indifferent, selfish and aggressive 'dark side' of human nature—represented by the **'Tiger'**—be both reconcilable with and derivative of the same force that created **'the lamb'** in all its innocence? As Paton identified, despite humans' denial of it, *the* great, fundamental, underlying question that **'has troubled the mind of man'** has always been, *are* humans part of God's **'work'**, part of **'his'** purpose and design, or *aren't we*? In other words, why don't humans live in accordance with the cooperative, loving integrative meaning of existence? Interestingly, the final words of Paton's acclaimed book about apartheid, *Cry, the Beloved Country*, allude to humanity's dream of one day finding this answer, and, through doing so, freeing itself from the terrible **'bondage of fear'** of our condition: **'But when that dawn will come, of our emancipation, from the fear of bondage and the bondage of fear, why, that is a secret'** (1948). Yes, humanity's hope and faith *has* always been that one day we would be able to explain the paradox of the human condition, and, as a result of doing so, liberate ourselves from **'the bondage'** of our species' immensely psychologically upsetting sense of unworthiness or guilt.

[935] So while religious assurances such as 'God loves you' could comfort us, ultimately we humans had to understand *why* we were lovable; there *had* to be a rational, first-principle-based biological explanation for humans' divisive condition and our responsibility as conscious beings was to find that explanation. The poet Alexander Pope was recognising this fundamental task and responsibility when he wrote, **'Know then thyself, presume not God to scan / The proper study of Mankind is Man'** (*Essay on Man*, 1733); as were the ancients when they had emblazoned across their temples the phrase, **'Man, know thyself.'** To refer again to the words of that pre-eminent philosopher of the twentieth century, Sir Laurens van der Post: **'Only by understanding how we were all a part of the same contemporary pattern** [of wars, cruelty and greed] **could we defeat those dark forces with a true understanding of their nature and origin'** (*Jung and the Story of Our Time*, 1976, p.24 of 275); and, **'Compassion leaves an indelible blueprint of the recognition that life so sorely needs between one individual and another; one nation and another; one culture and another. It is also valid for the road which our spirit should be building now for crossing the historical abyss that still separates us from a truly contemporary vision of life, and the increase of life and meaning that awaits us in the future'** (ibid. p.29). Jacob Bronowski observed the same truth in his book *The Ascent of Man*: **'We are nature's unique experiment to make the rational intelligence prove itself sounder than the reflex** [instinct]. **Knowledge is our destiny. Self-knowledge, at last bringing together the experience of the arts and the explanations of science, waits ahead of us'** (1973, p.437 of 448).

[936] Indeed, many of our most enduring mythologies, such as King Arthur's Knights' search for the 'Holy Grail', and Jason and the Argonauts' search for the 'Golden Fleece', are actually expressions of this all-important quest for the reconciling, redeeming and rehabilitating understanding of ourselves.

[937] Yes, the finding of the biological explanation of the human condition *has been* 'the holy grail' of the whole Darwinian revolution, the *central* quest in our human journey of conscious thought and enquiry—but, as this book has shown, that journey has been, and remains, lineball. When Richard Neville wrote **'We are locked in a race between self destruction and self discovery'** (*Good Weekend, The Sydney Morning Herald*, 14 Oct. 1986) he was recognising that *only* understanding of ourselves could bring an end to all the devastation, heartache and suffering that has plagued the human race since time immemorial and is now fast engulfing our planet in a terrible, terminal embrace. AND, as our designated vehicle for enquiry, it was

science (which, again, literally means 'knowledge') that was charged with this crucial task of finding the human-race-saving understanding of the human condition. Yes, despite presenting a patently evasive, dishonest explanation of the human condition, E.O. Wilson was being truthful when he wrote that **'The human condition is the most important frontier of the natural sciences'** (*Consilience*, 1998, p.298 of 374). AND, most wonderfully, through the advances made in science, particularly understanding of the difference in the way genes and nerves process information, science *has* finally made it possible for a scientist to explain the human condition. While virtually all scientists have had no choice but to operate from a basis of denial of any truths that led to confrontation with the issue of the human condition, the *practice* of science has provided the all-important insights into the mechanisms of the workings of our world that were needed for that explanation to be assembled—which means, as pointed out in par. 182, it is the practice of science as a whole that is humanity's liberator, its so-called 'saviour' or 'messiah'. Although extremely rare since Resignation became all but universal some 11,000 years ago, there have always been a few individuals who could confront and think truthfully about the human condition, but until the practice of science as a whole found understanding of the difference in the way genes and nerves process information such human-condition-confronting, truthful-thinking individuals were in no position to put together the explanation of the human condition. So it is the discipline of science that has finally made explanation of the human condition possible—and, by doing so, given the human race the means, the understanding, that enables all the psychological anger, egocentricity and alienation in humans to subside. Yes, with the human condition now explained, our previously inescapable need to live alienated from our beautiful soul, with all the horror that such a destructive, dishonest and shallow existence entailed, can end. The human race is at last in a position to return to the non-upset ideal state we have longed for, be it Heaven, Paradise, Eden, Nirvana, Utopia, Shangri-La, or whatever term we ascribed to it—the fundamental difference being that where we were once, as Moses said in the Bible, **'in the image of God'** (Gen. 1:27), instinctively orientated to the Godly integrative meaning of life, this time we'll return in a fully conscious, aware and understanding state. We will be **'like God, knowing** [understanding] **good and evil'** (ibid. 3:5). Again, as the poet T.S. Eliot wrote, **'We shall not cease from exploration and the end of all our exploring will be to arrive where we started and know the place for**

the first time' (*Little Gidding,* 1942). Buddha similarly looked forward to the time when **'In the future they will every one be Buddhas** [be free of psychosis and neurosis] / **And will reach Perfect Enlightenment** [reach understanding of the human condition]**'** (Buddha [Siddartha Gautama], *The Lotus Sutra*, c.560-480 BC, ch.9; tr. W.E. Soothill, 1987, p.148 of 275).

938 YES, with understanding of the human condition found, we can finally be upset-free, secure, sound, effective managers of our world and, as a result, the peace and happiness that humans throughout history have hardly dared to dream of can come into being. Christ's anticipated time that he appealed to us to pray for in his *Lord's Prayer*, when **'Your** [the Godly, ideal, integrated, peaceful] **kingdom come'** (Matt. 6:10 & Luke 11:2), can now become our reality. We can finally leave the dark cave of denial that Plato so accurately described—the cave in which we were imprisoned, having to hide from the condemning glare of all the confronting truths about our life (especially that most confronting truth of all of Integrative Meaning)—and stand free at last in the warm, healing sunlight of reconciling knowledge, and be part of, and see and feel, the world of staggering beauty that exists outside that debilitating 'cave' of denial. As William Blake famously prophesised in his appropriately titled poem *The Marriage of Heaven and Hell*, **'When the doors of perception are cleansed, man will see things as they truly are. For man has closed himself up, till he sees all things through narrow chinks of his cavern'** (1790). Again, Buddhist scripture accurately anticipated the form humans would take once the ameliorating understanding of the human condition arrived and was absorbed—the time, in the words of the scripture, when humans **'will with a perfect voice preach the true Dharma, which is auspicious and removes all ill'**. Of that time, the scripture says, **'Human beings are then without any blemishes, moral offences are unknown among them, and they are full of zest and joy. Their bodies are very large and their skin has a fine hue. Their strength is quite extraordinary'** (*Maitreyavyakarana*; tr. Edward Conze, *Buddhist Scriptures*, 1959, pp.238-242). The Bible similarly describes the time when **'Another book** [will be]**...opened which is the book of life** [the human-condition-explaining and thus humanity-liberating book]...[and] **a new heaven and a new earth** [will appear] **for the first heaven and the first earth** [will have]**...passed away...**[and the dignifying full truth about our condition] **will wipe every tear from...**[our] **eyes. There will be no more death or mourning or crying or pain, for the old order of things has passed away'** (Rev. 20:12, 21:1, 4). Yes, with understanding of the human condition now found, the human journey *can* have the happy ending we always

believed it would: **'The happy ending is our national belief'** (Mary McCarthy, *On the Contrary*, 1961).

939 *Achieving* that **'happy ending'**, however, obviously requires that we allow the truth to live in our lives, and that is an adjustment resigned humans will initially resist. While we at last have the means to ameliorate the human condition, there remains one last great problem to overcome (aside from the risk that science fails in its responsibility to support these understandings), and that is the difficult responsibility of facing the truth about ourselves. As will now be explained, while the finding of understanding of the human condition *is* wonderfully liberating, it is also fearfully confronting—but, thankfully, as will also be explained in this final chapter, there is an all-exciting and all-relieving way to overcome this last great hurdle of our fear of exposure of our now immensely corrupted state or condition.

Chapter 8:3 'Judgment day'

940 Humanity's journey thus far has been astonishing. In fact, as emphasised in chapter 7, the greatest, most heroic story ever told is our own. That saga, however, is not quite over. When the human-race-liberating, psychosis-addressing-and-solving, real explanation of the human condition arrives, which—as has been fully evidenced—it now has, all the denial that our species has had to employ to cope with the injustice of being unfairly condemned as bad when it was not, is suddenly exposed, and that exposure cannot come as anything other than a tremendous shock. The truth reveals the lies; the light of understanding suddenly floods the darkness of Plato's cave of denial where the human race has been hiding; yes, the blinds are suddenly drawn on all the lies and delusions we humans have been using to protect ourselves from exposure of our corrupted condition.

941 As has been revealed throughout this book, denial has played a dominant role in human life, even pervading science. Although, as stated above, science—through its painstaking discoveries of the mechanisms of the workings of our world—enabled the assembly of this explanation of the human condition, its practitioners have certainly not been rigorously objective and impartial in their 'scientific method'. No, as it (understandably) turns out, scientists—just like every other resigned human—have been committed to avoiding any insights that brought the

unbearably depressing subject of the human condition into focus, and there have been many, many such insights. There has been denial of the human condition itself; of Integrative Meaning; of the truth that our distant ancestors lived in an upset-free, innocent state; of the importance of nurturing in human development; of the true nature of consciousness; of our psychosis and the resulting differences in alienation—most especially between our distant ancestors and humans today, but also between individuals, races, genders, generations, countries, civilisations and cultures; and many, many other denials. Yes, there has been a great deal of *subjectivity* involved in science, and with the arrival of the truth about the human condition all that subjective dishonesty is suddenly revealed. The arrival of understanding of the human condition exposes not just all the denials that every human has been practising—but all the denials that science has been engaging in.

942 So while the arrival of understanding of the human condition liberates the human race from 2 million years of living with unjust condemnation, the immense change that this understanding brings cannot help but come as a *very* great shock, with the most difficult change to adjust to being the exposure of our corrupted condition—the *extent* of our accumulated anger, egocentricity and denial/falseness/dishonesty/alienation. In his famous 1970 book, *Future Shock*, Alvin Toffler was remarkably prescient in his anticipation of this time when understanding of the human condition would emerge and humans would suddenly be faced with (as he put it) **'the shattering stress and disorientation that we induce in individuals by subjecting them to too much change in too short a time'** (p.4 of 505). Yes, the truth about ourselves unavoidably and necessarily reveals the extent of our now immensely upset condition; it destroys the lies, our denials, our pretences and delusions, as it must, otherwise it wouldn't be the truth. Consider the Adam Stork analogy that was presented in chapter 2:4. If Adam could have explained why he had to carry out his search for knowledge when he was *first* criticised for doing so he would never have become upset—he would never have become defensively angry, egocentric and alienated. Or, if he had found the explanation after only a few days of carrying out that search, he would have accumulated very little anger, egocentricity and alienation to have to heal with understanding. But in *our* species' case, our conscious, self-managing, human-condition-afflicted state fully emerged some *2 million years ago* and we have only *just* found the understanding of why we became upset, which means

there is an absolute mountain of accumulated anger, unsatisfied ego, and denial in us humans to have to heal with this understanding. Certainly, we have learnt to restrain and conceal a great deal of the upset that exists within us. We have learnt to, as we say, 'civilise it', not let it show; for instance, we don't normally attack someone now the moment we become angry. Adult humans now exercise a great deal of self-control, but underneath our manufactured facade of restrained civility, even manufactured happiness, lies volcanic anger and immense frustrated egocentricity, which expresses itself in all the ferocious atrocities and vengeful acts of bloodshed we humans commit, and—albeit to a much lesser degree—in the smaller disputes that mar our everyday lives.

943 Yes, truth day *is* honesty day, exposure day, transparency day, revelation day. It is, in fact, the long-feared so-called **'judgment day'** referred to in the Bible (Matt. 10:15, 11:22, 24, 12:36; Mark 6:11; 2 Pet. 2:9, 3:7; 1 John 4:17). And although this **'judgment day'** is actually a day of compassionate understanding, not a day of condemnation—as an anonymous Turkish poet once said, judgment day is **'Not the day of judgment but the day of understanding'** (Merle Severy, 'The World of Süleyman the Magnificent', *National Geographic*, Nov. 1987)—it *is*, nevertheless, a day when the extent of our species' by now *extremely* psychologically upset condition is exposed. This paradox of being wonderfully liberated but at the same time agonisingly exposed was captured by the prophet Isaiah when he said that the liberation that **'gives you relief from suffering and turmoil and cruel bondage...will come with vengeance; with divine retribution...to save you. Then will the eyes of the blind be opened and the ears of the deaf unstopped...Your nakedness will be exposed'** (Bible, Isa.14:3; 35:4, 5; 47:3). Also referring to **'the Day of Reckoning'** (*The Koran*, ch.56) and **'the Last Judgement'** (ibid. ch.69), the prophet Muhammad provided a similar description of this paradoxical situation when he said, **'when the Trumpet is blown with a single blast and the earth and the mountains are lifted up and crushed with a single blow, Then, on that day, the Terror shall come to pass, and heaven shall be split...On that day you shall be exposed, not one secret of yours concealed'** (ibid). And in Zoroastrian religion, **'The universe was seen as a battleground in which these opposing forces contended, both in the sphere of political conquest and in the depths of each man's soul. But in time the light would shine out, scattering the darkness, and truth would prevail. A day of reckoning would arrive in which the blessed would achieve a heavenly salvation, while all others would find themselves roasting in fiery purgatory'** (*A Soaring Spirit, 600-400 BC*, Time-Life History of the World, 1988, p.37 of 176).

Another anticipation of this time when the liberating truth about our species' immensely heroic but also horrifically upset, exhausted condition is suddenly revealed is the word 'apocalypse'. The original name for the Book of Revelation in the Bible, **'apocalypse'** is **'Ancient Greek: meaning "un-covering"...translated literally from Greek,** [it] **is a disclosure of knowledge, hidden from humanity in an era dominated by falsehood and misconception, i.e., a lifting of the veil or revelation'** (Wikipedia, accessed Apr. 2013; see <www.wtmsources.com/127>); **'a cataclysm in which the forces of good triumph over the forces of evil'** (The Free Dictionary, accessed Apr. 2013; see <www.wtmsources.com/112>).

944 With regard to the exposure of the differences in alienation that exist *between* individuals (*and* races, genders, generations, countries, civilisations and cultures), immediately after describing how the arrival of the all-exposing, shocking truth about humans will come **'like the lightning, which flashes and lights up the sky from one end to the other'**, Christ described how **'two people will be in one bed; one will be taken** [revealed as sound, relatively free of upset and alienation] **and the other left** [revealed as being upset and alienated]**. Two women will be grinding corn together; one will be taken and the other left'** (Luke 17:24, 34, 35; see also Matt. 24:27, 40). Again, it has to be stressed that **'judgment day'** is *not* a time when some will be judged as deserving of being **'taken'** to heaven and others **'left'** behind, rejected as bad and 'evil', but a time of compassionate understanding of everyone's situation. Yes, the way **'the forces of good triumph over the forces of evil'** is through 'evil' finally being compassionately, lovingly understood as a heroic consequence of humanity's immensely courageous but psychologically upsetting battle to defeat ignorance—which is the point Sir Laurens van der Post was making when he wrote that **'True love is love of the difficult and** [historically] **unlovable'** (*Journey Into Russia*, 1964, p.145 of 319).

945 On this note, I should immediately stress that while the left-wing faction in politics, mechanistic science as a whole, and the scientists E.O. Wilson, Richard Wrangham, Victoria Wobber and Brian Hare, have been particularly targeted for exposure of their alienating denials in this book, *everyone's* denials are now also exposed. The fact is, the human-condition-confronting, truthful analysis of *anyone*'s situation would reveal similar levels of dishonesty. The only reason the left-wing political paradigm, mechanistic science, and the individuals listed above have been targeted for exposure is because *their* particular denials happen to be the most dangerous in terms of the threat of taking humanity to terminal alienation and extinction. In the case of the extreme danger

that denial-based mechanistic science poses to the world, I should include these honest words from Charles Birch, who, as I've mentioned before, I was fortunate enough to have as head of the biology faculty at Sydney University when I was studying science there: **'Reductionism or Mechanism...is the dominant mode of science and is particularly applicable to biology as it is taught today...**[it is] **A view or model of livingness that leaves out feelings and consciousness...**[and] **I believe it has grave consequences...In the name of scientific objectivity we have been given an emasculated vision of the world and all that is in it. The wave of anti-science...is an extreme reaction to this malaise...I believe biologists and naturalists have a special responsibility to put another image before the world that does justice to the unity of life and all its manifestations of experience—aesthetic, religious and moral as well as intellectual and rational'** ('Two Ways of Interpreting Nature', *Australian Natural History*, 1983, Vol.21, No.2). And this **'unity of life and all its manifestations of experience—aesthetic, religious and moral as well as intellectual and rational'** is *not* the fake one that the lord of lying, that master of keeping humanity *away* from any truth—indeed, the quintessential anti-Christ—E.O. Wilson put forward in his 1998 book, *Consilience: The Unity of Knowledge*, when he proposed that Evolutionary Psychology's alleged ability to explain the moral aspects of humans meant biology and philosophy, the sciences and the humanities, indeed science and religion, could at last be reconciled; he wrote of **'the attempted linkage of the sciences and humanities...of consilience, literally a "jumping together" of knowledge... to create a common groundwork of explanation'** (p.6 of 374). And nor is it Wilson's more recent—and even more dishonest—attempt to explain the origins of humans' moral nature and the human condition. No, it is *only* the psychosis-addressing-and-solving *real* explanation of the human condition that is being presented here in this book that brings about the realisation of Birch's belief in a **'malaise'**-of-**'science'**-ending, **'unity of life and all its manifestations'**. It is only *this* book that reconciles religion and science, instinct and intellect, conscience and conscious, idealism and realism.

946 Of course, the huge issue about this situation where all our **'nakedness will be exposed'** and, for example, the differences in alienation between **'two people** [and races, genders, generations, countries, civilisations and cultures]**'** will be revealed—this arrival of **'judgment day'**—is HOW IS THE HUMAN RACE, and all its institutions, such as our political and scientific bodies, SUPPOSED TO COPE with this ultimate **'future shock'**, the **'Terror'** and

'cataclysm' of having to face such a massive paradigm shift? The answer, which will now be explained, is that while there will be an initial, brief period of shock and overwhelmed bewilderment, this great change from denial to honesty is not only *not* difficult or painful, it is extremely easy and incredibly exciting, painless and joyful. Yes, there is not merely a way of *coping* with the truth about humans when it arrives, but an almost *unbearably exciting way of living* with it.

Chapter 8:4 The Transformed Lifeforce Way of Living

947 Exasperated by all the denial/dishonesty/falseness/'**phon**[iness]'/'**fake**[ness]'/delusion/pretence/artificiality/superficiality in human life, John Lennon, that creative genius and member of The Beatles band, desperately pleaded for some honesty in his 1971 song *Gimme Some Truth*: **'All I want is the truth, just gimme some truth. I've had enough of reading things by neurotic...politicians...I'm sick to death of seeing things from tight-lipped... chauvinists...I've had enough of watching scenes of schizophrenic...prima-donnas...I'm sick and tired of hearing things from uptight...hypocrites...All I want is the truth now, just gimme some truth NOW.'** Yes, we are all literally **'sick and tired of'** the dishonest denial the world has been drowning in; we have all completely had enough of the human condition and are all desperate to be free of it—BUT, while it is one thing to wish for the truth, it is quite another to cope with that truth, and the changes it heralds, when it actually arrives!

948 So *how is* the human race supposed to cope with having its 2 million years of accumulated denials, all our dishonest delusions and artificialities, and our pain and alienation, suddenly revealed? How are we to cope with **'judgment day'**? To arrive at the answer we need to summarise how we humans came to be in the situation we are now in, and how this situation changes with the arrival of understanding of the human condition—because doing so will provide the logical sequence of events that leads to the equally logical way that humans will be able to manage their transition from being a victim of the human condition to becoming a secure, sound, effective manager of our world.

949 As has been emphasised throughout this book, all humans are born instinctively expecting to encounter an unconditionally selfless, all-loving environment, but as a result of humanity's 2-million-year necessary search for knowledge we instead encountered an immensely

upset, human-condition-afflicted world where everyone was variously alienated from their true selves or souls with little or no unconditional love to give. As a result, we too became psychologically upset sufferers of the human condition. Then, as adolescents, we tried to understand why there was so much imperfection in the world around us and within ourselves, but unable to find an explanation we eventually, at around 15 years of age, had no choice but to resign ourselves to living in denial of the unbearably depressing subject of the horrendously corrupted state of the human condition. At this point, however, unable to *refute* the negative views of ourselves with understanding, we could only *counter* those negatives by focusing on, emphasising and developing whatever superficial positive views of ourselves we could find. In particular, we became preoccupied trying to find ways to feel good about ourselves, to validate ourselves, to find relief from the criticism we were living with that we weren't worthy beings by competing for whatever power, fame, fortune and glory was available, the result of which was that we became extremely self-preoccupied, ego-centric, competitive and selfish. This, of course, being a situation that could only be remedied through the first-principle-based explanation of the human condition itself, namely that we humans are conscious beings and as such it wasn't sufficient for us to be instinctively *orientated* to the world, we had to find *understanding* of the world if we were to master our conscious mind, a realisation that means we weren't 'bad' to have defied our instinctive self and become, as a result of that defiance, angry, egocentric and alienated. (The full explanation of the human condition appears in chapter 2.) What the arrival of this understanding means is that we no longer need to be angry, egocentric and alienated. But as pointed out, the problem is we humans needed this explanation when we *first* set out in search of knowledge some 2 million years ago, because if we had that explanation then we would never have become upset by the criticism coming from our instinctive self. We would have known *why* we had to defy our instincts—and thus we would have been secure about what we were doing. But unfortunately we didn't have that explanation, and so defensive and retaliatory anger, egocentricity and alienation resulted—behaviour that has been accumulating throughout those *2 million years*. So while the ameliorating understanding of why we became upset, namely the explanation of the human condition, *has* finally been found, to use it to heal all the upset that now exists within us would take a great deal of time and counselling—counselling that isn't presently available and time the world cannot afford at the moment. In fact, it will naturally take a number of generations for all the psychological upset within humans to be healed through the dissemination and absorption of the understanding of the human condition that is now available.

950 As mentioned in par. 176, Harry Prosen, a professor of psychiatry who has worked in the field for over 50 years, including chairing two departments of psychiatry and serving as president of the Canadian Psychiatric Association, has rightly said about the arrival of this psychosis-addressing-and-solving *real* explanation of the human condition that **'I have no doubt this biological explanation of the human condition is the holy grail of insight we have sought for the psychological rehabilitation of the human race.'** Yes, it *is* true that the **'psychological rehabilitation of the human race'** *is* now possible, and everyone *could* now retrace and analyse all the hurtful events in their life and replace all the defensive denials that they have had to employ to cope with that pain with truthful, compassionate understanding of those events and the feelings they produced—why their mother and father were preoccupied and unable to give them the unconditional love their instincts expected during their infancy and childhood, and why they are now so insecure, etc, etc—but, again, to undertake such psychological therapy and dismantle all the psychoses and neuroses would take an immense amount of time and supportive counselling, which just isn't possible at the moment. *And*, apart from extreme cases of psychosis and neurosis, *such therapy is not, in fact, necessary*, which is an absolutely fabulous situation that I will now explain.

951 While understanding of the human condition finally gives us the means to heal all our anger, egocentricity and alienation (which, in truth, are our psychoses [psyche or soul repressions] and neuroses [neuron or mind distresses]), we can't, in practice, hope to heal it in our lifetime, *however*—and this is all-important—we can immediately *know* that we humans are all fundamentally good and not bad, knowledge that puts each of us in a very powerful position *because it means we can legitimately decide not to think about and live in accordance with the upset within us*. We can reason that, 'Yes, I do have all these angry, egocentric and alienated feelings and thoughts, and I know that without undergoing a great deal of psychological therapy I can't hope to heal all the underlying upset that is producing these feelings and thoughts. But I *also* know that way of feeling and thinking is now obsoleted, so I'm simply going to

leave that way of feeling and thinking behind me and adopt the new way of feeling and thinking that is now available where I live in support of a reconciled, non-aggressive, non-competitive, truth-based new world for humans!'

952 Such transcendence of our real, upset self and adoption of a different way of thinking and behaving—a different attitude—may at first appear to be an extraordinary feat, but we humans are quite capable of being what we are not. For example, as explained in pars 677-679, when the human race became 'civilised' we chose not to live out our real, upset (in truth, volcanic) anger that now, after 2 million years of being unjustly condemned, exists within us, but to contain it; we chose not to let it express itself. We superimposed a different way of thinking and responding to our real, upset self's way of thinking and responding. As pointed out, we don't normally attack someone now the moment we become angry; instead, we transcend our real self's angry, defensive, destructive response and defer to and live by, or adopt, a different basis of thinking and behaving. In exactly the same way, we can study this explanation of the human condition and know that our upset condition has finally been explained and, therefore, that behaving in an upset way is no longer necessary, and on the basis of that logic prevent ourselves from behaving in an upset way by redirecting all our thoughts and all our energies to supporting the new human-condition-understood, reconciled world. Instead of letting our now obsoleted upset state rule our lives, we defer to and live through a different, now much more positive, way of thinking and behaving. We make a decision to shift our whole mindset from one state to another state. We mentally let go of one whole mindset and attach ourselves to another, and after a little practise the new mindset becomes automatic. This is how we learnt to civilise ourselves—we adopted a different response to our upset response and now, for the most part, we humans are very good at being restrained and civilised in the way we think and behave.

953 Yes, we can all now know that our insecure, egocentric need to validate ourselves through winning power, fame, fortune and glory is obsoleted. We can understand that it is no longer necessary to prove that we are good and not bad through demonstrations of our worth because our goodness has now been established at the most fundamental level through first-principle-based, biological explanation. We can know that with the arrival of understanding of the human condition our whole competitive,

self-preoccupied, selfish, must-win-power-fame-fortune-and-glory way of living has been made redundant. In fact, the fundamental situation we are now in is that the resigned, competitive way of living has become pointless, meaningless, because it doesn't progress the human race to greater knowledge as it once did because the key piece of knowledge that we needed to find of understanding of the human condition has been found. Furthermore, living out that self-preoccupied, competitive, selfish existence is now unnecessarily destructive of our relationships with our fellow humans, and of our world in general. We can now know that the resigned, egocentric, competitive, selfish and aggressive way of living is finished with, which means we can, and, in fact, should, simply cast aside that way of living, and take up, in its place, a new way of living that is available to humans now where we live in support of a reconciled, non-competitive, selfless, loving existence.

954 And, most wonderfully, what happens when we give up our old way of living and take up the new way of living that understanding of the human condition has made possible is we become *transformed* from a competitive and selfishly behaved human to a cooperative and selflessly behaved one. Yes, even though we are not yet free of the upset state of our own personal human condition, we *can immediately* have a change of attitude and decide not to live out that upset state that remains within us. The overall effect in our lives is that, despite our retention of the upset state of the human condition, we are effectively free from its hold and its influence, which is an absolutely fabulous transformation to have made in an instant—in one simple decision!

955 To elaborate, now that we know the reason why we humans had to set out in search of knowledge and defy our original ideal-behaviour-insisting instinctive self or soul, *all* the upset anger, egocentricity and alienation that unavoidably resulted from being unjustly criticised by our instincts is now rendered obsolete, unnecessary and meaningless. No longer do we have to retaliate against criticism of our upset/corrupted state because our upset/corrupted state has been defended with truthful, compassionate understanding at the most profound level. And no longer do we have to try to prove our worth because our worth has been established at the most fundamental level. And no longer do we have to deny any confronting truths about our immensely upset/corrupted condition because no longer are there any truths about our upset/corrupted state that condemn us. Our upset lives are explained and defended now, which

means we no longer have to be preoccupied compensating for that upset by finding forms of self-aggrandisement, by seeking self-distraction, or by chasing relief through materialistic forms of compensation for all the hurt we experienced growing up in an immensely human-condition-afflicted world. In other words, we no longer need to seek power, fame, fortune and glory to make ourselves feel good about ourselves because our goodness has now been established at the deepest, most profound level. Instead, we can simply leave our whole 'must-prove-our-worth, attack-and-deny-any-criticism' way of living behind as obsolete and redirect all our thoughts and all our energies into supporting and disseminating these human-race-saving understandings, and to repairing the world from all the damage our species' upset behaviour has caused—because with the human condition solved, all the upset that is causing the destruction of the planet can now end, which means it is at last possible to *properly* and permanently repair our environment. We *can*, as it were, put the issue of all our upsets/corruptions in a 'suitcase', attach a label to it saying, 'Everything in here is now explained and defended', and simply leave that suitcase behind at the entrance to what we in the World Transformation Movement call the Sunshine Highway and set out unencumbered by all those upset behaviours into a new world that is effectively free of the human condition. All our egocentric, embattled posturing to get a win out of life, all our strategising every minute of every day to try to find a way to compensate for feeling inadequate or imperfect or bad about ourselves, can suddenly end. We can leave Plato's dark cave where we have been hiding to escape the glare of the truth about our seemingly imperfect condition. Excitement and meaning—based on liberating, truthful, honest understanding of ourselves and our world—is what we have to sustain ourselves now.

956 Yes, the excitement of being effectively free of the human condition—the joy and happiness of being liberated from the burden of our insecurities and self-preoccupations; the awesome meaning and power of finally being aligned with the truth and participating in the magic true world; the wonderful empathy and equality of goodness and fellowship that understanding of the human condition now allows us to feel for our fellow humans; the freedom now to effectively focus on repairing the world; and, above all, the radiant aliveness from the optimism that comes with knowing our species' march through hell has finally ended—CAN NOW TRANSFORM THE WORLD.

957 The relief of being able to leave the issue of our upset state behind as dealt with, and the excitement of knowing a human-condition-free new world is coming and that we can all fully participate in bringing that about, is so absolutely incredibly relieving and exciting it has the power to transform *all* humans. From being a human-condition oppressed and depressed alienated person, all humans can now be TRANSFORMED into redeemed, liberated-from-the-human-condition, exhilarated, ecstatic, enthralled-with-existence, empowered, world-transforming LIFEFORCES. This exhilarated, ecstatic, enthralled-with-existence aspect is the 'Life' in 'Lifeforce', while the empowered, world-transforming aspect is the 'force' in 'Lifeforce', so LIFEFORCE covers both the personal benefit and the benefit to the world in one word. From being human-condition-afflicted, Plato's-cave-dwelling, effectively dead humans, we become Transformed Lifeforces or simply LIFEFORCES! *That* is the difference the arrival of understanding of the human condition makes to the human race!

958 Before recommending the reader watch some short videos in which people describe their transformation to a fabulous life in the new human-condition-understood world, more needs to be explained about this new Transformed Lifeforce Way of Living, especially how it differs in a most important and fundamental way to religious transformations, and to New Age Movement-type transcendences of self.

Chapter 8:5 <u>The Transformed Lifeforce Way of Living is not another religion</u>

959 So the reality is that while current generations of humans are unable to heal the psychological upset within themselves, what *is* possible is to let go of the resigned competitive mindset and take up support of the new Transformed Lifeforce Way of Living. And, as emphasised, humans *are* able to do this, decide to change from living for self to living a more selfless existence—*because we have been doing it throughout our history*. Chapter 7 just described the journey our species has been on as it attempted to find ways of living, and causes to support, that helped to contain and, in some cases, completely transcend our upset, resigned, self-centred, egocentric, selfish way of living. There was the just mentioned <u>self discipline</u>, where we 'civilised' our upset behaviour by superimposing more ideal principles for living upon our selfish and aggressive psychologically upset way of thinking and responding to life;

then there was the imposed discipline of our upset, where we lived by obeying specific laws of behaviour, such as Moses' Ten Commandments; then there was religion, where we deferred to and lived through our support of an exceptional denial-free thinking, sound prophet; and then there was the whole sequence of increasingly guilt-relieving attitudes that we adopted, from socialism/communism, to the New Age Movement, to feminism, to the environment or green movement, to the politically correct movement, to finally the completely guilt-stripped and totally dishonest, 'there-is-no-such-thing-as-truth' postmodern or deconstructionist movement. Of course, the degree to which we abandoned or left behind or let go or repressed or transcended our selfish and aggressive upset, real self and superimposed upon it or adopted a more selfless and cooperative attitude, varied enormously, but within the spectrum there existed the possibility of *completely* abandoning our upset way of behaving. In Christianity for example, this complete so-called 'conversion' from living for yourself to living for Christ is described as being **'born again'** (John 3:3). Similarly, while we humans can't change from being psychologically upset humans—which, as emphasised, is a process that will take a few generations—we *can immediately and completely* change our mind's *attitude* from living a selfish, self-preoccupied life and be, as it were, 'born again' to a consider-the-welfare-of-others-above-your-own-welfare, unconditionally selfless, soulful, pre-resigned-like way of thinking and living. While every resigned human naturally becomes extremely habituated to living for the relief of power, fame, fortune and glory, it *is* possible to completely relinquish that way of thinking and living and, in its place, adopt a completely different, unconditionally selfless way of thinking and behaving.

960 So while it is true that, as just described, the Transformed Way of Living is similar to a religion in that it involves completely letting go of or transcending our real, upset self and deferring to another way of living, that is where the similarity ends. The fundamental difference between the Transformed State and a religious conversion is that the Transformed State *is all about knowledge, not faith or dogma*—knowledge that, after a few generations, has the ability to eliminate all the upset in humans. Yes, as has been emphasised, dogma can't heal upset, only understanding can do that. As was explained in chapters 7:15 and 7:16, while religions *were* an incredibly effective means of containing the upset in humans while the search for understanding of that upset condition was being carried out,

the Transformed Way of Living, and the World Transformation Movement that promotes it, is, in complete contrast, concerned with what happens *after* that liberating understanding is found, which is the progression of the human race from a human-condition-afflicted state to a state completely free of that terrible affliction. The Transformed Lifeforce Way of Living, and the psychological amelioration of the human condition that it allows (given the time, the few generations, needed for it to take place), completely changes the human race from a state of troubled upset to a state of secure soundness. In fact, it metamorphoses—it matures—the human race from insecure adolescence to secure adulthood.

961 Religions were based on deferring to, and living through support of the *embodiment of the ideals* in the form of the soundness and truth of the unresigned, denial-free-thinking-and-behaving prophet around whom the religion was founded, whereas the Transformed Lifeforce State is based on deferring to and living through support of first-principle-based *understandings of the ideals and of our species' unavoidable historical lack of compliance with those ideals*. In fact, the Transformed Way of Living, where humans live in support of the understanding of the human condition that can, after only a few generations, eliminate all the psychological upset in the human race, represents the realisation of religion's hope and faith that that liberating understanding of the human condition would one day appear—it represents the end of faith and dogma and the beginning of knowing. Yes, and I will comment more on this shortly (in par. 1015), this is the end of religion and the beginning of the time of understanding that all the great prophets looked forward to—including Moses, who anticipated a time when we will **'be like God, knowing'** (Gen. 3:5).

962 It needs to be emphasised that while waiting for the understanding of the human condition to heal all the upset psychosis and neurosis in humans, all humans can adopt the Transformed Lifeforce Way of Living that leaves that upset state behind, effectively freeing humanity of its upset *straight away*. Freedom from the human condition is only a decision away now for all humans.

963 Another immense difference between this new Transformed Way of Living and a religion is that in the Transformed Way of Living there is no deity involved, or deference to any one personality; in fact, there is no worship of any kind. And best of all, unlike religion, there is no involvement or emphasis on guilt, because guilt—and the whole notion

of 'good and evil'—has been eliminated forever with the reconciling understanding of the human condition.

964 So while religion and the human-condition-resolved Transformed Lifeforce Way of Living both involve letting go of living through our corrupted self and deferring to something else, that is where the similarity ends. Yes, there is a *world of difference* between the religious state and the Lifeforce state, about which more will be explained shortly.

Chapter 8:6 Nor is the Transformed Lifeforce Way of Living another deluded, false start to a human-condition-free new world

965 And just as the Transformed Lifeforce Way of Living is not another religion, it is also not *another* deluded false start to a human-condition-free world. Again, as just described in par. 959 and at the end of chapter 7, there has been a litany of false starts to an ideally behaved, cooperative, selfless, loving new world for humans, and the reason they were all false starts was because for humans to achieve that cooperative, loving, peaceful state we first had to find the reconciling, healing understanding of our divisive human condition. It is true that the new Transformed Way of Living does involve artificially ridding ourselves of our upset state by simply leaving it behind or 'transcending' it, which is also the way all the false starts to an ideal new world for humans achieved their artificial elimination of upset, *but*, as just explained, the immense difference is that in the Transformed Way of Living such transcendence is only an interim measure—we are only doing so because it will take a few generations to actually rid the human race of upset. The all-important difference is that in the Transformed State we are leaving the old upset world behind because it is finished with—the upsetting search for knowledge, ultimately self-knowledge, understanding of the human condition, is complete. So the Transformed Lifeforce Way of Living is not another *false* start to a new world for humans but the *real* start to that world.

966 What supporters of all the false starts to an ideal new world for humans were doing was 'jumping the gun' on this job, abandoning humanity's critically important, heroic, upsetting search to find knowledge before that search had been completed, which means that what they were doing was *fundamentally irresponsible*. Certainly it could

be immensely relieving giving up the upsetting search for knowledge, but humanity's real freedom from upset depended on that search being carried out until understanding of the human condition was found. To refer again to the Adam Stork analogy, at any time Adam Stork could abandon his upsetting pursuit of knowledge and 'fly back on course' and obey his instinctive orientations and feel good as a result of doing so, but taking that option would mean he would never find the understanding he needed to actually *end* his upset existence once and for all. While giving up the upsetting search for knowledge while it was still to be completed was relieving, it *was* fundamentally irresponsible; however, with the upsetting search for self-knowledge now over it is no longer irresponsible to abandon that upsetting search. In fact, it is now irresponsible to *continue* it. Yes, it is not only legitimate now to leave the old resigned, embattled, competitive, aggressive and selfish egocentric way of living behind in favour of living in support of the new reconciled, human-condition-understood world, it is imperative; it is, in short, the *only responsible* way to behave.

967 Of course, when we humans became overly upset we had to do *something* to avoid behaving so destructively—as explained at length in chapters 7:15 and 7:16, as upset increased we first had to civilise the upset, then try to obey an imposed set of rules, then adopt a religion, and then, when we became so upset we could no longer bear the guilt-emphasising honesty in religion, we had to take up support of one of the pseudo idealistic causes to relieve ourselves of our insecure condition—but the fact remains that all these ways of coping *were* fundamentally irresponsible. As was also explained in chapters 7:15 and 7:16, while religions were by far the least irresponsible way of abandoning our upset, because the search for knowledge continued indirectly through our support of the honesty of the prophet around whom our religion was founded, it was still fundamentally an irresponsible pseudo idealistic strategy. We were still personally abandoning the upsetting search for knowledge.

968 So a very big difference with the new Transformed Way of Living is that it is not irresponsible at all; in fact, it is totally responsible and the only legitimate way to behave because the battle to find understanding has been won, so everyone *should* stop fighting, leave the battlefield, and, as they do at the end of a game of football, 'go to the showers and clean themselves up'!

969 Yes, this transformation from a human-condition-afflicted existence to an almost unbelievably wonderful and exciting, exhilarated and empowered Transformed Lifeforce State is not achieved by abandoning humanity's critical battle to find knowledge, ultimately self-knowledge, understanding of the human condition. In particular, it is not another fundamentally irresponsible 'New Age' or, more recently, 'A New Earth', 'alternative' movement where 'spiritual gurus' advocate we transcend our embattled conscious thinking egoic self when that battle hadn't been won. Nor is it another of those fundamentally irresponsible, superficial and ultimately ineffectual 'think positive', 'human potential', 'self development', 'self improvement', motivational programs where we tried to artificially defy the reality of the human condition by immersing ourselves in positive thoughts. Nor is our freedom from the agony of the human condition being achieved through the fundamentally irresponsible practice of deep, meditative extinction of our human-condition-distressed thinking mind as some religious practices taught. And nor does the new Transformed Lifeforce Way of Living involve the fundamentally irresponsible way of escaping the real issue before us as a species of our deeply troubled selves by adopting a focus-away-from-yourself, guilt-free, feel-good, pseudo idealistic cause like environmentalism or feminism or aboriginalism or multiculturalism. *And* our species' freedom is *certainly* not being achieved in the fundamentally irresponsible way of dogmatically imposing a deconstructed, good-and-evil-differentiation-free, politically-correct-but-human-reality-dishonest, ideal world, as the postmodern movement and, before it, the socialist and communist movements tried to do. No, the transformation is achieved through what is ultimately the only real and lasting and fundamentally responsible way it could be—through satisfying our conscious thinking mind with first-principle-based, biological understanding of *why* we humans are wholly worthwhile and meaningful beings.

970 There *had* to be a biological explanation for our species' non-ideal divisive, competitive, aggressive, angry, even-brutal-and-mean, selfish, self-obsessed, indifferent-to-others, arrogant, egocentric, deluded, defensive, escapist, superficial, artificial, alienated, despairing, lonely, depressed lives, and our fundamental responsibility as conscious beings was to find that ameliorating explanation. So while it *could* be immensely relieving giving up the all-important, upsetting search for knowledge, our fundamental responsibility as conscious beings was *not* to give up

that search. To include more of what Bronowski said in his concluding statement to his 1973 television series and book of the same name, *The Ascent of Man*: **'I am infinitely saddened to find myself suddenly surrounded in the west by a sense of terrible loss of nerve, a retreat from knowledge into – into what? Into...falsely profound questions about, Are we not really just animals at bottom; into extra-sensory perception and mystery. They do not lie along the line of what we are now able to know if we devote ourselves to it: an understanding of man himself. We are nature's unique experiment to make the rational intelligence prove itself sounder than the reflex** [instinct]**. Knowledge is our destiny. Self-knowledge, at last bringing together the experience of the arts and the explanations of science, waits ahead of us.'**

971 There certainly has been a **'terrible loss of nerve'**, a loss of commitment to humanity's great struggle to find **'self-knowledge'**. Irresponsible transcendence of the issue of self, thought repression, enforced dogma and escapism were all fundamentally a **'retreat from knowledge'**. But de-braining ourselves was never going to work; ultimately, we humans needed brain food not brain anaesthetic; we needed knowledge—specifically the dignifying, uplifting, healing, ameliorating, relieving, peace-bringing understanding of the human condition. Anything else *was* an abrogation of the fundamental responsibility that came with our greatest capacity and nature's greatest invention: our species' fully conscious, thinking, self-managing, self-adjusting mind. And, thankfully, that fabulous destiny and potential to progress from mere abstract, **'art**[istic]**'** *description* of the agonising, good-and-evil-afflicted dilemma of our human situation to reconciling, first-principle-based, scientific *understanding* of that dilemma and resulting amelioration, integration and unification of ourselves and of our species has finally arrived.

972 It should be mentioned here that all the deluded false starts to a new human-condition-free world have *so* severely discredited the possibility of the emergence of such a new world that it has made it easy for the unscrupulous to attempt to brand and dismiss our World Transformation Movement, which is promoting the *real*, human-condition-solved, human-race-saving transformation of humans, as just another of these deluded movements. Such an attack *is* unscrupulous because the transformation is based on assessable, understandable, first-principle-based biological explanation of the human condition that makes the practice of transcending our upset state not only legitimate but the only practice to adopt. Of course, the other evidence that this is not another deluded

dream of utopia is that, far from making people superficially feel good, these human-race-liberating understandings are *extremely* exposing, confronting and unsettling. As Christ warned, **'false prophets will appear and perform great signs and miracles to deceive even the elect—if that were possible...Wherever there is a carcass** [the extremely upset]**, there the vultures** [false prophet promoters of 'feel-good' delusion, escapism and denial] **will gather'**, whereas the real, human-condition-exposing liberation of the human race will come like terrifying rolling thunder: **'as the lightning comes from the east and flashes to the west...**[it will be very frightening as if] **the sun will be darkened...the stars will fall from the sky, and the heavenly bodies will be shaken'** (Matt. 24). Yes, the real love of our dark side is *nothing* like the superficial, artificial, trivial, transitory, sickly, weak 'do good to feel good' form of 'love' that ultimately only buried humans deeper in alienation. Again, the history of all the patently dishonest and thus cowardly attacks on our world-saving World Transformation Movement can be read about at <www.worldtransformation.com/persecution>.

Chapter 8:7 <u>But how does the Transformed Way of Living solve the problem of the unbearable, 'judgment day' exposure that understanding of the human condition unavoidably brings?</u>

973 What is yet to be explained is how are humans going to cope with having all our upsets that we have been living in denial of suddenly exposed; how are we going to manage the arrival of the liberating but at the same time all-exposing **'day of judgment'**, **'the Day of Reckoning'** when **'**[Our] **nakedness will be exposed'** and **'not one secret of** [ours will be] **concealed'**?

974 The simple answer to this question of how we cope is that we don't try to confront all the truth about our corrupted condition. Once we have investigated these understandings sufficiently to know that they have explained the human condition, we don't actually need to know any more than that to take up a life where we direct all our thoughts and energies into supporting these understandings, and to the repair of the world that they finally make possible. We don't need to know the full extent of the truth that this information reveals about the upset state of humanity; and we particularly don't need to know how it explains and reveals everything about our own human-condition-afflicted life.

975 Yes, we leave our upsets behind as dealt with for *two* reasons: firstly, for the reason already given, that healing all our upset would require a great deal of time and supportive counselling that simply isn't possible at the moment; and, secondly, because trying to face all the truth that is now revealed about our upset condition could be too confronting, exposing and depressing, and thus dangerously destabilising, an exercise.

976 The danger is that if we study this information beyond what our particular level of soundness and security of self can cope with we risk becoming overly confronted by the extent of our corrupted condition and dangerously depressed. As has been pointed out, the human race has coped this far by maintaining extreme levels of denial of many, many truths, so obviously we can't hope to dismantle all those denials overnight. Humans typically take months, if not years, to adjust to big changes in our lives, so in this *massive* paradigm shift, which takes us from a position where we didn't face *any* truth about our corrupted condition to facing the prospect of having *all* the truth about it exposed, we have to accept that it will be a process that will take generations, but that doesn't mean we can't support the truth while this adaption, and the digestion and healing that goes with it, takes place. We simply can't hope to, and mustn't, overly confront the truth while this long process of adjustment takes place, and, according to each person's level of upset, there will be a limit to how much truth each person can cope with; there will be a limit to how much they can listen to, read about and study these human-condition-confronting understandings. But, again, that doesn't mean that everyone can't *immediately* access and live in support of these human-race-liberating understandings.

977 Regarding the degree to which we should each investigate these explanations, obviously it is necessary to sufficiently verify to our own satisfaction that they are the liberating understandings of the human condition that the whole human race has been tirelessly working its way towards for some 2 million years—but we shouldn't risk investigating them to the extent that we start to become *overly* exposed and confronted by the truths they reveal. Having lived without any real understanding of human life it is natural to want to keep studying these explanations that finally make sense of the world, both within and around us, but, again, such analysis can lead to becoming overly confronted and depressed by the extent of our own corrupted state, and that of our world. The more intelligent and/or more educated in the human-condition-avoiding,

denial-based, mechanistic, reductionist paradigm, who pride themselves on being able to think and study and grasp new ideas, will initially be especially tempted to study these understandings beyond what their varying levels of security of self can cope with, but it won't be long before everyone learns that such an approach is both psychologically dangerous and irresponsible and unnecessary.

978 When Christ spoke of a time when **'the meek...inherit the earth'** (Matt. 5:5), and when **'many who are first will be last, and many who are last will be first'** (Matt. 19:30, 20:16; Mark 10:31; Luke 13:30), he was anticipating this time when understanding of the human condition would arrive and instead of the more intelligent and intellectual leading the way, as has been the case in almost every human situation to date, the more innocent and sound, the more soulful and instinctual and functional, the less upset or corrupted or dysfunctional, will do so. As the story of Adam Stork reveals, throughout the 2-million-year battle to find understanding, our instinctive self or soul was repressed because of its unjust condemnation of our intellect, but when understanding of the human condition is finally found this process is reversed, soul becomes sought-after. Our innocent, upset-free, original instinctive self or soul—soundness—has to lead us back home to soundness. It makes sense. Again, Christ gave the perfect description of this new situation when he said, **'The stone the builders rejected has become the capstone'** (Ps. 118:22; Matt. 21:42; Mark 12:10; Luke 20:17; Acts 4:11; 1 Pet. 2:7). Sir Laurens van der Post referred to this biblical analogy when he too anticipated this new situation: **'It is part of the great secret which Christ tried to pass on to us when He spoke of the "stone which the builders rejected" becoming the cornerstone of the building to come. The cornerstone of this new building of a war-less, non-racial world, too, I believe, must be...those** [more innocent, instinctual] **aspects of life which we have despised and rejected for so long'** (*The Dark Eye in Africa*, 1955, p.155 of 159). In his 1964 song *The Times They Are A-Changin'*, Bob Dylan, another prophet of our time, also anticipated the paradigm shift that the arrival of understanding of the human condition brings: **'Come gather round people wherever you roam / And admit that the waters around you have grown / And accept it that soon you'll be drenched to the bone** [the end play state of terminal alienation is upon us]**...keep your eyes wide, the chance won't come again** [for the denial-free understanding of the human condition to be found]**...the loser now will be later to win / For the times they are a-changin'...Don't stand in the doorway, don't block up the hall / For he that gets hurt will be he who has stalled / The battle outside ragin' /**

Will soon shake your windows and rattle your walls [the liberating, but also exposing, truth about humans is on its way] / **For the times they are a-changin' // Come mothers and fathers throughout the land / And don't criticise what you can't understand / Your sons and your daughters are beyond your command / Your old road is rapidly agein' / Please get out of the new one if you can't lend your hand / Oh the times they are a-changin' // The line it is drawn, the curse it is cast** [the exposing truth about humans is out] / **The slow one now will later be fast / As the present now will later be past / The order is rapidly fadin' / And the first one now will later be last / For the times they are a-changin'.'**

979 So as with having to abandon our obsoleted upset, resigned, competitive, power-fame-fortune-and-glory way of living, investigating the truth only to the degree we are sound enough to do so will initially be a difficult proposition for the denial-based world to accept, but it too is a proposition that can be understood and accepted as reasonable. It is the more secure in self, the least alienated, who have to develop these understandings of the human condition. If you are not sound enough to study these ideas to any great depth and you try to do so, you will end up in a state of fearful depression, or, worse still, mad. This means that it can be particularly dangerous for people with a history of psychological instability to study these understandings. Furthermore, if you *do* become overly confronted by what is being presented your natural reaction could be to try to attack and deny it in order to protect yourself—in effect, reinstate all your denials and those of others, which means advocating humanity's retreat *back into* Plato's dark cave of denial that the human race has been desperately trying to get out of for 2 million years! Yes, if you do become overly confronted, you could become, as Dylan warned, **'hurt'** and **'stalled'**, defensive, angry and retaliatory toward the information, and the consequence of such a response could be to, in effect, sabotage the efforts of every human who has ever lived to bring the human race to this dreamed-of moment of its liberation. As has been mentioned, we in the World Transformation Movement have endured years and years of this furiously angry, defensive reaction towards this information, attacks that were ultimately fruitless because this information is true and it won't be intimidated or oppressed: it is too precious to allow that. In short, the effect of overly studying this information, studying it more than your degree of security of self can cope with, can be both dangerous to you and dangerous to the human race, and no one should want, nor risk, either of those outcomes.

980 So while initially it will be difficult accepting this advice to only investigate all this amazing truth that is now available about humans to the degree that each person is sound enough to do so, with honesty it *can* be appreciated that it is a reasonable and necessary proposition. For instance, in the old human-condition-avoiding, denial-based world, academia limited those who could be involved in the pursuit of knowledge to the more intelligent, those with a high IQ (intelligence quotient). To enter university you had to pass entrance exams that assessed your IQ, your suitability to study complex subjects like higher mathematics and physics, because if you didn't have an adequate IQ you would make little progress in studying such subjects. Well, in the new human-condition-resolved, human-condition-confronting world we similarly need the most appropriate people to study its information, which are those with a high SQ (soul or soundness quotient). If you don't have an adequate SQ you simply won't be able to make any progress with the information involved. But, importantly, with the explanation of the human condition we can now appreciate that while everyone is necessarily variously upset/unsound/alienated, that upset/unsoundness/alienation is not something bad, just as in the old denial-based world those who lacked IQ weren't considered bad people, just not as able to think as effectively about certain complex subjects. As has been emphasised throughout this book, upset is not an evil, sinful, bad state—it is a heroic, good state, because it is an unavoidable product of humanity's heroic, all-important search for knowledge.

981 The overall situation then is that everyone has to measure and limit how much they can study these human-condition-confronting understandings against how much self-confrontation they can cope with. If you start to feel unbearably confronted then curtail your study of the information. Again, the human race cannot have lived in the darkness of both ignorance and denial for some 2 million years and then suddenly be able to face all the truth about ourselves individually; it is simply not a reasonable expectation. Everyone *has to* expect that there will be a limit to how deeply they can afford to look into this information. But thankfully, and most importantly, no one has to overly confront their old, upset self—everyone *can* leave that behind as dealt with and simply live for the new world and all its potential. As emphasised, although existing generations who grew up without these understandings will still suffer the effects of the human condition, they can still live effectively free of

the human condition because they no longer have to live preoccupied with it. Future generations are the ones who will be in the best position to further develop these understandings of the human condition because among current generations there are naturally very, very few individuals who remain secure enough in self to confront, study and develop them. The main point to remember is that once you know that this information has explained the human condition then you know all the upset in the world and within you is now explained and defended—which means you *can*, as described earlier, put the issue of all your upsets or corruptions in a 'suitcase', attach a label to it saying 'Everything in here is now explained and defended', and simply leave it behind as dealt with as you set out completely free and unencumbered into the new, human-condition-liberated world. You can join the Sunshine Army on the Sunshine Highway to the World In Sunshine.

982 Once you know this information is true, that the upset state of the human condition is explained and defended at the fundamental level, you can leave the issue of your own and the world's corruption behind as effectively dealt with and focus all your attention on developing a human-condition-free new world. You can preoccupy yourself with disseminating this information throughout the world and to a fresh generation, and with supporting all the projects that must now be undertaken to rehabilitate the world from the destructive effects of 2 million years of living under the duress of the human condition. In fact, it shouldn't even be necessary to talk about the whole issue of the human condition any more than what appears in this presentation and in the presentations on our World Transformation Movement website. Humanity moves on to an entirely new paradigm of reality now. We get the truth up, and we move on.

983 Yes, all that matters now is that the truth is kept alive and that it is disseminated to the world's population, because it alone can heal the human race and save the world. All everyone should do now is support the truth about the human condition and it will achieve everything everyone has ever dreamt of. If we look after this information it, in turn, will look after the world. That is the mantra of the new world that understanding of the human condition brings about.

984 We, the existing generations, are now *the conduit generations*, the generations who will connect the old dead world to the new living world. Having grown up 'in the dark' without understanding of the human

condition, our priority is to shine the light on these insights to ensure fresh generations won't have to adopt all the artificial ways of coping with the human condition—all the dishonest denials, all the egocentric means of seeking compensation for the injustice of being unfairly condemned, and all the angry retaliations needed against the unjust condemnation—that previous and existing generations have had to resort to. If children and pre-resigned adolescents can access these understandings they won't have to resign themselves to a life of living in denial of the human condition as their only means of coping with it—and not having to grow up with all the dishonesties, denials, delusions, artificialities and superficialities that current generations have had to endure, they will be like a new species of beings. They will have an inner freedom that is almost unimaginable to those who grew up without understanding of the human condition. As described in pars 48-52, Holden Caulfield, the central character in J.D. Salinger's seminal book, *The Catcher in the Rye*, said that **'the only thing that I'd really like to be'** is a **'catcher in the rye'** saving **'thousands of little kids'** from going **'over the cliff'** of Resignation. Well, we can *all* start living Holden's dream now; we can all become **'catcher**[s] **in the rye'**!

[985] Yes, finding understanding of the human condition is the dreamed-of breakthrough that can bring about the complete transformation—in fact, TRANSFIGURATION—of the entire human race and thus our world. The word transfiguration means **'a change that glorifies or exalts'**, **'a marked change in form or appearance; a metamorphosis'** (The Free Dictionary, accessed Dec. 2013; see <www.wtmsources.com/140>), so is a perfect description for what happens to humans when, through the application of dignifying, uplifting, relieving, ameliorating, reconciling, redeeming, healing and curing understanding, we are finally able to not only end our old egocentric, must-prove-that-we-are-not-bad, insecure, human-condition-afflicted existence, but enter an incredibly exciting state of glorious freedom, optimism and human-condition-liberated, empowered-to-repair-the-world capability.

[986] I should make the point here that evidence that the human race can survive the exposure of 'judgment day'—and not merely survive it but live with it in an absolutely wonderfully joyous and exciting way—is provided by the affirmations that have been given by people living in the Transformed Lifeforce State (which can be viewed at <www.worldtransformation.com/affirmations>), and by the existence of the many other people who have been aware of and have lived happily with these

understandings for decades. While a few people have become psychologically destabilised by these understandings, that is to be expected given the enormity and the novelty of the change that these understandings bring, and that, initially, there was always going to be a few people who would not heed the advice to study these understandings *only* to the degree that they could personally cope with them, and to stop if they started to become overly confronted and destabilised—advice that especially applies to people with a history of psychological instability. Of course, the ultimate evidence that it is possible for the human race to survive 'judgment day' is inherent in the fact that the human race would never have struggled so determinedly all these thousands of generations to find understanding of the human condition if we didn't intuitively know that we could free ourselves from the human condition when that understanding finally arrived.

Chapter 8:8 Plato's anticipation of the initial 'violently opposed', treated-as-'blasphemies' stage that the arrival of the understanding of the human condition has had to withstand

987 Before elaborating upon how understanding of the human condition has the power to transform the human race, more should be said about the initial overwhelming period of shock and bewilderment that is brought about by the arrival of the terrifying, rolling-thunder-like, stars-falling-from-the-sky, body-shaking truth about our condition.

988 Any meritorious new idea in science has typically gone through stages of resistance and even persecution before gaining universal acceptance—a process the philosopher Arthur Schopenhauer articulated when he **'said that the reception of any successful new scientific hypothesis goes through predictable phases before being accepted'**. First, **'it is ridiculed'** and **'violently opposed'**. Second, after support begins to accumulate, **'it is stated that it may be true but it's not particularly relevant'**. Third, **'after it has clearly influenced the field** [including members of the establishment quickly remodelling/plagiarising the ideas as their own discoveries, something we in the World Transformation Movement have, unfortunately, experienced] **it is admitted to be true and relevant but the same critics assert that the idea is not original'**. Finally, **'it is accepted as being self-evident'** (compiled from two references to Schopenhauer's quote—*New Scientist*, 15 Nov. 1984 and *PlanetHood*, B. Ferencz &

K. Keyes, 1988). (Note that each stage of recognition is achieved in a way that protects the ego of the onlookers. The extent of the insecurity caused by the human condition is very apparent.) The famous playwright George Bernard Shaw was another who warned of the true nature of progress when he wrote that **'All great truths begin as blasphemies'** (*Annajanska*, 1919). (The reason great truths so often began as **'blasphemies'** is because the search for truth was very often not a case of finding it but of uncovering it from years and years of denial—an exposure that seemed highly *blasphemous* to all those who had been practising the denial. We have seen how finding the great truths presented in this book of the explanation of the human condition, of the meaning of existence, and of the origin of our moral soul and conscious mind, all required the defiance of very entrenched denials, which means all these great truths *are* going to **'begin as blasphemies'**.)

989 But given the arrival of understanding of the human condition represents a *massive* paradigm shift for humans—especially a shift from denial to honesty—it was always an inevitability that this initial **'violently opposed'**, treated-as-**'blasphemies'** stage would have to be negotiated. Even though the Transformed Lifeforce Way of Living is underpinned by the compassionate understanding of the human condition, there were always going to be some who would try to remain hidden in Plato's dark cave of denial. Indeed, Plato has given us the best account of the nature of this initial **'violently opposed'**, treated-as-**'blasphemies'** stage that the arrival of understanding of the human condition would have to endure—which is not surprising in light of how extraordinarily honest and thus penetrating Plato's thinking was (recall in par. 59 how the acclaimed philosopher Alfred Whitehead described the history of philosophy as merely **'a series of footnotes to Plato'**). Since it will help the reader both reconnect with and appreciate the fact that the human condition does exist, that it can be solved, and that there are stages of acceptance that its explanation has to progress through, I have included the following summary of Plato's amazingly insightful description as it evidences each point. (And, impressively, this clear thinking was all done long, long ago, approximately 360 years before Christ and some 2,370 years before now, and the insightfulness is even more impressive when we consider how few people there were in the world when Plato lived compared to the billions now—yes, how asleep/alienated has the human race become since Plato's time!)

[990] As was briefly mentioned in par. 59, Plato wrote: **'I want you to go on to picture the enlightenment or ignorance of our human conditions somewhat as follows. Imagine an underground chamber, like a cave with an entrance open to the daylight and running a long way underground. In this chamber are men who have been prisoners there'** (c.360 BC; *Plato: The Republic*; tr. H.D.P. Lee, 1955, p.278 of 405). In this cave allegory Plato went on to describe how the cave's exit is blocked by a fire, such that if one of the prisoners were **'to stand up and turn his head and look and walk towards the fire; all these actions would be painful...he would** [have to] **turn back and take refuge'** in the cave of **'shadows'**, which are only an **'illusion'** of the real world outside the cave (pp.279-280). The allegory makes clear that while **'the light of the fire in the cave prison corresponds to the power of the sun'** (p.282), with **'the sun...making things we see visible'** (p.273), such that without it we can only **'see dimly and appear to be almost blind'** (p.272), having to hide in the **'cave'** of **'illusion'** and endure **'almost blind'** alienation was infinitely preferable to facing the **'painful'** light of the **'fire'**/**'sun'** that would make **'visible'** the unbearably depressing issue of **'the imperfections of human life'** (p.282).

[991] Having described how living in a cave-like state of **'almost blind'** alienated denial has, tragically, been absolutely necessary, Plato wrote that it was ultimately only by being **'illuminated by truth and reality'** (p.273) that **'the enlightenment...of our human conditions'** could be achieved and the cave prisoners be **'released from their bonds and cured of their delusions'** (p.279). He then described the initial resistance the cave prisoners would have to being **'released from their bonds and cured of their delusions'** when someone achieved **'the enlightenment...of our human conditions'** through being **'illuminated by truth and reality'**. To quote from a summary of the cave allegory that appeared in the *Encarta Encyclopedia*'s entry for 'Plato': **'Breaking free, one of the individuals escapes from the cave into the light of day. With the aid of the sun** [necessarily living free of denial of the glaring, burning, searing truth of Integrative Meaning and the issue it raises of the human condition—and assisted by the understandings that science has found of the differences in the way genes and nerves process information], **that person sees for the first time the real world and returns to the cave with the message that the only things they have seen heretofore are shadows and appearances and that the real world awaits them if they are willing to struggle free of their bonds. The shadowy environment of the cave symbolizes for Plato the physical world of appearances. Escape into the sun-filled setting outside the cave symbolizes the transition to the real world, the world of full and perfect**

being, the world of Forms, which is the proper object of knowledge' (written by Prof. Robert M. Baird, accessed Jul. 2008; see <www.wtmsources.com/101>). To return to *The Republic* and Plato's own words: **'if he** [the cave prisoner] **were made to look directly at the light of the fire** [again the fire represents the searing truth of Integrative Meaning and the issue of the human condition it raises], **it would hurt his eyes and he would turn back and take refuge in the things which he could see** [take refuge in all the denials and dishonest explanations and arguments that he has become attached and accustomed to], **which he would think really far clearer than the things being shown him. And if he were forcibly dragged up the steep and rocky ascent** [out of the cave of denial by the person who has broken free of the cave] **and not let go till he had been dragged out into the sunlight** [shown the truthful all-liberating—but at the same time all-exposing and confronting—explanation of the human condition], **the process would be a painful one, to which he would much object, and when he emerged into the light his eyes would be so overwhelmed by the brightness of it that he wouldn't be able to see a single one of the things he was now told were real** [this inability to absorb discussion of the human condition because it has been such an unbearable, off-limits subject for so long is what we in the WTM regularly encounter and refer to as the 'deaf effect'—as one reader of my books admitted, **'When I first read this material all I saw were a lot of black marks on white paper.'** The deaf effect was described in chapter 1:4, and you can read more about it at <www.worldtransformation.com/freedom-deaf-effect>]' (p.280). Plato continued: **'they would say that his visit to the upper world had ruined his sight** [they would treat the person who tries to deliver understanding of the human condition as if he was mad, which is how I have been treated for many years, as I talk more about below], **and** [they would say] **that the ascent** [out of the cave] **was not worth even attempting** [such assertions that it's not worth attempting have been regularly made against our work, with one of the architects of the mid-1990s public campaign of persecution against myself and the WTM saying, **'You know you are encroaching on the personal unspeakable inside people and you won't succeed'** (WTM records, 12 Feb. 1995), and the other architect of the attacks saying, **'You realise you are attempting the impossible, you will be fighting to have this material accepted right down to the last person on the planet'** (WTM records, 16 Feb. 1995)]' (p.281). Plato continued: **'And if anyone tried to release them and lead them up, they would kill him if they could lay hands on him'** (ibid). In fact, my persecutors did everything they could, short of physical attack, to **'kill'** me. (As I mentioned in par. 441, I ended up suffering from the extremely debilitating Chronic Fatigue

Syndrome from 1999 to 2009, which were 10 of what should have been the most productive years of my life, as a result of having to endlessly fight their attacks, and the WTM was so marginalised that there was virtually no one who would have anything to do with us or our project.) What they did that was *so* destructive of myself and the WTM was base their whole ferocious and malicious media fear campaign against me and our work at the WTM on the accusation that I am a deluded, megalomanic leader of a dangerous anti-social sect. But how could I possibly be so unsound as to be a deluded, megalomanic leader of a dangerous anti-social organisation when, as even the statements made by my attackers mentioned above acknowledge, I have been sound enough to look into the human condition? Quite simply, as Christ pointed out when he was similarly accused of being **'possessed by…the prince of demons'** (Mark 3:22) for his honest, human condition confronting and penetrating words, **'How can Satan drive out Satan?'** (3:23). Christ was making the same point when he said that **'A good tree cannot bear bad fruit, and a bad tree cannot bear good fruit'** (Matt. 7:18). The real motivation for the whole vicious media campaign was clearly prejudice against addressing the issue of the human condition.

992 Some idea of the prescience of Plato's anticipation of the *ferocity* of the attack that would be made on the person who **'escapes from the cave into the light of day'** and then **'dragged** [humanity] **out into the sunlight'** can be gained by the first ruling that the court made *against* my work in the defamation action we took to redress the horror media campaign. As was mentioned in par. 441, it ruled that my work is of **'such a poor standard that it has no support at all from the scientific community'**! It was certainly an immense relief then when, in 2010, the three judges in the New South Wales Court of Appeal unanimously recognised the real problem was that the primary judge clearly found my work unbearably confronting, ruling that he did **'not adequately consider' 'the nature and scale of its subject matter'**, in particular **'that the work was a grand narrative explanation from a holistic approach, involving teleological elements'**, and that other important submissions **'were not adequately considered by the primary judge'**, including that the work can make **'those who take the trouble to grapple with it uncomfortable'** because it **'involves reflections on subject-matter including the purpose of human existence which may, of its nature, cause an adverse reaction as it touches upon issues which some would regard as threatening to their ideals, values or even world views'**! Yes, the problem was *not* that I am a deluded monster, but that I am daring to address the human condition!

993 Since the subject of the human condition is both extremely novel and extremely confronting, the initial resistance stage that Schopenhauer said new ideas in science are typically subjected to—and which Plato anticipated the arrival of understanding of the human condition would be *particularly* subjected to—was always going to be exceptional, and it certainly has been. On this journey, however, we in the WTM believe that after having successfully fought the terrible public campaign of persecution against us in the law courts, which is the proper, civilised place to seek redress, that the initial **'violently opposed'**, treated-as-**'blasphemies'** stage is over and we can now move on to the next stage in the journey to acceptance where there will be scepticism as these ideas are evaluated for their accountability and thus truth. However, since these explanations of the human condition are *so* accountable and thus truthful, this last stage should be over as soon as these explanations are properly considered by the scientific establishment.

994 I should say that we have fought so determinedly against persecution because we have been well aware of what the philosopher John Stuart Mill wrote in his famous 1859 essay *On Liberty*, that **'the dictum that truth always triumphs over persecution, is one of those pleasant falsehoods which men repeat after one another till they pass into commonplaces, but which all experience refutes. History teems with instances of truth put down by persecution. If not suppressed for ever, it may be thrown back for centuries'** (ch.2). And in the case of these biological explanations of human nature, such **'persecution'** has already **'thrown back'** its recognition for 30 years! The science historian Thomas Kuhn similarly warned that the only guarantee truth will survive prejudice is the determination of its advocates, writing that **'In science...ideas do not change simply because new facts win out over outmoded ones...Since the facts can't speak for themselves, it is their human advocates who win or lose the day'** (Shirley Strum, *Almost Human*, 1987, p.164 of 297—Strum's references are to Thomas Kuhn's *The Structure of Scientific Revolutions*, 2nd edn, 1970). Interestingly, Kuhn also recognised that in science **'revolutions are often initiated by an outsider—someone not locked into the current model, which hampers vision almost as much as blinders would'** (ibid). Even Charles Darwin was **'a lone genius, working from his country home without any official academic position'** (Geoffrey Miller, *The Mating Mind*, 2000, p.33 of 538). While there are certainly advantages to not being **'hamper**[ed]**'** by **'the current model'**, the inherent danger of not being part of the establishment is that the **'outsider'** is an easy, undefended target for those in the establishment

who feel threatened by their new ideas. Again, the documentation of the 30 years of terrible persecution that we were subjected to can be read at <www.worldtransformation.com/persecution>.

[995] Incidentally, the extract from *The Republic* shows how extremely truthful and accurate Plato was in describing the human condition and its eventual resolution. Plato was certainly one of the soundest men in recorded history. Indeed, in *Freedom Book 1* I explain that Plato, along with the other two very great denial-free thinking prophets in history, namely Moses and Christ, made the most important contributions to humanity's great journey to enlightenment; specifically, that through the Ten Commandments Moses gave humanity the most effective form of Imposed Discipline for containing the ever-increasing levels of upset in humans; that Christ gave humanity the soundest and thus most effective corruption-and-denial-countering religion; and that Plato gave philosophy—the actual business of studying **'the truths underlying all reality'** (*Macquarie Dictionary*, 3rd edn, 1998), in particular studying and finding the all-important understanding of the human condition—the best possible orientation and assistance. So, we could say that 'the beauty and taste of roses, rice and potato saved the human race'!! And I probably should add 'a leg of lamb'—the prophet Abraham—to that world-saving feast because Abraham contributed the precious foundations of real, effective religion with his emphasis on monotheism, the need to revere the fundamental truth about our existence of there being one true 'God', which we can now acknowledge is Integrative Meaning. Indeed, in what has to be the most truth-filled and thus greatest poem ever written (extraordinarily insightful extracts from it are referred to throughout this book), *Intimations of Immortality*, the poet William Wordsworth probably best described the critical role denial-free thinking prophets have played in leading humanity by terming them the **'Eye**[s] **among the blind'**: **'Thou best Philosopher, who yet dost keep / Thy heritage, thou Eye among the blind / That, deaf and silent, read'st the eternal deep / Haunted for ever by the eternal mind / Mighty Prophet! Seer blest! / On whom those truths do rest / Which we are toiling all our lives to find / In darkness lost, the darkness of the grave'** (1807). The **'darkness lost, the darkness of the grave'** where no truth is accessible equates perfectly with Plato's cave of alienating denial where virtually the entire human race has had to live incarcerated against the glaring truth of our condition.

[996] So the options now for the human race are to continue living in the psychologically dead state of denial when it's no longer necessary—

an option that will lead to our species' extinction—or to take up the Transformed Lifeforce Way of Living and bring an end to our tortured, human-condition-afflicted existence. Of those two options, of either terminal alienation or transformation, the human race obviously *must* choose the latter. In the words of the *Encarta* summary of Plato's cave allegory, **'the proper object of knowledge'** is to achieve **'the transition to the real world'**—that is, end the alienated state of denial that humans have had to live in, get out of our species' horrible **'darkness of the grave'**, **'cave'** existence. Yes, the **'enlightenment'** of our **'imperfect'** **'human condition'** enables us to be **'released from'** the **'bonds'** of our **'cave'**-**'like'** **'prison'** of **'almost blind'** alienated denial and, as a result, be **'cured of'** our **'illusion'** and **'delusions'**—and obviously that is the path we *must* take. I should reiterate that Plato also emphasised how relieved the cave prisoner would be to be free of his old dishonest existence by saying, **'when he thought of his first home and what passed for wisdom there and of his fellow-prisoners, don't you think he would congratulate himself on his good fortune and be sorry for them?'** (*Plato: The Republic*, p.280). BUT, again, to achieve this wonderful **'transition to the real world'** the scientific establishment first has to fulfil its role of recognising this fully accountable biological explanation of the human condition, because that is the guidance the wider public needs for it to give serious consideration to the explanation. (I might mention that I have written more about Plato's extraordinarily insightful thinking at <www.worldtransformation.com/plato>.)

Chapter 8:9 <u>The utter magnificence of the Transformed Lifeforce State</u>

997 To reveal just how magnificent the Transformed Lifeforce Way of Living is, a further contrast can be made between it and Christianity—remembering that Christianity, although human-condition-*relieving*, was *not* human-condition-*resolving*.

998 It was St Paul who, more than anyone else, introduced the benefits of Christianity to the world, so consideration of his discovery of the beauty of Christianity, his conversion to the faith and his life as a Christian is especially indicative of what the new Transformed Lifeforce Way of Living offers the human race. Acts 8 and 9 of the Bible provide the account of Saul's (as St Paul was known before his conversion) journey to Damascus where he planned to persuade the authorities to

destroy the fledgling group of Christians, such was his fury towards the truth about the human condition that Christ, through his sound words and life, had dared to reveal and which this small group were living in courageous support of. However, while riding along on his donkey in a seething, **'murderous'** (Acts 9:1) rage towards Christ and his followers, Saul had an epiphany, the effect of which was so incredible that it was to change the world and basically save the human race from unbearable levels of upset.

999 With an appreciation now of what the human condition is, we can explain and thus understand what happened to Saul. The human condition is a terrifyingly confronting subject and, to cope with that terror, all that the upset human race has been able to do is live in denial of it, try to block out the whole issue of the imperfection in our lives. We humans attacked, denied and attempted to prove wrong any exposing criticism of our corrupted state. This response relieved our condition but there was, of course, a significant downside—a loathsome life of extreme anger, alienation and egocentricity. So in one sense our retaliation against the threat of exposure of our corrupted condition made us feel even *more* condemned. Defensive retaliation as a strategy for coping with the human condition, while relieving, also fuelled feelings of self-loathing and unworthiness. That was Saul's predicament—his anger was very great but so was the level of self-hatred he would have felt deep within himself for being such a brutally angry person. The whole issue of Christ's soundness that was making him *so* angry was also, at another level, making him extremely distressed and unhappy with himself. We can imagine, therefore, that, in a moment of full engagement in his mind with the horror of both sides of his situation, a thought would have occurred to him: *'If this man, Jesus, is so truthful and sound that he is provoking such anger in me, what would happen if I was to flip the situation around and instead of attacking him I supported him? Wouldn't I suddenly then be a force for enormous good—because I would be supporting someone who is the absolute opposite of my immensely corrupted and angry-with-soundness, truth-hating-and-denying self. I would be a force for good in the world, instead of a self-loathing monster. Wow, that would certainly turn my life around, wouldn't it!'* And, at that moment, as the metaphorical account goes, Saul fell off his donkey and was struck blind for three days; in more direct terms, he was completely overwhelmed by the sudden freedom he was feeling from all the pain

of his human-condition-afflicted life. Through his support of Christ, the agonising weight of the human condition had been lifted from his shoulders and he was able to live again. By siding with Christ, he was able to resurrect the truthful, soulful side of himself; he had been **'born again'** (John 3:3); **'he has crossed over from death to life'** (John 5:24). As Christ authoritatively said, **'I am the resurrection and the life** [through me, your soulful true self can live again]**'** (John 11:25).

1000 One way of measuring how much upset humans are preoccupied with, and how oppressed they are by the insecurity caused by the human condition, and thus how absolutely incredibly relieving it is to not have to be preoccupied with that condition, is to consider the experience of someone who has had a Near-Death Experience. For instance, as I explained in par. 800, mountain climbers who survive falls that they were convinced would be fatal (they were saved, perhaps, by landing in a snow drift) often report that during those near-death moments they experienced a state of extraordinary euphoria in which the world suddenly appeared utterly beautiful and radiant and that they were flooded with a feeling of ecstatic enthrallment. With understanding of the horror of the human condition we can appreciate how, in such cases, the mind would give up worrying, and that all the facades—in particular the denial that they adopted at Resignation—would become meaningless. If death is seemingly imminent, there is no longer any reason to worry or to pretend, at which point the struggle and agony of having to live under the duress of the human condition ceases and the true world of our all-sensitive soul suddenly surfaces. You suddenly discover what it is like to be free of the human condition. You suddenly have access to all the real beauty that exists in the world. You suddenly discover another seemingly magic world that is all-radiant and magnificent. This is the freedom Saul experienced when he abandoned his struggle with the human condition and deferred to Christ. He was suddenly 'born again' from a state of near death. All his embattled posturing to get a win out of life, all his focus on egocentrically building a castle of grandeur for himself, an edifice, a representation of glory around himself, all his strategising every minute of every day to try to find a way to compensate for feeling bad about himself, suddenly ceased. Realising the magnificence of the new way of living that Christianity offered, this **'way to be saved'** (Acts 16:17) from an effectively dead, resigned, alienated state (Christianity was actually originally called **'the Way'** (Acts 9:2, 19:9, 23, 22:4, 24:14, 22)), Saul took this breakthrough, life-saving idea to the four corners of the known world, and was thereafter known as Paul the Apostle and, following his canonisation, St Paul.

1001 To gain some further insight into the relieving nature of the new Transformed Lifeforce State, where humans' embattled struggle with the human condition similarly ends, you can, if you haven't already, watch the affirmations at <www.worldtransformation.com/affirmations>. You can especially sense the relief of the Transformed State in Doug Lobban's affirmation, but all the others are equally powerful. For example, Neil Duns says, **'The freedom this brings is like nothing humans have ever experienced'**; while Stacy Rodger says, **'This is all that we have ever, ever wanted'**; Tony Gowing says, **'This is like seeing the world for the first time, it's like waking up from a nightmare, it's like 100 tonnes being lifted off your shoulders'**; and John Biggs says, **'The relief and through-the-roof excitement that flows through your veins is incredible…It is beyond comprehension how much wonder and fun and excitement and happiness the human race has coming.'**

1002 So while leaving your baggage, your 'suitcase' of human-condition-embattled strategies and posturing behind by deferring to Christ *was* a marvellous solution to the problem of the horror of the human condition, as has been emphasised, it meant giving up your battle to prove yourself, giving up your particular participation in humanity's heroic struggle to overthrow the ignorance of our original instinctive self or soul, which, in effect, didn't appreciate our conscious mind's search for knowledge. As was explained in some detail in chapter 7:15, while supporting the prophet around whom religions were based did ensure the battle to find knowledge continued indirectly through that prophet's alienation-free, truthful words and sound life (this was the very great benefit of religion compared to the other ways of transcending your upset self), in becoming religious you had personally given up the battle. To use the Adam Stork analogy, you had stopped searching for knowledge and were flying 'back on course'. As mentioned in par. 745, in *The Simpsons* cartoon series the character Ned Flanders is a Christian, someone who has deferred to, and lives through, Christ. His neighbour Homer Simpson, on the other hand, is still living out the corrupting battle to overthrow the implication that we humans aren't fundamentally good. In one episode, Ned lends Homer his lawnmower, which Homer wrecks without remorse. But rather than getting angry or defensive, Ned simply accepts Homer's behaviour—he is the 'goody-goody' while Homer is one upset Adam

Stork, living out the battle to the full; he's massively angry, egocentric and alienated. If Homer could explain the situation, he would say to Ned: 'Listen Ned, you love Christ and he loves you, and you're a goody-goody, and I'm one upset, corrupted dysfunctional dude, but Ned, I'm still out there participating in humanity's heroic battle to find knowledge, so I'm a legend and you're a wimp.'

1003 The new Transformed Lifeforce State, where you leave all your upset baggage behind in favour of supporting these understandings of the human condition, offers a similar all-relieving and all-exciting, 'near-death-experience'-like, 'fall-off-your-donkey', 'go-blind-for-three-days'-type incredible freedom from the human condition that Christians experienced when they deferred to Christ. BUT—and, again, this is a *very, very* important difference—unlike Christianity or any other religion, *this freedom does not come at the cost of abandoning the battle because the battle is now won. It is not an act of irresponsibility or weakness to let go of the battle now, it is an act of strength*, because with the battle now won the illogical and thus irresponsible and thus weak thing to do would be to continue, or somehow re-engage with, that now-obsolete battle.

1004 So, if we return to the Homer and Ned example and imagine that Ned has taken up the new Transformed Lifeforce Way of Living, Homer would have no grounds to criticise him. With the battle to find knowledge, ultimately the understanding of the human condition, now won, there is no longer any good reason to keep living out the battle. In this scenario, the tables are turned: Ned would be in the position to censure Homer.

1005 Again, religions were a way of avoiding living out your upset when you became overly upset, but it meant giving up directly participating in the battle to find the knowledge needed to save the human race. Religion was a weak abandonment of the battle that still had to be fought. With the Transformed Lifeforce Way of Living there is no weakness involved because the battle has been won. In fact, taking up the Transformed Way of Living is the only strong course of action for a human to take now. What is weak is *not* taking it up. Again, this is the *very* important difference—and it is a most profound difference—between religion and the Transformed Lifeforce State.

1006 It is this fact that there is no longer any reason to keep living out the battle to champion the ego that has the potential now to change the world so rapidly from one of conflict and suffering to a world of peace and happiness. The explanations being presented in this book are all rational, there is no dogma or mysticism or abstract concept involved. As such, we can know if what is being presented is accountable and true or not, and since this information does explain the human condition there is no longer any justification for continuing the upsetting battle to champion the ego, so it follows that no one can argue against taking up the Transformed Way of Living, so when everyone *does* take up that way of living our world will be completely transformed. We are rational creatures, and so when all the logic says there is only one response we can make, namely the Transformed State, then that is the response the human race will take.

1007 And this is not a revolution dogmatically imposed by others upon us, as has pretty much always been the case with revolutions in the past; no, this is a revolution based on logic, on understanding, on knowledge, on information. Once someone is given this information there is really only one outcome in the end and that is that they take up the Transformed Way of Living. Of course, while it is the power of the logic, the rationale, that makes the Transformed Way of Living irresistible, the adoption of this way of living is also made irresistible by the absolutely wonderful transforming effect it has on people. Indeed, before long those still living in the old embattled, have-to-prove-your-worth way of living will feel like they have been caught wearing last year's fashions! There is such a stark difference between the free, expansive, enthralled-with-life, transformed existence and the old embattled egocentric existence where your mind is narrowly focused and preoccupied working all the angles every second of every day, defensively seeking validation, reinforcement, a 'win', that those living in the old way, in the past, will seem like a different species. Compared to the new expansive, all-exciting, free, Transformed State, those still stuck in the old embattled paradigm *will* be like sad, forgotten ghosts trapped in an obsoleted state.

1008 It is certainly true that people will initially find it difficult absorbing and taking in all these explanations because of the aforementioned 'deaf effect', the historical denials/blocks in the human mind to so many of the ideas being put forward; to add another example of this problem of the 'deaf effect' to the one mentioned in par. 991, a reader of my books wrote that **'This stuff is so head on it can be crippling, which, initially at least, can make it hard to get behind what's being said and access the profundity of where it's coming from.'** And there *will* be the initial stage of procrastination, what we in the WTM call the 'Mexican Standoff', where people deliberate over the choice that is in front of them, because they don't want to deny the truth and transforming potential of the explanations, but they also don't want to let go of their old, now-obsoleted, resigned power, fame, fortune and glory-seeking way of living that they have become so attached to. And there is also the problem of what we call 'Pocketing The Win', where people misuse the information's ability to make sense of those and the world around them to enable them to become *even more effective winners* of power, fame, fortune and glory—with, in the case of scientists, some even re-working the new insights in this book as their own! There is also the initial temptation to try 'Siding With The Angels', which involves avoiding self-confrontation through deluding yourself you are an unresigned innocent; or even a not fully resigned, relatively innocent 'Ship At Sea', someone who bravely refused to pull into a port and adopt denial when the storm of unbearable self-confrontation occurred during their adolescence. But despite these and other inevitable procrastinations that we in the WTM are all too familiar with, since *all* the logic and thus power lies with the new Transformed Lifeforce Way of Living, and because the transformation is so wonderful, eventually support and enthusiasm for the Transformed Way of Living *will* reach a critical mass, at which point a tidal wave of support for this information and the Transformed State it makes possible *will* sweep the world.

1009 So the only real impediment remaining to humanity's fabulous break-out from its 2-million-year horrible incarceration in Plato's cave of darkness is the one that this whole book has focused upon, which is that the scientific establishment continues to ignore this psychosis-addressing-and-solving, real explanation of the human condition instead of fulfilling its fundamental responsibility to recognise the truth, especially such critically important truth, when it arrives. As described in par. 756, the levels of block-out or alienation have increased so incredibly rapidly that the **'attention span'** of **'students'** is now so **'brief'** they can focus their minds for **'only eight minutes'**; that if they **'have to read...a book they all now groan'**; that they live in **'psychic saunas of superficial sensory stimulation'** from their **'iTunes'** and **'Facebook profiles'** that have resulted in **'a fathomless ocean of online ignorance and trivia that makes what is worthy and timeless more inaccessible than ever'**, **'decapitating our culture, trading in the ideas of some 3000 years of civilization for Buzz-Feed'**—the result of all of this being that **'We have a lot of competing noise**

for our attention these days, and it would be criminal to let that overwhelm our true potential, by masking [the] **useful information** [in Griffith's books] **with hideous noise.'** Yes, the world *is* becoming so superficial, so drowning in **'hideous noise'**, that it *is* rapidly approaching a situation where these deep-thinking-dependent, extremely profound, penetrating, confronting and 'deafening' understandings of human existence will **'overwhelm'** humans' **'true potential'** to understand their world, rendering the public at large incapable of recognising the substance and importance of this book, which means it is up to the scientific establishment to recognise its significance. To avoid terminal alienation and achieve our species' transformation from upset, science *has* to complete its job and recognise the substance and importance of this psychosis-addressing-and-solving true explanation of the human condition.

1010 But assuming science *does* recognise the accountability and importance of these understandings, then there is really nothing stopping interest in this information and the adoption of the Transformed Lifeforce Way of Living sweeping the world. Life for humans will change from a state of terrible turmoil, confusion and despair to one of immense relief, happiness and togetherness. The world for humans *will be* transformed from darkness into light. It really will be as if a light in the dark room we have been staggering around in has suddenly been turned on—everything becomes illuminated. As mentioned in par. 650, we humans had this awesome computer put in our heads, our fully conscious, thinking brain, but we weren't given the program for it and were instead left to wander this planet searching for the program in a terrifying darkness of confusion and bewilderment. Well, from that terrifyingly cold darkness we can now emerge into the warm sunshine of dignifying, redeeming, relieving and transforming understanding.

1011 A further very important difference between religions and the Transformed Lifeforce State is that there is no delusion involved in the Transformed State. To use *The Simpsons* analogy again: Ned has a 'goody-goody', self-satisfied, 'I-occupy-the-moral-high-ground' attitude, which drives the still-human-condition-embroiled heroic Homer crazy with frustration because he intuitively knows that Ned is deluding himself in thinking he occupies the moral high ground, that he is the more together, sound person and is on the right track, but Homer can't explain *why* Ned's view of himself is so extremely deluded and totally dishonest. Homer can't explain and thus reveal the truth that real idealism and the truly on-track, moral high ground lay with continuing the upsetting battle to find knowledge, and that Ned had become so upset, so unsound, that he had to abandon that all-important battle and leave it to others, including Homer, to fight. Worse, by effectively condemning those still upset and fighting the battle, Ned was basically siding *against* those still trying to win that battle, adding substantially to the opposition they had to overcome. In fact, it was the delusion and dishonesty that made giving up the battle particularly dangerous, because to uphold that position required constantly persuading yourself, and others, that you weren't being dishonest and deluded in what you were doing. (Indeed, I should say that in recent times, as the plight of the world has become more serious by the day and supporters of the left-wing in politics have had to struggle so much harder to justify, to themselves and others, their patently dishonest, fundamentally irresponsible and counter productive pseudo idealistic response to that crisis, the resulting shrill fanaticism that has emerged *has* become all but suffocating of those in the right-wing who have been trying to continue the all-important battle to find understanding of the human condition. Thank goodness the dangerous delusion and resulting fanaticism in left-wing politics has finally been exposed and made redundant by the arrival of understanding of the human condition.)

1012 So while religion was by far the least dishonest of all the forms of pseudo idealism that upset humans could take up (because of the honesty involved in acknowledging the truthful, sound life and words of the founding prophet), it still involved substantial dishonesty and delusion. In near total contrast to this situation, the Transformed Lifeforce Way of Living involves no delusion and virtually no dishonesty. By not fully confronting the extent of the upset within yourself and avoiding looking too deeply into all the truth about the human condition, you are practising some denial/dishonesty, but the compassionate full truth about the upset state of the human condition means you aren't deluded about the fact that you are an upset human. Indeed, you have to recognise and embrace that truth (which you can now safely do because the upset state is defended) to effectively be able to adopt the Transformed Lifeforce State.

1013 As briefly mentioned in par. 741, possibly the best sales pitch ever given for Christianity was one delivered by St Paul, as documented in the Bible in 2 Corinthians: **'Now if the ministry that brought death, which**

was engraved in letters on stone [Moses' Ten Commandments that were enforced by the threat of punishment], **came with glory** [because they brought society back from the brink of destruction from excessively upset behaviour]... **fading though it was** [there was no sustaining positive in having discipline imposed on you], **will not the ministry of the Spirit** [the positive mental state from being at last aligned with truth and soundness through your support of someone relatively free of upset and alienation] **be even more glorious? If the ministry that condemns men** [through punishment] **is glorious, how much more glorious is the ministry that brings righteousness** [that allows you to be part of the ideal state]**! For what was glorious has no glory now in comparison with the surpassing glory. And if what was fading away came with glory** ['fading' because it's hard to maintain attachment to a system merely out of fear of punishment]**, how much greater is the glory of that which lasts!** [compared with the relative ease of maintaining an attachment to a system that makes you feel that you are at last on the side of what is good, ideal and right]' (3:7-11). But if Christianity was considered the **'surpassing glory'** to living in fear of punishment, then the Transformed Lifeforce State is the surpassing of all surpassing glories. It is the best solution the upset human race has ever had by an immense, stupendous, incredible, absolutely wonderful, utterly magnificent country mile, because although it involves living in a deferential state of transition, we are only a few generations away from completely eradicating the human condition from the human race—and there is very little dishonesty and no irresponsibility, no weakness, no delusion, no deity, no worship, no focus on a personality, no faith, no dogma, no guilt of any sort involved in the process. What we have now is so relieving and so exciting that when this way of living catches on it will certainly sweep the world.

1014 So the Transformed Lifeforce Way of Living and World Transformation Movement that promotes it is *not* a religion; it is a *metamorphoser*, the proponents of a way of living and a movement that brings about a complete progression—the metamorphosis of the human race from a horrible human-condition-afflicted state to a completely liberated, human-condition-free state. Humanity is transformed from a state of insecure adolescence, where we humans searched for understanding of ourselves, to a state of secure adulthood, where we humans at last achieve understanding of ourselves.

1015 Finally, it should be emphasised that religions aren't being threatened by the arrival of dignifying understanding of the human condition—

they are being fulfilled. The whole purpose of religion was to be the custodian of the ideal state while the search for the liberating understanding of humans' 'fallen' condition was underway. Buddha, for instance, looked forward to the arrival of the amelioration of the human condition when he said that **'In the future they will every one be Buddhas** [meaning in the future everyone will be free of psychosis] / **And will reach Perfect Enlightenment / In domains in all directions / Each will have the same title** [there will be no more distorting alienation] / **Simultaneously on wisdom-thrones / They will prove the Supreme Wisdom'** (Buddha [Siddartha Gautama] 560–480 BC, *The Lotus Sutra*, ch.9; tr. W.E. Soothill, 1987, p.148 of 275). In the Bible, Moses similarly anticipated a time when we **'will be like God, knowing'** (Gen. 3:5). Christ also looked forward to the time when **'another Counsellor to be with you forever—the Spirit of truth** [first-principle-based, scientific understanding]**...will teach you all things and will remind you of everything** [all the denial-free truths] **I have said to you'** (John 14:16, 17, 26). Christ also said he looked forward to the time when, instead of being restricted to **'speaking figuratively'**, we **'will no longer use this kind of language but will** [be able to] **tell you plainly about my Father** [be able to explain the world of Integrative Meaning in denial-free, human-condition-reconciled, compassionate, understandable, rational, first principle, scientific terms]**'** (John 16:25). And again, the same anticipation of our species' liberation from the human condition is expressed in Revelations in the Bible where it states that **'Another book** [will be]**...opened which is the book of life** [the human-condition-explaining and humanity-liberating book]**...**[and] **a new heaven and a new earth** [will appear] **for the first heaven and the first earth** [will have]**...passed away...**[and the dignifying full truth about our condition] **will wipe every tear from...**[our] **eyes. There will be no more death or mourning or crying or pain** [insecurity, suffering or sickness]**, for the old order of things has passed away'** (20:12, 21:1, 4). Yes, as Isaiah hoped, there would come a time when humans **'will beat their swords into ploughshears...Nation**[s] **will...**[not] **train for war any more'** (Isa. 2:4). And what did that truth-saying prophet John Lennon **'imagine'**? A time when the human condition is resolved and **'the world will be as one'**, when there will be **'no heaven** [above us and] **no hell below us'**; when, in essence, there will be a world without the condemning differentiation of good and evil, a world liberated from the insecurity of the human condition and thus the need for religion, where, as Lennon sang, there will be **'Nothing to kill or die for, and no religion too...all the people living life in peace...No need for greed or hunger, a brotherhood of man...all the people sharing all the world'** (*Imagine*, 1971).

1016 The human race has *always* hoped and believed that some day, some where, at some time humanity's heroic search for and accumulation of knowledge would lead to, as Alan Paton hoped, **'the dawn...of our emancipation'**—the finding of the liberating understanding of the human condition—at which point every aspect of human life that was seemingly so inexplicable would suddenly make sense, and we can now see how true that hope and belief was. From an overwhelmingly complex and problematic existence a straight forward and totally effective, extremely-rapidly-repairing-of-human-life-and-the-Earth, way of living for humans is now able to emerge.

Chapter 8:10 How the Transformed Lifeforce Way of Living will quickly repair the world

1017 Not only does the Transformed Lifeforce Way of Living have the power to immediately transform every human's life from one that is tortured by the human condition to one that is characterised by a near-death-experience-like, the-world-suddenly-becomes-fabulously-radiant state of freedom from that condition—and thus the power to bring the whole human race back to life from its effectively dead state of alienation—it also has the power to repair the world we live in, to resuscitate planet Earth from all the brutal anger, selfish egocentricity and mindless alienated behaviour we have inflicted upon it.

1018 I briefly mentioned in par. 899 that I and others in the World Transformation Movement, including my brother Simon and fellow Patron of the WTM, Tim Macartney-Snape, are beneficiaries of Sir James Darling's extraordinary soul-rather-than-intellect-emphasising, Platonic education system, having attended Geelong Grammar School in Victoria, Australia. This is the school HRH The Prince of Wales, the heir to the British throne, was sent halfway around the world to attend for part of his education. (While referring to the soul-reinforcing influences on my life, and that of Prince Charles', I should also mention the exceptionally sound, soul-preserving inspiration of Sir Laurens van der Post. Sir Laurens' books about humanity's lost state of innocence helped me to hold onto the truth of the existence of another true world, and he is the author most often quoted in this book. Sir Laurens was also so important a person to Prince Charles that he was chosen to be godfather to Prince Charles' eldest son, and the future king,

Prince William, and there is **'A bronze bust of van der Post…in Prince Charles' garden at Highgrove'** ('Post, Sir Laurens Jan van der (1906-1996)', *Oxford Dictionary of National Biography*, Christopher Booker, 2004). A former British Prime Minister, Baroness Thatcher, no less, once described Sir Laurens as **'the most perfect man I have ever met'** (mentioned in J.D.F. Jones interview on *Late Night Live*, ABC Radio, 25 Feb. 2002), and Sir James Darling wrote that **'In the world of books there are, for me, Antoine de Saint-Exupéry, or Laurens van der Post'** (*The Education of a Civilized Man*, ed. Michael Persse, 1962, p.36 of 223). For those interested, more can be read about the visions of Sir James Darling [who was described as **'a prophet in the true biblical sense'** in his 3 November 1995 full-page obituary in *The Australian* newspaper], and Sir Laurens van der Post [who was similarly described as **'a prophet out of Africa'** in his 20 December 1996 full-page obituary in the *London Times*], at <www.worldtransformation.com/darling> and <www.worldtransformation.com/vanderpost>.) With regard to the inspired teaching of Sir James Darling, in one of his many extraordinary speeches, he said, **'selfishness is, as it has ever been, the ultimately destructive force in a society, and there are only two cures for selfishness—the regimented state which we all profess to dislike, and the change of heart, which we refuse to make. That is the choice, believe me, for each one of us, and we have not much time in which to make it. The need for decision** [to have a **'change of heart'** and live selflessly] **is serious and urgent, and the sands** [of time] **are running out'** (1950 GGS Speech Day address; Weston Bate, *Light Blue Down Under*, 1990, p.219 of 386).

1019 In this extract from his speech, Sir James identified selfishness as **'the ultimately destructive force'** and stated that historically there have only ever been **'two cures for selfishness—the regimented state which we all profess to dislike, and the change of heart, which we refuse to make'**. The **'regimented state'** is obviously a reference to the dogmatically imposed and strictly enforced selfless, cooperative, communal, social state of communism/socialism. As has been emphasised, being fully conscious, self-managing beings, having to subordinate our thinking mind to dogma was never going to work. De-braining ourselves was no real solution to our problems; ultimately we needed brain food, not brain anaesthetic. Our thinking mind needed understanding—specifically, understanding of why we humans are fundamentally good and not bad, evil, worthless beings. In short, we needed the psychologically dignifying, redeeming and transforming understanding of the human condition. But what Sir James has, in effect, also recognised in this passage is that, despite the urgency, we would **'refuse to make'** or adopt the other alternative, of having a **'change of heart'** and decide to live selflessly instead of selfishly, until humanity achieved that breakthrough. And he was right—until the liberating understanding of why humans are fundamentally good and not bad was found we *had no choice* but to keep on trying to achieve some relief from the insecurity of our condition by finding superficial ways to prove we were good and not bad, such as through winning power, fame, fortune and glory. We had no choice but to be 'ego-centric'—we had no choice but for our conscious thinking self (which again is how the dictionary defines **'ego'**) to be focused or 'centred' on trying to prove it/we were not bad, worthless and meaningless. It is *only now* with the fundamental, trustable, knowable, first-principle-based, dogma-free biological understanding of why we humans are not bad that we can afford to stop trying to prove ourselves all the time—that we can stop being egocentric; let go of being so self-worth-preoccupied; live for the human-condition-understood world free of that old insecure existence; transform from being selfish to selfless; have the **'change of heart'** that Sir James recognised was so **'serious and urgent'**.

1020 In his acclaimed 1969 BBC television documentary series *Civilisation*, the eminent historian Kenneth Clark mentioned that **'People who hold forth about the modern world often say that what we need is a new religion. It may be true but it isn't easy to establish'** (episode 'The Fallacies of Hope'). Saying that we need a new religion is really another acknowledgment of the fundamental need for the human race to have that **'change of heart'**, that change from living selfishly to selflessly, if we are to solve the world's problems. *And*, saying **'but it isn't easy to establish' 'a new religion'** is really, deep down, a recognition that what is needed for this change of heart to occur is for the daunting issue of the human condition to be confronted and solved. As has been explained, the Transformed Lifeforce Way of Living is not a religion, but, like a religion, it does bring about a change from living selfishly for yourself to living selflessly for something beyond self—so Clark's acknowledgment that we need a new religion was about as close as the old denial-based world could come to saying we needed to solve the human condition. In this regard, comments made by the Tibetan Buddhist leader, the Dalai Lama, in the following newspaper report show that he was getting closer to the truth of the need to solve the human condition and by so doing produce a universally acceptable 'God and man reconciled' new potential for humanity: **'The Dalai Lama believes secular ethics, not religion, is best placed to assist the "moral crisis" facing the world's people...**[because it] **respected all traditional faiths as well as non-believers. "Some people—some my friends—believe moral ethics must be based on religious faith," he said.** [But] **"No matter how wonderful a religion, (it will) never be universal. The crisis is universal—now the remedy must also be universal"'** (*The Australian*, 14 Jun. 2013).

1021 Plato also recognised the great transformation that occurs from living in the cave of denial where we are self-preoccupied trying to prove we are good and not bad all the time, to living free from all of that insecure 'baggage' in *selfless* support of the liberating understanding of *why* we humans are fundamentally good and not bad, when he posed the rhetorical question, **'Will our released** [cave] **prisoner hanker after these prizes or envy this power or honour? Won't he be more likely to feel, as Homer says, that he would far rather be "a serf in the house of some landless man", or indeed anything else in the world, than live and think as they** [the power and glory hungry] **do?'** (*Plato: The Republic*, tr. H.D.P. Lee, 1955, p.281).

1022 As to why we have **'hanker**[ed] **after these prizes or envy this power or honour'**, it has been explained that since materialistic luxuries, like glittering dresses, sparkling diamonds, bubbling Champagne, huge chandeliers, silver tea sets, big houses, swimming pools and pretentious cars, gave us (if we were lucky enough to afford them) the fanfare and glory we knew was due us, but which the world in its ignorance of our true goodness would not give us, materialism was one of the important means by which we could sustain our sense of self-worth while the upsetting search for the liberating understanding of why we are good and not bad was being carried out. In fact, when we *only* had our ability to win power, fame, glory and fortune (and, with that fortune/money/wealth/capital, acquire materialistic luxuries) to sustain our sense of self-worth during that upsetting search, those relieving artificial forms of reinforcement were the only engines driving the human-condition-afflicted, insecure world—the *only* rewards sustaining the all-important search for self-understanding. This is why socialism, which sought to replace the engine of greed with the idealism of selfless cooperativeness before we had found self-understanding, couldn't and didn't work. So, while the egocentric, self-centred, individualistic, selfish, greedy, competitive, power-fame-fortune-and-glory-seeking existence was the inequality-producing, human-suffering-causing, Earth-destroying,

'destructive force' that Sir James described it as, it *was also a 'constructive force'*. Yes, there was truth indeed in the saying **'Greed is good.'** The public demonstrations against greed and capitalism that occurred in 2011 and 2012 were actually naive, pseudo idealistic, make-yourself-feel-good-but-don't-solve-anything protests against the reality of the insecurity of life under the duress of the human condition. Of course, this destructive and constructive, 'bad and good' aspect of our behaviour is the core paradox of the human condition that we are now at last able to explain and understand and thus bring to an end.

1023 Nevertheless, while the **'selfishness'** of the old egocentric, self-centred, individualistic, selfish, greedy, capitalistic, competitive, materialistic way of living *was* the driving force that kept the all-important search for the relieving understanding of why we humans are good and not bad going, and was thus a 'constructive force', it is, as Sir James said, the **'ultimate' 'destructive force in a society'** because in the end *only* a selfless way of living is sustainable. *Ultimately* selfishness *is* destructive. No matter how much you try to control and regulate it, a society operating from a basis of selfishness will ultimately become dysfunctional. There was a limit to how long we could keep going under the drive of selfish greed. For human civilisation to survive, selflessness *had to* become the driving force in the world; *ultimately, there had to be*, Sir James' **'change of heart'**. Yes, as many people have pointed out over the years, **'it's not that humans lack the ability to fix the world, it's that they lack the will'**, the **'change of heart'**, the preparedness to live selflessly.

1024 So despite its precious contribution to the human race's progress towards finding understanding of ourselves, the ultimate truth is that selfishness, especially insatiable materialistic, capitalistic greed, was poised to destroy the world. It was 'insatiable' because, as was explained in par. 715, as an artificial form of reinforcement, materialism was never going to genuinely make us feel we were good and not bad—*only* understanding of our fundamental goodness could and now does achieve that. The celebrated leader of Indian nationalism, Mahatma Gandhi, was really making this point about the insatiability of trying to make ourselves feel good by surrounding ourselves with material luxury when he famously said, **'Earth provides enough to satisfy every man's need, but not every man's greed.'** And when the author F. Scott Fitzgerald wrote in *The Great Gatsby* that **'Living well is the best revenge'** (1925), he too was recognising that while we lacked the real **'revenge'** for the injustice of the human condition, namely reconciling understanding, materialism *was* **'the best revenge'** we had. So, it is going to come as an enormous relief for the planet and for humankind—in fact, it is going to make *all* the difference—that, through the finding of understanding of why we humans are good and not bad, we can finally let go of our selfish, egocentric, have-to-prove-our-worth, materialistic, power-fame-fortune-and-glory-seeking way of living and take up the transformed, human-condition-liberated, selfless existence.

1025 Just to illustrate the change that is going to come to our materialistic way of living, in front of me is a teaspoon—well, the monetary value of all the human-glorifying, egocentric content and effort that has gone into its ornate, embellished design, extravagant silver plating and competitive salesmanship and marketing to sell it to me, etc, etc, could feed a starving person for a week! Almost everything I see in front of me—my elaborate watch, my fancy shirt, my sophisticated pen—is, in truth, obscenely extravagant in a human-condition-free world. I should re-design all these items so they are not so extravagant, and it won't be long before I and everyone else in the world will be doing just that. In fact, I idealistically once designed and manufactured a full range of wooden furniture that was free of embellishment and artificial content before realising that for such integrity to be tolerated the human condition needed to be explained. Imagine if all the car makers in the world were to sit down together to design one extremely simple, embellishment-free, functional car that was made from the most environmentally-sustainable materials, how cheap to buy and humanity-and-Earth-considerate that vehicle would be. And imagine all the money that would be saved by not having different car makers duplicating their efforts, competing and trying to out-sell each other, and overall how much time that would liberate for all those people involved in the car industry to help those less fortunate and suffering in the world. Similarly, when each house is no longer designed to make an individualised, ego-reinforcing, status-symbol statement for its owners and all houses are constructed in a functionally satisfactory, simple way, imagine how much creative energy, labour, time and expense that will free up to care for the wellbeing of the less fortunate and the planet.

1026 Again, while we *needed* the individualistic, materialistic world to sustain our sense of self-worth while we couldn't establish it through understanding, now that we have established it that insecure way of living is obsoleted. If you watch Doug Lobban's affirmation of the Transformed State, which can be viewed at <www.worldtransformation.com/affirmations>, you will see how he let go of his egocentric mindset when he became transformed and that his only desire now is to fix the world, starting with his dysfunctional shower curtain! The finding of understanding of the human condition naturally brings about a whole new way of living. There's no dogma involved, like there was in socialism; as has been emphasised, this is about the *end of dogma* and also of faith, and the *beginning of understanding*.

1027 It also has to be remembered that our species' original, natural, instinctive way of behaving is in an unconditionally selfless, consider-the-larger-whole-above-yourself manner, so when we no longer have to behave inconsistently with that, then that original, natural way to behave comes thundering though. Our whole natural, excited, the-world-is-as-it-should-be self bursts forth, as you can see occurring in Doug Lobban, and in others who have provided affirmations. And all you have to do to experience this amazing human-condition-free state is let your unnatural, false way of living go, remembering also that letting it go is no longer an act of weakness, but a strong, fully responsible way to behave.

1028 The story of Lei Feng, a hero during China's cultural revolution, provides an example of the satisfaction and excitement and meaning that can be gained from selflessly serving the larger whole. Of course, as has been emphasised, communism was an extremely dangerous false start to the human-condition-free world, nevertheless, we can learn from Feng's example. During Mao Tse-tung's cultural revolution a Chinese newspaper published the posthumous diary of one of its readers, a young communist soldier (Feng) whose job it was to sweep one of the city's streets. The diary proved so powerful it was printed as a small book that was eventually circulated throughout China and the rest of the world. The gist of the diary was that Feng was happy doing his job because he saw himself as a cog in a giant machine, instrumental in helping the whole machine to work. He wrote, **'A man's usefulness to the revolutionary cause is like a cog in a machine. Though a cog is small, its use is beyond estimation. I am willing to be a cog'** (*The Australian*, 7 Mar. 1987). In essence, Feng wrote to tell of the satisfaction he felt at being able to serve. Serving *is* our original, natural way of behaving, and now that it has become the legitimate and responsible way to behave, it can again become the great joy in everyone's life.

1029 So while it will take time, in fact, a few generations, for humans to sufficiently absorb the understanding of the human condition for our underlying insecurity to be fully ameliorated and thus the need for some artificial reinforcement from materialism, and some artificial relief through escapism, self-distraction and sexual adventure, etc, etc, to completely disappear, once you adopt the Transformed State, while you haven't eliminated the insecure state of the human condition within yourself, you have completely changed your mind's focus from living an ego-embattled, selfish life to living selflessly; you have had the **'change of heart'**, the fundamental change of direction, that is so urgently needed to fix the world.

1030 Importantly, the more this transformation catches on, the easier it will be for people to move from the old insecure, ego-centric way of living to the new, secure, ego-less way of living. The egocentric way of living has had such a powerful hold on humans because while there was no understanding of the human condition all we had was the superficial reinforcement we gained from seeking power, fame, fortune and glory, but now that relieving understanding has been found, and assuming support for the understanding from the scientific establishment ensues, then that old egocentric way of living will quickly lose its power—so quickly, in fact, that everyone will be amazed by how swiftly their need for and attachment to the old artificial forms of reinforcement falls away.

1031 So, what I have been talking about is the transformation that the current, transitional generations can take up, from living a selfish, egocentric, have-to-prove-that-I-am-good-and-not-bad existence to embracing the new, selfless, egocentricity-obsoleted, don't-have-to-prove-that-I-am-good, liberated, exhilarated and empowered, Transformed Lifeforce Way of Living. Of course, in the future when human-condition-extinct generations appear who are *fully* integrated, harmonious, upset-free, all-sensitive and all-loving towards each other, the world and, indeed, the whole universe, they will be properly termed 'Universal Beings', at which point even greater relief will course through our planet. But the point is that, assuming support from the scientific community is forthcoming, the great **'change of heart'** is now on! The human race *is* on its way home, back to soundness, security of self and real love of one another. Imagine when we no longer have to dress to impress, deceive and disguise—especially when women no longer have to be preoccupied with being sex objects—how much freedom that is going to unleash, and how much time, energy and resources it will save? Imagine when communication technology is used only to spread reconciling truth rather than truthless, alienated, escapist, superficial drivel, as it is currently doing—how much relief that is going to bring to humans and thus our world? End the human condition and we end all the big problems of the world—and thus all the little problems too.

1032 So there are degrees of selflessness that we will now be capable of as we move from the old egocentric world to the new egocentricity-obsoleted, Transformed Lifeforce State, before our species eventually arrives at the future human-condition-extinct, Universal Beings State—*however*, it is the fundamental shift in direction, and the rapid changes that accompany it, that is all-important. The point overall is the immense power and potential of cooperative selflessness to solve *all* our problems. If we were all selfless we could solve the AIDS epidemic literally overnight because all it would take would be for everyone to agree to be tested for the AIDS virus and those who tested positive to agree to not have sexual relationships with anyone who doesn't have the virus. When everyone is selfless we will be able to solve the greenhouse gases, global warming problem almost overnight by everyone agreeing to hold every third breath or something—that's obviously not going to work, but the point is we will be able to do whatever it takes! Actually, the most effective mechanism we have available to control our environmental impact is our ability to curb our numbers by not reproducing as much, at least until we have stabilised all the threats to the world that our exploding population is causing. Under the duress of the human condition, we often had children for selfish reasons—to egocentrically perpetuate our family name; or to distract us from our own psychologically distressed lives; or, if we live in a dysfunctional society that isn't able to care for its elderly, to ensure we have someone to look after and provide for us in our old age—the result of which is that there *is* a plague of humans on Earth, in the sense that a 'plague' is **'a widespread affliction or calamity'** (*Free Online Dictionary*). With the control the contraceptive pill allows us to have over our reproduction it is obviously very irresponsible to have children if it means the destruction of our world. We have the means now, both practically and, with these understandings, psychologically, to love our world, so it's madness to choose self-destruction. The world needs—and we are now in a position to supply—love, not selfishness. Controlling our numbers will certainly be important in rehabilitating both humans and our environment. In fact, we in the WTM have already employed this device to help establish these understandings and the new world they can bring about. As mentioned in par. 442, we decided not to have children so that we would have all the resources we could possibly muster to ensure these precious understandings of the human condition would survive the initial onslaught of vilification, persecution and attempted repression for daring to address the issue of the human condition. Certainly our tiny band of 50 people would not have been able to undertake the legal actions we had to take against two of the biggest media organisations in Australia to redress their ferociously vilifying publications about us if we had children to nurture and support. It is the capacity to be selfless that has been missing from the human situation. Basically, as soon as humans no longer have to be selfishly preoccupied with having to artificially try to relieve themselves of the insecurity caused by their human condition through winning as much power, fame, fortune and glory as they possibly can, and are thus at last able to be fully concerned about the well-being of others, all the suffering in the world will be able to be brought to an end. The key rehabilitating effect for humanity and our planet is our ability now to fulfil Sir James Darling's instruction to transition from living selfishly to living selflessly, to have that **'change of heart'**.

1033 Understandably, this power of selflessness has not been something we humans wanted to acknowledge or think about because it has been too confronting of our present massively embattled, ego-hungry, desperately-needing-self-gratification-and-glorification, selfish, greedy, materialistic existence. We have *had to* live an alienated, escapist, materialistic life because we haven't been able to live a secure, honest, spiritualistic life, but now, with the human condition explained, we can. And now that we can afford to think about it, all manner of insights into the power of selfless cooperation become accessible, including those provided through observing ant, bee and termite colonies—truly, as King Solomon advised, **'Go to the ant...consider its ways and be wise'** (Bible, Prov. 6:6). The 2004 award-winning documentary *Ants—Nature's Secret Power* admitted the power of selfless cooperation when it concluded that **'The secret of ant societies is their cooperation...**[it's what has enabled them to] **act as a superorganism...**[and become] **nature's true world power'** (produced by Adi Mayer Films and ORF). In his book *The Soul of the White Ant*, Eugène Marais wrote that **'the termite...never rests or sleeps'** (1937, p.61 of 154).

Bees' extreme selflessness was also made apparent in a documentary on bee colonies, which reported that **'when bees become sick they sacrifice themselves and leave the hive to die to prevent infecting the rest of the colony'**, and how **'in the summer, the workers only live around 30 days because they literally work themselves to death'** (*Silence of the Bees*, National Geographic Channel, 2007). I'm certainly not suggesting that we go beyond self-sacrifice to self-elimination as a strategy for solving problems, but these examples do make it clear how powerful selflessness is as a force for solving problems. The main point is that our ability now to leave behind the insecure, self-preoccupied, ego-centric, selfish state of the human condition makes possible the *true* reparation of our whole world; it is us self-adjusting, conscious humans who—now that we no longer have to live an insecure, selfish existence and can live a selfless, cooperative existence—can now become **'nature's true world power'**.

1034 Basically, what everyone can now do and should do is adopt the Transformed Lifeforce Way of Living—and, when this fabulously empowering breakthrough explanation spreads, that objective will soon become the main objective of all nations. Then, with everyone's help, we each have to rehabilitate and make environmentally sustainable our realm—our home, our community, our region, our country, and then our world. For 2 million years humanity has strained with every fibre of its being to achieve this great breakthrough from ignorance to enlightenment, and for those of us living now our wonderful opportunity is to implement the rewards of that achievement. Here comes the golden sunshine revolution!

1035 As I have previously said, the main personal 'reward' for humans now is the anticipation and the excitement of being effectively free of the human condition—the joy and happiness of being liberated from the burden of your insecurities and self-preoccupations; the awesome meaning and power of finally being aligned with the truth and participating in the magic true world; the wonderful empathy and equality of goodness and fellowship that understanding of the human condition now allows you to feel for your fellow humans; the freedom now to effectively focus on repairing the world; and, above all, the radiant aliveness from the optimism that comes with knowing our species' march through hell has finally ended.

1036 Although this new world for humans, which the World Transformation Movement has been established to facilitate, will have a slow beginning because humans initially find it difficult even taking in or 'hearing' discussion about the human condition, if the scientific establishment fulfils its responsibility to recognise the truthful biological explanation of the human condition that we are presenting, then the WTM will be discovered and word of it and what it offers every human and the world will spread like wildfire—as the exceptional denial-free thinking prophet Teilhard de Chardin wrote, **'The Truth has to only appear once...for it to be impossible for anything ever to prevent it from spreading universally and setting everything ablaze'** (*Let Me Explain*, 1966; tr. René Hague et al, 1970, p.159 of 189). Indeed, all the quotes from prophets, songwriters and poets included in par. 113 testify to the nature and consequences of this information finally appearing in the world. Yes, before long we will be taking this information to the world—marketing our own human-condition-free, world-saving products, providing our own all-exciting and meaningful, denial-free, honest, human-condition-understood films, documentaries and books, launching our own TV station—and our website will be the biggest in the world, bigger than Google or any other existing site. And this is not deluded hubris, or wild exaggeration, it is simply the logical truth of what is able to happen when the dignifying, redeeming and healing understanding of the human condition is finally found.

1037 Sir Laurens van der Post once said that **'The carrier of new life, the carrier of renewal in societies, has always started with a lone voice, with one lonely individual somewhere taking upon himself in terms of his own life wherever it's challenged, the start of renewal and living the "being" with which he has been charged in a world of "having"; and it always starts like that before it becomes a group...**[and then] **a community—which is not yet, but which is coming'** (presentation by Laurens van der Post at the Institute of Contemporary Arts, London, 9 Dec. 1982). Yes, the bottom line truth is that *only* the finding of the reconciling understanding of the human condition and the **'change of heart'** from living by the doctrine of **'having'** to living by the transformed doctrine of **'being'** that the understanding finally makes possible can save the human race. The World Transformation Movement provides the *only* path forward for humankind—anything else represents ever-increasing, excruciating, unthinkable suffering and ultimately doom for our species. So, become a member of the World Transformation Movement!

Chapter 8:11 <u>Humanity's overall situation now that we have understanding</u>

1038 I drew the following picture to illustrate the situation the human race now faces.

Humanity's Situation: the Sunshine Highway to Freedom, the Abyss of Depression, our Cave-like Dead Existence and the Spiralling Pit of Terminal Alienation

1039 At the top of the picture humans are depicted revelling in the sun-drenched **'FREEDOM'** of the Transformed Lifeforce Way of Living. In the middle of the picture, however, lies a terrible abyss with the word **'DEPRESSION'** written prominently across its base, representing the depression humans suffered when they tried to confront the issue of the human condition. You can see that this abyss consumes two-thirds of the whole picture, which is an accurate reflection of how big a part depression has played in human life—basically, *all* our activities have been ruled by our fear of it. Yes, the human condition has essentially been our fear of the depressing implications of our non-ideal behaviour, of are we good or bad; are we actually worthwhile creatures or evil beings—and, although the human condition has finally been compassionately explained, the fact remains that it will still take a few generations to heal, so current generations will *still* be faced with that depression if they try to fully confront all the truth about themselves and the world. As has been explained, the arrival of truth 'day' or honesty 'day' is also exposure or 'judgment' 'day', the time when we are faced with too much truth to have to suddenly confront. On the lower side of the abyss lies a narrow strip of land that is coloured the grey-green of dead flesh. Featuring the words **'CAVE-LIKE DEAD EXISTENCE'**, it is populated by a throng of people walking around like zombies. This is a truthful depiction of the state of extreme block-out or denial and resulting alienation from our true selves or soul that humans currently live in. The bottom corner of the picture also contains a big black spiralling pit that has written beside it, **'Extinction of the human species from terminal levels of alienation.'** This describes the destiny of the human race if understanding of the human condition was not found—or if the scientific establishment continues to ignore its existence.

1040 In this situation that the human race is now in we obviously can't continue living a **'CAVE-LIKE DEAD EXISTENCE'**. The amount of suffering is way beyond obscene and unconscionable. What's more, if we remain where we are the human race will very soon go down the gurgler, into **'THE SPIRALLING PIT OF TERMINAL ALIENATION'**. There can be no doubt that our species has already reached the end play or end game point where it is about to plunge into that pit of perpetual darkness and extinction—indeed, the following condensation of the descriptions that were given in chapters 7:16B and 7:16C of what is occurring at the forefront of human progress, namely in the Western world, and in particular in the USA, makes it very clear how serious our species' plight is:

'today's children are probably the least loved generation of all', **'96 percent of American families are now dysfunctional'**; **'someone born since 1945** [is] **likely to be up to 10 times more depressed than their grandparents'**; **'Autism...now occurs at epidemic rates...According to California records, autism has increased 1,000% in the past 20 years!'**; there has been a **'41% increase in ADHD in US children during the past decade'**; **'there are more and more cases of ADD among children** [in the US], **one in five to be exact...So what is going on?...The brain** [of the ADHD child] **is busy, busy, dealing with the pain'** from **'childhood'**, **'where there was no** [real] **love...**[so] **the kid is agitated out of his mind, driven by agony inside...drowning in misery'**; **'while only two years ago students were able to sit through a half hour art documentary, I now know I lose them after only eight minutes'**; **'Alone and adrift in...our "psychic saunas" of superficial sensory stimulation, members of my** [Y] **generation lock and load our custom iTunes playlists, craft our Facebook profiles to self-satisfied perfection, and...bravely venture forth into life within glossy, opaque bubbles that reflect ourselves back to ourselves'**; **'The youth of today are living their lives one mile wide and one inch deep'**; **'the wisdom of the ages is...buried like lost treasure beneath a fathomless ocean of online ignorance and trivia that makes what is worthy and timeless more inaccessible than ever...**[The] **digital revolution...**[is] **decapitating our culture, trading in the ideas of some 3000 years of civilization for BuzzFeed'**; **'We have a lot of competing noise for our attention these days, and it would be criminal to let that overwhelm our true potential, by masking** [the] **useful information** [in Griffith's books] **with hideous noise'**; **'Generation Y is being ravaged by depression, anxiety and stress disorders.'** Young people now feel that **'Our innocence is lost...It scares me to death to think of what I have become...I feel so lost in this world...I scream to the sky but my words get lost along the way. I can't express all the hate that's led me here and all the filth that swallows us whole. I don't want to be part of all this insanity. Famine and death. Pestilence and war.** [Famine, death, pestilence and war are the 'Four Horsemen of the Apocalypse', the state of terminal alienation, described in Revelation 6 in the Bible.] **A world shrouded in darkness...Fear is driven into our minds everywhere we look...How do we save ourselves from this misery...So desperate for the answers...No one hears our cries. And no one sees us screaming...This is the end.'** The aforementioned *TIME* magazine story titled **'The Me Me Me Generation'** about the **'narcissism epidemic'** in the present Y-generation was essentially about how this **'80 million strong...biggest age group in American history'** exists in a state of terminal alienation. And the frightening question that the very existence of a Y-generation—

and the X-and-Z generations that flank it—poses is what exactly do we mean when we say the X, Y, Z generations? The answer is the end play state of terminal alienation. After all, what comes after Z?

1041 With the world **'so desperate for the answers'** that will **'save ourselves from this misery'** it is little wonder there is, as I've said, panic in officialdom, with the President of the United States announcing in 2013 a **'Brain Initiative'**, giving **'$100 million initial funding'** to mechanistic science to find **'the underlying causes of...neurological and psychiatric conditions'** afflicting humans; and in 2012, an American billionaire pledging $200 million to **'accomplished scholars whose collective mission is... greater understanding of the human condition'**; and in 2013, the European Commission announcing the launch of the **'Human Brain Project with a 2013 budget of €54 million (US$69 million)'** with a **'projected billion-euro funding over the next ten years'** with the goal of providing **'a new understanding of the human brain and its diseases'** to **'offer solutions to tackling conditions such as depression'**; and there are similar initiatives taking place at Oxford and Cambridge universities—this despite the mechanistic scientists in these institutions being committed to living and practising their craft in complete denial of the human condition!

1042 *Clearly* humanity can't stay living on the 'narrow strip of land that is coloured the grey-green of dead flesh' a moment longer. The pain of our tortured condition HAS TO STOP!! And we have **'the answers'** that we are **'so desperate for'** to **'save ourselves from this misery'**—the explanation of the human condition! However, as emphasised, trying to confront all these now terminal levels of upset inside us humans is also not an option. Enrico, a reader of my books, described how difficult it is trying to face down the terrible Abyss of Depression when he wrote: **'Diving into a sea of truth where everything is completely transparent one can't but ask, "how will anybody cope with this; how in the world can anybody cross such darkness to reach light?!"'** (Email to the WTM, 24 Feb. 2011). Yes, one of the people I have drawn looking down into the Abyss of Depression, holding their head in horror, could be a mother with an autistic child trying to fully confront the truth that the main reason for autism is a lack of alienation-free, unconditional, true love. As emphasised in chapter 7:16C, that mother is now able to understand that autism is *not* the fault of parents, but of the ever-increasing upset in the human race as a whole, which in turn is due to our species' heroic search for knowledge. The difficulty, however, is that while the mother *can* access some relief from that guilt in that truthful, fully compassionate understanding, if she dwells at all on the truth of how much upset exists within her and all the pain that upset has inadvertently caused her child, she will still experience unbearable feelings of depression. Clearly she is going to have to maintain a degree of block-out or denial of such deeper thinking. And virtually every human is in the same situation, where it will be impossible for them to fully immerse themselves in and confront the truth about their corrupted condition and the suffering and destruction it has caused. So virtually all humans *will* have to take the **'SUNSHINE HIGHWAY'** bridge *over* the Abyss of Depression: adopt the Transformed Lifeforce Way of Living, where they support the truth while avoiding overly confronting it. Incidentally, the only reason I made the bridge over the abyss in the drawing so narrow is so that all the letters of the word 'depression' below it are visible. The bridge should really be a mile wide because it is, in truth, a wide open and easy path to take. But the point is that there *is* a way **'to cross such darkness to reach light'**, and the evidence for that is all the people who have *already* crossed that darkness to reach light. People are already coping with the problem of exposure, and coping in an extremely relieved and joyous way, through having adopted the Transformed Lifeforce Way of Living.

1043 The examples are endless of the shock of exposure or 'judgment day', and the changes it will bring. With understanding of the human condition and the ability to be truthful that it makes possible, all the books on psychology are going to have to be re-written. What particularly changes the way psychology has been practised is the truth that we humans are born with cooperatively-orientated, all-loving instinctive expectations, and, therefore, how compromising of an infant's healthy development it is when these expectations for unconditional love are not met. It particularly reveals how easily children, in their trusting naivety, blame themselves for any shortfall in love, and how devastatingly hurtful such conclusions that they are a bad, unworthy person are for them, and, therefore, how much children have to block out any thinking about that terrible conclusion, and, therefore, how alienated they become from their true self or soul, and how that lack of access to their true self compromises their ability to nurture the next generation with real love when they in turn became parents. It is our all-loving and thus all-trusting soul that has not been acknowledged in psychology; in fact, it has not been acknowledged in any aspect of human life.

1044 Indeed, the human race has been living in denial of so many truths that are now revealed that nearly all academic books will have to be re-written. Science as a whole will be faced with changing from complying with the practice of denial to having to be honest, which will mean its whole evasive, intellectualism-emphasising-instinctualism-denying structure, including its traditional universities and academic departments and millions of denial-drenched research papers (as has been described and evidenced, the human condition is such that humans have been incapable of being objective), will be faced with obsolescence. A stark indication of the immense change that comes with the arrival of understanding of the human condition is that with the exception of just a handful of titles (the main ones being the great religious texts), virtually all the world's books have been deeply denial-compliant in their content and so will, therefore, be largely obsoleted by the new denial-free paradigm. Basically, it won't be long before all our libraries and academic institutions become museums; indeed, if it wasn't for the Transformed Lifeforce Way of Living that allows everyone to fully participate in the human-condition-understood new world, the whole, current, denial-saturated human race itself would become a museum piece! Interestingly, in his classic book, *Nineteen Eighty-Four*, George Orwell predicted that **'By 2050—earlier, probably—all real knowledge of Oldspeak will have disappeared. The whole literature of the past will have been destroyed'** (1949). While Orwell was referring to the emergence of a completely thought-oppressing, pseudo-idealistic state, there is some intuitive realisation deep within what he has written of a time when all dishonest thinking will **'disappear'**; a time when, as predicted in the Bible, revelation about our human condition will lead to a situation where **'the old order of things has passed away'** (Rev. 21:4).

1045 It is true; when understanding of the human condition arrives, as it has, every denial-based aspect of the whole world becomes redundant; basically, the world as we know it comes to an end, and virtually overnight—well, within a few generations, which is still very fast. For teachers at schools or universities, all the truth-avoiding, immensely superficial, basically dishonest, mechanistic ways of viewing the world will soon no longer be taught. The arrival of the actual understanding of the reason why we humans are good and not bad will mean that materialistic goods that have been designed to make ourselves feel superficially better about our corrupted condition will, before long, be seen as meaningless and worthless. The whole basis of materialism has been undermined, made pointless and unnecessary. All our posturing and promenading and pretending will eventually lose its power to impress. Our whole artificial and superficial existence has been exposed and made redundant. Going to church will, in time, be superseded by this reconciling understanding of our previously God-fearing existence. The business of politics, of having left and right-wing factions, is similarly made redundant by the reconciling understanding of the human condition. Basically, all our denials are exposed, made transparent; they no longer work. So yes, before long, within a few generations, that whole denial-based existence will come to an end. Life as we have known it will be over—and, most relievingly for humankind, with its passing all the suffering and destruction that went with it will also go. A whole new world now begins to open up across the world based on reconciling understanding and the Transformed Lifeforce Way of Living it makes possible. As I've said in every major document I have written since I started writing about the human condition nearly 40 years ago, **'soon from one end of the horizon to the other will appear an army in its millions to do battle with human suffering and its weapon will be understanding'**.

1046 Humanity's situation now is actually perfectly depicted by Hans Christian Andersen's 1837 fable, *The Emperor's New Clothes*, in which a child breaks the spell of deception that the emperor is beautifully clothed when he discloses the truth of the emperor's nakedness. The first reaction of the emperor and his entourage is to try even harder to maintain the deception, maintain the denial, but it is futile. To quote from the fable: **"But he has got nothing on," said a little child. "Oh, listen to the innocent," said its father. And one person whispered to the other what the child had said. "He**

has nothing on—a child says he has nothing on!" "But he has nothing on!" at last cried all the people. The Emperor writhed, for he knew it was true. But he thought "The procession must go on now." So he held himself stiffer than ever, and the chamberlains held up the invisible train' (*Andersen's Fairy Tales*, tr. E.V. Lucas & H.B. Paull, 1963, p.243 of 311). Yes, the spell is broken, our delusions and pretences have been stripped of their power.

1047 It is true that we humans are afraid of change almost as much as we are of self-confrontation, and this information does bring about phenomenal change, a massive paradigm shift, but this change is entirely manageable, most especially because it is a positive paradigm shift, not a negative one—we are going from an extremely negative, absolutely awful paradigm, to an extremely positive, fabulously wonderful paradigm. We can procrastinate and try to stay in the old, essentially dead, hell-on-earth world; stay in the egocentric castles we have built, or are trying to build, for ourselves based on our pursuit of power, fame, fortune and glory; think **'But...The procession must go on...**[and hold up our **'invisible train**[s]'] **stiffer than ever'**—or we can join the new all-exciting world that will be opening up right across the planet. There is now so much to do in support of the liberated new world that the range of new opportunities, needs, positions, industries, services, etc, etc, will be endless—and the best part of it all is that this new world for humans will be filled with immense excitement, happiness, meaning and fellowship. With understanding of the human condition found there is no longer *any* justification for staying in the old, lonely, effectively dead world. Assuming science fulfils its responsibility to recognise this psychosis-addressing-and-solving, real biological explanation of the human condition—which is what is needed for the public at large, which is presently preoccupied with superficial **'hideous noise'**, to stop and take an interest in, and discover the truth and beauty of these deeply profound and penetrating explanations of human existence—then before long *everyone* will be rolling their sleeves up and excitedly going to work for the absolutely fabulous, human-condition-understood, peaceful, reconciled, all-loving new world for humans.

1048 At this point I might address the concern that in **'a thoroughly transparent world'**, which understanding of the human condition does produce, our **'species would...suffer spiritual anemia and perhaps terminal boredom'** (Frank Tippett, 'Essay', *TIME*, 17 Jan. 1983). This often-stated perception that **'the price of paradise will be boredom'**, and even that we should, therefore, stay living in our alienated state of denial of the human condition, is what I call 'black speak', blindness preaching blindness, because the alienated state is such a numb, seared and exhausted state that to argue that we should stay alienated is effectively arguing that we should stay dead! If a bucket of water represents the depth of sensitivity humans are capable of then our alienated state is equivalent to living life on the thin surface meniscus of that water. So to maintain that we should remain in an empty, insecure, immensely unhappy and deeply, deeply distressed, effectively dead state that can hardly feel or know anything when we can finally, at last, savour the magnificence of our world is simply alienation trying to build a positive out of a negative. It is an understandable defensive reaction that people have had to make themselves feel better about their human-condition-afflicted, empty lives, but clearly it is an absurd betrayal of all that the human race has worked towards. While our world will, for example, lose some of its superficial variety or 'colour' when the different states of alienation disappear and everyone has similar personalities, the incredible sensitivity and happiness that will come to humans from being able to access the world of our soul again will mean our lives have a depth and potential that we have hardly dared to dream of. For example, each person will be able to immerse themselves in whatever aspect of sensitivity they choose, with the number of different aspects of sensitivity available in our world to savour being innumerable. Some people might spend a whole lifetime perfecting, and sharing with others, what it's like to feel what a certain mood feels like, or what a certain animal or plant feels like, or what a certain time of the day truly feels like. There are awarenesses and feelings and knowledge and thoughts and imaginings that we haven't even begun to tap into in our human-condition-preoccupied, virtually dead state. Once people allow themselves to appreciate what life will be like free of the human condition they will never make such sad, blind, defensive comments as suggesting a human-condition-understood future will be 'boring'.

1049 Yes, the fact is, having lived in such a dead, alienated, dark, cave-like state of dishonest denial and resulting estrangement from our true soulful self for so long, coming out into the truthful world of liberating sunshine of understanding will be *so* redeeming and exalting it *will be* almost more exciting, transfiguring, rapturous and exhilarating than the human body is capable of handling! To finally be free from having to—and we did *have to*—live such a world-destroying, selfish, mean and egocentric existence for so long is such an *immense* relief. To now be able to live a life of transformed freedom from the dark, effectively dead state of denial and be able to participate fully in the true, human-condition-liberated, soul-resuscitating, loving world with all its beauty and sensitivity is something *so* exciting that appreciating just how exciting it *will be* is almost beyond our present powers of imagination. So let me paint the picture as starkly as I possibly can. Firstly, the following is a re-inclusion of a powerful description that was given in par. 851 of the situation the materialism-driven world has arrived at. It was explained that with communication technology—especially television and the internet—letting everyone in the world know what is going on elsewhere in the world, everyone is now acutely aware of the financial wealth and material luxury that some people and cultures have created, and they envy it; they ask themselves, 'Why shouldn't I have all this?' As the psychiatrist Clancy McKenzie observed, **'While visiting Machu Picchu in Peru in 1979 I noted very poor persons, living in the mountains, who had only the clothes they wore and perhaps a lama or two, but had beautiful, warm smiles and seemed content and happy. Days later I was in Bogota in Colombia. It was a very hot day and we asked the driver to stop at an outdoor tavern to buy cold beer. The people were very impoverished, but there was a TV playing and they were able to view the "outside world" where everyone seemed to have more, and luxury was abundant. I offered to go in with the driver and he urged me to wait in the car. I soon learned why. The absolute hatred was so intense that it was palpable. These people did not have less than those in Machu Picchu but they saw others who had more, and their needs were intensified.'** Yes, the result of this awareness of what materialism offers is that there is now rampant greed everywhere; all those populations who haven't had the material luxury that is so prevalent in the Western world see themselves as being on catch-up—they want a big house, a car, a fridge, a suit, a gown, a fancy watch, Nikes, Coca-Cola and Big Macs, a beach holiday, etc. The effect is that our original instinctive self or soul's natural inclination to be selfless and loving has been thrown by the wayside and everywhere cynicism reigns—a 'cynic' being **'one who doubts or denies the goodness of human motives'**. The following is an account of the situation in Kenya, but it very much applies to what is happening throughout Africa, South America, China, India, south-east Asia and in parts of the Middle East. Written by the Kenyan photographer

Mirella Ricciardi, it describes Kenya's current **'chaos and confusion, the crime and lawlessness, the greed, the bribery and corruption at all levels, the senseless wildlife killings, the environmental destruction, the mindless power games, the hopeless destitution and hunger, the lack of concern and forward thinking...The outrageous greed and disregard for anything and anyone other than themselves and their immediate families in the top ranks of governments, have created a financial drain on the country that filters down and affects whole populations in varying degrees, threatening the fragile infrastructure of the land. All those, without exception, who have somehow attained a position of command and authority, however humble, have but one idea in mind, to emulate the example set by the head of state – to milk their positions for all they are worth, for as long as it lasts. Because finances are siphoned off to Swiss bank accounts, salaries are so low they need topping up by whatever means. As a consequence bribery and corruption at every level is the only option. The police force and judiciary, the ministers, the administrators, the teachers, the doctors and the heads of government divisions, extract whatever they can, by fair means or foul, under the guise of their uniforms and titles, while the country collapses and the masses starve.'** So, the 'developing' 'poor' parts of the world are fast approaching **'collapse'** and **'hopeless destitution'**—while the so-called 'developed' 'affluent' parts of the world are fast approaching a state of terminal alienation with anxiety disorders such as ADHD and autism rampant and the **'biggest age group in American history'** in the grip of an extremely insecure **'narcissism epidemic'**. After presenting this analysis in par. 851, I summarised that it is basically all over for the insecure, human-condition-avoiding, escapist, power-fame-fortune-and-glory-seeking, materialism-driven, greedy, never-satisfied, selfish existence. Yes, that world *has* reached its calamitous end, BUT THE FABULOUS NEWS IS THAT AT THAT BRINK THE WORLD IS ABLE TO BE SAVED! The God of Mammon and Greed *is able to be replaced by the God of Soul and Love*, humanity *can* have a **'change of heart'**—because the insecurity-ending understanding of the human condition has been found. Humanity is finally able to come home. While we had no other form of reinforcement other than materialism, we built it up in our mind to be something wonderful, but the truth has always been that materialism really offers nothing; it is a desperately sad, shallow and unsatisfactory way of living—a very poor substitute for our lack of spiritualism. With understanding of the human condition finally found, however, what is on sale now is nothing less than heaven on Earth, a way of living that is *incredibly* exciting and happy. Between the materialistic way of living and the Transformed Lifeforce life there is just no contest, NONE AT ALL. The way forward now is back to our soul's true world. The gates have been flung open to paradise. Materialism and greed is old baggage now, discarded rubbish to be left on the garbage tip. No one is going to buy it anymore. Its power to seduce is gone because the most magnificent alternative has arrived to transport everyone to a world of soul, real love, real meaning, real togetherness and real happiness. Everyone, every Peruvian, every Kenyan, every Chinese person, every Indian, every American, every European, *all* people on Earth, can now join the great exodus from the, in truth, terrible human-condition-afflicted, materialistic, greedy, selfish existence. NO ONE WILL WANT THAT ANYMORE WHEN THE TRUTH GETS OUT ABOUT WHAT IS ON OFFER NOW!

1050 Yes, 'the cavalry is coming'—well, that is the phrase used in 'Cowboys and Indians' folklore; in more contemporary 'Star Wars' terms, 'the Force' *is* with us—everything *is* possible now, help *is* on its way to every person in every corner of the world. To paraphrase U2's fabulous anticipation of our species' liberation from the agony of the human condition, 'love has come to town' in its absolute purest form, namely as relieving understanding. Everyone in every situation and predicament can now rise up as radiant new beings from their corpse-like state. Everyone can now come back to life—can wake up from their human-condition-afflicted torpor and look outwards and see each other and the world for the first time, and move across and help each other, and do anything and everything that needs to be done to end the suffering and pain that plagues this planet.

1051 We humans *have*, in truth, all been asleep, owned by so much pain and suffering. And, certainly, we are going to be in shock for a little while absorbing the realisation that we have finally won our freedom from the agony of the human condition—but it's on, the great awakening, the rising up of the human race from its deep slumber. From the festering, stalled state it has been in for far too long, waiting for these liberating understandings of the human condition, the human race is finally on its way! As Doug Lobban says in his aforementioned affirmation, **'I've got this image of marching straight forward with incredible purpose, straight out of a massive swamp into long green grass with flowers and butterflies and sunshine, and never ever looking back, just keep marching forward. Everyone together, we**

are going to have so much fun.' Yes, the gateway *is* wide open and sunlight is now streaming into our world. We can all wake up now; we can all break out of our chrysalis, our human-condition-afflicted straitjacket, where we have been constrained, letting the wings of our great potential unfold in the sunshine of relieving understanding to reveal the fabulously beautiful creatures we really are.

Chapter 8:12 The 'pathway of the sun'

[1052]The other aspect of *The Emperor's New Clothes* fable that needs explaining is the significance of the **'innocent' 'child'** breaking the spell of all the denial and delusion in the emperor's court. The overall situation is that we humans have been practising denial for some 2 million years, ever since we became fully conscious and had to defend ourselves without understanding of why we are good and not bad. This means that this practice of denial has been refined and added to for *so* long now that it has achieved a state of near complete block-out of any truths that bring the issue of the human condition into focus, which, as has been revealed in this book, is a phenomenal amount of truth. It was as if our planet had become enclosed in a solid steel casing through which no light, no truth, could enter. Every time some light crept through a crack, the great welding machines of denial would come out and weld a slab of steel over the offending fracture. In the end it was a truly massive edifice of denial, so strongly reinforced that it seemed no one would ever be able to break through it, leaving the human race to perish inside its terrible darkness. The great fortress of denial *did* seem impenetrable, unassailable, BUT NOT SO TO OUR SOUL, which an **'innocent' 'child'** is the perfect representation of, for it has always had so much love that it was able to lie in wait for the time when science would find sufficient understanding of the mechanisms of the workings of our world for it, that little innocent child of loving, denial-free soundness, to rise up and break through that steel case and liberate the human race from its incarceration.

[1053]The imagery of a child one day liberating the human race occurs in many mythologies. The biblical story of David and Goliath is actually a metaphorical description of this situation where the whole army of humanity is besieged by the monstrous, all-pervading and all-powerful hold that denial has on the world, which Goliath represents, when David, who is **'only a boy'** (Sam. 17:33, 42), **'come**[s] **against you [Goliath] in the name of the Lord Almighty** [in the form of Integrative Meaning-accepting, denial-free thinking]**'** (17:45) **'and kill**[s] **him'** (17:50), allowing everyone to celebrate their liberation **'with singing and dancing'** (18:6).

1054 In the great European legend of King Arthur, the wounded (alienated) king whose realm was devastated (humans unavoidably made their world an expression of their own madnesses), could only have his wound healed, and his realm restored, by the arrival in his kingdom of a simple, naive boy. The boy's name is Parsifal, which, depending on which of the numerous online sources you refer to, means either **'guileless fool'** or **'pure fool'**. To the alienated, only a naive **'fool'** would dare approach and grapple with the confronting truths about our divisive condition. The Jungian analyst Robert A. Johnson offered this rendition of the legend, saying firstly that **'Alienation is the current term for it** [the wounded state of humans today]**. We are an alienated people, an existentially lonely people; we have the Fisher King wound'** (*He, Understanding Masculine Psychology*, 1974, p.12 of 97). He then described how **'The court fool had prophesied that the Fisher King would be healed when a wholly innocent fool arrives in the court. In an isolated country a boy lives with his widowed mother** [since the male ego can be especially oppressive of the souls of children, saying that the mother is widowed provides recognition that the child wasn't oppressed by an especially egocentric father]**...His mother had taken him to this faraway country and raised him in primitive** [**isolated**, innocent, unconditionally-loving, alienation-free, sound, soulful] **circumstances. He wears homespun clothes, has no schooling, asks no questions. He is a simple, naive youth'** (p.90). Johnson went on to recount that in the myth it is this boy, Parsifal, who, when he becomes an adult, is able to heal the Fisher King's wound of alienation, so that **'the land and all its people can live in peace and joy'** (p.94).

1055 In an astonishingly prophetic 1940 essay titled *The Almond Trees*, the literature Nobel Laureate Albert Camus also recognised this truth that to find understanding of the human condition and **'heal'** humanity's **'Fisher King wound'** of **'alienation'** was going to require an **'innocent'** person, symbolised by a **'boy'** or **'youth'**, from **'an isolated' 'faraway country'** that had not been overly exposed to the horrifically upset, **'alienated'**, **'existentially lonely'** almost universal state of humans now. Camus wrote that, despite all the corrupted upset in the world, there are still **'those shining** [innocent] **lands where so much strength is still untouched'** that can produce the innocent and sound **'whiteness and its sap' 'strength of character'** needed to **'stand up to'** all the alienated, fraudulent evasion and denial and last **'just long enough to prepare the fruit'**—find the world-saving liberating understanding of the human condition. The following is an extract from what he wrote (underlinings are my emphasis): **'All we then need to know is what we want. And what indeed we want is never again to bow down before the sword, never more to declare force to be in the right when it is not serving the mind. This, it is true, is an endless task. But we are here to pursue it. I do not have enough faith in reason to subscribe to a belief in progress, or to any philosophy of History. But I do at least believe that men have never ceased to grow in the knowledge of their destiny. We have not overcome our condition, and yet we know it better. We know that we live in contradiction, but that we must refuse this contradiction and do what is needed to reduce it. Our task as men is to find those few first principles that will calm the infinite anguish of free souls. We must stitch up what has been torn apart, render justice imaginable in the world which is so obviously unjust, make happiness meaningful for nations poisoned by the misery of this century. Naturally, it is a superhuman task. But tasks are called superhuman when men take a long time to complete them, that is all.**

[1056] **Let us then know our aims, standing steadfast on the mind, even if force dons the mask of ideas or of comfort to lure us from our task. The first thing is not to despair. Let us not listen too much to those who proclaim that the world is ending. Civilizations do not die so easily, and even if this world were to collapse, it will not have been the first. It is indeed true that we live in tragic times. But too many people confuse tragedy with despair. "Tragedy", Lawrence said, "ought to be a great kick at misery." This is a healthy and immediately applicable idea. There are many things today deserving of that kick.**

[1057] **When I lived in Algiers, I would wait patiently all winter because I knew that in the course of one night, one cold, pure February night, the almond trees of the Vallée des Consuls would be covered with white flowers. I was then filled with delight as I saw this fragile snow stand up to all the rain and resist the wind from the sea. Yet every year it lasted, just long enough to prepare the fruit.**

[1058] **This is not a symbol. We shall not win our happiness with symbols. We shall need something more weighty. All I mean is that sometimes, when life weighs too heavily in this Europe still overflowing with its misery, I turn towards those shining lands where so much strength is still untouched. I know them too well not to realize that they are the chosen lands where courage and contemplation can live in harmony. The contemplation of their example then teaches me that if we would save the mind we must pass over its power to groan and exalt its strength and wonder. This world is poisoned by its misery, and seems to wallow**

in it. It has utterly surrendered to that evil which Nietzsche called the spirit of heaviness. Let us not contribute to it. It is vain to weep over the mind, it is enough to labour for it.

1059 **But where are the conquering virtues of the mind? This same Nietzsche listed them as the mortal enemies of the spirit of heaviness. For him they are the strength of character, taste, the "world", classical happiness, severe pride, the cold frugality of the wise. These virtues, more than ever, are necessary today, and each can choose the one that suits him best. Before the vastness of the undertaking, let no one in any case forget strength of character. I do not mean the one accompanied on electoral platforms by frowns and threats. But the one that, through the virtue of its whiteness and its sap, stands up to all the winds from the sea. It is that which, in the winter for the world, will prepare the fruit'** (*Summer*, 1954, pp.33-35 of 87).

1060 To examine what Camus has so beautifully articulated, he began by stating that the fundamental priority and responsibility of humanity is to solve our human condition, liberate the human mind from its underlying upset and by so doing replace the need for **'force'** to control our upset, troubled natures with the ability to explain, understand and thus pacify that upset. He acknowledged the human condition—that **'we live in contradiction'**—and that we have to live in denial of this condition—that **'we must refuse this contradiction'**. He also acknowledged the reality of the current extremely upset, alienated, soul-disconnected, depressed human state—**'nations poisoned by the misery of this century...**[a world] **utterly surrendered to that evil which Nietzsche called the spirit of heaviness'**—and went on to emphasise the need for the clarifying, first principle reconciling biological explanation to **'overcome our condition'**, saying we need **'to find those first few principles'** that will **'stitch up what has been torn apart'**.

1061 Again, of special significance here is Camus' acknowledgment that these answers were not going to come from the ivory towers of intellectualdom, but from outlying realms where there is still sufficient innocence, **'strength still untouched'**, **'whiteness and its sap'**, to overcome all the evasion, denial and dishonesty, to **'stand up to all the winds from the sea'**, and find the reconciling understanding of the human condition, **'prepare the fruit'**. And, about where the answers about ourselves would emerge, Camus wrote, **'I turn towards those shining lands where so much strength is still untouched. I know them too well not to realize that they are the chosen lands where courage** [to defy all the dishonest denial in the world] **and contemplation** [denial-free, honest thought] **can live in harmony.'**

[1062]Australia has been—at least up until very recent times when communication technology especially, but also immigration and multiculturalism, have ended our isolation and spread **'alienation'** everywhere here as well—just such a **'shining land...where so much strength is still untouched'**. And, in fact, **'courage and contemplation'** have been able to **'live in harmony'** here **'just long enough to prepare the fruit'**, to find the understanding of the human condition that liberates the human race! And, moreover, the anticipation of this fantastic breakthrough occurring here has been very strong in Australian mythology.

[1063]Although Australia is a young modern civilisation, we nevertheless have our own powerful contemporary mythology, at the heart of which is Banjo Paterson's 1895 poem *The Man From Snowy River*; in fact, in recognition of the poem's significance to us, Australia's $10 note features Paterson's image and, in microprint, *all* the many words to *The Man From Snowy River*. Ostensibly, the poem is about a great and potentially dangerous ride undertaken by mountain horsemen to recapture an escaped thoroughbred that had joined the brumbies (wild horses) in the mountain ranges, but what the poem is *really* recognising is that in Australia's isolation and relative innocence there would emerge sufficient soundness to defeat denial and put together the liberating explanation of the human condition. In the poem, the character Clancy persuades the station owner, Harrison, to let a **'stripling' 'lad'**—a boy—on his **'hardy mountain pony'** join their expedition to retrieve the escaped thoroughbred; Clancy argues, **'I warrant he'll be with us when he's wanted**

at the end.' Again, a boy is the embodiment of the innocence that is needed **'at the end'** of humanity's heroic journey of accumulating sufficient scientific understandings of the mechanisms of our world to assemble the denial-free explanation of the human condition. The poem describes how the boy goes beyond where the rest of the horsemen (the resigned, alienated adults) dare to go, and follows the brumbies down the **'terrible descent'** of a steep mountain where (if you weren't sound) **'any slip was death'** (to confront the unconfrontable issue of the human condition) and recaptures the thoroughbred from the impenetrable mountains (retrieves the escaped truth from the depths of denial). As the poem says, the boy **'ran them** [the brumbies]**...till their sides were white with foam / He followed like a bloodhound on their track / Till they halted, cowed and beaten—then he turned their heads for home'** (he fought all the alienation and its denial that has been enslaving this world to a standstill until it finally gave up the truth).

1064 Paterson was even more explicit in his anticipation of the liberating understanding of the human condition emerging from the Australian bush (innocent countryside) in another of his works from 1889, his aptly titled poem *Song of the Future*. Using the analogy of the pioneers who finally forged the path through our eastern coastal mountain range (appropriately enough called 'The Great Dividing Range') that had barred the way to Australia's interior during the early days of white settlement, Paterson envisaged **'the future'** heroic, Australian-led expedition humanity would take from the alienated bondage of the human condition to the fertile, sun/understanding-drenched freedom of a human-condition-resolved new world. These are the poem's key verses: **'Tis strange that in a land** [Australia] **so strong, so strong and bold in mighty youth** [innocence]**, we have** [in 1889] **no poet's voice of truth to sing for us a wondrous song** [explain the human condition]**...**[However,] **we yet may find achievements grand within the <u>bushman's quiet life</u>. Lift ye your faces to the sky, ye far blue mountains of the west...Tis hard to feel that years went by before the pioneers broke through your rocky heights and walls of stone, and made your secrets all their own** [broke through the wall of denial blocking access to the truth about the human condition]**. For years the fertile western plains were hid behind your sullen** [alienated] **walls, your cliffs and crags...Between the mountains and the sea, like Israelites with staff in hand, the people waited restlessly: They looked towards the mountains old and saw the sunsets come and go with gorgeous golden afterglow that made the west a fairyland, and marvelled what that**

west might be of which such wondrous tales were told...At length the hardy pioneers by rock and crag found out the way, and woke with voices of today, a silence kept for years and years [brought an end to the silence of the resigned world of denial]**...The way is won! The way is won! And straightway from the barren coast there came a westward-marching host, that aye and ever onward prest with eager faces to the west along the pathway of the sun...Could braver histories unfold than this bush story, yet untold—the story of their westward march** [liberation from the human condition]**...And it may be that we who live in this new land apart, beyond the hard old world grown fierce and fond and bound by precedent and bond** [bound up in sophisticated, intellectual denial], **may read the riddle** [of the human condition] **right** [synthesise the liberating truth from science's hard won insights into our world] **and give new hope to those who dimly see** [those who are embedded in blind denial/alienation], **that all things may be yet for good and teach the world at length to be one vast united brotherhood.'** Banjo Paterson was certainly an extraordinarily prophetic and gifted writer! (I might mention that the gifted editor of my writing, World Transformation Movement [WTM] Founding Member Fiona Cullen-Ward, is related to Banjo.)

1065 Henry Lawson, who was introduced in par. 45 when his poem about Resignation, *The Voice from Over Yonder*, was included, is another of Australia's greatest poets, and he too anticipated that the answer to the human condition would be found here. Making the same insights, employing the same comparisons and using the same imagery as Paterson's poems, Lawson wrote in his 1892 poem, the also aptly titled *When the Bush Begins to Speak*, that **'They know us not in England yet, their pens are overbold. We're seen in fancy pictures that are fifty years too old. They think we are a careless race—a childish race, and weak. They'll know us yet in England, when the bush begins to speak** [when innocence makes its contribution]**..."The leaders that will be", the men of southern destiny, are not all found in cities that are builded by the sea. They learn to love Australia by many a western creek** [while Australia as a whole has been relatively sheltered and thus innocent, it is from the Australian inland countryside or 'bush', rather than from the cities, that exceptional innocence will appear]**. They'll know them yet in England, when the bush begins to speak...All ready for the struggle, and waiting for the change, the army of our future lies encamped beyond the** [coastal] **range. Australia, for her patriots, will not have far to seek, they'll know her yet in England when the bush begins to speak...We'll find the peace and comfort that our fathers could**

not find, or some shall strike the good old blow that leaves a mark behind. We'll find the Truth and Liberty [the truth about the human condition that brings liberating understanding to humanity] **our fathers came to seek, or let them know in England when the bush begins to speak.'**

1066 And in yet another extraordinarily prescient work (which again has the same insights, comparisons and imagery), the Australian poet A.D. Hope also wrote of the role of the relatively innocent continent of Australia in delivering the liberating understanding of the human condition. In his 1931 poem, actually titled *Australia*, Hope wrote of **'A nation of trees, drab green and desolate grey...Without songs, architecture, history...And her five cities, like five teeming sores, each drains her, a vast parasite robber-state, where second-hand Europeans pullulate timidly on the edge of alien shores. Yet there are some like me turn gladly home from the lush jungle of** [alienated, dishonest, intellectual] **modern thought, to find the Arabian desert of the human mind, hoping, if still from the deserts the prophets** [innocent, unresigned, denial-free, truthful, profound, effective thinkers] **come. Such savage and scarlet as no green hills dare, springs in that waste, some spirit which escapes the learned doubt, the** [dishonest, intellectual as opposed to the honest, instinctual] **chatter of cultured apes which is called civilization over there** [in England].**'**

1067 In more recent Australian mythology, the 1998 Australian-made animated film *Babe: Pig in the City* tells the story of a little pig who, in order to save his farm, which symbolises the true world, has to take his innocence into the very heart of alienation, the city, and defy and defeat the alienation to release the captured soul of humanity, represented by the innocent animals being held captive there. To symbolise that the hazardous undertaking is going to require toughness as well as innocence, the pig wears a spiked collar into battle, the collar being donated by the bulldog who recognises he hasn't got the sufficient innocence/soundness to do the job himself. Interestingly, the Australian production company that made the *Babe* movies, Kennedy Miller, also produced the late 1970s and early 1980s *Mad Max* trilogy which features the same theme of unrecognised (because the world practises evading and denying truthful unevasiveness) innocence taking on the alienated world and leading humanity out of bondage from the darkness of that blind, alienated world. To quote a review of the film *Mad Max* (or *The Road Warrior* as the film was titled in the USA), Max is **'a hero in the classical tradition—a figure whose origins lie in the ancient myths; his role,**

in common with classical heroes, is of a man from nowhere destined to lead society into the next generation' (*Sunday Telegraph*, 21 Mar. 1982). Yes, again, these liberating answers were not going to come from the ivory towers of the truth-avoiding mechanistic scientific establishment, but from deep left field, from way out on the periphery of the great battle humanity has been waging against ignorance.

1068 Sir Laurens van der Post, that exceptional denial-free-thinking prophet of our time, clearly recognised the truth of innocence lying in wait for science to do its job so that it could lead humanity home when he wrote that **'Whatever happens, I shall be there in the end, for I, child that I am, am mother of your future self'** (*Jung and the Story of Our Time*, 1976, p.167 of 275). Yes, it is guidance from the long-repressed innocent, soulful, conscience-infused clarity that humans had before the upset, denial-based, alienated human condition developed that was needed to synthesise the liberating truth from science's hard won insights into the workings of our world. Evasive, whole-view-avoiding, reductionist, mechanistic science had to complete the difficult task of finding all the 'pieces of the jigsaw' of the explanation of the human condition before soul-guided, whole-view-confronting, denial-free thinking could come along **'in the end'** (or, as Paterson wrote, **'at the end'**) and piece the 'jigsaw' together. As emphasised in pars 182 and 937, science is the real 'messiah' or liberator of humanity; it made the explanation of the human condition that is presented in chapter 2 possible. Historically, people have talked about a 'second coming' of innocence, but in the vast spectrum of alienation that inevitably developed in humanity's heroic battle to defeat the ignorance of our instinctive self or soul and establish that we humans are good and not bad, there have always been an extremely rare few individuals who have been sufficiently nurtured with alienation-free, unconditional love in their infancy and childhood to be sound and secure enough in self to confront and think truthfully about the human condition, but until the discipline of science found understanding of the difference in the way genes and nerves process information, such upset-free, innocent individuals could not make their final contribution and assemble the explanation of the human condition from those understandings. Science was tasked with the tedious, hard work of finding the clues that make explanation of the human condition possible—and it has to be remembered that science is only the peak expression of the courageous effort that every human has exerted to defeat ignorance; that, in reality, it is 'on the shoulders' of 2 million years of human struggle against our soul's oppression that understanding of our species' fundamental goodness has finally been found—but our soul *was* needed **'in the end'** to assemble the actual explanation of the human condition from those clues. It is like a game of gridiron football where the team as a whole, with one exception, does all the hard work, gaining yardage down the field. Finally, when the side gets within kicking distance of the goal posts, a specialist kicker, who until then has played no part, is brought onto the field. While he—in his unsoiled attire—kicks the winning goal, the win clearly belongs to the team of exhausted players who did all the ground work. It is science, backed up by humanity as a whole, that is the real liberator of the human race from ignorance as to our species' fundamental goodness.

1069 The point here is that it is true that exceptional innocence has had a crucial concluding role to play but, most importantly, such an innocent is no more special or worthy or wonderful than any other human. If anything, they are less worthy because they haven't been involved in humanity's heroic battle as much as everyone else; they are yet to encounter upset—but, in any case, viewing those who have been more involved in the battle as more worthy is not accurate either, because no one could choose where they were going to be born/positioned on that battlefield. Yes, the fundamental insight that understanding of the human condition gives us is that in the epic battle to defeat ignorance that the human race has been waging for some 2 million years, there was going to be a vast spectrum of exhaustion/alienation but in this great army of warriors ALL HUMANS ARE EQUALLY GOOD, SPECIAL AND WONDERFUL. Since upset resulted from fighting humanity's battle against ignorance, upset, exhausted alienation is a heroic state, not a bad, inferior or lesser state, and, by the same logic, alienation-free innocence that resulted from having escaped encounter with all the upset on that battlefield is not in any way better, or in some way more special or more deserving or more wonderful than the upset, embattled, alienated state. The differentiation where some people are viewed as bad and others as good has gone forever from our discourse; it doesn't exist anymore; it has no basis of truth or fact. Understanding of the human condition completely and permanently eliminates the concept of 'good and evil'. Again, while instinct and intellect had particular concluding roles to play (the soul/instinct had to synthesise the explanation of the human condition from the insights into the workings of our world that science/knowledge/intellect had to first find), the truth is that it is the human race as a whole, *all the efforts of every human who has ever lived*, that has liberated our species from the horror of the human condition. Understanding the human condition allows us to know that we are all absolutely wonderful, utterly sublime, completely lovable, and that having all fully contributed to humanity's successful battle against ignorance we are *all* fabulous heroes of the story of life on Earth. So now we, THE HUMAN RACE AS A WHOLE, should give ourselves the biggest party ever. Everyone everywhere is going to be hoppin' and boppin', rompin' and stompin', hollerin' and howlin', movin' and groovin', rollickin' and rollin', hootin' and tootin', jumping and jiving, jolting and somersaulting, skipping and skating, shaking and shimmering, hugging and laughing, embracing and gyrating, twisting and shouting, dancing and singing, slipping and sliding, jamming and slamming, ripping and roaring, whirling and twirling and reelin' and rockin'. Yes, **'allow freedom to ring…from every village and every hamlet, from every state and every city'** because **'all of God's children, black men and white men, Jews and Gentiles, Protestants and Catholics'**, the more innocent and the more upset, the short and the tall, the big and the small, those who are left-handed and those who are right-handed, the butcher, the baker and the candlestick-maker, EVERYONE, can **'join hands and sing', 'Free at last! Free at last! Thank God Almighty, we are free at last** [from unjust condemnation]**!'** (Martin Luther King Jr, 'I Have A Dream' speech, 28 Aug. 1963).

[1070]In about 1969 (when I was around 23 years old) I wrote a poem that clearly anticipated the excitement of this time of humanity's liberation. The poem indicates just how strong my vision has always been of one day being able to find the liberating understanding of the human condition: **'This is a story you see, just a story—but for you / Um—I remember a long time ago in the distant future a timeless day / a sunlit cloudless day when all things were fine / when we all slow-danced our way to breakfast in the sun // You see the day awoke with music / Can you imagine one thousand horses slow galloping towards you across a vast plain / and we loved that day so much / We all danced like Isadora Duncan through the morning light // We skipped and twirled and spun about / Fairies were there like dragonflies over a pool / Little girls with wings they hovered and flew about / their small voices you could hear / You see it was that kind of morning // When the afternoon arrived it was big and bold and beautiful / In worn out jeans and bouncing breasts we began / to fight—our way—into another day / into something new—to jive our way into the night / from sunshine into a thunderstorm // We all took our place, rank upon rank we came / as an army with Hendrix out in front / and the music busted the horizon into shreds / By God we broke the world apart / The pieces were of different colours and there were so many people / We danced in coloured dust, we left in sweat no room at all / We had a ball in gowns of grey and red / There were things that happened that nobody knew / Bigger and better, I had written on my sweater / Where there was sky there was music, huge clouds of it / and there were storms of gold with coloured lights / It was so good we cried tears into our eyes / In a tug of war of love we had no strength left at all / Dear God we cried but he only sighed and / whispered strength through leaves of laughter // On and on we came in bold ranks of silvered gold / to lead a world that didn't know to somewhere it didn't care / It couldn't last, it had to end and yet it had an endless end / We were so happy in balloons of coloured bubbles that wouldn't bust / and we couldn't, couldn't quench our lust / There we were all together for ever and ever / and tomorrow had better beware because / when we've wept and slept we will be there to shake its bloody neck.'**

[1071]I should explain how, when we were young children and still thinking completely honestly and thus effectively, we knew the truth about our destiny; we didn't know exactly what would happen in our adult lives, but we did know the general form our future would take as a result of how hurt our original instinctive self or soul was in our infancy and early childhood. When we are children we have a clear awareness of the imperfections or otherwise of our circumstances and think truthfully about the consequences of those circumstances. If they were not ideal, as was the case in nearly everyone's lives, then we rapidly began to stop thinking about those unhappy consequences, but the point is there was a time in everyone's life when we knew the basic path our life was going to take. In that brief time in our early childhood of total honesty and thus insightfulness we all knew of the immense problem facing humanity of the dilemma of the human condition, the issue of the extreme imperfection of human life and, knowing that crux problem, we made an assessment of the extent and nature of our contribution to its solution. For the very rare exceptionally fortunate, those exceptionally loved and nurtured in their infancy and early childhood, *they* knew they could make an exceptional contribution and the precise nature of it. When we spoke of people being driven by a 'vision' this is essentially what we were recognising—an awareness in someone of them having the opportunity to make a critical contribution to the underlying battle humanity has been engaged in of bringing understanding to the human condition. And because people with such guiding visions were carrying such a strong awareness of what they could and must do from such a young age, it was as if they were owned or possessed by their vision and were thus very difficult to deter from their path. This is why, despite extreme resistance to the point of almost unbearable persecution, I have so tenaciously held on to my vision of being able to deliver understanding of the human condition. Yes, everyone is born a truthful, denial-free thinking and thus insightful prophet—as mentioned in par. 519, R.D. Laing recognised this when he said, **'Each child is a new beginning, a potential prophet'**—but few could afford to stay thinking so truthfully and insightfully; most had to forget what they could see because it was too confronting and depressing and just get on with their life as best they could. So while the main resignation to the imperfections of life typically occurred when people were about 15 years of age, there were many mini-resignations prior to that major one as people variously adjusted and eventually acquiesced to life under the duress of the human condition.

1072 Sir Laurens van der Post recognised the, in truth, least heroic and special, but critical, concluding role soul/conscience-led innocence had to play when he wrote these beautiful words: **'One of the most moving aspects of life is how long the deepest memories stay with us. It is as if individual**

memory is enclosed in a greater which even in the night of our forgetfulness stands like an angel with folded wings ready, at the moment of acknowledged need, to guide us back to the lost spoor of our meanings' (*The Lost World of the Kalahari*, 1958, p.62 of 253). Sir Laurens' writings about the relatively innocent Bushmen race of the Kalahari desert in Africa were so incredibly precious to me as a young man trying to stand up to all the dishonesty in the world because what he wrote about the Bushmen confirmed for me that there *is* another alienation-free, true world—and Sir Laurens' great vision was that he knew exactly that this is what all his beautiful descriptions of the relatively innocent life of the Bushmen could achieve, writing that **'I had a private hope of the utmost importance to me. The Bushman's physical shape combined those of a child and a man: I surmised that examination of his inner life might reveal a pattern** [of child-like innocence, soundness and truth] **which reconciled the spiritual opposites in the human being and made him whole...it might start the first movement towards a reconciliation'** (*The Heart of the Hunter*, 1961, p.135 of 233). Yes, what Sir Laurens wrote did **'start the first movement towards a reconciliation'**; it gave me the support I needed to synthesise the explanation of the human condition. In short, his writings prepared the way for this book and our species' freedom.

The author (left), and fellow World Transformation Movement Patron Tim Macartney-Snape AM OAM, with Sir Laurens van der Post in London in 1993, a few years before Sir Laurens' death in 1996

1073 Interestingly, in a 2006 interview with the Australian television presenter Andrew Denton, Bono, the prophetic lead singer of the rock band U2, said, **'You do get the feeling in Australia that there's...something going on down here, a new society being dreamt up...**[that in Australia there is] **the opportunity to lead the world'** (*Enough Rope*, ABC, episode 97, 13 Mar. 2006). As well as the story of David and Goliath, the Bible contains Isaiah's description of how innocence is needed to defeat the denial that is blocking access to the truth and, by doing so, lead humanity home, with Isaiah describing how **'a little child will lead them'** to the state where the concepts of 'evil' and 'good', and, in the process, the more corrupted and the more innocent, will be reconciled; to where, he says, the **'wolf will live with the lamb'** (Isa. 11:6). Isaiah also foresaw the wonderful effects of this reconciliation when he prophesised there would come a time when humans **'will beat their swords into ploughshears...Nation**[s] **will...**[not] **train for war any more'** (2:4).

1074 So that is what has happened—the truth about humans now exists, the spell has been broken, the proverbial 'elephant in our living rooms' has finally been acknowledged and exposed. Denial no longer works; its hold has been released. The great citadel of lies has been stormed. The steel casing of denial enveloping Earth has been wrenched apart. Yes, the truth *is* out, and thank goodness it is because the suffering simply could not go on. This is no exaggeration—we humans currently live in a place of such horrendous pain and suffering (which can be measured by the degree we block it out; by our astronomical levels of alienation) that it is as though we are living in a smouldering, toxic, sulphurous cauldron. BUT, there is a path out of there now to a world of soothing, healing sunshine, and amazing freedom, tranquility and beauty. All we have to do is cross the **'SUNSHINE HIGHWAY'** bridge: adopt the Transformed Lifeforce Way of Living. There is nothing stopping every human now from escaping the horror of the human condition, except their now completely obsoleted habituation to living in what is literally HELL ON EARTH. Our freedom comes as such a shock that we initially won't quite know how to take it up, but the truth is there is absolutely nothing that justifies *not* taking it up; there really is *no* impediment at all to taking it up, none at all—the gateway to freedom *is* now WIDE open to everyone.

1075 Sir Laurens van der Post, who, again, wrote *so* beautifully, gave us this wonderful description of how, in **'the midnight hour of the crashing darkness'** of the time we are living in, we had to build **'a bridge'** to **'cross**

from one side to the other' of **'that split between night and day in ourselves'**. He wrote: **'We must shut our eyes and turn them inwards, we must look far down into that split between night and day in ourselves until our head reels with the depth of it, and then we must ask: 'How can I bridge this self? How cross from one side to the other?' If we then allow that question to become the desire for its own answer, and that desire to become a bridge across the chasm, then, and only then, from high above on this far peak of our conscious self, on this summit so far above the snow-line of time, in this cold, sharp, selected moment, clearly and distinctly we shall see a cross. A gulf bridged makes a cross; a split defeated is a cross. A longing for wholeness presupposes a cross, at the foundations of our being, in the heart of our quivering, throbbing, tender, lovely, lovelorn flesh and blood, and we carry it with us wherever we journey on, on unto all the dimensions of space, time, unfulfilled love, and Being-to-be. That is sign enough. After that the drum can cease from drumming, the beating and troubled heart have rest. In the midnight hour of the crashing darkness, on the other side of the night behind the cross of stars, noon is being born'** (*Venture to the Interior*, 1952, p.229 of 239). It could be that Sir Laurens was intuitively envisaging Australia as the place where the human condition would be solved when he talked of **'noon'** being **'born'** beneath the Southern Cross, which is the constellation that forms the centrepiece of our Australian flag. I know Sir Laurens was very impressed with the soundness of the Australians he was held alongside as a prisoner of war, and that he knew of the exceptional isolation of Australia from all the upset in the world, **'the bushman's quiet life'** that Paterson referred to, and that he also knew of Australia's natural soulfulness—that Australia is like Africa where our species grew up—writing that **'When I first went to Australia...my senses told me at once that here, beyond rational explanation, was a land physically akin to Africa'** (*The Dark Eye in Africa*, 1955, p.35 of 159). His emphasis on **'a cross'** being **'sign enough'**, also invokes the involvement of Christ in 'saving the day', and there is truth in that, in the sense that Christ was exceptionally sound and, as has been explained, such innocent soundness is necessarily what was required to be able to **'turn...inwards'** and **'look far down into that split between night and day in ourselves until our head reels with the depth of it'**, namely look into the human condition. As Sir Laurens has also written, **'He who tries to go down into the labyrinthine pit of himself, to travel the swirling, misty netherlands below sea-level through which the harsh road to heaven and wholeness runs, is doomed to fail and never see the light where night joins day unless he goes out of love in search of love'** (*The Face Beside the Fire*, 1953, p.290 of 311).

[1076] Yes, as Billy Joel wrote and sang in his 1993 song *River of Dreams* about the overwhelming difficulty virtually all humans have of crossing the metaphorical **'river so deep'** to find the human-race-liberating-and-rehabilitating understanding of the human condition, **'In the middle of the night I go walking in my sleep, from the mountains of faith...through the valley of fear...through the jungle of doubt...through the desert of truth...to the river so deep...that is runnin' to the promised land...but the river is wide and it's too hard to cross...I try to cross to the opposite side so I can finally find what I've been looking for...I've been searching for something taken out of my soul.'** While *only* understanding of our devastated condition could rehabilitate the human race—as Tracy Chapman wrote and sang in her 1994 song *New Beginning*, **'The world is broken into fragments and pieces that once were joined together in a unified whole...The whole world's broke and it ain't worth fixing. It's time to start all over, make a new beginning... change our lives and paths; create a new world...There's too much fighting, too little understanding...We need to...make a new** [truthful] **language. With these we'll redefine the world and start all over'**—to **'start all over'** and **'create a new world'** required uncorrupted, alienation-free, sound, soul-full **'love'** because only it could **'cross' 'the river so deep'**; **'travel the swirling, misty netherlands below sea-level through which the harsh road to heaven and wholeness runs'**.

[1077] Indeed, for U2's wonderfully named and aforementioned song, *When Love Comes To Town*, Bono wrote these exciting lyrics that anticipated humanity's liberation from the human condition: **'I was a sailor, I was lost at sea / I was under the waves...But I did what I did before love** [the ultimate expression of which is truth] **came to town...I've seen love conquer the great divide** [between 'good and evil', yin and yang, idealism and realism, instinct and intellect, conscience and conscious, altruism and egotism, spiritualism and materialism, young and old, women and men, blacks and whites, religion and science, the left-wing and the right-wing, socialism and capitalism, country and city—and, most especially, between the **'split between night and day in ourselves'**]**...When love comes to town I'm gonna jump that train'** (1987). Yes, yes, yes, when the reconciling understanding of the human condition arrives we should *all* **'jump that train'**, get on board, *choose transformation NOT terminal alienation for the human race*. To include more anticipations of our species' liberation—from Bono and the other prophets of our time—consider these lyrics from U2's song *Love Rescue Me*, which Bono wrote: **'I've conquered my past, the future is here**

at last, I stand at the entrance to a new world I can see. The ruins to the right of me, will soon have lost sight of me' (1988). Yes, **'The ruins to the right of'** us **'will soon have lost sight of'** us because we have **'conquered'** our **'past'**, found the reconciling understanding of our corrupted condition, which means **'the future is here at last'** and we can leave that old, now dealt with and thus obsoleted, upset existence behind us forever. And what did Bob Dylan say? **'The present now will later be past…For the times they are a-changin''** (*The Times They Are A-Changin'*, 1963). Cat Stevens/Yusuf Islam similarly foresaw a time when **'Yesterday has past, now let's all start the living for the one that's going to last'** (*Changes IV*, 1971). So, yes, let's go, let's get out of this dead place we have all been living in for far too long—join the Sunshine Army on the Sunshine Highway to the World in Sunshine!

1078 I might mention that not only did I have the good fortune to be born in Australia, I also grew up in the post-war 1960s period of relative innocence and idealism. After a terrible bloodletting like the Second World War, which amounts to a valving off of upset, there always emerges a period of enormous relief and freshness, and it was all this innocence and its soundness that made the 1960s a **'golden age'** (Bono, *God Part II*, 1988), for not only did our generation conquer outer space by sending man to the moon, we have now conquered *inner* space by finding the truth about the human condition. And, moreover, we actually believed we were going to find that world-saving insight because it was around that time that 'rock and roll' music was created, for what is 'rock and roll' if not totally optimistic, all-out, rock-solid *'determination and resilience'* to achieve freedom from our species' historic state of unjust condemnation—determination to, as Bono sang, **'kick the darkness till it bleeds** [the] **daylight'** (ibid) of the truth about us humans and end the damned condemnation of our species FOREVER! What did Bob Dylan famously say about Elvis Presley—**'Hearing him for the first time was like busting out of jail.'** John Lennon famously reiterated the sentiment, saying, **'Before Elvis there was nothing'**; there was not all-out determination and optimism, there was no rock and roll, there was just endless decades and epochs and ages of resigned, lonely music—which, sadly, we have now returned to, but in an even *more* lonely state, with today's terminally alienated, head-banging, autistic, psychotic music.

1079 Jerry Lee Lewis, Chuck Berry and Little Richard were also locked onto the immensely excited driving beat that lay at the heart of rock and roll of anticipation of our species' liberation from the horror of the human condition, *especially* that belt-it-out, blast-out-of-here, boiling-with-excitement, completely-raging supernova from Ferriday, Louisiana, Jerry Lee Lewis—who was rightly the first performer inducted into the Rock and Roll Hall of Fame. In the 1992 documentary *Mojo Working: The Making of Modern Music*, which contains a wonderful collection of footage and commentary about Jerry Lee Lewis, writer Charles 'Dr Rock' White reported that **'John Lennon came into Jerry Lee Lewis' dressing room… and he walked over to Jerry Lee and…bent down and kissed Jerry Lee's feet…** [and then he] **walked out speechless'** (directed by Mark Neal). I wholeheartedly agree with Lennon's gesture; to me no one's music channelled the excitement of the anticipation of the liberation of the human race from the human condition as purely as Jerry Lee's did. If you listen to a live recording of Jerry Lee's April 5, 1964 performance at the Star Club in Hamburg, Germany, especially the tracks *Long Tall Sally* and *Hound Dog*, you will hear what **'is regarded by many music journalists as one of the wildest and greatest rock and roll concert albums ever'** and that Jerry Lee **'sounds possessed'** and was **'rocking harder than anybody had before or since… words can't describe the music'** (Wikipedia, accessed Dec. 2010; see <www.wtmsources.com/125>). Jerry Lee's performances were just drenched in the excitement of breaking free from the dungeon of our species' tortured condition. In fact, my vision is of a hysteria of millions and millions of excited people with Jerry Lee's piano being held aloft out in front and Jerry Lee standing on top of it, flicking his hair back and hitching his pants up, as he used to do—and filling the air is the musical build-up to humanity's great breakthrough to its freedom in *Prologue/Crunchy Granola Suite*, from Neil Diamond's 1972 *Hot August Night* album. But then, to actually take us through the portals of the new world that understanding of the human condition now makes possible, instead of Diamond's *Crunchy Granola Suite* vocals coming in, Jerry Lee would start singing 'Great Balls of Fire', Let's get out of here, LET'S GO!, to an immense roar of unbelievable relief and excitement from the flood of humanity bursting through that doorway to its freedom. One of our WTM Founding Members, Tony Gowing, has actually written a song titled *LET'S GO* that he sings with our WTM band, The Denialators—you can watch a rendition of this song at <www.worldtransformation.com/denialators>.

[1080] (Many more excited anticipations of the human-condition-reconciled new world, especially from 1950s and 1960s music, can be found in *Freedom Book 1* at <www.worldtransformation.com/freedom-anticipations>.)

Genevieve's Sun poster, copies of which are freely available at
<www.worldtransformation.com/posters>

1081 This picture of a tiny figure standing with upraised arms in front of an immense sun as it rises over the horizon represents the arrival of understanding and the dawn of humanity's all-magnificent freedom from the darkness and horror of the human condition. Created by WTM Founding Member Genevieve Salter in January 1998, this image has become the emblem of the World Transformation Movement.

1082 Finally, bringing these understandings to the world—an immense undertaking that will not have been achieved until science gives these insights into human behaviour the recognition they deserve and need—has been an enormous effort for the tiny band that is the 50 Founding Members of the WTM. It has been such a struggle and such a team effort that in the more than 20 years that most of us have been involved there has hardly been a week go by when one of us hasn't been prompted to say 'we couldn't have survived that attack on us', or 'we couldn't have accomplished that critical task', 'if we were even one less in number'. We have fought so hard and for so long and been through so much together it really is like we are one organism—and, given all that we have managed to achieve together based on our love of this information and what it can do for the world, it is a *superorganism*, with each member—Annabel Armstrong, Susan Armstrong, Sam Belfield, John Biggs, Richard Biggs, Anthony Clarke, Lyn Collins, Steve Collins, Lachlan Colquhoun, Eric Crooke, Emma Cullen-Ward, Fiona Cullen-Ward, Anthony Cummins, Neil Duns, Sally Edgar, Anna Fitzgerald, Brony FitzGerald, Connor FitzGerald, Tony Gowing, Jeremy Griffith, Simon Griffith, Damon

Isherwood, Felicity Jackson, Charlotte James, Lee Jones, Monica Kodet, Anthony Landahl, Doug Lobban, Tim Macartney-Snape, Manus McFadyen, Ken Miall, Tony Miall, Rachel O'Brien, James Press, Stacy Rodger, Marcus Rowell, Genevieve Salter, Will Salter, Nicholas Shaw, Wendy Skelton, Pete Storey, Ali Watson, Polly Watson, Prue Watson, Tess Watson, Tim Watson, James West, Stirling West, Prue Westbrook, Annie Williams—acting as beacons in **'the crashing darkness'** in this **'winter for the world'**, lining **'the pathway of the sun'** to humanity's freedom for *everyone*.

""" The Rising Sun """
Founding Members of the World Transformation Movement with Supporters and Friends, behind Jill Griffith's Memorial Tree, Sydney, 13 December 2008

L to R: Back Row: Sean Makim*, John Biggs, Tim Macartney-Snape, Will Salter, Tony Gowing, Sam Belfield, Neil Duns, Anna Fitzgerald, Anthony Cummins, Tess Watson, Anthony Landahl. 3rd Row: Monica Kodet, Prue Watson, Andy Colquhoun*, James Press, Tony Miall, Ken Miall, Richard Biggs, Pete Storey, Connor FitzGerald, Emma Cullen-Ward, Nicholas Shaw, Lee Jones, James West, Eric Crooke, Tim Watson. 2nd Row: Felicity Jackson, Fiona Cullen-Ward, Marie McNamara*, Damon Isherwood, Ali Watson, Anna Zilioli*, Stirling West, Susan Armstrong, Polly Watson, Doug Lobban, Dave Downie*, Sally Edgar, Charlotte James, Manus McFadyen, Annabel Armstrong, Genevieve Salter, Marcus Rowell. Front Row: Katrina Taylor*, Simon Griffith, Jeremy Griffith, Annie Williams, Lachlan Colquhoun, Anthony Clarke, Wendy Skelton, Brony FitzGerald, Prue Westbrook, Sarah Colquhoun*, Lyn Collins, Steve Collins, Amanda Purdy*, Stacy Rodger, Simon Mackintosh*. Sitting on grass: Tess Colquhoun*, Meika Collins*, Hebe Colquhoun*, Katie Collins*. Also present in spirit, our adored friend Rachel O'Brien. (*Supporter or Friend of the WTM.)

1083 So please visit our World Transformation Movement website and join us—because be assured that this book epitomises what the whole human journey has come down to: this book represents the final great battle, the war of the worlds, the battle of Armageddon, between the Christ and the anti-Christ; between the truth and the lies; between what

I represent and what E.O. Wilson represents; between the responsibility of our conscious mind to deliver knowledge, ultimately self-knowledge, the honest explanation of the human condition, and the abrogation of that responsibility, which is Wilson's dishonest explanation of the human condition; basically between transformation and terminal alienation for the human race. We, the human race, everyone, *must* make sure transformation wins.

1084 (Video presentations where I and others in the WTM answer questions about and describe the Transformed Lifeforce Way of Living can be viewed at <www.worldtransformation.com/transformation-videos>. And a more complete description of all aspects of the transformation can be found in *Freedom Book 2* at <www.worldtransformation.com/freedom-book2>.)

1085 In conclusion, the image below is a still frame from a documentary about the artist Paul Gauguin (*Gauguin: The Full Story*, directed by Waldemar Januszczak, 2005). The beauty and happiness of the dancing girl captures something of the joy and excitement that can now occur for the whole world. What a celebration it should be! More of this girl's dancing and some other wonderful footage of excited, transformed-new-world-like Pacific Island dancing from another documentary about Gauguin (*Palette Collection – Gauguin*, directed by Alain Jaubert, 2003) can be seen in this YouTube clip at <www.wtmsources.com/108>.

[1086] As a tribute to the pure, unconditional, selfless love I received from my parents, Norman and Jill Griffith, which created the determined and excited inspiration in this book, I include this photograph of them. It was taken under the wisteria beside our home on our sheep station, 'Totnes', near the town of Mumbil in the central west of New South Wales in 1959. You can see something of the unresigned, human-condition-defying, unaccepting-of-corrupted-reality, belief-in-another-true-world, relatively innocent, sound, secure, upright strength in my mother that nurtured in me the love of, and belief in, the authentic, soulful, true world.

About the Author

The following is taken from Professor Prosen's Introduction to this book, where a much more complete biography of the author is presented.

Jeremy Griffith (1945-) is an Australian biologist who has dedicated his life to bringing fully accountable, biological understanding to the dilemma of the human condition—the underlying issue in all human life of our species' extraordinary capacity for what has been called 'good' and 'evil'. While it's undeniable that humans are capable of great love and empathy, we also have an unspeakable history of greed, hatred, rape, torture, murder and war; a propensity for deeds so shocking and overwhelming that the eternal question of 'Why?' has seemed depressingly inexplicable. Even in our everyday behaviour, why, when the ideals of life are to be cooperative, selfless and loving, are we so ruthlessly competitive and selfish that human life has become all but unbearable and our planet near destroyed? How could we humans possibly be considered good when all the evidence seems to unequivocally indicate that we are a deeply flawed, bad, even 'evil' species?

For most people, trying to think about this ultimate of questions of whether we humans are fundamentally good or not has been an unbearably self-confronting exercise. Indeed, the issue of the human condition has been so depressing for virtually all humans that only a rare few individuals have been sound and secure enough in self to go anywhere near the subject. Nurtured by a sheltered upbringing in the Australian bush (countryside), Jeremy is one of those rare few. His soundness and resulting extraordinary integrity and thus clarity of thought, coupled with his training in biology, has enabled him to successfully grapple with this most foreboding of all subjects of the human condition and produce the breakthrough, human-behaviour-demystifying-and-ameliorating explanation of it, which is presented here in his summa masterpiece, *IS IT TO BE Terminal Alienation or Transformation For The Human Race?*

Jeremy Griffith, Sydney, Australia, November 2013

Printed in Australia
AUOC02n1217100614
261570AU00001B/1/P

9 781741 290172